VOLUME NINETY

ADVANCES IN FOOD AND NUTRITION RESEARCH

Functional Food Ingredients from Plants

VOLUME NINETY

ADVANCES IN FOOD AND NUTRITION RESEARCH

Functional Food Ingredients from Plants

Edited by

ISABEL C.F.R. FERREIRA
Centro de Investigação de Montanha (CIMO), Instituto Politécnico de Bragança, Campus de Santa Apolónia, Bragança, Portugal

LILLIAN BARROS
Centro de Investigação de Montanha (CIMO), Instituto Politécnico de Bragança, Campus de Santa Apolónia, Bragança, Portugal

ELSEVIER

ACADEMIC PRESS
An imprint of Elsevier

Academic Press is an imprint of Elsevier
50 Hampshire Street, 5th Floor, Cambridge, MA 02139, United States
525 B Street, Suite 1650, San Diego, CA 92101, United States
The Boulevard, Langford Lane, Kidlington, Oxford OX5 1GB, United Kingdom
125 London Wall, London, EC2Y 5AS, United Kingdom

First edition 2019

Notices

Knowledge and best practice in this field are constantly changing. As new research and experience broaden our understanding, changes in research methods, professional practices, or medical treatment may become necessary.

Practitioners and researchers must always rely on their own experience and knowledge in evaluating and using any information, methods, compounds, or experiments described herein. In using such information or methods they should be mindful of their own safety and the safety of others, including parties for whom they have a professional responsibility.

To the fullest extent of the law, neither the Publisher nor the authors, contributors, or editors, assume any liability for any injury and/or damage to persons or property as a matter of products liability, negligence or otherwise, or from any use or operation of any methods, products, instructions, or ideas contained in the material herein.

ISBN: 978-0-12-816567-6
ISSN: 1043-4526

For information on all Academic Press publications
visit our website at https://www.elsevier.com/books-and-journals

Publisher: Zoe Kruze
Acquisition Editor: Sam Mahfoudh
Editorial Project Manager: Letícia M. Lima
Production Project Manager: Denny Mansingh
Cover Designer: Mark Rogers

Typeset by SPi Global, India

Contents

Contributors

Ryszard Amarowicz
Institute of Animal Reproduction and Food Research, Polish Academy of Sciences, Olsztyn, Poland

Begoña Ayuda-Durán
Grupo de Investigación en Polifenoles (GIP-USAL), Universidad de Salamanca, Salamanca, Spain

Nicolas Bordenave
School of Nutrition Sciences, Faculty of Health Sciences, University of Ottawa, Ottawa, ON, Canada

Adelar Bracht
Postgraduate Program in Food Science, Department of Biochemistry, Laboratory of Biochemistry of Microorganisms and Food Science, State University of Maringa, Maringá, Paraná, Brazil

Bruna L. Buzati Pereira
Interdisciplinary School of Health Sciences, Faculty of Health Sciences, University of Ottawa, Ottawa, ON, Canada; Internal Medicine Department, Botucatu Medical School, UNESP—Univ Estadual Paulista, Botucatu, Brazil

María Ciudad-Mulero
Department of Nutrition and Food Science, Faculty of Pharmacy, Complutense University of Madrid, Madrid, Spain

Rúbia Carvalho Gomes Corrêa
Postgraduate Program in Food Science, Department of Biochemistry, Laboratory of Biochemistry of Microorganisms and Food Science, State University of Maringa, Maringá, Paraná, Brazil

Vanesa Gesser Correa
Postgraduate Program in Food Science, Department of Biochemistry, Laboratory of Biochemistry of Microorganisms and Food Science, State University of Maringa, Maringá, Paraná, Brazil

Francesco Di Gioia
Department of Plant Science, Pennsylvania State University, University Park, PA, United States

Corrine C. Dobson
School of Nutrition Sciences; Interdisciplinary School of Health Sciences, Faculty of Health Sciences, University of Ottawa, Ottawa, ON, Canada

Virginia Fernández-Ruiz
Department of Nutrition and Food Science, Faculty of Pharmacy, Complutense University of Madrid, Madrid, Spain

Jéssica Amanda Andrade Garcia
Postgraduate Program in Food Science, Department of Biochemistry, Laboratory of Biochemistry of Microorganisms and Food Science, State University of Maringa, Maringá, Paraná, Brazil

Susana González-Manzano
Grupo de Investigación en Polifenoles (GIP-USAL), Universidad de Salamanca, Salamanca, Spain

Ana M. González-Paramás
Grupo de Investigación en Polifenoles (GIP-USAL), Universidad de Salamanca, Salamanca, Spain

Riadh Hammami
School of Nutrition Sciences, Faculty of Health Sciences, University of Ottawa, Ottawa, ON, Canada

Cecilia Jiménez López
Nutrition and Bromatology Group, Department of Analytical and Food Chemistry, Faculty of Food Science and Technology, University of Vigo—Ourense Campus, Ourense; Nutrition and Food Science Group, Department of Analytical and Food Chemistry, CITACA, CACTI, University of Vigo—Vigo Campus, Vigo, Spain

Mª Cruz Matallana-González
Department of Nutrition and Food Science, Faculty of Pharmacy, Complutense University of Madrid, Madrid, Spain

Patricia Morales
Department of Nutrition and Food Science, Faculty of Pharmacy, Complutense University of Madrid, Madrid, Spain

Walid Mottawea
School of Nutrition Sciences, Faculty of Health Sciences, University of Ottawa, Ottawa, ON, Canada

Taofiq Oludemi
Mountain Research Center (CIMO), Polytechnic Institute of Bragança, Bragança, Portugal

Ronald B. Pegg
Department of Food Science & Technology, The University of Georgia, Athens, United States

Rosane Marina Peralta
Postgraduate Program in Food Science, Department of Biochemistry, Laboratory of Biochemistry of Microorganisms and Food Science, State University of Maringa, Maringá, Paraná, Brazil

Spyridon A. Petropoulos
Department of Crop Production and Rural Environment, University of Thessaly, Volos, Greece

Jovana Petrović
Department of Plant Physiology, Institute for Biological Research "Siniša Stanković", University of Belgrade, Belgrade, Serbia

Krista A. Power
School of Nutrition Sciences, Faculty of Health Sciences, University of Ottawa, Ottawa, ON, Canada

M.A. Prieto
Nutrition and Bromatology Group, Department of Analytical and Food Chemistry, Faculty of Food Science and Technology, University of Vigo—Ourense Campus, Ourense; Nutrition and Food Science Group, Department of Analytical and Food Chemistry, CITACA, CACTI, University of Vigo—Vigo Campus, Vigo, Spain

Alexane Rodrigue
School of Nutrition Sciences; Interdisciplinary School of Health Sciences, Faculty of Health Sciences, University of Ottawa, Ottawa, ON, Canada

Celestino Santos-Buelga
Grupo de Investigación en Polifenoles (GIP-USAL), Universidad de Salamanca, Salamanca, Spain

Jesus Simal-Gandara
Nutrition and Bromatology Group, Department of Analytical and Food Chemistry, Faculty of Food Science and Technology, University of Vigo—Ourense Campus, Ourense, Spain

Marina Soković
Department of Plant Physiology, Institute for Biological Research "Siniša Stanković", University of Belgrade, Belgrade, Serbia

Dejan Stojković
Department of Plant Physiology, Institute for Biological Research "Siniša Stanković", University of Belgrade, Belgrade, Serbia

Tatiane Francielli Vieira
Postgraduate Program in Food Science, Department of Biochemistry, Laboratory of Biochemistry of Microorganisms and Food Science, State University of Maringa, Maringá, Paraná, Brazil

Preface

There has been a growing interest in functional foods and in the incorporation of bioactive extracts into food products, thus obtaining foods with beneficiary health effects. This fact is due to raising consumer's awareness for chemopreventive nutrition, which is linked to the use of chemical or biological substances for the prevention of diseases, being highly receptive to functional foods with specific components. This concept has stimulated consumers' interest in food products, such as functional foods, with specific bioactive molecules.

Plants are an excellent example of such products, having remarkable medicinal properties due to different compounds, which have a defined action on a diverse of bioactivities. In addition, plant extracts have several beneficial physiological effects, many related to the high content in polysaccharides, polyphenols, pigments, vitamins, glucosinolates, steroids, saponins, phytoestrogens, dietary fiber, terpenes, terpenoids, among other compounds. Most of these molecules are known for being strong scavengers of free radicals, which have key roles in aging and various other diseases, such as coronary heart disease, cancer, or neurodegenerative diseases. Plant phytochemicals have effectively been the subject of several studies in recent years, which shows its great potential to act as a functional food ingredient.

Therefore, the use of natural resources is crucial, especially if these sources include bioactive products, which certainly add value to food products because of their functional properties and health benefits. The aim of this book, *Functional Food Ingredients From Plants*, is to explore different classes of phytochemicals, as possible molecules to be added to food products, in order to enhance their bioactive properties and therefore constitute a functional food, exploiting this concept.

Isabel C.F.R. Ferreira
Lillian Barros

CHAPTER ONE

Natural antioxidants of plant origin

Ryszard Amarowicz[a,*], Ronald B. Pegg[b]
[a]Institute of Animal Reproduction and Food Research, Polish Academy of Sciences, Olsztyn, Poland
[b]Department of Food Science & Technology, The University of Georgia, Athens, United States
*Corresponding author: e-mail address: r.amarowicz@pan.olsztyn.pl

Contents

Advances in Food and Nutrition Research, Volume 90
ISSN 1043-4526
https://doi.org/10.1016/bs.afnr.2019.02.011

Abstract

Interest in the content of natural antioxidants in plant-based foods can be from the human health perspective, in terms of how these compounds might help promote one's health and wellness, or from the storage point-of-view, as the endogenous antioxidant constituents aid to extend a foodstuff's shelf-life. This chapter reports essential information about the mechanism of antioxidant action and methods employed for determination of their activity, classes of phenolic compounds (phenolic acids, flavonoids, lignans, stilbenes, tannins), sources of plant antioxidants (oil seeds, cereals, legumes, plants of the Lamiaceae family, tea and coffee, tree nuts, fruits, and berries), extraction strategies of phenolic compounds from plant material, and the influence of processing and storage on the content of natural antioxidants in foods and their antioxidant activity. Thermal processing, if not releasing bound phenolics from the structural matrices of the food, tends to decrease the antioxidant potential or, in the best case scenario, has no significant negative impact. Gentler sterilization processes such as high-pressure processing tend to better retain the antioxidant potential of a foodstuff than thermal treatments such as steaming, boiling, or frying. The impact of processing can be assessed by determining the antioxidant potential of foodstuffs either at the point of formulation or after different periods of storage under specified conditions.

1. Introduction

The natural phenolic compounds that are present in plants are responsible for antioxidant activity. This activity has been confirmed in numerous *in vivo* and *in vitro* studies. Phenolic compounds also have other important biological activities, which makes them applicable as alternatives to synthetic additives. For food technologies, it is important to understand the chemistry of antioxidants and the analytical methods that are used in the determination of those antioxidants. From a practical point of view, the results of research on the influence of processing and storage on natural antioxidants are valuable for practice.

2. Mechanisms of action of natural phenolic antioxidants

Antioxidant can be classified as either "primary antioxidants" or "secondary antioxidants." Primary antioxidants actively inhibit oxidation reactions, whereas secondary antioxidants act in an indirect way; for example, they react with pro-oxidants or are able to scavenge oxygen (Craft, Kerrihard, Amarowicz, & Pegg, 2012; Shahidi & Wanasundara, 1992).

Phenolic compounds, as primary antioxidants, act according to two mechanisms: hydrogen-atom transfer (HAT) or single-electron transfer (SET). The HAT mechanism occurs when an antioxidant compound quenches free radical species by donating hydrogen atoms:

$$(n)\,RO_2^{\cdot} + ArOH \rightarrow (n)\,ROOH + ArO^{\cdot} \tag{1}$$

The free radical formed in this reaction is much more stable than $RO_2^{\cdot}$.

The SET mechanism occurs in cases where an antioxidant transfers a single electron to aid in the reduction of potential target compounds:

$$(n)\,RO_2^{\cdot} + ArOH \rightarrow (n)\,RO_2{-} + [ArOH]^{\cdot +} \tag{2}$$

The resultant radical-cationic antioxidant compound is then deprotonated by interacting with water:

$$[ArOH]^{\cdot +} + H_2O \rightleftarrows ArO^{\cdot} + H_3O^{+} \tag{3}$$

$$RO_2{-} + H_3O^{+} \rightleftarrows ROOH + H_2O \tag{4}$$

Experimental investigations of Leopoldini, Marino, Russo, and Toscano (2004) and Wright, Johnson, and Di Labio (2001) assert that HAT and SET chemical processes can occur simultaneously as a sequential proton-loss electron transfer (SPLET), which is also termed as a proton-coupled electron transfer (PCET) (Huang, Ou, & Prior, 2005). The reaction schemes below illustrate a SPLET mechanism (Klein & Lukeš, 2006):

$$ArOH \rightarrow ArO^{-} + H^{+} \tag{5}$$

$$ArO^{-} + ROO^{\cdot} \rightarrow ArO^{\cdot} + ROO^{-} \tag{6}$$

$$ROO^{-} + H^{+} \rightleftarrows ROOH \tag{7}$$

The presence of ionic metals such as copper and iron in a system can promote the production of hydroxyl radicals by the Fenton reaction:

$$Fe^{2+} + H_2O_2 \rightarrow Fe^{3+} + {}^{\cdot}OH + {}^{-}OH \tag{8}$$

Phenolic compounds can operate as "secondary" antioxidants in a chelation process by inhibiting oxidation without directly interacting with oxidative species (Mira et al., 2002). A high chelation activity is often characteristic of phenolic compounds that have a 5-OH and/or 3-OH moiety with a 4-oxo group in the A/C ring structure. Positive effects on the chelation activity of flavonoids are associated with the presence of 3′-4′ and/or 7′-8′-*o*-dihydroxyphenyl groups on the B- and A-rings (Khokhar & Owusu Apenten, 2003).

3. Methods used for the determination of antioxidant activity

3.1 ORAC (oxygen radical absorbance capacity) assay

In the ORAC assay (Adom & Liu, 2005; Glazer, 1990; Huang, Ou, Hampsch-Woodhill, Flanagan, & Deemer, 2002; Ou, Hampsch-Woodhill, & Prior, 2001) $RO_2^{\cdot}$ is generated by thermal degradation of (AAPH). The generated peroxyl radicals react with fluorescein or its derivative by decreasing the fluorescence signal at an excitation/emission wavelength pair of 493/515 nm. The HAT reactions of antioxidants with peroxyl radicals protect against the disappearance of fluorescein in the sample. The following reaction scheme illustrates this process:

$$2RO_2^{\cdot+}(FL)OH\,(\text{fluorescence at }515) \rightarrow 2ROOH + (FL)O^{\cdot}(HAT) \quad (9)$$

3.2 Photochemiluminescence (PCL) assay

In the first step of the PCL assay (Popov & Lewin, 1994, 1996), luminol (5-amino-2,3-dihydro-1,4-phthalazinedione) is photodegradated resulting in the production/quenching of $O_2^{\cdot-}$:

$$\text{Luminol} + h\nu_1\,(UV) \rightarrow L^* + {}^3O_2 \rightarrow [L^*O_2] \rightarrow L^{\cdot+} \rightarrow L^{\cdot+} + O_2^{\cdot-} \quad (10)$$

Once the $O_2^{\cdot-}$ and luminol radicals are generated, they proceed through a series of reactions that result in the production of blue luminescence:

$$L^{\cdot+} + O_2^{\cdot-} \rightarrow N_2 + AP^{*2-} \rightarrow AP^{2-} + h\nu_2\,(\text{blue at }360\,\text{nm}) \quad (11)$$

In the reaction presented above, AP^{*2-} is an excited aminophthalate anion, and AP^{2-} is the aminophthalate anion at the ground state. The antioxidant species that is present in the reaction mixture will out-compete the luminol radical in the HAT reaction. The production of blue luminescence will be produced until the concentration increases.

3.3 FRAP (ferric reducing antioxidant power) assay

The FRAP assay was developed by Benzie and Strain (1996) to measure the ferric reducing power of human plasma. Pulido, Bravo, and Saura-Calixto (2000) adapted this method to quantify the ferric reducing antioxidant power of plant extracts. Dragsted et al. (2004) used the FRAP assay in a microtiter plate reader in a 96-well format. In FRAP, the assay reaction involves the reduction of Fe^{3+}—TPTZ (iron[III]-2,4,6-tripyridyl-*S*-triazine) to Fe^{2+}—TPTZ through SET with an antioxidant compound. The result of this reaction is an intense blue color:

$$Fe^{3+} - \text{TPTZ} + \text{ArOH} \rightarrow Fe^{2+} - \text{TPTZ}\,(\text{blue at 595 nm}) + [\text{AROH}]^{\cdot +}\,(\text{SET}) \tag{12}$$

3.4 CUPRAC (cupric reducing antioxidant capacity) assay

Redox reactions with copper are often faster than redox reactions with iron. Similar to iron, copper ions coordinate with nitrogen-containing chelating agents such as 2,2′-bipyridine or 1,10-phenanthroline and its derivatives. The CUPRAC method (Apak, Güçlü, Özyürek, & Karademir, 2004; Özyürek et al., 2011) involves the reduction of free copper(II) to copper(I) in the presence of neocuproine (NC) (2,9-dimethyl-1,10-phenanthroline), which results in the coordinated complex Cu(I)—NC at a ratio of 2:1 according to the following reaction scheme:

$$Cu^{2+} + \text{ArOH} - 2\text{NC} \rightarrow Cu^{+} - (\text{NC})_2(\text{blue at 450 nm}) + [\text{ArOH}]^{\cdot +} \tag{13}$$

3.5 TEAC (Trolox equivalent antioxidant capacity) assay

The TEAC assay was developed by Miller, Rice-Evans, Davies, Gopinathan, and Milner (1993) for the measurement of the antioxidant capacity of human plasma based on the scavenging of the free-radical cation ($\text{ABTS}^{\cdot +}$) by antioxidants. This method was modified by Re et al. (1999) for the direct generation of ($\text{ABTS}^{\cdot +}$) without radical intermediates. The TEAC assay is generally accepted as a SET assay. However, an ABTS radical cation can be neutralized by SET and HAT mechanisms:

$$\text{ABTS}^{\cdot +}(\text{green at 734 nm}) + \text{ArOH} \rightarrow \text{ABTS}\,(\text{colorless}) + [\text{ArOH}]^{\cdot +}(\text{SET}) \tag{14}$$

$$\text{ABTS}^{\cdot+}(\text{green at } 734\,\text{nm}) + \text{ArOH} \rightarrow \text{ABTS(H)}\ (\text{colorless}) + \text{ArO}^{\cdot}\,(\text{HAT}) \tag{15}$$

3.6 DPPH˙ (2,2′-diphenyl-1-picrylhydrazyl radical) assay

The DPPH˙ assay (Blois, 1958; Bondet, Brand-Williams, & Berset, 1997; Brand-Williams, Cuvelier, & Berset, 1995) is often run due to its relative inexpensiveness. DPPH˙ undergoes a HAT (Brand-Williams et al., 1995; Litwinienko & Ingold, 2003), SET (Foti, Daquino, & Geraci, 2004; Huang et al., 2005) or mixed (Schaich, 2006) mechanisms according to the following reaction schemes:

$$\text{DPPH}^{\cdot}(\text{violet at } 515\,\text{nm}) + \text{ArOH} \rightarrow \text{DPPH(H)}\ (\text{colorless}) + \text{ArO}^{\cdot}\,(\text{HAT}) \tag{16}$$

$$\text{DPPH}^{\cdot}(\text{violet at } 515\,\text{nm}) + \text{ArOH} \rightarrow \text{DPPH}^{-}(\text{colorless}) + \text{ArO}^{\cdot+}(\text{SET}) \tag{17}$$

3.7 β-carotene-linoleic acid (linoleate) assay

In this assay (Miller, 1971), an emulsion system is used, the pentadienyl free-radical formed from linoleic acid (Frankel, 2005) attacks highly unsaturated β-carotene molecules, and the characteristic orange color of the emulsion disappears. Phenolic antioxidants can protect β-carotene against destruction by "neutralizing" the linoleate free radical. This process can be monitored spectrophotometrically by measuring the absorbance of the sample at 470 nm.

3.8 Critical opinion on antioxidant methods

At the end of this paragraph we want to emphasize that the antioxidant activity of phenolic compounds determined *in vitro* using the above described methods cannot be generalized to their role *in vivo*. Very critical opinions about antioxidant methods are presented by Harnly (2017). He cited the opinion of the U.S. Department of Agriculture (USDA), which removed its ORAC database from the internet in 2012 "due to mounting evidence that the values indicating antioxidant capacity have no relevance to the effects of specific bioactive compounds, including polyphenols on human health." Harnly summarized his article by stating, that "antioxidant" is a marketing term of questionable health and analytical value.

4. Classification of natural phenolic compounds

4.1 Phenolic acids

Phenolic acids are derivatives of benzoic and cinnamic acids. Fig. 1 depicts the chemical structures of the main phenolic acids that are found in plants and foods of plant origin. The chemical structure of chlorogenic acid, the esters of caffeic acid and (−)-quinic acid, called chlorogenic acid is depicted in Fig. 2.

Phenolic acids exhibit antimicrobial activity (Gyawali & Salam, 2014). Their hydroxy groups can interact with the cell membrane of bacteria to

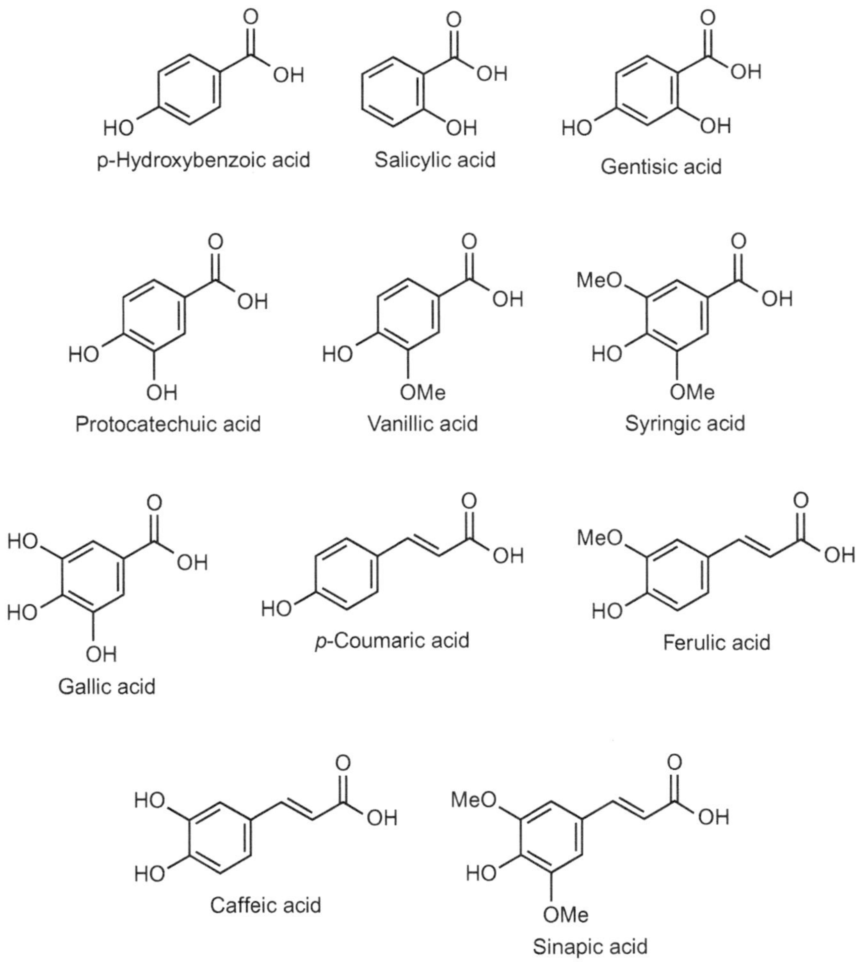

Fig. 1 Chemical structure of phenolic acids.

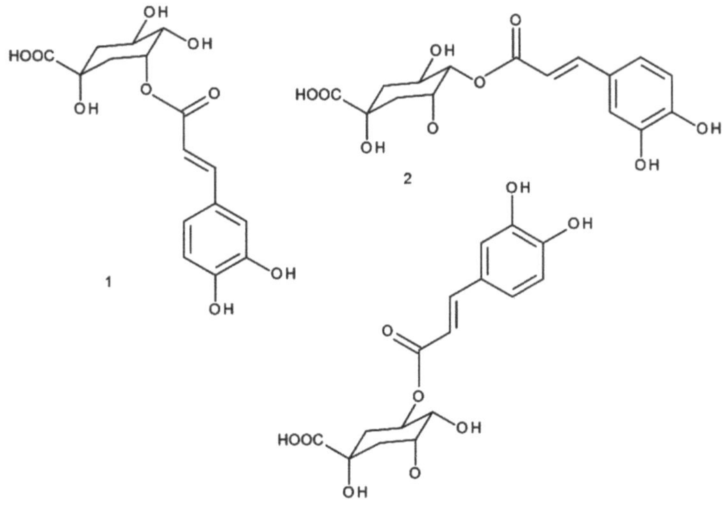

Fig. 2 Chemical structure of chlorogenic acids. **1**–3-*O*-caffeoylquinic acid; **2**–4-*O*-caffeoylquinic acid; **3**–5-*O*-caffeoylquinic acid.

disrupt membrane structures (Lai & Roy, 2004; Xue, Davidson, & Zhong, 2013). Gallic acid induced the apoptosis of cancer cells (Lu et al., 2010) and possessed anti-inflammatory properties (Maggi-Capeyron et al., 2001). Syringic and vanillic acids acted as suppressors of immune-mediated liver inflammation (Itoh et al., 2009). Ferulic acid exhibited chemopreventive activity against oral cancer (Mori et al., 1999). In experiments on rats, sinapic acid exhibited an antihyperglycemic effect (Kanchana, Shyni, Rajadurai, & Periasamy, 2011) and a protective effect against arsenic-induced toxicity (Pari & Jalaludeen, 2011).

Several authors confirmed the antioxidant activity of phenolics using such methods as ABTS radical cation, DPPH radical, β-carotene-linolenic acid, inhibition of lipid peroxidation in rat brain homogenates, and inhibition of lipid peroxidation induced by the superoxide anion radical (Graff, 1992; Karamać, Buciński, Pegg, & Amarowicz, 2005; Karamać, Koleva, Kancheva, & Amarowicz, 2017; Koroleva et al., 2014; Nenandis, Zhang, & Tsimidou, 2003; Re et al., 1999). It was observed that the radical-scavenging activities of phenolic acids depend on the number of hydroxy moieties that are attached to the aromatic ring of the benzoic or cinnamic acid molecules. Two methoxy moieties attached to the aromatic ring at positions 3 and 5 increased the radical-scavenging activity of phenolic acids.

4.2 Flavonoids

Flavonoids are widely distributed in plants and foods of plant origin and consist of two outer aromatic rings with a three-carbon bridge. According to their chemical structure flavonoids have been classified into several subgroups (Fig. 3).

Due to their biological activity, flavonoids are considered as an indispensable component in many nutraceuticals (Panche, Diwan, & Chandra, 2016). Plants rich in flavonoids have been applied for the preparation of functional foods.

Flavonoids possess many biochemical properties, such as antioxidant, antibacterial, antiviral, anti-inflammatory, anticancer, and hepatoprotective activities (Kumar & Pandey, 2013). The mechanism of antioxidant activity in flavonoids can be characterized by the direct scavenging of oxygen free radicals or excited oxygen species, the inhibition of oxidative enzymes, and chelation properties (Heim, Tagliaferro, & Bobilya, 2002; Korkina & Afanas'ev, 1997; Terao, 2009).

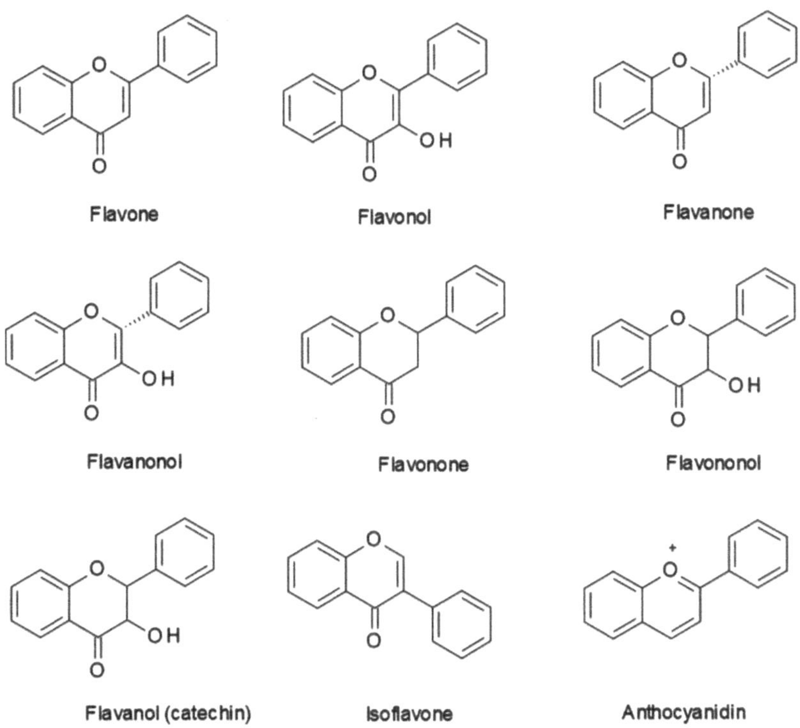

Fig. 3 Chemical structure of flavonoids.

The flavonoid derivatives have properties of inhibitory activities to acetylcholinesterase (Sheng et al., 2009). The inhibition of this enzyme is one of the central focuses for the development of drugs against Alzheimer's and Parkinson's diseases. The results of Hügel, Jackson, May, Zhang, and Xue (2016) indicated that dietary flavonoids are associated with a decreased risk of hypertension and cardiovascular disease. The antidiabetic activity of flavonoids was shown in an experiment with rats (Waisundara, Hsu, Tan, & Huang, 2009).

4.3 Lignans

Lignans are diphenol compounds that are formed from phenylalanine *via* dimerization of substituted cinnamic alcohols. Flax and sesame seeds are the main sources of lignans in the human diet. The chemical structures of flaxseed and sesame lignans are depicted in Figs. 4 and 5.

Under the activity of human intestinal bacteria, flaxseed lignans are metabolized to enterodiol and enterolactone (Touré & Xueming, 2010). Because some of the effect of estrogen are reduced by enterodiol and enterolactone, lignans are commonly referred to as phytoestrogens. Due to their structural similarity to 17-β-oestradiol, at normal oestradiol levels

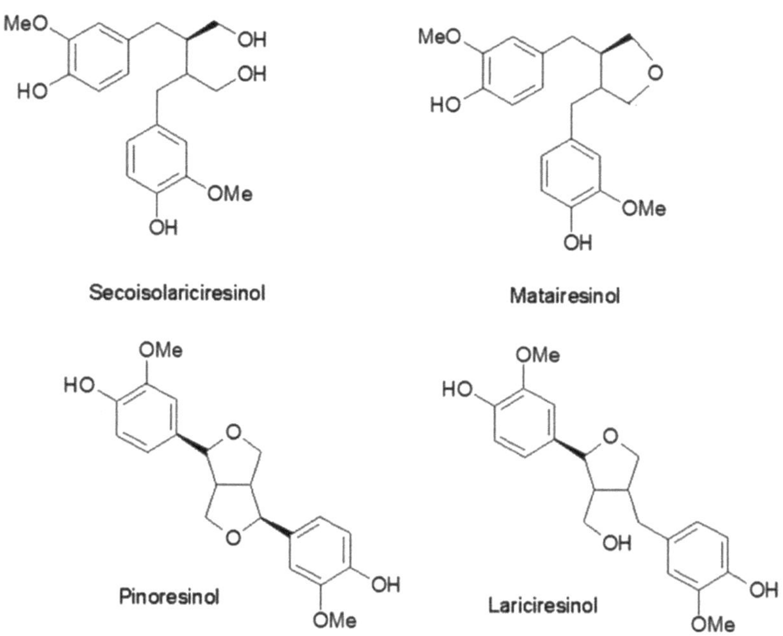

Fig. 4 Chemical structure of flaxseed lignans.

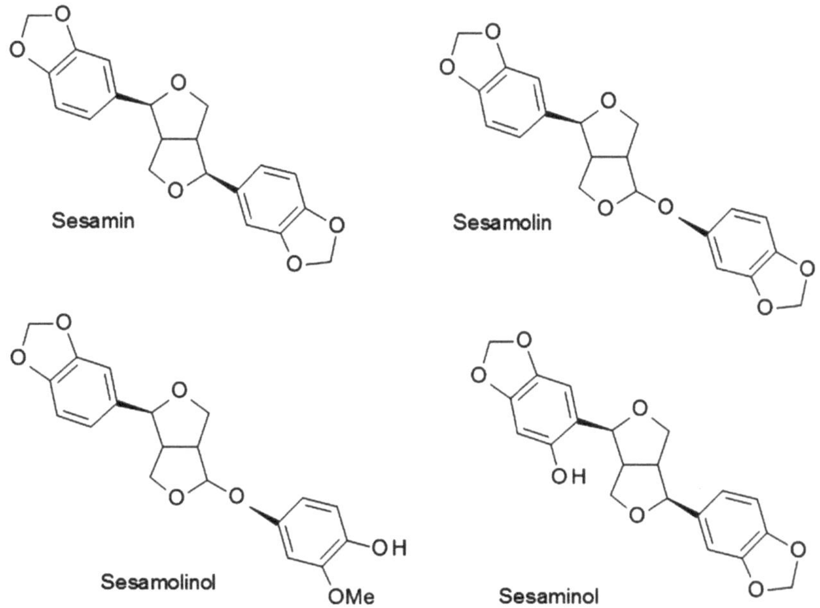

Fig. 5 Chemical structure of sesame lignans.

in organisms, lignans are an antagonist of this hormone (Hutchins & Slavin, 2003). The activity of lignans against breast and prostate cancer was confirmed by several studies (Adlercreutz et al., 1992; Demark-Wahnefried et al., 2001, 2004; Saarinen, Penttinen, Smeds, Hurmerinta, & Mäkelä, 2005).

Numerous experiments have shown the antioxidant activity of lignans. For example, secoisolariciresinol (SECO) and SDG were active in bulk oil and emulsion systems (Slavova-Kazakova, Karamać, Kancheva, & Amarowicz, 2016) and in liposome systems (Hu, Yuan, & Kitts, 2007). In the experiments of Hosseinian, Muir, Westcott, and Krol (2006), SECO and SDG retarded the degradation of canola oil in a concentration-dependent manner. SDG was an active scavenger of hydroxy radicals (Prasad, 1997) and DPPH radicals (Eklund et al., 2005). SECO and matairesinol exhibited a ferric-reducing antioxidant power that was greater than that of ascorbic acid (Niemeyer & Metzler, 2003).

The antioxidant activities of sesamin and sesamolin were confirmed by using FRAP, ORAC, DPPH radical, and β-carotene-linoleate model systems. Sesamin enhanced the radical scavenging ability of γ-tocopherol against the DPPH radical by three-fold (Mahendra Kumar & Singh, 2015). Sesame lignans enhanced the antioxidant activity of vitamin E in lipid peroxidation systems (Ghafoorunissa, Hemalatha, & Rao, 2004). Papadopoulos, Nenadis, and Sigalas (2016) proposed, for the first time, that sesamin and sesamolin may present antioxidant activity through a hydrogen atom transfer mechanism. The addition of sesame lignan compounds protected sunflowers against auto-oxidation at 60 °C and thermal oxidation at 180 °C. (Lee, Kim, & Choe, 2007). Sesamin and sesamolin protected methyl linoleate against autoxidation at 60 °C for 18 h in the dark. Oxidation was monitored by its conjugated dienoic acid content and *p*-anisidine value (Lee & Choe, 2006). Osawa (1999) found that sesaminol inhibits oxidative damage in DNA.

4.4 Stilbenes

Stilbenes are a class of phenolic compounds that are typical in plants such as berries, grapevines, and peanuts. In grapes, resveratrol accumulates in the skins. In the stilbene molecule, two phenyl groups are joined *via* an ethene double bond (Leopoldini, Russo, & Toscano, 2011). Fig. 6 depicts representatives of stilbenes. The most well-known and best-characterized stilbene is resveratrol (3,4′,5-trihydroxystilbene).

This stilbene exhibits cardio-protective, neuro-protective, anticancer, antidiabetic, and antiaging capabilities (El Khawand, Courtois, Valls,

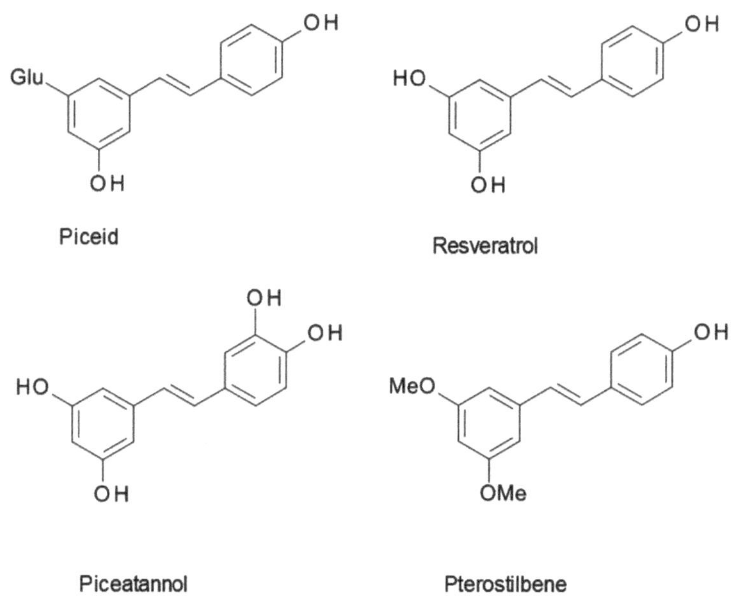

Fig. 6 Chemical structure of stilbenes.

Richard, & Krisa, 2018; Pandey & Rizvi, 2011). The results of several studies show that resveratrol has the capability to protect lipids and proteins against oxidation induced under conditions that challenge the body's redox status (Markus & Morris, 2008; Pandey & Rizvi, 2009, 2010).

The prevalence of resveratrol in red wine is often considered to play a major role in the so-called "French Paradox": The French have a relatively low incidence of coronary heart disease, while consuming a diet that is relatively rich in saturated fatty acids (Ferrieres, 2004; Orallo, 2006).

The dominance of piceid (the glycosylated form) over resveratrol (the aglycon form) was reported in such vines as Cabernet Sauvignon, Chardonnay Merlot, and Riesling (Lee & Rennaker, 2007). The stilbene chemistry in the *Vitis genus* and in wine has been reviewed by Pawlus, Waffo-Téguo, Shaver, and Mérillon (2012).

4.5 Tannins

According to the chemical structure, tannins are divided into two subclasses: condensed tannins and hydrolysable tannins. Condensed tannins (proanthocyanidins-PAC) are biopolymers that are based on flavan-3-ols (Fig. 7 (1)). At high temperatures in alcohol solutions of strong mineral acids, these tannins release anthocyanidins and catechins as terminal end groups. The chemical structure of procyanidin B1, which is a molecule with a $4 \rightarrow 8$ bond (epicatechin-($4\beta \rightarrow 8$)-epicatechin), is presented in Fig. 7 (2).

Hydrolyzable tannins are classified into simple gallic and ellagic acid derivatives, namely gallotannins and ellagitannins. The hydrolysis of gallotannins yields gallic acid, whereas that of ellagitannins yields ellagic acid (Fig. 7 (3)). Gallotannins are natural polymers that are formed by the subsequent esterification of hydroxyl groups of D-glucose and gallic acid in polymeric chains (Fig. 7 (4)). Ellagitannins are esters of hexahydroxydiphenic acid and polyols: glucose or quinic acid (Fig. 7 (5)).

The results of several investigations showed strong antimicrobial activities of condensed tannins against such bacteria as *Micrococcus luteus*, *Proteus mirabilis*, *Bacillus licheniformis*, *Nocardia asteroids*, *Salmonella typhimurium*, *Staphylococcus aureus*, and *Bacillus subtilis* (Hatano et al., 2005; Jabri et al., 2016; Shahwar, Raza, Mughal, Abbasi, & Ahmad, 2010). Tannins separated from red bean (*Phaseolus vulgare* L.), buckwheat (*Fagopyrum esculentum* Moench), walnuts (*Junglas regia* L.), and hazelnuts (*Corylus avellana* L.) showed antibacterial activities against *Listeria monocytogenes*, *Staphylococcus aureus*, *Escherichia coli* O157:H7, *Brochothrix thermosphacta*, *Pseudomonas fragi*,

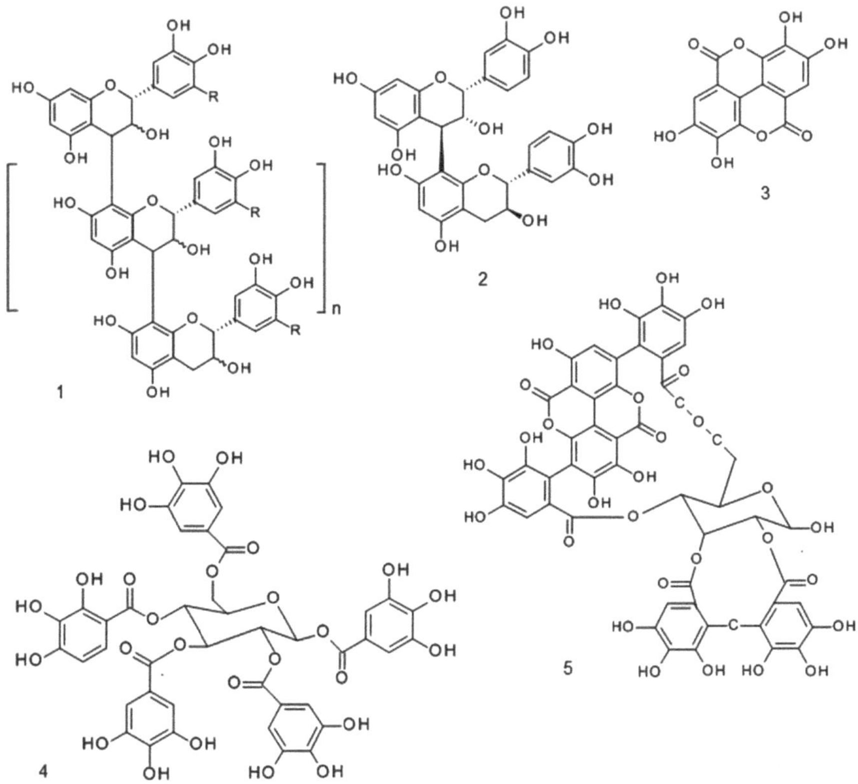

Fig. 7 Chemical structure of tannins. **1**—condensed tannins (proanthocyanidins); **2**—procyanidin B2; **3**—ellagic acid; **4**—gallotanin isolated from red maple; **5**—punicalagin (ellagotannin from pomegranate).

Salmonella typhimurium, and *Lactobacillus plantarum* (Amarowicz, Dykes, & Pegg, 2008). Gallotannins exhibited anticancer, anti-angiogenic, antioxidant, anti-inflammatory, and anti-ulcerative activities (Karas, Ulrichová, & Valentová, 2017).

Proanthocyanidins can donate hydrogen atoms or electrons as typical primary antioxidants and act as secondary antioxidants (Amarowicz, 2007). These compounds can chelate Fe(II) (Karamać, Kosińska, & Amarowicz, 2006) and inhibit the activity of cyclooxygenase (Zhang, De Witt, Murugesan, & Nair, 2004). Proanthocyanidins that were separated from green tea and from bearberry (*Arctostaphylos uva-ursi*) leaves exhibited antioxidant activity in a meat model system (Amarowicz, Pegg, & Barl, 2001; Amarowicz, Pegg, Dykes, Troszyńska, & Shahidi, 2005; Pegg, Amarowicz, & Naczk, 2005). The tannin fraction that separated from the crude extracts of red bean, adzuki bean, lentil, faba bean, and broad bean exhibited much stronger antioxidant activity than the fractions of flavonoids and phenolic acids that were separated from the same extracts (Amarowicz, Estrella, Hernández, & Troszyńska, 2008; Amarowicz, Estrella, et al., 2009; Amarowicz et al., 2010; Amarowicz, Karamać, Dueñas, & Pegg, 2017; Amarowicz & Shahidi, 2017, 2018). In a cited investigation, the authors employed DPPH radical, ABTS radical cation, and reducing-power assays, as well as the β-carotene-linoleate model system. The antiradical activity of procyanidins B1 and B3 from adzuki beans against peroxyl radicals was reported by Ariga and Hamano (1990). The correlation between TEAC values and the content of condensed tannins in extracts obtained from broad bean, adzuki bean, faba bean, green lentil, red lentil, and read bean was described by Amarowicz, Troszyńska, Baryłko-Pikielna, and Shahidi (2004).

Ellagitannin sanguine H-6 is an important antioxidant of raspberries (Mullen et al., 2002). According to Borges, Degeneve, Mullen, and Crozier (2010) this ellagitannin is responsible for 44.7% of the total antioxidant capacity of raspberries. Ellagic acid, 4-acetylarabinosylellagic acid, 4-arabinosylellagic acid, and 4-acetylxylellagic acid contributed to the antioxidant potential of raspberry jam (Zafrilla, Ferreres, & Tomás-Barberán, 2001). Gallotannins with higher degrees of galloylation exhibited stronger antioxidant activities than those with low degrees of galloylation (Tian, Li, Ji, Zhang, & Luo, 2009). The antioxidant properties of gallotannins increased after thermal hydrolysis. The product of such processes exhibited a synergistic antioxidant effect with citric acid, ascorbyl palmitate, and α-tocopherol in a bulk oil system or edible oil-based system (Terán-Hilares, Chirinos, Pedreschi, & Campos, 2018).

5. Sources of natural antioxidants

The main sources of natural antioxidants include oil seeds, cereals, legumes, plants of the Lamiaceae family, tea and coffee, tree nuts, fruits, and berries. The potential antioxidant capacity of plant materials, or the antioxidant activity of their derived extracts depends on the content of phenolic compounds in the plants or extracts.

In laboratory practice, two methods are used for the evaluation of these parameters: the total phenolics content (TPC) and total flavonoids content (TFC). The determination of phenolic compounds in plant material includes an extraction procedure, followed by a colorimetric reaction.

For the TPC determination, the extracted phenolics react under alkaline conditions with Folin-Ciocalteu's phenol reagent (Singleton, Orthofer, & Lamuela-Raventhós, 1999). The results are reported as equivalent of the standard mass per unit of raw material or extract. Gallic acid, as standard compounds, gallic acid, (+)-catechin, tannic acid, ferulic acid, and sinapic acid have been used as standard compounds. For the TFC determination, flavonoid-aluminum chloride ($AlCl_3$) complexation was applied (Christ & Müller, 1960). The results are expressed as (+)-catechin or rutin equivalents.

The TPC and TFC methods have several limitations. The reaction of Folin-Ciocaleu's reagent used in the TPC reaction, is subject to some great interferences, particularly any readily reducible component present within the assay mixture. Ascorbic acid is the major interference in the case of most fruits (Craft et al., 2012). Moreover, regarding TFC, chelation of flavones/flavonols with $AlCl_3$ do not react uniformly with selected standards, indicating these methods as inadequate for the estimation of total flavonoid content in unknown samples. In plants, flavonols and flavones exist as glycosides, and the presence of sugar moieties hamper proper chelation with $AlCl_3$. Any blockage of the hydroxyl groups by glycosylation in carbons of positions 3, 5, 3′ or 4′ prevents chelation with $AlCl_3$ and the bathochromic shift toward 415/420 nm (Mammen & Mammen, 2012). In a research work of Pękal and Pyrzyńska (2014), the flavone luteolin, formed complexes that showed a strong absorption at 405–420 nm, while the λ_{max} for the complexes formed by chrysin and apigenin, did not show the catechol moiety in B ring, at 377 nm.

Nevertheless, these methods are still used to estimate the total quantities of phenolic compounds and flavonoids in plants. The selected concentrations of the total phenolics and flavonoids in plant material are presented in Tables 1–7.

5.1 Oil seeds

Among oil seeds, rapeseed and canola exhibit the highest content of phenolic compounds (Table 1). The content of flavonoids is very low, and the content of total flavonoids in rapeseed has not been reported. The main phenolic compounds of rapeseed are sinapine (choline ester of sinapic acid), sinapic acid as well as, is esters, and glucosides (Szydłowska-Czerniak, 2013). In rapeseed hulls, the presence of condensed tannins has been reported (Naczk, Amarowicz, Pink, & Shahidi, 2000). In crude rapeseed/canola oil obtained from roasted rapeseed/canola seeds, the main phenolic

Table 1 Content of total phenolics and total flavonoids in oil seeds.

Plant material	Total phenolic compounds		Total flavonoids		Reference
	Units	Content	Unit	Content	
Rapeseed flour	μg SAE/g	5310–6940			Vuorela, Meyer, and Heinonen (2004)
Rapeseed	mg SAE/g DW				Zago et al. (2015)
Seeds		10.80			
Cake		16.39			
Meal		19.68			
Canola seeds	mg SAE/g defatted seeds	7.73; 8.75; 11.95			Khattab, Eskin, Aliani, and Thiyam (2010)
Rapeseed crude extract	mg SAE/ 100 g extract	1577; 1705			Siger, Czubinski, Dwiecki, Kachlicki, and Nogala-Kalucka (2013)
Soybean seeds	mg GAE/kg	253			Cho et al. (2010)
Soybean seeds	mg GAE/g	5.6			Lee, Hwang, Son, and Cho (2019)
Soybean seeds					Alu'datt, Rababah, Ereifej, and Alli (2013)
Full fat	mg GAE/g	0.45–2.75			
Defatted		0.67–3.05			
Soybean seeds	mg GAE/g	6.10 ± 0.10			Yao, Cheng, Wang, Wang, and Ren (2011)
Soybean CO_2 extract	mg GAE/ 100 g extract	0.9–16.0	mg QE/ 100 g extract	0.0–65.0	Alvarez, Cabred, Ramirez, and Fanovich (2019)
Flaxseed	mg GAE/ 100 g	109.9–246.9			Deng et al. (2017)
Flaxseed					Alu'datt et al. (2013)
Full fat	mg GAE/g	0.92–1.90			
Defatted		1.09–1.59			

Continued

Table 1 Content of total phenolics and total flavonoids in oil seeds.—cont'd

Plant material	Total phenolic compounds		Total flavonoids		Reference
	Units	Content	Unit	Content	
Seasame cake extract	mg GAE/g	1.72			Nadeem et al. (2014)
Sunflower seeds	mg GAE/g	11.2±0.5			Herbello-Hermelo et al. (2018)
Pumpkin seeds		0.387±0.138			

GAE, gallic acid equivalents; SAE, sinapic acid equivalents; QE, quercetin equivalents; DW, dry weight.

Table 2 Content of total phenolics and total flavonoids in cereal grains.

Plant material	Total phenolic compounds		Total flavonoids		Reference
	Unit	Content	Unit	Content	
Wheat	mg GAE/g DW	2.86±0.07			Deng et al. (2012)
Oat		2.83±0.16			
Corn		1.97±0.06			
Buckwheat		4.48±0.46			
Black rice		9.47±0.48			
Millet		2.05±0.13			
Sorghum		1.92±0.05			
Barley	μg GAE/g DW	1929–2917			Suriano et al. (2018)
Blue highland	mg GAE/ 100g DW	336–453	mg CE/ 100g DW	37.9–45.3	Yang, Dang, and Fan (2018)
Barley (12 cultivars)					
Corn water extract	mg TAE/g extract	66.9			Yogesh, Jha, and Ahmad (2014)
Wheat bran	mg GAE/ 100g DW	460–845	mg CE/ 100g DW	988–1861	Smuda, Mohsen, Olsen, and Hassan (2018)
Rice bran		954–1012		556–1012	
Corn bran		1538–1925		677–1235	
Wheat germ		164–321		317–682	
Rice germ		272–674		180–674	
Corn germ		228–624		257–546	

GAE, gallic acid equivalents; TAE, tannic acid equivalents; CE, catechin equivalents; DW, dry weight.

Table 3 Content of total phenolics and total flavonoids in legume seeds.

Plant material	Total phenolic compounds: Unit	Total phenolic compounds: Content	Total flavonoids: Unit	Total flavonoids: Content	Reference
Black turtle clipse bean	mg GAE/g seeds	6.98 ± 0.48	mg CE/g seeds	3.30 ± 0.11	Xu, Yuan, and Chang (2007)
Black turtle T-39 bean		3.37 ± 0.15		2.51 ± 0.12	
Navy bean		0.57 ± 0.05		0.92 ± 0.02	
Pinto bean		3.76 ± 0.06		2.99 ± 0.12	
Red kidney bean		4.05 ± 0.05		3.39 ± 0.09	
Pink bean		3.77 ± 0.19		3.65 ± 0.13	
Small red bean		5.76 ± 0.38		4.24 ± 0.10	
Grass pea (30 cultivars)	mg CE/g extract	1.88–7.12			Rybiński, Karamać, Sulewska, Börner, and Amarowicz (2018)
	mg CE/ 100 g seeds	20.3–70. 3			
Mung bean water extract	mg TAE/g extract	248.6			Yogesh et al. (2014)
Lima bean	mg GAE/g	4.72 ± 0.23			Yao et al. (2011)
Broad bean		6.43 ± 0.71			
Common bean		8.59 ± 0.11			
Pea		4.87 ± 0.14			
Jack bean		3.77 ± 0.34			
Goa bean		2.44 ± 0.20			
Adzuki bean		2.68 ± 0.19			
Hyacinth bean		6.28 ± 0.23			
Chicking vetch		1.58 ± 0.14			
Garbanzo bean		1.04 ± 0.24			
Dral		7.95 ± 0.29			
Cow bean		3.94 ± 0.05			
Rice bean		4.88 ± 0.11			
Mung bean		8.14 ± 0.21			

Continued

Table 3 Content of total phenolics and total flavonoids in legume seeds.—cont'd

Plant material	Total phenolic compounds		Total flavonoids		Reference
	Unit	Content	Unit	Content	
Blue lupin	mg GAE/ 100g DW	313.7; 394.2			Grela et al. (2017)
White lupin		289.3; 278.4			
Yelow lupin		328.9; 273.5			
Pea		225.6; 189.2			
Chickpea		259.7			
Lentil		398.3			
Grass pea		249.7; 294.1			
Bean		418.4			
Broad bean		366.7			

GAE, gallic acid equivalents; TAE, tannic acid equivalents; CE, catechin equivalents; DW, dry weight.

Table 4 Content of total phenolics and total flavonoids in plants of Lamiaceae family.

Plant material	Total phenolic compounds		Total flavonoids		Reference
	Unit	Content	Unit	Content	
Extract of:	mg SAE/g extract				Amarowicz, Żegarska, et al. (2009)
Thyme		203			
Oregano		288			
Marjoram		254			
Oregano	mg GAE/g	2.1–4.2			Santos-Zea, Antunes-Ricardo, Gutierrez-Uribe, García-Pérez, and Benedito (2018)
Oregano	mg GAE/g DM	22.87; 40.74; 51.26	mg QE/g dm	10.44; 10.48; 11.80	Gutiérrez-Grijalva, Angulo-Escalante, León-Félix, and Heredia (2017)
Oregano extract	mg GAE/g extract	41			Timothy, Vishnu Priya, and Gayathri (2018)

Table 4 Content of total phenolics and total flavonoids in plants of Lamiaceae family.—cont'd

Plant material	Total phenolic compounds		Total flavonoids		Reference
	Unit	**Content**	**Unit**	**Content**	
Thyme ethanolic extract	g GAE/100g	20.31	g CA/100 g	11.39	Wisam, Nahla, and Tariq (2017)
Thyme water extract		20.3		10.31	
Thyme	mg TAE/g DM	62.40 ± 0.03	mg QE/g dm	8.55 ± 0.04	Tohidi, Rahimmalek, and Arzani (2017)
Sage extract	mg GAE/100 g extract	47.92	mg GAE/100 g extract	20.47	Gantner et al. (2018)
Sage extract	mg RAE/100 g extract	5514–7787	mg RAE/100 g extract	1199–1905	Dent, Kovačević, Bosiljkov, and Dragović-Uzelac (2017)
Basil	mg GAE/g	57			Alnahdi, Ayaz, and Danial (2011)
Rosemary		55			

GAE, gallic acid equivalents; TAE, tannic acid equivalents; RAE, rosmarinic acid equivalents; QE, quercetin equivalents; DM, dry matter.

Table 5 Content of phenolic compounds in tea and coffee.

Material	Compound	Unit	Content	Reference
Green tea ($n=95$)	Total phenolics	g/100 g	17.5 (11.9–25.2)	Astill, Birch, Dacombe, Humphrey, and Martin (2001)
	Catechins (HPLC)		13.3 (7.1–20.8)	
Black tea ($n=55$)	Total phenolics		14.4 (7.3–21.9)	
	Catechins (HPLC)		2.1 (0.7–8.8)	

Continued

Table 5 Content of phenolic compounds in tea and coffee.—cont'd

Material	Compound	Unit	Content	Reference
Teas prepared according to package instructions	Catechins	mg/L		Astill et al. (2001)
Green tea ($n=15$)			591 (287–825)	
Black tea ($n=8$)			999 (801–1299)	
Green tea (different time and temperature for preparation)	Total phenolics	mg GAE/g DW	68.13–131.31	Balci and Özdemir (2016)
	Total flavonoids	mg CE/g DW	17.97–32.04	
	C	mg/g DW	8.91–17.09	
	EC		4.29–9.55	
	EGC		28.03–59.42	
	ECG		8.02–14.61	
	EGCG		38.05–69.66	
	GCG		5.53–18.53	
	GC		2.30–12.92	
	CG		0.04–1.74	
Green tea (water extract)	Total phenolics	mg GAE/mL	26.33 ± 1.73	Nibir, Sumit, Akhand, Ahsan, and Hossain (2017)
Black tea (water extract):				
Flowery broken orange pekoe			6.78 ± 0.55	
Broken orange pekoe			8.84 ± 0.50	
Red dust			8.20 ± 0.49	
Green tea (water extract)		mg CE/mL	50.12 ± 0.60	
Black tea (water extract):				

Table 5 Content of phenolic compounds in tea and coffee.—cont'd

Material	Compound	Unit	Content	Reference
Flowery broken orange pekoe			13.93 ± 1.08	
Broken orange pekoe			17.7 ± 0.82	
Red dust			19.12 ± 0.33	
Green tea	Total phenolics	mg GAE/g DW%	86.3 (65.8–106.2)	Khokhar and Magnusdottir (2002)
	contribution of catechins in total phenolics		50.4–98.0	
	EGC	mg/g	16.2–34.6	
	C		0–1.3	
	EC		4.4–9.5	
	EGCG		20.3–42.6	
	ECG		3.7–8.5	
Black tea	Total phenolics	mg GAE/g DW%	103.0 (80.5–134.9)	
	contribution of catechins in total phenolics		10.1–37.3	
	EGC	mg/g	0.2–6.3	
	C		0–1.7	
	EC		1.4–5.6	
	EGCG		2.7–25.2	
	ECG		0.5–8.6	
Black tea	Total phenolics	g GAE/100 g	7.52–8.29	Serpen et al. (2012)
	C	mg/100 g	59.3–98.3	
	EC		61.9–78.8	
	ECG		89.5–11.5	
	EGC		1038–1141	
	EGCG		102–155	

Continued

Table 5 Content of phenolic compounds in tea and coffee.—cont'd

Material	Compound	Unit	Content	Reference
	GC		524–650	
	Theaflavin		121–298	
	Theaflavin 3,3′-digallate		12.5–15.9	
	Thearubigins		5940–6830	
Green tea	Total phenolics	mg GAE/g DW	31.6 ± 0.31	Bizuayehu, Atlabachew, and Ali (2016)
Black tea			21.3–25.0	
Green tea	Total flavonoids	mg CE/g DW	23.2 ± 0.68	
Black tea			8.17–14.7	
Green tea	Tannins	mg TAE/g DW	7.45 ± 0.23	
Black tea			5.64–6.90	
Coffee bean:				Farah, de Paulis, Trugo, and Martin (2005)
Coffea arabica (Brazil)	CQA	g/100 g DW	4.2	
	FQA		0.28	
	diCQA		0.77	
	TCGA		5.25	
C. arabica (Ethiopia)	CQA		4.6	
	FQA		0.60	
	diCQA		1.37	
	TCGA		5.73	
C. canephora (Uganda)	CQA		5.77	
	FQA		0.47	
	diCQA		1.34	
	TCGA		7.58	
C. arabica (wild)	CQA	g/100 g DW	3.26	Ky, Louarn, Guyot, Hamon, and Noirot (2001)
C. canephora (wild)	FQA		0.19	
	diCQA		0.60	
	TCGA		4.10	

Table 5 Content of phenolic compounds in tea and coffee.—cont'd

Material	Compound	Unit	Content	Reference
	CQA		7.66	
	FQA		1.43	
	diCQA		2.31	
	TCGA		11.30	
Commercial ground roasted coffee	CQA	g/100g DW	0.38–1.25	Monteiro and Trugo (2005)
	FQA		0.06–0.22	
	diCQA		0.09–0.24	
	TCGA		0.47–1.72	
Ground roasted coffee:				Farah et al. (2005)
Coffea arabica (Brazil)	CQA	g/100g DW	2.15	
	FQA		0.17	
	diCQA		0.14	
	TCGA		2.46	
	CQL		0.36	
	FQL		0.04	
	diCQL		0.01	
	CoQL		0.01	
	TCQL		0.41	
	CQA		1.65	
	FQA		0.15	
	diCQA		0.13	
	TCGA		1.93	
C. arabica (Ethiopia)	CQL		0.33	
	FQL		0.44	
	diCQL		0.01	
	CoQL		0.01	
	TCQL		0.38	
C. canephora (Uganda)	CQA		2.76	

Continued

Table 5 Content of phenolic compounds in tea and coffee.—cont'd

Material	Compound	Unit	Content	Reference
	FQA		0.34	
	diCQA		0.23	
	TCGA		3.3	
	CQL		0.3	
	FQL		0.03	
	diCQL		0.03	
	CoQL		–	
	TCQL		0.45	
Instant coffee:				Nogueira and Trugo (2003)
Non-decaffeinated	CQA	g/100 g DW	2.41; 1.30	
	FQA		0.27; 0.14	
	diCQA		0.09; 0.04	
	TCGA		2.77; 1.48	
Decaffeinated	CQA		4.73; 3.33	
	FQA		0.84; 0.60	
	diCQA		0.28; 0.17	
	TCGA		5.85; 4.10	

GAE, gallic acid equivalents; TAE, tannic acid equivalents; CE, catechin equivalents; DW, dry weight; C, catechin; EC, epicatechin; EGC, epigallocatechin; ECG, epicatechin gallate; EGCG, epigallocatechin gallate; GCG, gallocatechin gallate; GC, gallocatechin; CG, catechin gallate; CQA, caffeoyilquinic acid; FQA, feruloyiquinic acid; diCQA, dicaffeoyilquinic acid; TCGA, total chlorogenic acids; CQL, caffeoylquinic lactone; FQL, feruloylquinic lactone; diCQL, dicaffeoylquinic; CoQL, total caffeoylquinic lactone; TCQL, total coumaroylquinic lactone.

Table 6 Content of total phenolics in nuts.

Plant material	Total phenolic compounds		Reference
	Unit	Content	
Walnuts	mg GAE/100 g	1625	Kornsteiner, Wagner, and Elmadfa (2006)
Almonds		239	
Brazil nuts		112	
Pine nuts		32	
Pistachios		867	

Table 6 Content of total phenolics in nuts.—cont'd

Plant material	Total phenolic compounds: Unit	Total phenolic compounds: Content	Reference
Cashew		137	
Macadamia		46	
Peanuts		420	
Pecan		1284	
Hazelnuts (raw)	mg GAE/g	2.07 ± 0.0597	Herbello-Hermelo et al. (2018)
Hazelnuts (toasted)		0.960 ± 0.024	
Walnuts		13.9 ± 2.4	
Almonds (raw)		0.654 ± 0.087	
Almonds (toasted)		5.43 ± 0.305	
Brazil nuts		0.723 ± 0.016	
Pine nuts		1.05 ± 0.08	
Pistachios		1.86 ± 0.07	
Cashew		1.12 ± 0.10	
Macadamia		2.23 ± 0.17	
Peanuts		1.73 ± 0.09	
Pecan		4.40 ± 0.68	
Chestnuts		2.69 ± 0.01	
Walnuts	mg GAE/g	15.5 ± 4.1	Wu, Boecher, Holden, and Haytowitz (2004)
Almonds		4.18 ± 0.84	
Brazil nuts		3.10 ± 0.96	
Pine nuts		0.68 ± 0.25	
Pistachios		16.6 ± 1.2	
Cashew		2.74 ± 0.39	
Macadamia		1.56 ± 0.29	
Peanuts		3.96 ± 0.54	
Pecan		20.2 ± 1.0	
Almond:	mg CE/g extract or fraction		Amarowicz, Troszyńska, and Shahidi (2005)
Crude extract		16.1 ± 0.4	
LMW fraction		7.14 ± 0.2	
HMW fraction		80.4 ± 2.1	

GAE, gallic acid equivalents.

compound with antioxidant activity is 4-ethenyl-2, 6-dimethoxyphenol (other names: canolol, 4-vinyl-2,6-dimethoxyphenol, 4-vinylsyringol) (Galano, Francisco-Márquez, & Alvarez-Idaboy, 2011; Shrestha, Stevens, & de Meulenaer, 2012; Wakamatsu et al., 2005). This

Table 7 Content of total phenolics and total flavonoids in fruits and berries.

Plant material	Total phenolic compounds		Total flavonoids		Reference
	Unit	Content	Unit	Content	
Apple (three cultivars)	mg GAE/kg FW				Lin, Peng, Yang, and Zou (2018)
Whole fruit		671; 862; 1455			
		2363; 3763; 3624			
Peel		160; 825; 1098			
		1437; 1406; 1395			
Flesh					
Pomace					
Apple juice (three apple cultivars)	mg GAE/L	880; 880; 1590			Lin et al. (2018)
African star apple	mg GAE/g extract		mg CE/g extract		Oboh, Adebayo, Ejakpovi, Ogunsuyi, and Boligon (2018)
Flesh pulp		8.07 ± 0.04		4.44 ± 0.11	
Seed coat		6.21 ± 0.10		5.56 ± 0.07	
Back coat		6.13 ± 0.09		4.14 ± 0.10	
Pears of Turkish cultivars	mg GAE/g	2.40–5.87	mg RE/g	3.50–5.47	Ozrenk, Erez, Altintas, and Inal (2018)

Asian pear juice powder	mg GAE/100 g	50.44–223.54			Lee, Ahmed, Jiang, and Eun (2017)
Plum skin	μg GAE/g powder	82.69 ± 0.83	μg CE/g powder	29.39 ± 2.84	Coman et al. (2018)
Peach seeds extract	mg TAE/g	1.92 ± 0.04	mg CE/g	0.81 ± 0.01	
Strawberries (60 varieties)	mg GAE/100 g	8.45–208.58			Nowicka, Kucharska, Sokół-Łętowska, and Fecka (2019)
Red navel orange	mg GAE/100 g DW	652–792			Lu, Lv, Peng, Zhu, and Pan (2018)
Orange	mg GAE/100 g pulp	11.52–31.88			Do Couto, de Souza, Morgado, Ogata, and Cunha Júnior (2018)
Lemon slices	mg GAE/g	3.35 ± 0.6			Fu et al. (2017)
Yellow passion fruit:	mg GAE/100 g DW				Dos Reis, Facco, Flôres, and Rios (2018)
Pulp					
Peel		1297 ± 13			
Seeds		1062 ± 25			
Purple passion fruit:		347 ± 6			
Pulp					

Continued

Table 7 Content of total phenolics and total flavonoids in fruits and berries.—cont'd

Plant material	Total phenolic compounds		Total flavonoids		Reference
	Unit	Content	Unit	Content	
Peel					
Seeds		789 ± 4			
Orange passion fruit:		1571 ± 27			
Pulp		326 ± 1			
Peel					
Seeds					
		1559 ± 5			
		2585 ± 97			
		429 ± 1			
Seeds of:	mg GAE/kg				İnan, Özcan, and Aljuhaimi (2018)
Mandarin		204–287			
Orange		204–230			
Lemon		152–212			
Blackberry	mg GAE/100 g DW	2835.9 ± 63.8			Lee et al. (2015)
Black currant		2382.4 ± 60.8			
Blueberry		2706.7 ± 96.0			

Chokeberry juice	mg GAE/L	4772			Daskalova et al. (2015)
Goji berrie	mg GAE/g	14.1 ± 0.455			Herbello-Hermelo et al. (2018)
Elderberry fruits	μg GAE/g powder	474 ± 8	μg CE/g powder	164 ± 2	Coman et al. (2018)
Elderberry skin		1005 ± 54		535 ± 15	
Elderberry skin and seeds		122 ± 5		47 ± 1	
Black raspberry seed extract	mg GAE/g extract	11.8 ± 0.3			Luther et al. (2007)
Chardonnay grape seed extract		99.8 ± 3.7			
Grapes (30 varieties)	mg GAE/g FW	0.294–1.407	mg CE/g FW	0.082–0.132	Liu et al. (2018)
Italian red grape skin	μg GAE/g powder	275.39 ± 1.65	μg CE/g powder	140.74 ± 9.66	

GAE, gallic acid equivalents; CE, catechin equivalents; RE, rutin equivalents; FW, fresh weight.

compound is produced by the decarboxylation of sinapic acid. Improvement in the oxidative stability of canola and mustard seeds oils is due to the appearance of canolol after the roasting process, which was confirmed by Wijesundera, Ceccato, Fagan, and Shen (2008). Morley, Grosse, Leischa, and Lau (2013) showed that sinapic acid decarboxylase (SAD) produced canolol with an overall yield of 3.0 mg per g of canola meal.

Isoflavones are the dominant phenolic constituents of soybean. Based on their chemical structure, three aglycon forms of isoflavones that are commonly found in soybeans are daidzein, genistein, and glycitein. These three isoflavones can be conjugated by glucose (daidzin, geinstin, glycitin, respectively), malonylglucose (malonyldaidzin, malonylgeinstin, malonylglycitin), and acetylglucose (acetyldaidzin, acetylgeinstin, acetylglycitin) (Luthria, Biswas, & Natarajan, 2007). The content of the total isoflavones in soybeans ranges from 1.2 to 2.4 mg/g (Rostagno, Palma, & Barroso, 2004).

Lignans are the main phenolic constituent of flaxseed. In seeds of this plant, phenolics are present in the form of the lignan macromolecule (LM), which is composed of five SDG residues that are interconnected by hydroxymethylglutaric acid (HMGA). Additional phenolic compounds present in LM are 4-*O*-β-glucopyranosyl-*p*-coumaric acid (CouAG), 4-*O*-β-glucopyranosyl-ferulic acid (FeAG), 4-*O*-β-glucopyranosyl-caffeic acid (CaAG), and herbacetin diglucoside (flavonoid) (Kosińska, Penkacik, Wiczkowski, & Amarowicz, 2011; Strandås, Kamal-Eldin, Andersson, & Åman, 2008; Struijs, Vincken, Verhoef, Voragen, & Gruppen, 2008). According to Struijs, Vincken, Doeswijk, Voragen, and Gruppen (2009), the molecular weight of LM as determined by using MALDI-TOF MS, ranged from 1500 to 4300.

The SDG content in 29 cultivars grown in Sweden and in Denmark ranged from 11.7 to 24.1 mg/g in defatted flaxseed flour and from 6.1 to 13.3 mg/g in whole flaxseed (Johnsson, Kamal-Eldin, Lundgren, & Åman, 2000). In 32 varieties of Chinese flaxseed, the content of SDG was between 11.37 and 37.31 mg/g (Deng et al., 2017).

The predominant phenolic compound in the sunflower kernel is one of the chlorogenic acid isomers: 5-*O*-caffeoylquinic acid (Aramendia et al., 2000). This compound comprises 43–73% of the phenolic compounds that are extracted from kernels (Weisz, Kammerer, & Carle, 2009). Karamać, Kosińska, Estrella, Hernández, and Dueñas (2012) additionally found new isomers of coumaroylquinic acid (probably 3-*O*-*p*-coumaroylquinic and 4-*O*-*p*-coumaroylquinic acids) and dicaffeoylquinic acids (probably 1,3-di-*O*-caffeoylquinic and 1,4-di-*O*-caffeoylquinic acids).

5.2 Cereals

As presented in Table 2, phenolic acids are the chief phenolic constituents of cereals. Ferulic acid was reported to be the most dominant phenolic acid in wheat; the content of vanillic, *p*-coumaric, sinapic, and caffeic acids was significantly lower. Purple wheat possessed a higher content of vanillic and ferulic acid than did other colored wheat grains. The content of flavonoids in purple, yellow, and red wheat was much higher than that in white wheat (Verma, Hucl, & Chibbar, 2008).

Caffeic, *p*-coumaric, ferulic, and sinapic acids were the dominant phenolic acids determined in the caryopses of two cultivars of wheat, rye, and triticale. The majority of phenolic acids were found in the form of soluble esters (Weidner, Amarowicz, Karamać, & Dąbrowski, 1999). In the study by Pihlava et al. (2015), the major phenolic acid in the esterified fraction of rye was sinapic acid, followed by much lower amounts of ferulic and caffeic acids.

Alkylresorcinols (also known as resorcinolic lipids) are typical phenolic compounds occurring in cereal grains. These compounds are composed of a single phenolic ring with an alkyl side chain containing 13–27 carbon atoms (Kozubek & Tyman, 1999). From research by Ross et al. (2003), alkylresorcinols were found in wheat (489–1429 μg/g), rye (720–761 μg/g), triticale (439–647 μg/g), and barley (42–51 μg/g), but not in rice, oats, maize, sorghum, or millet.

For rice and rice bran, the presence of γ-oryzanol is very characteristic, which is a ferulate ester of triterpene alcohols and plant sterols (Patel & Naik, 2004). Cycloartenyl ferulate, 24-methylenecycloartanyl ferulate, and campesteryl ferulate are the three major components and account for 80% of γ-oryzanol (Xu, Godber, & Xu, 2001).

In the last decade, the interest of food scientists has been focused on pseudocereals as sources of natural antioxidants. In the seeds of common buckwheat (cultivars from western, central and southeastern Europe grown in the Balkan area), quercetin-3-*O*-rutinoside, isoorientin (luteolin-6-*C*-glucoside), vitexin (apigenin-8-*C*-glucoside), caffeic acid-pentoside, procyanidin trimer, and epiafzelechin–epicatechin were found to be the main phenolic compounds (Kiprovski et al., 2015).

The chemical structures of kaempferol-3-*O*-(2-β-glucopyranosyl)-α-L-rhamnopyranoside-7-*O*-α-L-rhamnopyranoside and kaempferol-3-*O*-(4-β-xylopyranosyl)-α-L-rhamnopyranoside -7-*O*-α-L-rhamnopyranoside were identified from canihua by Peñarrieta, Alvarado, Åkesson, and Bergenståhl (2008). The content of flavonols in quinoa seeds that were cultivated in Japan ranged from 130 to 193 mg/100 g fresh weight (FW) (Hirose, Fujita, Ishii, & Ueno, 2010). The presence of cinnamic acid derivatives of quinoa seeds was reported by Cutillo, Dellagreca, Gionti, Previtera, and Zarrelli (2006). The content of *p*-coumaric acid in quinoa samples from Peru ranged from 2.26 to 27.5 mg/100 g (Repo-Carrasco-Valencia, Hellström, Pihlava, & Mattila, 2010).

5.3 Legumes

Legumes are characterized by a relatively high content of total phenolics and flavonoids (Table 3). Some legumes are also rich in condensed tannins (Vaz Patto et al., 2015).

In broad bean, 14 compounds, namely, phenolic acids (*p*-coumaric and ferulic acid), catechins (epicatechin, epicatechin glucoside, and epicatechin gallate), procyanidin gallate, prodelphidin dimer, gallate procyanidin dimer, and digallate procyanidin dimer, were identified by Amarowicz and Shahidi (2017). Gallate procyanidin dimer, gallate procyanidins, and acetylated kaempferol hexose were the major phenolic compounds present in the extract of faba bean (Amarowicz & Shahidi, 2018). There were 20 compounds (hydroxycinnamates, procyanidins, gallates, flavonols, dihydroflavonols, dihydrochalcones), which were identified in the crude extract of red bean (Amarowicz et al., 2017). In green lentils, catechin and epicatechin glucosides, procyanidin dimers, quercetin diglycoside, and *p*-coumaric acid were the dominant phenolic compounds, while in red lentil, quercetin diglycoside, catechin, digallate procyanidin, and *p*-hydroxybenzoic were the dominant phenolic molecules (Amarowicz, Estrella, et al., 2009; Amarowicz et al., 2010). The adzuki bean extract was characterized by a high content of catechin and epicatechin glucosides, procyanidin dimers, myricetin, and protocatechuic acid (Amarowicz, Estrella, et al., 2008).

Legume seeds are also a source of lignans. The content of isolariciresinol, lariciresinol, secoisolariciresinol, pinoresinol, and matairesinol in legumes was reported by Durazzo, Turfani, Azzini, Maiani, and Carcea (2013). Green lentil exhibited the highest content of secoisolariciresinol and pinoresinol levels of approximately 75 μg/100 g dry matter. A high content of lariciresinol (177 μg/100 g dry matter) was determined in red lentils. The content of lignans in bean, chickpeas, and lentils were low, as reported by Thompsom, Boucher, Liu, Cotterchio, and Kreiger (2006). In the cited study, the greatest content of 29.9 μg secoisolariciresinol/100 g was found in white beans.

5.4 Plants of the Lamiaceae family

Plants belonging to Lamiaceace family are not only a source of essential oils but also phenolic compounds (Table 4). Very typical for these plants is the presence of rosmarinic acid, a caffeic acid ester of 3-(3,4-dihydroxyphenyl) lactic acid.

The phenolic profile of sage (*Salvia officinalis* L.) was characterized by a presence of rosmarinic acid and 16 flavonoids (10 flavones and 6 flavanones) including apigenin, luteolin, hispidulin, and nepetin. Some flavanones were glucosylated at the 5- or 7-position of the flavonoid structure (Lee et al., 2018).

The main phenolic constituents of rosemary (*Rosmarinus officinalis* L.) (mg/100 g dry matter) were as follows: rosmarinic acid (1286), epirosmanol (1113), carnosol (806), carnosic acid (655), caffeic acid (278), and catechin (255) (Shan, Cai, Sun, & Corke, 2005).

Sonmezdag, Kelebek, and Selli (2018) identified 21 phenolic compounds in thyme, of which 9 were phenolic acids (rosmarinic acid, rosmarinic acid-glucoside, chlorogenic acid, caffeic acid, lithospermic acid, gallic acid, 3,4-dihydroxyphenyl acetic acid, protocatechuic acid-hexoside, protocatechuic acid) and 12 were flavonoids. The flavonoids included rutin, luteolin, and derivatives of luteolin, apigenin, kaempferol, and hesperetin.

The most abundant phenolic compound in oregano (*Origanum vulgare* L.) was rosmarinic acid (12.8 mg/g of plant), followed by chlorogenic acid (2.10 mg/g). Among flavonoids, hyperoside was found in the largest amount (1.05 mg/g), followed by isoquercitrin (0.71 mg/g) (Oniga et al., 2018).

Rosmarinic acid was the dominant polyphenol (1.7–1.4 mg/g of fresh product) in three cultivars of basil (*Ocimum basilicum* L.). Much lower contents of caffeic acid, gallic acid, chlorogenic acid, quercetin, rutin, and apigenin were determined (Fratianni et al., 2017).

Waller et al. (2017) found 4-hydroxybenzoic acid, caffeic acid, chlorogenic acid, hesperetin, and rutin as the key phenolic compounds of marjoram (*Origanum majorano* L.).

5.5 Tea and coffee

The main polyphenolic compounds that are present in green tea are (−)-epigallocatechin-3-gallate (EGCG), (−)-epigallocatechin (EGC), (−)-epicatechin-3-gallate (ECG), and (−)-epicatechin (EC) (Fig. 8). EGCG accounts for 50–70% of catechins (Khan & Mukhtar, 2013). The literature data on the content of catechins in green tea are presented in Table 5. The

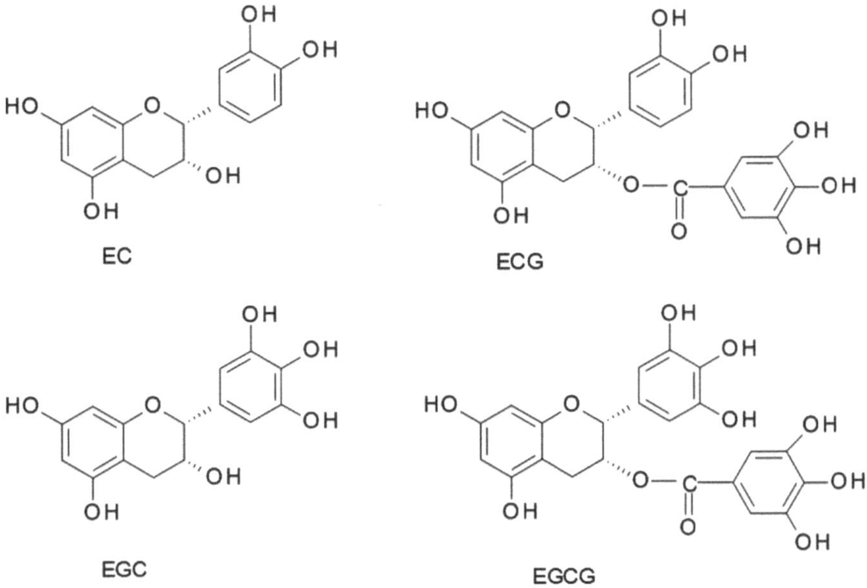

Fig. 8 Chemical structure of catechins. EC—(−)-epicatechin; ECG—(−)-epicatechin-3-gallate; EGC—(−)-epigallocatechin; EGCG—(−)-epigallocatechin-3-gallate.

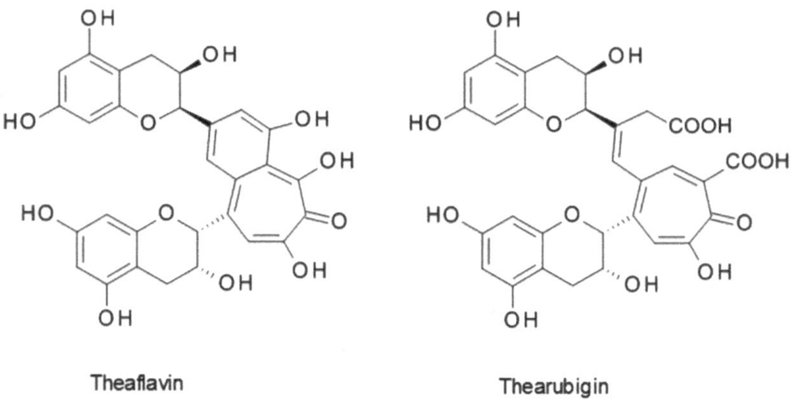

Fig. 9 Chemical structure of black tea polyphenolics.

content of individual catechins, which are expressed as mg/100 mL, are as follows: EC, EGC, ECG, and EGCG in an infusion that was obtained from 10 green teas that originated from countries, as follows: 28.6–49.4 (EC), 76.7–200 (EGC), 51.5–86.2 (ECG), and 107–124 (EGCG) (Koch et al., 2018).

In black tea, the dominant phenolic compounds are thearubigins and theaflavins (Fig. 9). These phenolics are formed from the oxidation and condensation of flavan-3-ols in tea leaves during the enzymatic oxidation of black tea (Takemoto & Takemoto, 2018). The content of thearubigins and theaflavins in black tea from India ranged from 3.63 to 7.86 g/100 g and from 0.08 to 0.47 g/100 g (Khanum, Faiza, Sulochanamma, & Borse, 2017). In black tea (beverage), the content of theaflavin, theaflavin-3-gallate, theaflavin-3′-gallate, and theaflavin-3-3′-digallate was 0.95, 0.73, 0.60, and 0.59 mg/100 mL infusion, respectively (Lee, Kim, Park, Kim, & Kim, 2016).

Chlorogenic acids (CGAs) and their derivatives are the main phenolic compounds of green coffee beans, reaching levels up to 14% dry matter. CGAs include such phenolics as caffeoylquinic acids, dicaffeoylquinic acids, feruloylquinic acids, *p*-coumaroylquinic acids and mixed diesters of caffeic and ferulic acids with quinic acid (Farah & Donangelo, 2007). During roasting, chlorogenic acids undergo such processes as isomerization, hydrolysis, and degradation of CGA into low-molecular-weight compounds (Clifford, 1999; Farah et al., 2005). At high temperatures of roasting, a part of CGA can also be transformed into quinolactones and melanoidins (Moreira, Nunes, Domingues, & Coimbra, 2012). The content of individual CGA in coffee is presented in Table 5.

The extraction of CGA into the beverage is affected by the grind of the coffee, the ratio of coffee to water, the brewing method, the water temperature and time of beverage preparation when coffee is in contact with water. Domestic preparation of Arabica and Robusta coffee results in the extraction of 70–200 and 70–350 mg CGA per 200 mL cup, respectively (Farah & Donangelo, 2007).

5.6 Tree nuts

Tree nuts are a rich source of phenolic compounds (Table 6) that belong to several classes. Fanali et al. (2018) confirmed the presence of catechin, epicatechin, epicatechin 3-gallate, and two procyanidins in hazelnut kernels.

Rusu et al. (2018) detected 18 polyphenols in the extracts of walnuts: chlorogenic acid, caftaric acid, ferulic acid, gentisic acid, caffeic acid, *p*-coumaric acid, sinapic acid, isoquercitrin, rutozid, myricetol, fisetin, quercitrin, quercetin, luteolin, kaempferol, patuletin, hyperoside, and apigenin.

In almond skins, the most prolific compounds are flavan-3-ols. Among phenolic acids, the greatest was procatechuic acid; among flavanols, the most abundant was isorhamnetin-3-*O*-rutinoside; and among flavanones, it was naringenin-7-*O*-glucoside (Pasqualone et al., 2018).

Ma et al. (2014) used high-performance liquid chromatography coupled with electrospray ionization mass spectrometry and determined a large variety of phenolic compounds in peanut skins, including phenolic acids and their esters, stilbenes (*trans*-resveratrol and *trans*-piceatannol), flavan-3-ols, isoflavones, flavanols, flavone biflavonoids, and proanthocyanidins.

Pecans were characterized by the presence of such phenolic compounds as gallic acid, ellagic acid pentose, epicatechin gallate, ellagic acid, valoneic acid dilactone hydrate, ellagic acid galloyl pentose, ellagic acid galloyl pentose/dimethyl ellagic acid pentose, methyl ellagic acid galloyl pentose, methyl ellagic acid galloyl pentose, and methyl ellagic acid galloyl pentose (Robbins, Greenspan, & Pegg, 2016).

5.7 Fruits and berries

Results reported in Table 7 confirmed that fruits and berries are a rich source of phenolic compounds in the human diet. Several classes of phenolic compounds are present in these plants.

According to Vrhovsek, Rigo, Tonon, and Mattivi (2004), in apples representing eight of the most widely cultivated varieties in western Europe, flavanols were a major class of polyphenols (71–90%), followed by hydroxycinnamates (4–18%), flavonols (1–11%), dihydrochalcones (2–6%) and, in red apples, anthocyanins (1–3%). Tsao, Yang, Xie, Sockovie, and Khanizadeh (2005) found that polyphenols were fivefold more prevalent in the skin than in the flesh of the apples. According to McGhie, Hunt, and Barnett (2005), almost 46–50% of the polyphenolics in whole apples are located in the skin.

Flavonols, flavanols, procyanidins, dihydrochalcones, and hydroxycinnamates were determined in the peel of Golden Delicious apples by Chinnici, Bendini, Gaiani, and Riponi (2004), with epicatechin, procyanidin B2, and phloridzin as the most abundant compounds.

A total of 23 bioactive compounds were identified in plums by UPLC-MS, including gallic acid, rutin, resorcinol, chlorogenic acid, catechin, and ellagic acid, and the antioxidant capacity can be attributed to these species (Hernández-Ruiz et al., 2018).

Profiles of 52 phenolic compounds (proanthocyanidins, flavonoids, phenolic acids, hydroxychalcones) in apples and plums were obtained by UPLC-DAD-ESI-MS, as reported by Navarro et al. (2018). Among the phenolic compounds that were identified in pear skin were rutin, (+)-catechin, daidzein, 5,7,3′,5′-tetrahydroxyflavanone, quercetin-3-*O*-(3″-*O*-galloyl)-α-l-rhamnopyranoside, apigenin, and quercetin were the phenolic compounds identified in pear skin (Qiu et al., 2018).

The analysis by LC/DAD of peach samples allowed for the identification and quantification of 14 non-colored phenolics (hydroxybenzoic acids, hydroxycinnamic acids, flavan-3-ols, flavanols) and 3 anthocyanins (unknown, cyanidin-3-*O*-glucoside, and cyanidin-3-*O*-rutinoside) (Bentoa, Gonçalvesa, & Silva, 2018).

The citrus flavonoids include a class of glycosides, namely, hesperidin and naringin, and another class of *O*-methylated aglycones of flavones, such as nobiletin and tangeretin, which are relatively two common polymethoxylated flavones (Li, Wang, Guo, Zhao, & Ho, 2014; Rafiq et al., 2018). Flavones rutin, nobiletin and tangeretin, and flavanones hesperidin, narirutin and eriocitrin were identified and quantified in all organic and conventional orange juices by Mesquita and Monteiro (2018).

Phenolic compounds of berries (strawberry, raspberry, blackberry, blueberry, and cranberry) include flavonoids, such as anthocyanins (i.e., cyanidin glucosides and pelargonidin glucosides), flavonols (quercetin, kaempferol, myricetin), flavanols (catechins and epicatechin), phenolic acids, and hydrolyzable tannins, such as ellagitannins (Skrovankova, Sumczynski, Mlcek, Jurikova, & Sochor, 2015). Among berries, blueberries and blackberries are the richest sources of anthocyanins (Kalt, Forney, Martin, & Prior, 1999).

The phenolic composition of grapes and their derived products is based on flavonoids, such as flavonols, flavanols, anthocyanins, stilbenes, such as *trans*-resveratrol, and phenolic acids, such as gallic, caffeic acids vanillic, syringic, and ellagic (Teixeira, Eiras-Dias, Castellarin, & Gerós, 2013). *trans*-Resveratrol is located mostly in grape skin.

6. Extraction strategies of phenolic compounds from plant material

In phytochemistry, mixtures of water with organic solvents, such as methanol, ethanol, acetone, propanol, dimethylformamide, ethyl acetate, and propanol, are used for the extraction of phenolic compounds from plant material (Antolovich, Prenzler, Robards, & Ryan, 2000; Kozłowska, Rotkiewicz, Zadernowski, & Sosulski, 1983; Luthria & Mukhopadhyay, 2006; Naczk & Shahidi, 2006; Robards, 2003; Zadernowski, Naczk, & Nesterowicz, 2005). Khattab et al. (2010) showed that 70% methanol was the most efficient solvent in extracting phenolic compounds from canola seed compared to either 70% ethanol or 70% isopropanol. The content of the total phenolic compounds in rapeseed meal for 70% methanol, 70% ethanol, and water was 6580, 5310, and 5960 μg sinapinic acid equivalents per g of meal, respectively (Vuorela et al., 2004). The study of Liang et al. (2018) demonstrated a 70% extraction of phenolic compounds from hempseed cake (*Cannabis sativa* L.).

The extraction time and the ratio of the solvent-to-sample (R) play important roles in the recovery of polyphenols from plant material. During longer extraction, phenolic compounds can be oxidized. The addition of reducing agents can protect phenolics against this process (Krygier, Sosulski, & Hogge, 1982; Naczk & Shahidi, 2006). However, according to Deshpande and Cheryan (1985), the optimum time for extraction of phenolic compounds from bean seeds was 50–60 min. The extraction of main phenolics from hempseed cake was positively affected by the time of exposure (Liang et al., 2018). Changing R from 1:5 to 1:10 increased the extraction yield of phenolic compounds and condensed tannins from commercial canola meals when using 70% (v/v) acetone (Naczk & Shahidi, 2006).

Luthria and Mukhopadhyay (2006) showed that the extraction of phenolics from eggplant was influenced by shaker, rotary shaker, stirring, sonication, or reflux applied for extract preparation based on the sample preparation parameters.

For the separation of free phenolic acids from esters and glucosides present in the extract, alkaline and acidic hydrolyses were applied (Acosta-Estrada, Gutiérrez-Uribe, & Serna-Saldívar, 2014; Krygier et al., 1982). Aglycons of flavonoids were obtained by using enzymatic or acidic hydrolysis. Phenolic compounds bound to plant walls were liberated after chemical (alkaline or acidic) or enzymatic hydrolysis. To release phenolic acids in cereals, α-amylase, cellulose, and commercial enzymes (Thermamyl, Ultrafo L. Viscozyme, Lallzyme) have been reported (Bartolomé & Gómez-Cordovés, 1999; Yu, Vasathan, & Temelli, 2001; Zupfer, Churchill, Rasmusson, & Fulcher, 1998).

Subcritical water extraction (SWE) has become a popular green chemistry method for the extraction of phenolic compounds from plant material. This is a simple, inexpensive, convenient and environmentally friendly approach that is easily coupled with other extraction and purification methods. The parameters affecting the efficiency of subcritical water extraction of phenolic compounds are the temperature, flow rate, extraction mode, matrix composition, pH, pressure, modifiers, and additives (Khoshnoudi-Nia, Niakosari, & Tahsiri, 2017).

The polarity of water under pressure changes with the temperature. At lower temperatures, the water is much more polar than at higher temperatures ($\geq$250 °C). Under this condition, the polarity of the pressurized water is similar to that of polar organic solvents (Herrero, Cifuentes, & Ibañez, 2006). Using SWE, the polar analytes are selectively extracted at lower temperatures, while fewer polar analytes are extracted at higher temperatures.

Subcritical water extraction has been applied recently for isolation of the phenolic compounds from grape seeds and American oak wood (Marchante et al., 2019), olive pomace (Seçmeler, Güçlü Üstündağ, Fernández-Bolaños, & Rodríguez-Gutiérrez, 2018), stems, leaves and berries of *Aronia melanocarpa* Medik. (Cvetanović et al., 2018), onion (*Allium cepa* L.) and onion juice product (Kim & Lim, 2018), pistachio (*Pistacia vera* L.) hulls (Erşan, Güçlü Üstündağ, Carle, & Schweiggert, 2018), black carrot (Aşkin Uzel, 2017), pomegranate (*Punica granatum* L.) (Yan, Cao, & Zheng, 2017), and spent coffee grounds (*Coffea arabica* L.) (Xu, Wang, Liu, Yuan, & Gao, 2015).

Enhancement of the total phenolics compounds in the extract from plant material can be obtained by microwave pretreatment. In an experiment by Papoutsis et al. (2017), the microwave pretreatment of lemon pomace significantly affected the total phenolics content, total flavonoids, and proanthocyanidins, as well as the antioxidant activity of the extract. The listed properties increased as the microwave radiation time and power increased.

7. Antioxidant capacity of plant and plant extracts—*In vitro* assays and model systems

The procedure for determining antioxidant capacity includes extraction, evaporation of organic solvent, freeze-drying of residual water, and choice of appropriate antioxidant activity assay. The results can be expressed in relation to the extract or by starting plant material. The procedure can be shortened when the antioxidant assay is used just after extraction, without sample concentration. In this case, the results are reported only in relation to the plant material.

7.1 *In vitro* assays

In laboratory practice, antioxidant assays are used as described in part 3 of this article. The selected results of *in vitro* assays are reported in Table 8.

Table 8 Antioxidant potential of selected plants.

Material	Method	Unit	Results	Reference
Rapeseed crude extract	FRAP	μmol TE/g	20.21; 27.05	Siger et al. (2013)
Rapeseed extract	DPPH	Scavenging % (concentration 0.5 mg/mL)	20	Vuorela et al. (2004)
	Liposome model	Inhibition of conjugated dienes (%; concentration 8.4 μg/mL)	69.3–90.9	
		Inhibition hexanal (%; concentration 8.4 μg/mL)	97.3–99.4	
Soybean CO_2 extract	DPPH	μmol TE/100 g extract	0.3–12.0	Alvarez et al. (2019)
Soybean	DPPH	μmol TE/g seeds	15.17 ± 0.93	Yao et al. (2011)
Flaxseed	DPPH	mg TE/100 g	32.56–46.22	Deng et al. (2017)
	ABTS	mmol TE/g	14.22–36.14	
	FRAP	mg TE/g	0.58–1.08	
Extracts of 12 common beans from Italy	DPPH	EC_{50} (mg/mg DPPH•)	39–2810	Ombra et al. (2016)
Black turtle Eclipse	DPPH	μmol TE/g seeds	18.95 ± 0.03	Xu et al. (2007)
Black turtle T-39			14.49 ± 0.14	
Navy bean			1.48 ± 0.04	
Pinto bean			13.79 ± 0.03	
Red kidney			16.81 ± 0.11	
Pink bean			15.49 ± 0.17	
Small red			17.90 ± 0.13	
Black turtle Eclipse	FRAP	mmol Fe^{2+}/100 g seeds	9.70 ± 0.35	
Black turtle T-39			6.05 ± 0.20	
Navy bean			1.27 ± 0.03	
Pinto bean			7.24 ± 0.28	
Red kidney			7.93 ± 0.47	
Pink bean			4.07 ± 0.08	
Small red			4.53 ± 0.19	

Table 8 Antioxidant potential of selected plants.—cont'd

Material	Method	Unit	Results	Reference
Black turtle Eclipse	ORAC	μmol TE/g seeds	92.73 ± 4.99	
Black turtle T-39			48.91 ± 2.04	
Navy bean			13.30 ± 0.55	
Pinto bean			51.13 ± 3.64	
Red kidney			70.48 ± 6.99	
Pink bean			90.85 ± 1.92	
Small red			70.58 ± 3.24	
Grass pea (30 cultivars)	ABTS	mmol TE/g extract	0.015–0.037	Rybiński et al. (2018)
		mmol TE/100g seeds	0.158–0.372	
	FRAP	mmol Fe^{2+}/g extract	0.045–0.120	
		mmol Fe^{2+}/100g seeds	0.487–1.189	
Lima bean	DPPH	μmol TE/g seeds	36.25 ± 1.02	Yao et al. (2011)
Broad bean			37.15 ± 2.14	
Common bean			46.83 ± 1.75	
Pea			31.92 ± 2.46	
Jack bean			37.81 ± 2.33	
Goa bean			37.15 ± 2.01	
Adzuki bean			18.08 ± 1.94	
Hyacinth bean			28.01 ± 1.17	
Chickling vetch			15.39 ± 1.48	
Garbanzo bean			1.28 ± 0.06	
Dal			37.93 ± 1.32	
Cow bean			37.27 ± 2.48	
Rice bean			35.36 ± 1.99	
Mung bean			45.36 ± 1.27	

Continued

Table 8 Antioxidant potential of selected plants.—cont'd

Material	Method	Unit	Results	Reference
Wheat	ABTS	μmol Trolox/g DW	3.69 ± 0.25	Deng et al. (2012)
Oat			1.57 ± 0.20	
Corn			4.52 ± 0.10	
Buckwheat			9.43 ± 0.35	
Black rice			30.03 ± 1.10	
Millet			1.88 ± 0.11	
Sorghum			1.03 ± 0.12	
Wheat	FRAP	μmol Fe(II)/g DW	9.28 ± 0.70	
Oat			16.15 ± 1.06	
Corn			11.66 ± 0.40	
Buckwheat			31.20 ± 0.93	
Black rice			126.19 ± 2.91	
Millet			11.29 ± 1.19	
Sorghum			11.70 ± 0.22	
Barley	DPPH	μmol TE/g DW	8.20–13.40	Suriano et al. (2018)
	ABTS	μmol TE/g DW	12.51–13.40	
Blue highland barley (12 cultivars)	DPPH	mg TE/100g DW	1337–1640	Yang et al. (2018)
	FRAP	mg TE/100g DW	817–1292	
	ABTS	mg TE/100g DW	639–1041	
Wheat bran	FRAP	mM $FeSO_4$	21.5–36.1	Smuda et al. (2018)
Rice bran			20.6–40.6	
Corn bran			12.6–34.7	
Wheat germ			28.8–39.2	
Rice germ			19.1–38.8	
Corn germ			16.3–37.3	
Oregano	DPPH	μM TE/g DW	225; 373; 500	Gutiérrez-Grijalva et al. (2017)
	ABTS	μmol TE/g DW	351; 350; 249	
	ORAC	μmol TE/g DW	754; 812; 349	
Oregano	FRAP	μmol TE/g	15.4–26.4	Santos-Zea et al. (2018)

Table 8 Antioxidant potential of selected plants.—cont'd

Material	Method	Unit	Results	Reference
Thyme	DPPH	IC_{50} (μg/ml)		Tohidi et al. (2017)
Thyme ethanolic extract	DPPH	IC_{50} (μg/ml)	12.1	Köksal et al. (2017)
Thyme water extract	ABTS		3.4	
Thyme ethanolic extract	DPPH		50.08	
Thyme water extract	ABTS		40.03	
Sage extract	FRAP	μmol TE/100 g extract	170–174	Dent et al. (2017)
	FRAP	μmol Fe^{2+} /100 g extract	457–604	
Basil extract	DPPH	Scavenge effect (%)	76	Alnahdi et al. (2011)
Rosemary extract			76	
Walnuts	ORAC	μmol TE/g	131 ± 35	Wu et al. (2004)
Almonds			43 ± 9	
Brazil nuts			8.62 ± 2.06	
Pine nuts			4.43 ± 1.11	
Pistachios			75.6 ± 10.5	
Cashew			15.2 ± 2.0	
Macadamia			14.4 ± 2.3	
Peanuts			28.9 ± 2.4	
Pecan			175 ± 10	
Yellow cashew	ABTS	mmol TE/100 g DW	3.322	Moo-Huchin et al. (2015)
		mg vit. C/100 g DW	0.970	
	DPPH	mmol TE/100 g DW	1.579	
		mg vit. C/100 g DW	0.340	

Continued

Table 8 Antioxidant potential of selected plants.—cont'd

Material	Method	Unit	Results	Reference
Red cashew	ABTS	mmol TE/100 g DW	3.050	
		mg vit. C/100 g DW	0.890	
	DPPH	mmol TE/100 g DW	1.593	
		mg vit. C/100 g DW	0.343	
Apple (3 cultivars) Whole fruit	DPPH	μmol TE/kg FW	5310; 5290; 10,210	Liu et al. (2018)
Peel			38,490; 26,850; 19,850	
Flesh			4860; 7330; 1040	
Pomace			7060; 6800; 15,090	
Whole fruit	ORAC	μmol TE/kg FW	12,980; 19,050; 21,540	
Peel			38,300; 62,810; 45,270	
Flesh			940; 17,400; 15,870	
Pomace			20,400; 23,870; 23,550	
Apple juice (3 apple cultivars)	DPPH	μmol TE/L	3750; 4000; 8680	Liu et al. (2018)
	ORAC	μmol TE/L	11,740; 12,600; 14,190	
African star apple extract Flesh pulp	DPPH	IC_{50} (mg/mL)	2.27 ± 0.14	Oboh et al. (2018)
Seed coat			3.21 ± 0.11	
Back coat			2.50 ± 0.09	
Flesh pulp	Scavenge of $^{\bullet}OH$	IC_{50} (mg/mL)	0.38 ± 0.07	
Seed coat			0.39 ± 0.11	
Back coat			0.48 ± 0.08	

Table 8 Antioxidant potential of selected plants.—cont'd

Material	Method	Unit	Results	Reference
Yellow passion fruit Pulp	DPPH	IC_{50} (mg/100 mL)	0.20 ± 0.03	Dos Reis et al. (2018)
Peel			1.69 ± 0.03	
Seed			1.18 ± 0.03	
Yellow passion fruit Pulp			3.32 ± 0.02	
Peel			6.98 ± 0.20	
Seed			6.30 ± 0.08	
Orange passion fruit Pulp			2.41 ± 0.01	
Peel			2.45 ± 0.03	
Seed			2.68 ± 0.03	
Yellow passion fruit Pulp	ABTS	IC_{50} (mg/100 mL)	0.82 ± 0.03	
Peel			2.22 ± 0.01	
Seed			3.84 ± 0.08	
Yellow passion fruit Pulp			4.59 ± 0.01	
Peel			9.37 ± 0.05	
Seed			4.76 ± 0.03	
Orange passion fruit Pulp			3.72 ± 0.05	
Peel			2.95 ± 0.02	
Seed			3.87 ± 0.00	
Seeds of: Mandarin	ABTS	mmol TE/kg	0.52–2.66	İnan et al. (2018)
Orange			1.49–1.83	
Lemon			1.79–1.86	
Orange	FRAP	μmol ferrous sulfate/g pulp	6.15–9.79	Do Couto et al. (2018)

Continued

Table 8 Antioxidant potential of selected plants.—cont'd

Material	Method	Unit	Results	Reference
Red navel orange	DPPH	μmol AAE/g DW	9.2–14.6	Lu et al. (2018)
	ORAC	μmol TE/g DW	271–306	
Strawberries (60 varieties)	DPPH	μmol TE/100 g	391–1287	Nowicka et al. (2019)
	ABTS	μmol TE/100 g	943–2254	
Blackberry	ABTS	mg AAE/100 g DW	6125.7 ± 176.3	Lee et al. (2015)
Black currant			4618.3 ± 188.8	
Blueberry			4814.6 ± 166.0	
Chokeberry juice	ORAC	μmol TE/L	55,307	Daskalova et al. (2015)
Black raspberry seed extract	ORAC	μmol TE/g extract	95.8 ± 4.5	Luther et al. (2007)
Chardonnay grape seed extract			662.5 ± 39.4	
Grapes (30 varieties)	ABTS	μmol TE/g FW	0.339–4.839	Liu et al. (2018)
	FRAP	μmol Fe^{2+}/g FW	1.289–11.767	
Green tea (different time and temperature for preparation)	DPPH	IC_{50} (mg DW/mg DPPH)	0.48–1.16	Balci and Özdemir (2016)
Tea infusion: Green	ABTS	IC_{50} (μg/mL)	6.72; 6.85; 6.73	Konieczyski, Viapiana, and Marek Wesolowski (2017)
Black			6.88; 6.42, 6.60	
Oolong			6.70	
Green tea	DPPH	mg AAE/g DW	80.0 ± 0.63	Bizuayehu et al. (2016)
Black tea			28.8–52.3	
Coffee fruit extract	ORAC	μmol Trolox/g	15,246; 6097	Mullen et al. (2011)
Coffee fruit powder			823; 735	
Coffee silverskin	DPPH	Antiradical effect (%)	24.1–40.6	Bessada, Alves, Costa, Nunes, and Oliveira (2018)
	FRAP	mM FSE	4.35–20.5	
Coffee beans ($n = 21$; different countries)	Total content of antioxidants (amperometric method)	mg QE/g	20.1–32.1	Yashin, Yashin, Wang, and Nemzer (2013)
		mg QE/cup	147.1–224.7	

TE, Trolox equivalents; AAE, ascorbic acid equivalents; QE, quercetin equivalents; DM, dry matter; FW, fresh weight; FSE, ferrous sulfate equivalents; IC_{50}/EC_{50}, concentration of antioxidant scavenging 50% of free radical in sample.

7.2 Model systems

Antioxidant activity of plant extracts, their fractions or pure phenolic compounds separated from the extract can also be investigated in model systems. In this case, the extracts, fractions or pure compounds are added to food, and the antioxidant activity is reported to have a protection effect against lipid or protein oxidation.

7.2.1 Oil system

Antioxidants that are added to edible oils should protect unsaturated fatty acids and increase their stability to thermal degradation, as well as demonstrate good thermal stability. Extracts from herbal plants, such as thyme, rosemary, sage, marjoram, and oregano, are a rich source of natural antioxidants (Chrpová et al., 2010; Mekinić et al., 2014; Oliveira, Ribeiro-Santos, Ramos, Castilho, & Sanches-Silva, 2018). According to the results of Kozłowska and Gruczyńska (2018), the oxidative stability of sunflower oil samples enriched with oregano extracts and soybean oil supplemented with thyme extracts was improved compared to samples without the addition of herbal plant extracts and synthetic antioxidants.

Combinations of different herbal plant extracts and synthetic antioxidants may also have synergistic effects in preventing hydroperoxide formation compared to samples containing only herbal plant extracts or only synthetic antioxidants (Hraš, Hadolin, Knez, & Bauman, 2000). In a study carried out by Ramalho and Jorge (2008), it was noted that rosemary extract added to soybean oil showed a positive effect on its oxidative and thermal stability.

Moreover, Abdalla and Roozen (1999) showed that thyme and lemon balm extracts are natural antioxidants in sunflower oil and oil-in-water emulsion. Suja, Abraham, Thamizh, Jayalekshmy, and Arumughan (2004) demonstrated that sesame cake extract could be used as a substitute for synthetic antioxidant to protect soybean, sunflower, and safflower oils.

In a bulk-stripped corn oil system, almond extracts at a concentration level of 200 ppm were able to reduce the formation of hexanal 82–93%. The inhibition of hexanal formation by the additives that were used decreased in the order of green shell cover extract ~>brown skin extract ~>whole seed extract. The TBARS values of the control shoved a five-fold increase at the end of a 7-day storage period; with the addition of almond extracts, there was only a two- to three-fold increase (Wijeratne, Amarowicz, & Shahidi, 2006).

In several studies, antioxidant properties of plant extracts were investigated in fish oil. After 4 days of storage, a significant decrease in the TBARS values of fish oil from yellowfin tuna (*Thunnus albacares* Bonnaterre) by the *Stevia rebaudiana* stem water extract, compared to the control group, was reported by Yu et al. (2017). The results indicate that *Stevia rebaudiana* waste can be exploited as a strong natural antioxidant material to inhibit fish oil oxidation.

Chardonnay grape and raspberry seed flour extracts were able to suppress lipid oxidation and rancidity development in menhaden fish oil. Chardonnay grape seed flour extract at 7.4 mg flour equivalents/mL exhibited the same suppression of lipid oxidation in fish oil as 130 ppm mixed tocopherols (Luther et al., 2007).

The findings of Raudoniute et al. (2011) indicate that strawberry leaf extract may retard the production of volatile oxidation products of bluefish (*Pomatomus saltatrix*), which agrees with the results of the measurement of hexanal.

In an experiment by Sekhon-Loodu, Sumudu, Warnakulasuriyaa, Rupasinghe, and Shahidi (2013), apple peel phenolics were incorporated at a phenolic concentration of 200 μg/mL into bulk fish oil. TBARS values revealed that at this concentration, the crude extract prevented fish oil oxidation better than α-tocopherol and BHT.

7.2.2 Butter system

Several authors used butter to investigate the antioxidant properties of natural antioxidants in a lipid-water-lipid emulsion system. The addition of green tea and rosemary extracts at a concentration of 0.02% to butter increased its oxidative stability, which was conducted under Rancimat and Oxidograph test conditions at 110 °C (Gramza-Michałowska, Korczak, & Reguła, 2007).

In an experiment by Amarowicz, Żegarska, et al. (2009), the protective effect of the addition of a thyme extract (0.05 and 0.01%) was observed during the first 27 days of storage at 60 °C. Oxidation of butter was monitored by the changes in PV. The observed induction period of clarified butter fat treated with rosemary extract and stored at 60 °C was 26 days (addition 0.05%) and 31 days (addition 0.1%). The result of the control without extract was only 13 days (Żegarska, Rafałowski, Amarowicz, Karamać, & Shahidi, 1998).

Ghee (butter oil) incorporated with orange peel extract showed a storage period of 21 days (6, 32, and 60 °C), with the least development of the PV, TBARS and FFA control samples (Asha et al., 2015).

Sumac (*Rhus coriora* L.) ethyl acetate extract (addition of 0.2 and 0.5%) exhibited significant positive effects on oxidative stability of Turkish Yayik butter stored for 120 days regarding both PV and TBARS values (Sert, Arslan, Ayar, & Özcan, 2015).

An alcoholic extract of rosemary at a concentration of 400 mg/kg improved the oxidative stability of butter at temperatures of 60 and 110 °C, as monitored by the formation and degradation of peroxides (Santos, Shetty, & da Silva Miglioranza, 2014).

Results obtained by Nadeem et al. (2014) showed that an ethanolic extract of sesame cake at concentrations of 50, 100 and 150 ppm can be used for the long-term preservation of olein butter, with acceptable sensory characteristics. In this research, lipid oxidation was evaluated using PV, the concentration of conjugated dienes, and the loss of oleic acid.

7.2.3 Meat system

Addition of plant materials containing natural phenolic antioxidants can prevent the oxidation of lipids and proteins in meat and poultry products (Ahmad, Gokulakrishnan, Giriprasad, & Yatoo, 2015; Oswell, Thippareddi, & Pegg, 2018; Shah, Bosco, & Mir, 2014). On the list of used plants are hawthorn, blackberry, strawberry, dog rose and extracts obtained from peels (Rodríguez-Carpena, Morcuende, Andrade, Kylli, & Estévez, 2011; Rodríguez-Carpena, Morcuende, & Estévez, 2011). The extracts of alma, lychee seeds, mustard leaf, and green tea were useful for extending the refrigerated storage of ground meat products (Kumar & Langoo, 2016; Lee et al., 2010; Qi, Huang, Huang, Wang, & Wei, 2015).

Shan, Cai, Brooks, and Corke (2009) listed a long list of commercially available extracts. The list includes extracts of grape skin and seed, pine bark, coffee, oregano, adzuki bean, carob fruit, green tea, and rosemary. In the meat model systems, in addition to scavenging free radicals, natural phenolic compounds can chelate metal ions and inhibit the Fenton reaction (Estévez & Heinonen, 2010).

The most-used natural antioxidant in the meat industry is rosemary extract. According to FSIS Directive 7120.1, rosemary extracts are "Safe and suitable ingredients used in the production of meat, poultry, and egg products, which explicitly allows the use of rosemary extract as a component of an antioxidant blend" (Oswell et al., 2018).

In the cooked comminuted pork model system, whole almond seed, almond brown skin extract, and almond green shell cover extract were used. Almond extracts at 100 and 200 ppm levels inhibited the formation of

TBARS, hexanal, and total volatiles by 2–36 and 22–74%, 20–44 and 54–76%, and 1–23 and 42–70%, respectively (Wijeratne et al., 2006). The water extracts obtained from sprouted mung bean (raw and sprouted), chick pea (raw and sprouted), bean, corn, and fenugreek effectively protected raw chicken meat against lipid oxidation (Yogesh et al., 2014). The lowest values of TBARS after 24 h storage at 4 °C storage were obtained for the chick pea extract. Raw chicken ground meat treated with water extracts of thuja cones (*Thuja occidentalis*) and peach seeds during an 8-day storage at 4 °C showed significantly ($P < 0.01$) lower amounts of TBARs compared to the control group (Yogesh & Ali, 2014).

The results of Karpińska-Tymoszczyk (2013) show that an oil-soluble rosemary extract is effective in delaying lipid oxidation of cooked turkey meatballs that are stored at 4 °C. The author suggests this could be contributed to the synergistic effect of sodium erythorbate and oil-soluble rosemary extract used in the model system. The addition of a sage extract delayed the formation of lipid-derived products of oxidation throughout the storage of turkey meatballs. A significant effect of the addition of sage to turkey meat on the levels of TBARS was observed on the 6th day of storage (Gantner et al., 2018).

Grape seed extracts reduced lipid oxidation in cooked beef and pork patties (overwrapped in PVC, 8 days, 4 °C) (Rojas & Brewer, 2007); pre-cooked, frozen, reheated beef sausage (overwrapped in PVC, then frozen at for 4 months, −18 °C) (Kulkarni, DeSantos, Kattamuri, Rossi, & Brewer, 2011); beef frankfurters (vacuum packed, 90 days 4 °C) (Ozvural & Vural, 2012); raw and cooked goat meat (9 days, 5 °C) (Rababah et al., 2011); restructured mutton slices (aerobic and vacuum packaging, 14 and 28 days, 4 °C) (Reddy et al., 2013); cooked, frozen, reheated beef patties (overwrapped with PVC film, 6 months, −18 °C) (Colindres & Brewer, 2011); and aerobically packaged raw pork (9 days, room temperature) (Shan, Cai, Brooks, & Corke, 2009).

The antioxidant effects of various plant extracts have been evaluated and tested in different seafood model systems (Maqsood, Benjakul, Abushelaibi, & Alam, 2014). In an experiment by Özen and Soyer (2018), pomegranate rind extract at a level of 100 ppm was an excellent antioxidant in inhibitory both the lipid and protein oxidation of mackerel (*Scomber scombrus*) mince during frozen storage. The oxidation process of lipids and proteins was monitored by the presence of TBARS and protein carbonyls. In a fish (salmon) model system, peanut skin extract prevented oxidation in non-irradiated and gamma-irradiated samples by up to 63 and 37%, respectively. TBARS were used in this research to measure lipid oxidation (De Camargo et al., 2017).

Both PV and TBARS analyses showed that the extract of polyphenols of quince (*Cydonia oblonga*) was able to retard lipid oxidative degradation in mackerel (*Scomber scombrus*) fillets stored at 4 °C when compared to the control lot at days 1, 5 and 11 (Fattouch, Sadok, Raboudi-Fattouch, & Ben, 2008).

The oxidation indices (TBARs, PV, CD, conjugated dienes and trienes, *p*-AV, free fatty acids, Totox values) of Atlantic salmon, Atlantic halibut, and blue shark showed that the active packaging with barley husk ethyl acetate extract slowed down lipid hydrolysis and oxidation, with a concentration-dependent effect (Pereira de Abreu, Losada, Maroto, & Cruz, 2010; Pereira de Abreu, Paseiro Losada, Maroto, & Cruz, 2010, 2011).

8. Influence of processing and storage on the content of natural antioxidants in food and their antioxidant activity

Some mechanical processes can influence the content of phenolic compounds in plant material. For instance, dehulling reduces the content of condensed tannins in leguminous seeds. Alonso, Aguirre, and Marzo (2000) reported that this process decreased tannin levels in faba and kidney beans by 92.3 and 93.3%, respectively. In the case of pea, the decrease ranged only between 11.2% and 28.7% (Alonso, Orúe, & Marzo, 1998). Micronization, an example of another processing technique, markedly reduced the content of total phenolics of red bean, but for pinto bean an opposite effect was observed (Oomah, Kotzeva, Allen, & Bassinello, 2014). According to Khattab and Arntfield (2009), micronization resulted in a 6% reduction in the content of tannins in red kidney beans. On the other hand, homogenization increased the concentration of cyaniding-3-*O*-glucoside and cyanidin-3-*O*-rutinoside. The major anthocyanins in a juçara, banana and strawberry smoothie (de Oliveira Ribeiro et al., 2018).

The content of total phenolic compounds in high pressure treated (600 MPa) strawberry purées increased from 855 mg gallic acid equivalents (GAE)/100 g DW to 939 mg GAE/100 g DW (Patras, Brunton, da Pieve, & Butler, 2009). The content of the naringenin and hesperetin (flavanones) in orange juice subjected to high pressure processing increased due to the pressure treatment (400 MPa, 1 min, 40 °C) by 20.2 and 39.9% respectively (Sánchez-Moreno, Plaza, De Ancos, Martin, & Cano, 2005).

In the study of Błaszczak, Amarowicz, and Górecki (2017), fresh and untreated aronia juice possessed a significantly higher total phenolics content compared to juices that had been treated with pressures of 200, 400, and 600 MPa for 15 min. However, after 20, 40 and 60 days of storage, the concentration of the total polyphenols in the high pressure-treated juices was found to be 5–10% greater than that of untreated juice counterpart samples. After 40 and 60 days of storage, the concentration of cyanidin-3-*O*-xyloside was almost twofold higher for pressurized juices compared to the untreated samples. After the same periods, the concentration of cyanidin-3-*O*-arabinoside in pressurized juice was higher by 58% and 10% compared to the untreated juices.

More than 43% of ferulic acid and *p*-coumaric acid was lost after pressurization of tomato purée at 450 MPa for 5 min. The decreases in isochlorogenic and sinapic acids were 79 and 72%, respectively. The contents of (−)-epicatechin and rutin in the purée pressurized at 650 MPa for 15 min were reduced by 52 and 76%, respectively (Jeż, Wiczkowski, Zielińska, Białobrzewski, & Błaszczak, 2018). Jayathunge, Grant, Linton, Patterson, and Koidis (2015) demonstrated that the content of total phenolic compounds of tomato juice decreased by 26% after 1 min exposure at 600 MPa.

The increase in the total phenolics content of grape by-products following high pressure processing, ultrasonics, and pulsed electric field was reported by Corramles, Toepfl, Butz, Knorr, and Tausche (2008). As reported by Tao et al. (2016), the phenolic contents of the wine treated together with French oak chips, by high pressure at 250, 450, and 600 MPa for up to 45 min, increased due to an enhanced extraction of phenolics from the oak chips.

The high pressure treatment of raw and roasted buckwheat groats decreased the contents of total phenolic compounds by 12.6 and 8%, respectively (Błaszczak, Zielińska, Zieliński, Szawara-Nowak, & Fornal, 2013).

Nemzer, Vargas, Xia, Sintara, and Feng (2018) studied the effects of hot-air drying (AD), freeze drying (FD), and refractance window drying (RWD) on the retention of anthocyanins, phenolics, flavonoids, and antioxidant capacity (oxygen radical absorbance capacity {ORAC}) in blueberries, tart cherries, strawberries, and cranberries, as well as concentrations of proanthocyanidins in cranberries and chlorogenic acid and catechins in blueberries. The freeze-dried products exhibited higher ORAC values and a greater content of anthocyanins and total phenolics than fruits processed by AD and RWD. The RWD yielded samples with a lower antioxidant potential and lower retention of total phenolics and anthocyanins. AD-dried fruits were also characterized by a significantly lesser quality retention, as determined by the various quality indices measured in the study.

In the study of Vashisth, Singh, and Pegg (2011), the different drying technologies used for muscadine pomace and time–temperature treatments resulted in a varying content of total phenolics in the dried muscadine products. The trend observed for the retention of phenolics in processed samples was as follows: freeze drying > vacuum belt drying >> hot air drying.

From the study of Siddhuraju (2006), the content of condensed tannins during dry heating (open hot plate at 125 °C for 25 min) of moth bean (*Vigna aconitifolia* (Jacq.) Marechal) seeds decreased from 1.91 to 1.31 g/100 g.

Soaking of leguminous seeds can change the content of the endogenous phenolic compounds. The contents of total phenolics and condensed tannins in faba beans and kidney beans after soaking were reduced (Alonso et al., 2000). A reversed effect was, however, observed by Huber, Brigide, Bretas, and Canniatti-Brazaca (2014): after 10 h of maceration, the content of phenolic acids and flavonoids in the extract of white beans was greater than that in extract of untreated seeds. Based on the research of Chau and Cheung (1997), soaking of beans indigenous to China decreased the content of condensed tannins. The migration of tannins from the legume's cotyledons into the soaking water was observed by Mkanda, Minnaar, and Kock (2007).

Several researchers have reported the effect of cooking on the phenolics of legumes. According to Gujral, Sharma, Gupta, and Wani (2013), cooking red kidney beans reduced the content of total phenolics, but increased the content of total flavonoids. Reduction of the content of total phenolics in beans was also noted by Rocha-Guzmán, González-Laredo, Ibarra-Pérez, Nava-Berúmen, and Gallegos-Infante (2007). In seed coats, this content decreased by ~90%. The authors confirmed diffusion of phenolic compounds out from the seed coats during cooking to the cooking water, and from there to the cotyledons. An increase in total phenolic compounds by 20% in pinto beans after microwave treatment was described by Oomah et al. (2014). Turkmen, Sari, and Velioglu (2005) reported higher extractability of phenolic compounds from legumes after cooking due to partial liberation of phenolic compounds bound to the cell walls.

The cooking of cereals (wheat, pearl millet, rice, maize, sorghum) enhanced selected phenolic, flavonoid and flavonol contents when evaluated by *in vitro* digestion and chemical extraction (Prajapati, Patel, Parekh, & Subhash, 2013). Fares, Platani, Baiano, and Menga (2010) reported a decrease in free phenolic acids in cooked wheat pasta samples. The authors were of the opinion that the reduction in phenolics after heat treatment was mainly due to a decrease in *p*-hydroxybenzoic acid. In the cited work, an increase in phenolic acids liberated from bound forms during *in vitro* digestion was observed.

Domestic cooking of potatoes (boiling, frying, microwaving) resulted in a partial loss of flavonols (quercetin-3-*O*-rutinoside, quercetin-3-*O*-diglucoside, and quercetin-3-*O*-glucosylrutinoside) and caffeic acid derivatives. The highest retention of caffeic acid derivatives was observed for steam-cooking: roughly 50% of their initial contents was retained after steam-cooking, whereas only ~33% of the initial contents was determined in potatoes that had been boiled and fried (Tudela, Cantos, Espin, Tomas-Barberan, & Gil, 2002).

A reduction in the content of total phenolic compounds in beans during extrusion was reported by Alonso et al. (2000). In the experiments of Korus, Gumul, and Czechowska (2007), the effect of extrusion on phenolic compounds of beans was not clear; that is, one variety showed an increase in the quantity of phenolic compounds in extrudates compared to the raw seeds, while the other varieties exhibited a decrease. Extrusion preconditioning and a drying treatment greatly reduced the contents of condensed tannins in faba bean, field pea, and chickpea (Adamidou, Nengas, Grigorakis, Nikolopoulou, & Jauncey, 2011).

In the experiments of Xu and Chang (2008) with black beans, steaming resulted in a greater retention of the total phenolic compounds than boiling did. Noteworthy is that the greatest decrease in the content of total phenolics was observed by pressure steaming. By this technique, the content of condensed tannins in immature faba beans was diminished by 42.3% compared to the corresponding raw material. Boiling of the same material resulted in a decrease of condensed tannins by 41.3% (Boukhanouf, Louaileche, & Perrin, 2016). After pasteurization, the content of anthocyanins like perlagonidin-3-*O*-glucoside in the juçara, banana and strawberry smoothie was reduced (de Oliveira Ribeiro et al., 2018).

The findings of Dos Reis et al. (2018) showed positive effects of pasteurization processing on orange passion-fruit juice and improved the bioaccessibility of the bioactive compounds. The pasteurized juice possessed higher concentrations of phenolic compounds (i.e., 15% > retention at days 0 and 4), epigallocatechin gallate (40% > retention at day 0 and 27% at day 4), and antioxidant capacity (12% > retention at day 0 and 7% at day 4). The fresh juice retained greater levels of quercetin (79% at day 0 and 245% at day 4).

Germination of legumes effects the content and composition of phenolic compounds. According to Alonso et al. (2000), the content of ferulic, vanillic, *p*-hydroxybenzoic, and *p*-coumaric acids decreased in beans after germination. In bean seeds, the presence of flavonol glycosides (i.e., quercetin-3-*O*-rutinoside, quercetin-3-*O*-ramnoside, kaempferol-3-*O*-rutinoside, and kaempferol-3-*O*-glucoside) was detected, but only after germination (López-Amorós, Hernández, & Estrela, 2006). Germination decreases the content of condensed tannins in beans (Alonso et al., 2000). For kidney beans, germination decreased the levels of flavan-3-ols and anthocyanins, while enhancing the contents of flavones and flavonols (Dueñas, Martínez-Villaluenga, Limón, Peñas, & Frias, 2015).

Oats fermented for 72h by *Aspergillus oryzae* var. *effuses*, *Aspergillus oryzae* and *Rhizopus oryzae* showed much higher contents of total phenolics and flavonoids compared to the unfermented starting material (Cai et al., 2014). Similar effects were noted for oat solid-state fermentation (Cai et al., 2012). In the ethyl acetate fraction from the crude extract of oats fermented by *A. oryzae*, the contents of caffeic and ferulic acids were ~3 and 9 times higher, respectively, when compared to native oats. The positive effect of fermentation was also observed for oat antioxidant capacity as determined by the ORAC assay and cyclic voltammetry.

Soybean koji fermented with various filamentous fungi, namely *Aspergillus oryzae*, *Aspergillus sojae*, *Aspergillus awamori*, *Actinomucor taiwanensis*, and *Rhizopus sp.*, exhibited higher contents of total phenolic compounds and antioxidant potential, as determined by the DPPH radical scavenging, reducing power, and Fe^{2+}-chelating capability *in vitro* assays in relation to the control. The soybean koji that had been fermented with *Aspergillus awamori* exhibited the greatest antiradical activity against the DPPH radical, and yielded the best results for the reducing power and Fe^{2+}-chelating capability assays (Lin, Wei, & Chou, 2006). Fungal fermentation (tempeh) of common beans led to a reduction in the seeds antioxidant potential, as determined by the DPPH radical and low-density lipoprotein oxidation assays (Gamboa-Gómez et al., 2016). It was found that application of a mixed culture of tempeh (*Lactobacillus plantarum* and *Rhizopus microspores*) enhanced the antioxidant capacity of unhulled common beans, as determined by the DPPH radical assay (Starzyńska-Janiszewska, Stodolak, & Mickowska, 2014).

From the research of Limón et al. (2015), liquid-state fermentation with *Bacillus subtilis* increased the content of several phenolic acids and flavonoids in extracts of kidney bean. However, fermentation with *Lactobacillus plantarum* showed a reversed effect. During *Rhizopus oligosporus* fungal fermentation of *Pinto durani* and *Bayo victoria* beans for tempeh, Gamboa-Gómez et al. (2016) observed a positive effect on the content of total phenolic compounds.

Changes in the phenolic compounds of plants have been reported by numerous researchers. The storage of "Galaxy" apples under a dynamic controlled atmosphere (DCA) and ultralow oxygen (ULO) conditions resulted in a higher content of total phenolic compounds in the flesh, higher contents of chlorogenic acid and procyanidin B_1 in the peel, and of chlorogenic acid, epicatechin and catechin in the flesh (Stanger et al., 2018). The "Royal Delicious" apple wedges treated with anti-browning agents, namely 4-hexylresorcinol, ascorbic acid, calcium chloride, and cysteine, as well as edible coatings (i.e., carboxyl methylcellulose (CMC) of *Aloe vera* gel) alone as well as in combination and then packed in polypropylene trays showed a reduction in the loss of phenolic compounds and antioxidant activity during storage compared to untreated control samples (Kumar, Sethi, Sharma, Singh, & Varghese, 2018).

In the experiment of Saba and Sogvar (2016), the application of CMC as a fruit coating in combination with anti-browning agents was effective in maintaining the total phenolics level in apple slices during storage. The above mentioned results are in agreement with previous findings reported by Gonzalez-Aguilar et al. (2005), that anti-browning treatments reduced the loss of phenolics content in fresh-cut fruits.

Improved storability of gaseous chlorine dioxid strawberries packed in perforated clamshell during long-term storage (12 days at 2 °C), in terms of their anthocyanins contents and antioxidant capacity was reported by Chiabrando, Giuggioli, Maghenzani, Peano, and Giacalone (2018).

The positive effects of N_2 packaging on dried lemon slices with regards to the retention of total phenolics, hesperidin, and antioxidant capacity determined by the ORAC assay was described by Fu et al. (2017). In this experiment, samples were stored at room temperature for 7 weeks.

During technological processing, the antioxidant potential of raw plant material and food of plant origin can be changed. Soaking of leguminous seeds increased their antioxidant potential when investigated using the ABTS radical cation and DPPH radical assays and decreased the antiradical potential against the hydroxy radical (Chakraborty & Bhattacharyya, 2014).

Pressure and microwave cooking were found to decrease the antiradical potential of red kidney beans against the DPPH radical and increased the antiradical potential against ABTS radical cation. After the microwave treatment of pinto beans, their antioxidant capacity as measured by the ORAC assay was greater by 18% in relation to untreated seeds (Oomah et al., 2014).

The radical-scavenging activity of the extracts obtained from bean seed coats was related to their cooking time (Rocha-Guzmán et al., 2007).

Boiling and steaming of beans exhibited significantly lower antioxidant potential than raw beans (Xu & Chang, 2008). In this study, DPPH radical scavenging and ORAC methods were used. Ombra et al. (2016) observed a marginal impact of cooking on the antioxidant activities of extracts prepared for different bean types. The sterilization of white bean reduced the antioxidant activity (as measured by the ABTS radical cation and DPPH radical assays) of the extract obtained from seeds (Huber et al., 2014). Extrusion of dry beans decreased their antioxidant potential, as determined by EPR studies, and using an emulsion system.

Cooking of cereals such as wheat, pearl millet, rice, maize, and sorghum resulted in an increase in their total antioxidant capacity (Prajapati et al., 2013).

The antioxidant potential of roasted groats, evaluated by a photochemiluminescence and the DPPH radical assays, decreased nearly tenfold in comparison with that of raw groats (Błaszczak et al., 2013). A statistically significant decrease of the Trolox equivalent antioxidant capacity (TEAC) values of the roasted buckwheat sample was also noted by Zieliński, Michalska, Amigo-Benavent, Del Castillo, and Piskula (2009). According to Craft, Kosińska, Amarowicz, and Pegg (2010), the oil-roasting process better retained the antioxidant capacities of peanut kernel phenolics than dry roasting, but both of these processes yielded similar results for the majority of the antioxidant assays conducted.

Pressurized (500 MPa, 150 s) plum purée exhibited a decreased antioxidant capacity (by 13%) compared to the untreated sample (Gonzales-Cebrino, Duran, Delgado-Adamez, Contador, & Ramirez, 2013). Barba, Esteve, and Frigola (2013) observed a reduction in the antioxidant capacity (8–16%) of blueberry juice treated by high pressure at 400 MPa (15 min) and 600 MPa (5–15 min); the ABTS radical cation assay was used to assess the antioxidant activity.

The results of the ABTS radical cation assay for high-pressure treated (200, 400, 600 MPa, 15 min) aronia juices were on average 8% lower compared to the data obtained for the untreated juices. However, the reductions in TEAC and FRAP values during storage of the pressurized juices were lower, when compared to the changes observed for the untreated juices over the same time period (Błaszczak et al., 2017). Contrasting findings were reported by Keenan et al. (2010), who demonstrated that fruit smoothies pressurized at 450 MPa for 5 min and stored for 20 and 30 days showed that 3% and 7% lower reducing activity were evaluated by the FRAP assay, compared to the findings of the untreated samples.

High pressure processing of wine in the presence of oak chips resulted in an increase in antioxidant activity, as determined by the ABTS radical cation assay. Wine samples treated by high pressure at 250, 450 and 650 MPa for 45 min increased in their antioxidant activity from 23.1 to 27.2, 26.4, and 26.8 mmol Trolox equivalents/L, respectively (Tao et al., 2016).

The results of Roldan-Martin, Sanchez-Moreno, Lioria, de Ancos, and Cano (2009) indicate that high pressure treatment (100–400 MPa) resulted in a decrease of the antioxidant activity of onions, as determined by the DPPH radical assay. In contrast, McInerney, Seccafien, Stewart, and Bird (2007) reported that green beans treated for 2 min by high pressure (400–600 MPa) augmented their antioxidant activity capacity, as determined by the FRAP assay.

Sánchez-Moreno, Plaza, de Ancos, and Cano (2006) indicated that anti-radical activity of the extracts of tomato against the DPPH radical was unaffected by high-pressure treatment of 400 MPa for 15 min at 25 °C. The high-pressure treatment (450, 550, and 650 MPa for 5, 10, and 15 min) of tomato purées induced irregular changes in the antioxidant activity of the extracts, as determined by FRAP, photochemiluminescence, and cyclic voltammetry assays (Jeż et al., 2018).

The antioxidant capacity (as measured by photochemiluminescence, ABTS radical cation, and DPPH radical *in vitro* assays), of roasted buckwheat treated with high pressure (200 MPa for 2, 4, and 9 min) was significantly lowered compared to untreated samples (Błaszczak et al., 2013).

9. Conclusions and future perspectives

Consumers are becoming more savvy these days and are reading food labels very carefully. As the push for clean labels becomes more prevalent, the employment of natural antioxidants in foodstuffs will continue to grow. It is therefore important to have an understanding of the possible mechanisms by which these natural antioxidants function in food systems, which of course is linked to the class of compound in question and its chemical structure. As outlined in this chapter, phenolic compounds are the key contributors in plant-based foods or plant-based ingredients that bestow antioxidant activity. Many types/classes of phenolic compounds exist; furthermore, there are even more possible *in vitro* and *in vivo* assays available for their biological characterization. Before any such assay can be performed, extraction strategies and the impact of processing on these natural antioxidative constituents must be ascertained. As described in the latter part of this work, there are numerous ways in which to carry out extractions and evaluations of antioxidant efficacy. The underlying problem, however, is that there is no conclusive means by which to confirm exhaustive extraction of the phenolic constituents from a food or plant material, nor can one ensure that partial degradation of the bioactives have not occurred. Even if one does his/her best job at performing an extraction of the antioxidant constituents, the ultimate question will then be how reliable are some of the *in vitro* techniques at providing insight to what might happen *in vivo* in human cells. If one is not prepared to carry out randomized clinical trials to evaluate the effect of consuming antioxidant-rich foods in a diet, a logical stream of testing of antioxidant capacity will evolve from chemical proof of antioxidant content (i.e., test tube assays), through biological testing via cell-based bioassays. This is not to say that existing *in vitro* assays have no place in the evaluation scheme of a food antioxidant potential; quite the contrary, they offer a great starting point in helping to decide which foodstuff might be worthy of further studies by cell-based or *in vivo* assays. Outside of an expensive clinical trial, this seems to be the future direction in which antioxidant capacity of natural phenolic compounds from foods is heading.

References

Abdalla, A., & Roozen, J. P. (1999). Effect of plant extracts on the oxidative stability of sunflower oil and emulsion. *Food Chemistry*, *64*, 323–329.

Acosta-Estrada, B. A., Gutiérrez-Uribe, J. A., & Serna-Saldívar, S. O. (2014). Bound phenolics in foods, a review. *Food Chemistry*, *152*, 46–55.

Adamidou, S., Nengas, I., Grigorakis, K., Nikolopoulou, D., & Jauncey, K. (2011). Chemical composition and antinutritional factors of field peas (*Pisum sativum*), chickpeas (*Cicer arietinum*), and faba beans (*Vicia faba*) as affected by extrusion preconditioning and drying temperatures. *Cereal Chemistry*, *88*, 80–86.

Adlercreutz, H., Mousavi, Y., Clark, J., Hockerstedt, K., Hämälainen, E., Wähälä, K., et al. (1992). Dietary phytoestrogens and cancer: In vitro and in vivo studies. *Journal of Steroid Biochemistry and Molecular Biology*, *41*, 331–337.

Adom, K. K., & Liu, R. H. (2005). Rapid peroxyl radical scavenging capacity (PSC) assay for assessing both hydrophilic and lipophilic antioxidants. *Journal of Agriculture and Food Chemistry*, *53*, 6572–6580.

Ahmad, S. R., Gokulakrishnan, P., Giriprasad, R., & Yatoo, M. A. (2015). Fruit-based natural antioxidants in meat and meat products: A review. *Critical Reviews in Food Science and Nutrition*, *55*, 1503–1513.

Alnahdi, H. S., Ayaz, N. O., & Danial, E. N. (2011). Screening of some medicinal plants for antioxidant and antimicrobial activity and their phenolic contents. *Australian Journal of Basic and Applied Sciences*, *5*, 808–815.

Alonso, R., Aguirre, A., & Marzo, F. (2000). Effects of extrusion and traditional processing methods on antinutrients and *in vitro* digestibility of protein and starch in faba and kidney beans. *Food Chemistry, 68*, 159–165.

Alonso, R., Orúe, E., & Marzo, F. (1998). Effects of extrusion and conventional processing methods on protein and antinutritional factor contents in pea seeds. *Food Chemistry, 63*, 505–512.

Alu'datt, M. H., Rababah, T., Ereifej, K., & Alli, I. (2013). Distribution, antioxidant and characterisation of phenolic compounds in soybeans, flaxseed and olives. *Food Chemistry, 139*, 93–99.

Alvarez, M. V., Cabred, S., Ramirez, C. L., & Fanovich, M. A. (2019). Valorization of an agroindustrial soybean residue by supercritical fluid extraction of phytochemical compounds. *Journal of Supercritical Fluids, 143*, 90–96.

Amarowicz, R. (2007). Tannins: The new natural antioxidants? *European Journal of Lipid Science and Technology, 109*, 549–551.

Amarowicz, R., Dykes, G. A., & Pegg, R. B. (2008). Antibacterial activity of tannin constituents from *Phaseolus vulgaris, Fagoypyrum esculentum, Corylus avellana* and *Juglans nigra. Fitoterapia, 79*, 217–219.

Amarowicz, R., Estrella, I., Hernández, T., & Troszyńska, A. (2008). Antioxidant activity of extract of adzuki bean and its fractions. *Journal of Food Lipids, 15*, 119–136.

Amarowicz, R., Estrella, I., Hernández, T., Troszyńska, A., Kosińska, A., & Pegg, R. B. (2009). Antioxidant activity of a red lentil extract and its fractions. *International Journal of Molecular Sciences, 10*, 5513–5527.

Amarowicz, R., Estrella, I., Hernández, T., Troszyńska, A., Kosińska, A., & Pegg, R. B. (2010). Free radical-scavenging capacity, antioxidant activity, and phenolic composition of green lentil (*Lens culinaris*). *Food Chemistry, 121*, 705–711.

Amarowicz, R., Karamać, M., Dueñas, M., & Pegg, R. B. (2017). Antioxidant activity and phenolic composition of a red bean (*Phasoelus vulgaris*) extract and its fractions. *Natural Product Communications, 12*, 541–544.

Amarowicz, R., Pegg, R. B., & Barl, B. (2001). Antioxidant activity of extracts of phenolic compounds from selected plant species. In W. Pfannhauser, C. R. Fenwick, & S. Khokhar (Eds.), *Biologicaly-active phytochemicals in food* (pp. 345–3040). Cambridge, UK: Royal Society of Chemistry.

Amarowicz, R., Pegg, R. B., Dykes, G. A., Troszyńska, A., & Shahidi, F. (2005). Antioxidant and antibacterial properties of extracts of green tea polyphenols. *ACS Symposium Series, 909*, 94–106.

Amarowicz, R., & Shahidi, F. (2017). Antioxidant activity of broad bean seed extract and its phenolic composition. *Journal of Functional Foods, 38*, 656–662.

Amarowicz, R., & Shahidi, F. (2018). Antioxidant activity of faba bean extract and fractions thereof. *Journal of Food Bioactives, 1*, 112–118.

Amarowicz, R., Troszyńska, A., Baryłko-Pikielna, N., & Shahidi, F. (2004). Polyphenolics extracts from legume seeds: Correlations between total antioxidant activity, total phenolics content, tannins content and astringency. *Journal of Food Lipids, 11*, 278–286.

Amarowicz, R., Troszyńska, A., & Shahidi, F. (2005). Antioxidant activity of almond seed extract and its fractions. *Journal of Food Lipids, 12*, 344–358.

Amarowicz, R., Żegarska, Z., Rafałowski, R., Pegg, R. B., Karamać, M., & Kosińska, A. (2009). Antioxidant activity and free radical-scavenging capacity of ethanolic extracts of thyme, oregano, and marjoram. *European Journal of Lipid Science and Technology, 111*, 1111–1117.

Antolovich, M., Prenzler, P., Robards, K., & Ryan, D. (2000). Sample preparation in the determination of phenolic compounds in fruits. *Analyst, 125*, 989–1009.

Apak, R., Güçlü, K., Özyürek, M., & Karademir, S. E. (2004). Novel total antioxidant capacity index for dietary polyphenols and vitamins C and E, using their cupric ion reducing capability in the presence of neocuproine: CUPRAC method. *Journal of Agriculture and Food Chemistry, 52*, 7970–7981.

Aramendia, M. A., Garcia, I. M., Lafont, F., Lizaso, J., Marinas, J. M., & Urbano, F. J. (2000). Rapid determination of chlorogenic acid and related compounds in sunflower seeds by high-performance liquid chromatography/atmospheric pressure chemical ionization mass spectrometry. *Rapid Communications in Mass Spectrometry, 14*, 1019–1022.

Ariga, T., & Hamano, T. (1990). Radical scavenging action and its mode in procyanidins B-1 and B-3 from adzuki beans to peroxyl radicals. *Agricultural and Biological Chemistry, 54*, 2499–2504.

Asha, A., Manjunatha, M., Rekha, R. M., Surendranath, B., Heartwin, P., Rao, J., et al. (2015). Antioxidant activities of orange peel extract in ghee (butter oil) stored at different storage temperatures. *Journal of Food Science and Technology, 52*, 8220–8227.

Aşkin Uzel, R. (2017). A practical method for isolation of phenolic compounds from black carrot utilizing pressurized water extraction with in-site particle generation in hot air assistance. *Journal of Supercritical Fluids, 120*, 320–327.

Astill, C., Birch, M. R., Dacombe, C., Humphrey, P. G., & Martin, P. T. (2001). Factors affecting the caffeine and polyphenol contents of black and green tea infusions. *Journal of Agriculture and Food Chemistry, 49*, 5340–5347.

Balci, F., & Özdemir, F. (2016). Influence of shooting period and extraction conditions on bioactive compounds in Turkish green tea. *Food Science and Technology, 36*, 737–743.

Barba, F. J., Esteve, M. J., & Frigola, A. (2013). Physicochemical and nutritional characteristics of blueberry juice after high pressure processing. *Food Research International, 50*, 545–549.

Bartolomé, A., & Gómez-Cordovés, C. (1999). Barley spent grain: Release of hydroxycinnamic acids (ferulic and *p*-coumaric acids) by commercial enzyme preparations. *Journal of the Science of Food and Agriculture, 79*, 435–439.

Bentoa, C., Gonçalvesa, A. C., & Silva, B. (2018). Assessing the phenolic profile, antioxidant, antidiabetic and protective effects against oxidative damage in human erythrocytes of peaches from Fundão. *Journal of Functional Foods, 43*, 224–233.

Benzie, I. F. F., & Strain, J. J. (1996). The ferric reducing ability of plasma (FRAP) as a measure of "antioxidant power": The FRAP assay. *Analytical Biochemistry, 239*, 70–76.

Bessada, S. M. F., Alves, R. C., Costa, A. S. G., Nunes, M. A., & Oliveira, M. B. P. P. (2018). *Coffea canephora* silverskin from different geographical origins: A comparative study. *Science of the Total Environment, 645*, 1021–1028.

Bizuayehu, D., Atlabachew, M., & Ali, M. T. (2016). Determination of some selected secondary metabolites and their *in vitro* antioxidant activity in commercially available Ethiopian tea (*Camellia sinensis*). *Springerplus, 5*(1), 412.

Błaszczak, W., Amarowicz, R., & Górecki, A. R. (2017). Antioxidant capacity, phenolic composition and microbial stability of aronia juice subjected to high hydrostatic pressure processing. *Innovative Food Science and Emerging Technologies, 39*, 141–147.

Błaszczak, W., Zielińska, D., Zieliński, H., Szawara-Nowak, D., & Fornal, J. (2013). Antioxidant properties and rutin content of high pressure-treated raw and roasted buckwheat groats. *Food and Bioprocess Technology, 6*, 92–100.

Blois, M. S. (1958). Antioxidant determination by the use of a stable free radical. *Nature, 181*, 1199–1200.

Bondet, V., Brand-Williams, W., & Berset, C. (1997). Kinetics and mechanisms of antioxidant activity using the DPPH˙ free radical method. *Lebensmittel-Wissenschaft und Technologie, 30*, 609–615.

Borges, G., Degeneve, A., Mullen, W., & Crozier, A. (2010). Identification of flavonoid and phenolic antioxidants in black currants, blueberries, raspberries, red currants, and cranberries. *Journal of Agricultural and Food Chemistry, 58*, 3901–3909.

Boukhanouf, S., Louaileche, H., & Perrin, D. (2016). Phytochemical content and *in vitro*antioxidant activity of faba bean (*Vicia faba* L.) as affected by maturity stage and cooking practice. *International Food Research Journal, 23*, 954–961.

Brand-Williams, W., Cuvelier, M. E., & Berset, C. (1995). Use of a free radical method to evaluate antioxidant activity. *Lebensmittel-Wissenschaft und Technologie, 28*, 25–30.

Cai, S., Gao, F., Zhang, X., Wang, O., Wu, W., Zhu, S., et al. (2014). Evaluation of γ-aminobutyric acid, phytate and antioxidant activity of tempeh-like fermented oats (*Avena sativa* L.) prepared with different filamentous fungi. *Journal of Food Science and Technology, 51*, 2544–2551.

Cai, S., Wang, O., Wu, W., Zhu, S., Zhou, F., Ji, B., et al. (2012). Comparative study of the effects of solid-state fermentation with three filamentous fungi on the total phenolics content, flavonoids, antioxidant activities of subfractions from oats (*Avena sativa* L.). *Journal of Agricultural and Food Chemistry, 60*, 507–513.

Chakraborty, A., & Bhattacharyya, S. (2014). Thermal processing effects on *in vitro* antioxidant activities of five common Indian pulses. *Journal of Applied Pharmaceutical Science, 4*, 65–70.

Chau, C.-F., & Cheung, P. C.-K. (1997). Effect of various processing methods on antinutrients and in vitro digestibility of protein and starch of two Chinese indigenous legume seeds. *Journal of Agricultural and Food Chemistry, 45*, 4773–4776.

Chiabrando, V., Giuggioli, N., Maghenzani, M., Peano, C., & Giacalone, G. (2018). Improving storability of strawberries with gaseous chlorine dioxide in perforated clamshell packaging. *Polish Journal of Food and Nutrition Sciences, 68*, 41–148.

Chinnici, F., Bendini, A., Gaiani, A., & Riponi, C. (2004). Radical scavenging activities of peels and pulps from cv. Golden delicious apples as related to their phenolic composition. *Journal of Agricultural and Food Chemistry, 52*, 4684–4689.

Cho, K. M., Lee, J. H., Yun, H. D., Ahn, B. Y., Kim, H., & Seo, W. T. (2010). Changes in phytochemical constituents (isoflavones, flavanols, and phenolic acids) during cheonggukjang soybeans fermentation using potential probiotics *Bacillus subtilis* CS90. *Journal of Food Composition and Analysis, 24*, 402–410.

Christ, B., & Müller, K. H. (1960). Zur serienmaessigen bestimmung des gehaltes an flavonol-derivaten in Drogen. *Archiv der Pharmazie, 293*, 1033–1042.

Chrpová, D., Kouřimská, L., Gordon, M. H., Heřmanová, V., Roubíčková, I., & Pánek, J. (2010). Antioxidant activity of selected phenols and herbs used in diets for medical conditions. *Czech Journal of Food Science, 28*, 317–325.

Clifford, M. N. (1999). Chlorogenic acids and other cinnamates. Nature, occurrence and dietary burden. *Journal of the Science of Food and Agriculture, 79*, 362–372.

Colindres, P., & Brewer, M. S. (2011). Oxidative stability of cooked, frozen, reheated beef patties: Effect of antioxidants. *Journal of the Science of Food and Agriculture, 91*, 963–968.

Coman, M. M., Oancea, A. M., Verdenelli, M. C., Cecchini, C., Bahrim, G. E., Orpianesi, C., et al. (2018). Polyphenol content and in vitro evaluation of antioxidant, antimicrobial and prebiotic properties of red fruit extracts. *European Food Research and Technology, 244*, 735–745.

Corramles, M., Toepfl, S., Butz, P., Knorr, D., & Tausche, B. (2008). Extraction of anthocyanins from grape by-product assisted by ultrasonics, high hydrostatic pressure or pulsed lectric fields: A comparison. *Innovative Food Science and Emerging Technologies, 9*, 85–91.

Craft, B. D., Kerrihard, A. L., Amarowicz, R., & Pegg, R. B. (2012). Phenol-based antioxidants and the *in vitro* methods used for their assessment. *Comprehensive Reviews in Food Science and Food Safety, 11*, 148–173.

Craft, B. D., Kosińska, A., Amarowicz, R., & Pegg, R. B. (2010). Antioxidant properties of extracts obtained from raw, dry-roasted, and oil-roasted us peanuts of commercial importance. *Plant Foods for Human Nutrition, 65*, 311–318.

Cutillo, F., Dellagreca, M., Gionti, M., Previtera, L., & Zarrelli, A. (2006). Phenols and lignans from *Chenopodium album*. *Phytochemical Analysis*, *17*, 344–349.

Cvetanović, A., Zengin, G., Zeković, Z., Švarc-Gajić, J., Ražić, S., Damjanović, A., et al. (2018). Comparative in vitro studies of the biological potential and chemical composition of stems, leaves and berries *Aronia melanocarpa*'s extracts obtained by subcritical water extraction. *Food and Chemical Toxicology*, *121*, 458–466.

Daskalova, E., Delchev, S., Peeva, Y., Vladimirova-Kitova, L., Kratchanova, M., Kratchanov, C., et al. (2015). Antiatherogenic and cardioprotective effects of black chokeberry (*Aronia melanocarpa*) juice in aging rats. *Evidence-Based Complementary and Alternative Medicine*, *2015*, 717439.

De Camargo, A. C., Regitano-d'Arce, M. A. B., Rasera, G. B., Canniatti-Brazaca, S. G., Do Prado-Silva, L., Alvarenga, V. O., et al. (2017). Phenolic acids and flavonoids of peanut by-products: Antioxidant capacity and antimicrobial effects. *Food Chemistry*, *237*, 538–544.

De Oliveira Ribeiro, L., Almeida, A. C. S., de Carvalho, C. W. P., Borguini, R. G., Freitas, S. P., & da Matta, V. M. (2018). Effect of processing on bioactive compounds, physicochemical and rheological characteristics of juçara, banana and strawberry smoothie. *Plant Foods for Human Nutrition*, *73*, 222–227.

Demark-Wahnefried, W., Price, D. T., Polascik, T. J., Robertson, C. N., Anderson, E. E., Paulson, D. F., et al. (2001). Pilot study of dietary fat restriction and flaxseed supplementation in men with prostate cancer before surgery: Exploring the effects on hormonal levels, prostate-specific antigen, and histopathologic features. *Urology*, *58*, 47–52.

Demark-Wahnefried, W., Robertson, C. N., Walther, P. J., Polascik, T. J., Paulson, D. F., & Vollmer, R. T. (2004). Pilot study to explore effects of low-fat, flaxseed-supplemented diet on proliferation of benign prostatic epithelium and prostate-specific antigen. *Urology*, *63*, 900–904.

Deng, G.-X., Xu, X.-R., Guo, Y.-J., Xia, E.-Q., Li, S., Wu, S., et al. (2012). Determination of antioxidant property and their lipophilic and hydrophilic phenolic contents in cereal grains. *Journal of Functional Foods*, *4*, 906–914.

Deng, Q., Yuc, X., Mad, F., Xu, J., Huang, F., Huang, Q., et al. (2017). Comparative analysis of the *in-vitro* antioxidant activity and bioactive compounds of flaxseed in China according to variety and geographical origin. *International Journal of Food Properties*, *20*, S2708–S2722.

Dent, M., Kovačević, D. B., Bosiljkov, T., & Dragović-Uzelac, V. (2017). Polyphenolic composition and antioxidant capacity of indigenous wild dalmatian sage (*Salvia officinalis* L.). *Croatica Chemica Acta*, *90*, 451–459.

Deshpande, S. S., & Cheryan, M. (1985). Evaluation of vanillin assay for tannin analysis of dry beans. *Journal of Food Science*, *50*, 905–910.

Do Couto, C. A., de Souza, E. R. B., Morgado, C. M. A., Ogata, T., & Cunha Júnior, L. C. (2018). *Citrus sinensis* cultivars: Alternatives for diversification of Brazilian orchards. *Revista Brasileira de Fruticultura*, *40*(5), e-097.

Dos Reis, L. C. R., Facco, E. M. P., Flôres, S. H., & Rios, A. D. O. (2018). Stability of functional compounds and antioxidant activity of fresh and pasteurized orange passion fruit (*Passiflora caerulea*) during cold storage. *Food Research International*, *106*, 481–486.

Dragsted, L. O., Pedersen, A., Hermetter, A., Basu, S., Hansen, M., Haren, G. R., et al. (2004). The 6-a-day study: Effects of fruit and vegetables on markers of oxidative stress and antioxidative defense in healthy nonsmokers. *American Journal of Clinical Nutrition*, *79*, 1060–1072.

Dueñas, M., Martínez-Villaluenga, C., Limón, R. I., Peñas, E., & Frias, J. (2015). Effect of germination and elicitation on phenolic composition and bioactivity of kidney beans. *Food Research International*, *70*, 55–63.

Durazzo, A., Turfani, V., Azzini, E., Maiani, G., & Carcea, M. (2013). Phenols, lignans and antioxidant properties of legume and sweet chestnut flours. *Food Chemistry, 140*, 666–671.

Eklund, P. C., Langvik, O. K., Warna, J. P., Salmi, T. O., Willfor, S. M., & Sjoholm, R. E. (2005). Chemical studies on antioxidant mechanisms and free radical scavenging properties of lignans. *Organic & Biomolecular Chemistry, 3*, 3336–3347.

El Khawand, T., Courtois, A., Valls, J., Richard, T., & Krisa, S. (2018). A review of dietary stilbenes: Sources and bioavailability. *Phytochemistry Reviews, 17*, 1007–1029.

Erşan, S., Güçlü Üstündağ, Ö., Carle, R., & Schweiggert, R. M. (2018). Subcritical water extraction of phenolic and antioxidant constituents from pistachio (*Pistacia vera* L.) hulls. *Food Chemistry, 253*, 46–54.

Estévez, M., & Heinonen, M. (2010). Effect of phenolic compounds on the formation of α-amoniadipic and γ-glutamic semialdehydes from myofibrillar proteins oxidized by copper, iron, and myoglobin. *Journal of Agricultural and Food Chemistry, 58*, 4448–4455.

Fanali, C., Tripodo, G., Russo, M., Della Posta, S., Pasqualetti, V., & De Gara, L. (2018). Effect of solvent on the extraction of phenolic compounds and antioxidant capacity of hazelnut kernel. *Electrophoresis, 39*, 1683–1691.

Farah, A., de Paulis, T., Trugo, L. C., & Martin, P. R. (2005). Effect of roasting on the formation of chlorogenic acid lactones. *Journal of Agriculture and Food Chemistry, 53*, 1505–1513.

Farah, A., & Donangelo, C. M. (2007). Phenolic compounds in coffee. *Brazilian Journal of Plant Biology, 18*, 23–36.

Fares, C., Platani, C., Baiano, A., & Menga, V. (2010). Effect of processing and cooking on phenolic acid profile and antioxidant capacity of durum wheat pasta enriched with debranning fractions of wheat. *Food Chemistry, 119*, 1023–1029.

Fattouch, S., Sadok, S., Raboudi-Fattouch, F., & Ben, S. M. (2008). Damage inhibition during refrigerated storage of mackerel (*Scomber scombrus*) fillets by a presoaking in quince (*Cydonia oblonga*) polyphenolic extract. *International Journal of Food Science and Technology, 143*, 2056–2064.

Ferrieres, J. (2004). The French paradox: Lessons for other countries. *Heart, 90*, 107–111. Fitoterapia, 79, 217–219.

Foti, M. C., Daquino, C., & Geraci, C. (2004). Electron-transfer reaction of cinnamic acids and their methyl esters with the DPPH˙ radical in alcoholic solutions. *Journal of Organic Chemistry, 69*, 2309–2314.

Frankel, E. N. (2005). *Lipid oxidation*. Bridgewater, Dundee: The Oily Press. Chapter 1.

Fratianni, F., Cefola, M., Pace, B., Cozzolino, R., Giulio, D., Cozzolino, A., et al. (2017). Changes in visual quality, physiological and biochemical parameters assessed during the postharvest storage at chilling or non-chilling temperatures of three sweet basil (*Ocimum basilicum* L.) cultivars. *Food Chemistry, 229*, 752–760.

Fu, M., Xiao, G., Wu, J., Chen, Y., Yu, Y., Chen, W., et al. (2017). Effects of modified atmosphere packaging on the quality of dried lemon slices. *Journal of Food Processing and Preservation, 41*(4), e13043.

Galano, A., Francisco-Márquez, M., & Alvarez-Idaboy, J. R. (2011). Canolol: A promising chemical agent against oxidative stress. *Journal of Physical Chemistry B, 115*, 8590–8596.

Gamboa-Gómez, C. I., Muñoz-Martínez, A., Rocha-Guzmán, N. E., Gallegos-Infante, J. A., Moreno-Jiménez, M. R., González-Herrera, et al. (2016). Changes in phytochemical and antioxidant potential of tempeh common bean flour from two selected cultivars influenced by temperature and fermentation. *Journal of Food Processing and Preservation, 40*, 270–278.

Gantner, M., Brodowska, M., Horczyczak, E. G., Kalinowska, I. W., Najda, A., Pogorzelska, E., et al. (2018). Antioxidant effect of sage (*Salvia officinalis* L.) extract on Turkey meatballs packed in cold modified atmosphere. *CyTA Journal of Food*, *16*, 628–636.
Ghafoorunissa, Hemalatha, S., & Rao, V. V. M. (2004). Sesame lignans enhance antioxidant activity of vitamin E in lipid peroxidation systems. *Molecular and Cellular Biochemistry*, *262*, 195–202.
Glazer, A. N. (1990). Phycoerythrin fluorescence-based assay for reactive oxygen species. *Methods in Enzymology*, *186*, 161–168.
Gonzales-Cebrino, F., Duran, R., Delgado-Adamez, J., Contador, R., & Ramirez, R. (2013). Changes after high-pressure processing on physicochemical parameters, bioactive compounds, and polyphenol oxidase activity of red flesh and peel plum puree. *Innovative Food Science and Emerging Technologies*, *20*, 34–41.
Gonzalez-Aguilar, G. A., Ruiz-Cruz, S., Soto-Valdez, H., Vazquez-Ortiz, F., Pacheco-Aguilar, R., & Wang, C. Y. (2005). Biochemical changes of fresh-cut pineapple slices treated with anti-browning agents. *International Journal of Food Science and Technology*, *40*, 377–383.
Graff, E. (1992). Antioxidant potential of ferulic acid. *Free Radical Biology and Medicine*, *13*, 435–448.
Gramza-Michałowska, A., Korczak, J., & Reguła, J. (2007). Use of plant extracts in summer and winter season butter oxidative stability improvement. *Asia Pacific Journal of Clinical Nutrition*, *16*(Suppl.1), 85–88.
Grela, E. R., Kiczorowska, B., Samolińska, W., Matras, J., Kiczorowski, P., Rybiński, W., et al. (2017). Chemical composition of leguminous seeds: Part I—Content of basic nutrients, amino acids, phytochemical compounds, and antioxidant activity. *European Food Research and Technology*, *243*, 1385–1395.
Gujral, H. S., Sharma, P., Gupta, N., & Wani, A. A. (2013). Antioxidant properties of legumes and their morphological fractions as affected by cooking. *Food Science and Biotechnology*, *22*, 187–194.
Gutiérrez-Grijalva, E. P., Angulo-Escalante, M. A., León-Félix, J., & Heredia, J. B. (2017). Effect of *in vitro* digestion on the total antioxidant capacity and phenolic content of 3 species of oregano (*Hedeoma patens*, *Lippia graveolens*, *Lippia palmeri*). *Journal of Food Science*, *2*(8), 2832–2839.
Gyawali, R., & Salam, I. A. (2014). Natural products as antimicrobial agents. *Food Control*, *46*, 412–429.
Harnly, J. (2017). Antioxidant methods. *Journal of Food Composition and Analysis*, *2017*(64), 145–146.
Hatano, T., Kusuda, M., Inada, K., Ogawa, T. O., Shiota, S., Tsuchiya, T., et al. (2005). Effects of tannins and related polyphenols on methicillin-resistant *Staphylococcus aureus*. *Phytochemistry*, *66*, 2047–2055.
Heim, K. E., Tagliaferro, A. R., & Bobilya, D. J. (2002). Flavonoid antioxidants: Chemistry, metabolism and structure-activity relationships. *Journal of Nutrition and Biochemistry*, *13*, 572–584.
Herbello-Hermelo, P., Lamas, J. P., Lores, M., Domínguez-González, R., Bermejo-Barrera, P., & Moreda-Piñeiro, A. (2018). Polyphenol bioavailability in nuts and seeds by an *in vitro* dialyzability approach. *Food Chemistry*, *254*, 20–25.
Hernández-Ruiz, K. L., Ruiz-Cruz, S., Cira-Chávez, L. A., Gassos-Ortega, L. E., Ornelas-Paz, J. J., Del-Toro-Sánchez, C. L., et al. (2018). Evaluation of antioxidant capacity, protective effect on human erythrocytes and phenolic compound identification in two varieties of plum fruit (*Spondias* spp.) by UPLC-MS. *Molecules*, *23*(12), 3200.
Herrero, M., Cifuentes, A., & Ibañez, E. (2006). Sub- and supercritical fluid extraction of functional ingredients from different natural sources: Plants, food-by-products, algae and microalgae-A review. *Food Chemistry*, *98*, 136–148.

Hirose, Y., Fujita, T., Ishii, T., & Ueno, N. (2010). Antioxidant properties and flavonoid composition of *Chenopodium quinoa* seeds cultivated in Japan. *Food Chemistry, 119*, 1300–1306.

Hosseinian, F. S., Muir, A. D., Westcott, N. D., & Krol, E. S. (2006). Antioxidant capacity of flaxseed lignans in two model systems. *Journal of the American Oil Chemists' Society, 83*, 835–840.

Hraš, H. A., Hadolin, M., Knez, Ž., & Bauman, D. (2000). Comparison of antioxidative and synergistic effects of rosemary extract with α-tocopherol, ascorbyl palmitate and citric acid in sunflower oil. *Food Chemistry, 71*, 229–233.

Hu, C., Yuan, Y. V., & Kitts, D. D. (2007). Antioxidant activities of the flaxseed lignan secoisolariciresinol diglucoside, its aglycone secoisolariciresinol and the mammalian lignans enterodiol and enterolactone *in vitro*. *Food and Chemical Toxicology, 45*, 2219–2227.

Huang, D., Ou, B., Hampsch-Woodhill, M., Flanagan, J. A., & Deemer, E. K. (2002). Development and validation of oxygen radical absorbance capacity assay for lipophilic antioxidants using randomly methylated β-cyclodextrin as the solubility enhancer. *Journal of Agriculture and Food Chemistry, 50*, 1815–1821.

Huang, D., Ou, B., & Prior, R. L. (2005). The chemistry behind antioxidant capacity. *Journal of Agriculture and Food Chemistry, 53*, 1841–1856.

Huber, K., Brigide, P., Bretas, E. B., & Canniatti-Brazaca, S. G. (2014). Effect of thermal processing and maceration on the antioxidant activity of white beans. *PLOS ONE, 9*(7), e99325.

Hügel, H. M., Jackson, N., May, B., Zhang, A. L., & Xue, C. C. (2016). Polyphenol protection and treatment of hypertension. *Phytomedicine, 23*, 220–231.

Hutchins, A. M., & Slavin, J. L. (2003). Effects of flaxseed on sex hormone metabolism. In L. U. Thompson & S. C. Cunnane (Eds.), *Flaxseed in human nutrition* (pp. 126–149). Champaign: AOCS Press AOCS Press.

İnan, Ö., Özcan, M. M., & Aljuhaimi, F. (2018). Effect of location and *Citrus* species on total phenolic, antioxidant, and radical scavenging activities of some *Citrus* seed and oils. *Journal of Food Processing and Preservation, 42*(3), e13555.

Itoh, A., Isoda, K., Kondoh, M., Kawase, M., Kobayashi, M., Tamesada, M., et al. (2009). Hepatoprotective effect of syringic acid and vanillic acid on concanavalin induced liver injury. *Biological and Pharmaceutical Bulletin, 32*, 1215–1219.

Jabri, K. M.-A., Rtibi, K., Ben-Said, A., Aouadhi, C., Hosni, K., Sakly, M., et al. (2016). Antidiarrhoeal, antimicrobial and antioxidant effects of myrtle berries (*Myrtus communis* L.) seeds extract. *Journal of Pharmacy and Pharmacology, 68*, 264–274.

Jayathunge, K. G. L. R., Grant, I. R., Linton, Patterson, M. F., & Koidis, A. (2015). Impact of long-term storage at ambient temperatures on the total quality and stability of high-pressure processed tomato juice. *Innovative Food Science & Emerging Technologies, 32*, 1–8.

Jeż, M., Wiczkowski, W., Zielińska, D., Białobrzewski, I., & Błaszczak, W. (2018). The impact of high pressure processing on the phenolic profile, hydrophilic antioxidant and reducing capacity of purée obtained from commercial tomato varieties. *Food Chemistry, 261*, 201–209.

Johnsson, P., Kamal-Eldin, A., Lundgren, L. N., & Åman, P. (2000). HPLC method for analysis of secoisolariciresinol diglucoside in flaxseeds. *Journal of Agricultural and Food Chemistry, 48*, 5216–5219.

Kalt, W., Forney, C. F., Martin, A., & Prior, R. L. (1999). Antioxidant capacity, vitamin C, phenolics, and anthocyanins after fresh storage of small fruits. *Journal of Agricultural and Food Chemistry, 47*, 4638–4644.

Kanchana, G., Shyni, W. J., Rajadurai, M., & Periasamy, R. (2011). Evaluation of antihyperglycemic effect of sinapic acid in normal and streptozotocin-induced diabetes in albino rats. *Global Journal Pharmacology, 2011*(5), 33–39.

Karamać, M., Buciński, A., Pegg, R. B., & Amarowicz, R. (2005). Antioxidant and antiradical activity of ferulates. *Czech Journal of Food Science*, *32*, 64–68.
Karamać, M., Koleva, L., Kancheva, V. D., & Amarowicz, R. (2017). The structure-antioxidant activity relationship of ferulates. *Molecules*, *22*(4), 527.
Karamać, M., Kosińska, A., & Amarowicz, R. (2006). Chelating of Fe(II), Zn(II) and cu(II) by tannin fractions separated from hazelnuts, walnuts and almonds. *Bromatologia i Chemia Toksykologiczna*, *39*, 257–260.
Karamać, M., Kosińska, A., Estrella, I., Hernández, T., & Dueñas, M. (2012). Antioxidant activity of phenolic compounds identified in sunflower seeds. *European Food Research and Technology*, *235*, 221–230.
Karas, D., Ulrichová, J., & Valentová, K. (2017). Galloylation of polyphenols alters their biological activity. *Food and Chemical Toxicology*, *105*, 223–240.
Karpińska-Tymoszczyk, M. (2013). The effect of oil-soluble rosemary extract, sodium erythorbate, their mixture, and packaging method on the quality of Turkey meatballs. *Journal of Food Science and Technology*, *50*, 443–454.
Keenan, D. F., Brunton, N. P., Gormley, T. R., Butler, F., Tiwari, B. K., & Patras, A. (2010). Effect of thermal and high hydrostatic pressure processing on antioxidant activity and colour of fruit smoothies. *Innovative Food Science and Emerging Technologies*, *11*, 551–556.
Khan, N., & Mukhtar, H. (2013). Tea and health: Studies in humans. *Current Pharmaceutical Design*, *19*, 6141–6147.
Khanum, H., Faiza, S., Sulochanamma, G., & Borse, B. B. (2017). Quality, antioxidant activity and composition of Indian black teas. *Journal of Food Science and Technology*, *54*, 1266–1272.
Khattab, R. Y., & Arntfield, S. D. (2009). Nutritional quality of legume seeds as affected by some physical treatments 2. Antinutritional factors. *LWT-Food Science and Technology*, *42*, 1113–1118.
Khattab, R., Eskin, M., Aliani, M., & Thiyam, U. (2010). Determination of sinapic acid derivatives in canola extracts using high-performance liquid chromatography. *Journal of the American Oil Chemist's Society*, *87*, 147–155.
Khokhar, S., & Magnusdottir, S. G. (2002). Total phenol, catechin, and caffeine contents of teas commonly consumed in the United Kingdom. *Journal of Agricultural and Food Chemistry*, *50*, 565–570.
Khokhar, S., & Owusu Apenten, R. K. (2003). Iron binding characteristics of phenolic compounds: Some tentative structure–activity relations. *Food Chemistry*, *81*, 133–140.
Khoshnoudi-Nia, S., Niakosari, M., & Tahsiri, Z. (2017). Subcritical water technology for extraction phytochemical compound. *Journal of Medicinal Plants*, *16*, 94–107.
Kim, D.-S., & Lim, S.-B. (2018). Quality properties of subcritical water extracts of onion and onion juice product. *Journal of the Korean Society of Food Science and Nutrition*, *47*, 750–758.
Kiprovski, B., Mikulic-Petkovsek, M., Slatnar, A., Veberic, R., Stampar, Malencic, D., et al. (2015). Comparison of phenolic profiles and antioxidant properties of European *Fagopyrum esculentum* cultivars. *Food Chemistry*, *185*, 41–47.
Klein, E., & Lukeš, V. (2006). DFT/B3LYP study of the substituent effect on the reaction enthalpies of the individual steps of single electron transfer – proton transfer and sequential proton loss electron transfer mechanisms of phenols antioxidant action. *Journal of Physical Chemistry A*, *110*, 12312–12320.
Koch, W., Kukula-Koch, W., Komsta, Ł., Marzec, Z., Szwerc, W., & Głowniak, K. (2018). Green tea quality evaluation based on its catechins and metals composition in combination with chemometric analysis. *Molecules*, *23*(7), 1689.
Köksal, E., Bursal, E., Gülçin, I., Korkmaz, M., Çağlayan, C., Gören, A. C., et al. (2017). Antioxidant activity and polyphenol content of Turkish thyme (*Thymus vulgaris*)

monitored by liquid chromatography and tandem mass spectrometry. *International Journal of Food Properties*, *20*, 514–525.

Konieczyski, P., Viapiana, A., & Marek Wesolowski, M. (2017). Comparison of infusions from black and green teas (*Camellia sinensis* L. Kuntze) and erva-mate (*Ilex paraguariensis* A. St.-Hil.) based on the content of essential elements, secondary metabolites, and antioxidant activity. *Food Analysis Methods*, *10*, 3063–3070.

Korkina, L. G., & Afanas'ev, I. B. (1997). Antioxidant and chelating properties of flavonoids. *Advances in Pharmacology*, *38*, 151–163.

Kornsteiner, M., Wagner, K.-H., & Elmadfa, I. (2006). Tocopherols and total phenolics in 10 different nut types. *Food Chemistry*, *98*, 381–387.

Koroleva, A., Torkova, A., Nikolaev, I., Khrameeva, E., Fedorova, T., Tsentalovich, M., et al. (2014). Evaluation of the antiradical properties of phenolic acids. *International Journal of Molecular Sciences*, *15*, 16351–16380.

Korus, J., Gumul, D., & Czechowska, K. (2007). Effect of extrusion on the phenolic composition and antioxidant activity of dry beans of *Phaseolus vulgaris* L. *Food Technology and Biotechnology*, *45*, 139–146.

Kosińska, A., Penkacik, K., Wiczkowski, W., & Amarowicz, R. (2011). Presence of caffeic acid in flaxseed lignan macromolecule. *Plant Foods for Human Nutrition*, *66*, 270–274.

Kozłowska, M., & Gruczyńska, E. (2018). Comparison of the oxidative stability of soybean and sunflower oils enriched with herbal plant extracts. *Chemical Papers*, *72*, 2607–2615.

Kozłowska, H., Rotkiewicz, D. A., Zadernowski, R., & Sosulski, F. W. (1983). Phenolic acids in rapeseed and mustard. *Journal of the American Oil Chemists' Society*, *60*, 1119–1123.

Kozubek, A., & Tyman, J. H. P. (1999). Resorcinolic lipids, the natural non-isoprenoid phenolic amphiphiles and their biological activity. *Chemical Reviews*, *99*, 1–25.

Krygier, K., Sosulski, F., & Hogge, L. (1982). Free, esterified, and insoluble-bound phenolic acids. 1. Extraction and purification procedure. *Journal of Agricultural and Food Chemistry*, *30*, 330–334.

Kulkarni, S., DeSantos, F. A., Kattamuri, S., Rossi, S. J., & Brewer, M. S. (2011). Effect of grape seed extract on oxidative, color and sensory stability of pre-cooked, frozen, re-heated beef sausage model system. *Meat Science*, *88*, 139–144.

Kumar, Y., & Langoo, B. A. (2016). Effects of aloe, green tea, and amla extracts on microbiological and oxidative parameters of refrigerated raw meat batter. *Agricultural Research*, *5*, 81–88.

Kumar, S., & Pandey, A. K. (2013). Chemistry and biological activities of flavonoids: An overview. *Scientific World Journal*, *2013*, 162750.

Kumar, P., Sethi, S., Sharma, R. R., Singh, S., & Varghese, E. (2018). Improving the shelf life of fresh-cut "Royal Delicious" apple with edible coatings and anti-browning agents. *Journal of Food Science and Technology*, *55*, 3767–3778.

Ky, C.-L., Louarn, D. S., Guyot, B., Hamon, S., & Noirot, M. (2001). Caffeine, trigonelline, chlorogenic acids and sucrose diversity in wild *Coffea arabica* L. and *C. canephora P. accessions*. *Food Chemistry*, *75*, 223–230.

Lai, P. K., & Roy, J. (2004). Antimicrobial and chemopreventive properties of herbs and spices. *Current Medicinal Chemistry*, *11*, 1451–1460.

Lee, C. G., Ahmed, M., Jiang, G. H., & Eun, J. B. (2017). Color, bioactive compounds and morphological characteristics of encapsulated Asian pear juice powder during spray drying. *Journal of Food Science and Technology*, *59*, 2717–2727.

Lee, J., & Choe, E. (2006). Extraction of lignan compounds from roasted sesame oil and their effects on the autoxidation of methyl linoleate. *Journal of Food Science*, *71*, C430–C436.

Lee, M.-A., Choi, J.-H., Choi, Y.-S., Han, D.-J., Kim, H.-Y., Shim, S.-Y., et al. (2010). The antioxidative properties of mustard leaf (*Brassica juncea*) kimchi extracts on refrigerated raw ground pork meat against lipid oxidation. *Meat Science*, *84*, 498–504.

Lee, J. H., Hwang, C. E., Son, K. S., & Cho, K. M. (2019). Comparisons of nutritional constituents in soybeans during solid state fermentation times and screening for their glucosidase enzymes and antioxidant properties. *Food Chemistry*, *272*, 362–371.

Lee, J., Kim, M., & Choe, E. (2007). Antioxidant activity of lignan compounds extracted from roasted sesame oil on the oxidation of sunflower oil. *Food Science and Biotechnology*, *16*, 981–987.

Lee, S.-H., Kim, H.-W., Lee, M.-K., Lee, M.-K., Kim, Y. J., Asamenew, G., et al. (2018). Phenolic profiling and quantitative determination of common sage (*Salvia plebeia* R. Br.) by UPLC-DAD-QTOF/MS. *European Food Research and Technology*, *244*, 1637–1646. 2018.

Lee, L.-S., Kim, Y.-C., Park, J.-D., Kim, Y.-B., & Kim, S.-H. (2016). Changes in major polyphenolic compounds of tea (*Camellia sinensis*) leaves during the production of black tea. *Food Science and Biotechnology*, *25*, 1523–1527.

Lee, J., & Rennaker, C. (2007). Antioxidant capacity and stilbene contents of wines produced in the Snake River valley of Idaho. *Food Chemistry*, *105*, 195–203.

Lee, S. G., Terrence, M., Vance, T. M., Nam, T.-G., Kim, D.-O., Koo, S. I., et al. (2015). Contribution of anthocyanin composition to total antioxidant capacity of berries. *Plant Foods for Human Nutrition*, *70*, 427–432.

Leopoldini, M., Marino, T., Russo, N., & Toscano, M. (2004). Antioxidant properties of phenolic compounds: H-atom versus electron transfer mechanism. *Journal of Physical Chemistry A*, *108*, 4916–4922.

Leopoldini, M., Russo, N., & Toscano, M. (2011). The molecular basis of working mechanism of natural polyphenolic antioxidants. *Food Chemistry*, *125*, 288–306.

Li, S., Wang, H., Guo, L., Zhao, H., & Ho, C. T. (2014). Chemistry and bioactivity of nobiletin and its metabolites. *Journal of Functional Foods*, *6*, 2–10.

Liang, J., Zago, E., Nandasiri, R., Khattab, R., Eskin, N. A. M., Eck, P., et al. (2018). Effect of solvent, preheating temperature, and time on the ultrasonic extraction of phenolic compounds from cold-pressed hempseed cake. *Journal of the American Oil Chemists' Society*, *95*, 1319–1327.

Limón, R. I., Peñas, E., Torino, M. I., Martínez-Villaluenga, C., Dueñas, M., & Frias, J. (2015). Fermentation enhances the content of bioactive compounds in kidney bean extracts. *Food Chemistry*, *172*, 343–352.

Lin, L., Peng, A., Yang, K., & Zou, Y. (2018). Monomeric phenolics in different parts of high-acid apple (*Malus sieversii* f. *niedzwetzkyana* (Dieck) Langenf): A promising source of antioxidants for application in nutraceuticals. *International Journal of Food Science and Technology*, *53*, 1503–1509.

Lin, C. H., Wei, Y. T., & Chou, C. C. (2006). Enhanced antioxidative activity of soybean koji prepared with various filamentous fungi. *Food Microbiology*, *23*, 628–633.

Litwinienko, G., & Ingold, K. U. (2003). Abnormal solvent effects on hydrogen atom abstractions. 1. The reactions of phenols with 2,2-diphenyl-1picrylhydrazyl (dpph˙) in alcohols. *Journal of Organic Chemistry*, *68*, 3433–3438.

Liu, Q., Tang, G.-Y., Zhao, C.-N., Feng, X.-L., Xu, X.-Y., Cao, S.-Y., et al. (2018). Comparison of antioxidant activities of different grape varieties. *Molecules*, *23*(10), 2432.

López-Amorós, M. L., Hernández, T., & Estrela, I. (2006). Effect of germination on legume phenolic compounds and their antioxidant activity. *Journal of Food Composition and Analysis*, *19*, 277–283.

Lu, Y., Jiang, F., Jiang, H., Wu, K., Zheng, X., Cai, Y., et al. (2010). Gallic acid suppresses cell viability, proliferation, invasion and angiogenesis in human glioma cells. *European Journal of Pharmacology*, *2010*(641), 102–107.

Lu, Q., Lv, S., Peng, Y., Zhu, C., & Pan, S. (2018). Characterization of phenolics and antioxidant abilities of red navel orange "Cara Cara" harvested from five regions of China. *International Journal of Food Properties*, *21*, 1107–1116.

Luther, M., Parry, J., Moore, J., Meng, J., Zhang, Y., Cheng, Z., et al. (2007). Inhibitory effect of chardonnay and black raspberry seed extracts on lipid oxidation in fish oil and their radical scavenging and antimicrobial properties. *Food Chemistry, 104*, 1065–1073.

Luthria, D. L., Biswas, R., & Natarajan, S. (2007). Comparison of extraction solvents and techniques used for the assay of isoflavones from soybean. *Food Chemistry, 105*, 325–333.

Luthria, D. L., & Mukhopadhyay, S. (2006). Influence of sample preparation on assay of phenolic acids from eggplant. *Journal of Agricultural and Food Chemistry, 54*, 41–47.

Ma, Y., Kosińska-Cagnazzo, A., Kerr, W. L., Amarowicz, R., Swanson, R. B., & Pegg, R. B. (2014). Separation and characterization of phenolic compounds from dry-blanched peanut skins by liquid chromatography-electrospray ionization mass spectrometry. *Journal of Chromatography A, 1356*, 64–81.

Maggi-Capeyron, M.-F., Ceballos, P., Cristol, J.-P., Delbose, S., Le Doucen, C., Pons, M., et al. (2001). Wine phenolic antioxidants inhibit AP-1 transcriptional activity. *Journal of Agricultural and Food Chemistry, 49*, 5646–5652.

Mahendra Kumar, C., & Singh, S. A. (2015). Bioactive lignans from sesame (*Sesamum indicum* L.): Evaluation of their antioxidant and antibacterial effects for food applications. *Journal of Food Science and Technology, 52*, 2934–2941.

Mammen, D., & Mammen, D. (2012). A critical evaluation on the reliability of two aluminum chloride chelation methods for quantification of flavonoids. *Food Chemistry, 135*, 1365–1368.

Maqsood, S., Benjakul, S., Abushelaibi, A., & Alam, A. (2014). Phenolic compounds and plant phenolic extracts as natural antioxidants in prevention of lipid oxidation in seafood: A detailed review. *Comprehensive Reviews in Food Science and Food Safety, 13*, 1125–1140.

Marchante, L., Izquierdo-Cañas, P. M., Gómez-Alonso, S., Alañón, S., García-Romero, M. E., Pérez-Coello, M. S., et al. (2019). Oenological potential of extracts from winery and cooperage by-products in combination with colloidal silver as natural substitutes to sulphur dioxide. *Food Chemistry, 276*, 485–493.

Markus, M. A., & Morris, B. J. (2008). Resveratrol in prevention and treatment of common clinical conditions of aging. *Clinical Interventions in Aging, 3*, 331–339.

McGhie, T. K., Hunt, M., & Barnett, L. E. (2005). Cultivar and growing region determine the antioxidant polyphenolic concentration and composition of apples grown in New Zealand. *Journal of Agricultural and Food Chemistry, 53*, 3065–3070.

McInerney, J. K., Seccafien, C. A., Stewart, C. M., & Bird, A. R. (2007). Effect of high pressure processing on antioxidant activity, and total carotenoid content and availability, in vegetables. *Innovative Food Science & Emerging Technologies, 8*, 543–548.

Mekinić, G. I., Skroza, D., Ljubenkov, I., Šimat, V., Možina, S., & Katalinić, V. (2014). *In vitro* antioxidant and antibacterial activity of Lamiaceae phenolic extracts: A correlation study. *Food Technology and Biotechnology, 52*, 119–127.

Mesquita, E., & Monteiro, M. (2018). Simultaneous HPLC determination of flavonoids and phenolic acids profile in Pêra-Rio orange juice. *Food Research International, 106*, 54–63.

Miller, H. E. (1971). A simplified method for the evaluation of antioxidants. *Journal of the American Oil Chemists' Society, 48*, 91.

Miller, N. J., Rice-Evans, C., Davies, M. J., Gopinathan, V., & Milner, A. (1993). A novel method for measuring antioxidant capacity and its application to monitoring the antioxidant status in premature neonates. *Clinical Science, 84*, 407–412.

Mira, L., Fernandez, T. M., Santos, M., Rocha, R., Florêncio, M. H., & Jennings, K. R. (2002). Interactions of flavonoids with iron and copper ions: A mechanism for their antioxidant activity. *Free Radical Research, 36*, 1199–1208.

Mkanda, A. V., Minnaar, A., & Kock, H. L. (2007). Relating consumer preferences to sensory and physicochemical properties of dry beans (*Phaseolus vulgaris* L.). *Journal of the Science of Food and Agriculture, 87*, 2869–2879.

Monteiro, M. C., & Trugo, L. C. (2005). Determination of bioactive compounds in Brazilian roasted coffees. *Química Nova*, *28*, 637–641.

Moo-Huchin, V. M., Moo-Huchin, M. I., Estrada-León, R. J., Cuevas-Glory, L., Estrada-Mota, I. A., Ortiz-Vázquez, E., et al. (2015). Antioxidant compounds, antioxidant activity and phenolic content in peel from three tropical fruits from Yucatan, Mexico. *Food Chemistry*, *166*, 17–22.

Moreira, A. S. P., Nunes, F. M., Domingues, M. R., & Coimbra, M. A. (2012). Coffee melanoidins: Structures, mechanisms of formation and potential health impacts. *Food and Function*, *3*, 903–915.

Mori, H., Kawabata, K., Yoshimi, N., Tanaka, T., Murakami, T., Okada, T., et al. (1999). Chemopreventive effects of ferulic acid on oral and rice germ on large bowel carcinogenesis. *Anticancer Research*, *19*, 3775–3778.

Morley, K. L., Grosse, S., Leischa, H., & Lau, P. C. K. (2013). Antioxidant canolol production from a renewable feedstock *via* an engineered decarboxylase. *Green Chemistry*, *15*, 3312–3317.

Mullen, W., McGinn, J., Lean, M. E., MacLean, M. R., Gardner, P., Duthie, G., et al. (2002). Ellagitannins, flavonoids, and other phenolics in red raspberries and their contribution to antioxidant capacity and vasorelaxation properties. *Journal of Agricultural and Food Chemistry*, *50*, 5191–5196.

Mullen, W., Nemzer, B., Ou, B., Stalmach, A., Hunter, J., Clifford, M., et al. (2011). The antioxidant and chlorogenic acid profiles of whole coffee fruits are influenced by extraction procedures. *Journal of Agricultural and Food Chemistry*, *59*, 3754–3762.

Naczk, M., Amarowicz, R., Pink, D., & Shahidi, F. (2000). Insoluble condensed tannins of canola/rapeseed. *Journal of Agricultural and Food Chemistry*, *48*, 1758–1762.

Naczk, M., & Shahidi, F. (2006). Phenolics in cereals, fruits and vegetables: Occurrence, extraction and analysis. *Journal of Pharmaceutical and Biomedical Analysis*, *41*, 1523–1542.

Nadeem, M., Situ, C., Mahmud, A., Khalique, A., Imran, M., Rahman, F., et al. (2014). Antioxidant activity of sesame (*Sesamum indicum* L.) cake extract for the stabilization of olein based butter. *Journal of the American Oil Chemists' Society*, *91*, 967–977.

Navarro, M., Moreira, I., Arnaez, E., Quesada, S., Azofeifa, G., Vargas, F., et al. (2018). Polyphenolic characterization and antioxidant activity of *Malus domestica* and *Prunus domestica* cultivars from Costa Rica. *Foods*, 7(2), 15.

Nemzer, B., Vargas, L., Xia, X., Sintara, M., & Feng, H. (2018). Phytochemical and physical properties of blueberries, tart cherries, strawberries, and cranberries as affected by different drying methods. *Food Chemistry*, *262*, 242–250.

Nenandis, N., Zhang, H.-Y., & Tsimidou, M. Z. (2003). Structure-antioxidant activity relationship of ferulic acid derivatives: Effect of carbon side chain characteristic groups. *Journal of Agricultural and Food Chemistry*, *51*, 1874–1879.

Nibir, Y. M., Sumit, A. F., Akhand, A. A., Ahsan, N., & Hossain, M. S. (2017). Comparative assessment of total polyphenols, antioxidant and antimicrobial activity of different tea varieties of Bangladesh. *Asian Pacific Journal of Tropical Biomedicine*, 7, 352–357.

Niemeyer, H. B., & Metzler, M. (2003). Differences in the antioxidant activity of plant and mammalian lignans. *Journal of Food Engineering*, *2003*(56), 255–256.

Nogueira, M., & Trugo, L. C. (2003). Chlorogenic acid isomers, caffeine and trigonellin contents in Brazilian coffee. *Ciência e Tecnologia de Alimentos*, *23*, 296–299.

Nowicka, A., Kucharska, A. Z., Sokół-Łętowska, A., & Fecka, I. (2019). Comparison of polyphenol content and antioxidant capacity of strawberry fruit from 90 cultivars of *Fragaria* × *ananassa* Duch. *Food Chemistry*, *270*, 32–46.

Oboh, G., Adebayo, A. A., Ejakpovi, I. I., Ogunsuyi, O. B., & Boligon, A. A. (2018). Phenolic profiling and *in vitro* antioxidant, anticholinesterase, and antimonoamine oxidase properties of aqueous extract of African star apple (*Chrysophyllum albidum*) fruit parts. *Journal of Food Biochemistry*, *42*(4), e12568.

Oliveira, A. S., Ribeiro-Santos, R., Ramos, F., Castilho, M. C., & Sanches-Silva, A. (2018). UHPLC-DAD multi-method for determination of phenolics in aromatic plants. *Food Analytical Methods*, *11*, 440–450.

Ombra, M. N., d'Acierno, A., Nazzaro, F., Riccardi, R., Spigno, P., Zaccardelli, M., et al. (2016). Phenolic composition and antioxidant and antiproliferative activities of the extracts of twelve common bean (*Phaseolus vulgaris* L.) endemic ecotypes of southern Italy before and after cooking. *Oxidative Medicine and Cellular Longevity*, *2016*, 1398298.

Oniga, I., Pus, C., Silaghi-Dumitrescu, R., Olah, N.-K., Sevastre, B., Marica, R., et al. (2018). *Origanum vulgare ssp. vulgare*: Chemical composition and biological studies. *Molecules*, *23*(8), 2077.

Oomah, B. D., Kotzeva, L., Allen, M., & Bassinello, P. Z. (2014). Microwave and micronization treatments affect dehulling characteristics and bioactive contents of dry beans (*Phaseolus vulgaris* L.). *Journal of the Science of Food and Agriculture*, *94*, 1349–1358.

Orallo, F. (2006). Comparative studies of the antioxidant effects of *cis*- and *trans*-resveratrol. *Current Medicinal Chemistry*, *13*, 87–98.

Osawa, T. (1999). Protective role of dietary polyphenols in oxidative stress. *Mechanisms of Aging and Development*, *111*, 133–139.

Oswell, N. J., Thippareddi, H., & Pegg, R. B. (2018). Practical use of natural antioxidants in meat products in the U.S.: A review. *Meat Science*, *145*, 469–479.

Ou, B., Hampsch-Woodhill, M., & Prior, R. L. (2001). Development and validation of an improved oxygen radical absorbance capacity assay using fluorescein as the fluorescent probe. *Journal of Agriculture and Food Chemistry*, *49*, 4619–4626.

Özen, Ö. B., & Soyer, A. (2018). Effect of plant extracts on lipid and protein oxidation of mackerel (*Scomber scombrus*) mince during frozen storage. *Journal of Food Science and Technology*, *55*, 120–127.

Ozrenk, K., Erez, M. E., Altintas, S., & Inal, B. (2018). The comparison of phenolic compounds content, antioxidant capacity and molecular analysis of some selected Turkish pear genotypes. *Fresenius Environmental Bulletin*, *27*, 584–589.

Ozvural, E. B., & Vural, H. (2012). The effects of grape seed extract on quality characteristics of frankfurters. *Journal of Food Processing and Preservation*, *36*, 291–297.

Özyürek, M., Güçlü, K., Tütem, E., Başkan, K. S., Erçağ, E., Çelik, S. E., et al. (2011). A comprehensive review of CUPRAC methodology. *Analytical Methods*, *3*, 2439–2453.

Panche, A. N., Diwan, A. D., & Chandra, S. R. (2016). Flavonoids: An overview. *Journal of Nutrition Science*, *5*, e47.

Pandey, K. B., & Rizvi, S. I. (2009). Protective effect of resveratrol on formation of membrane protein carbonyls and lipid peroxidation in erythrocytes subjected to oxidative stress. *Applied Physiology Nutrition and Metabolism*, *34*, 1093–1097.

Pandey, K. B., & Rizvi, S. I. (2010). Resveratrol may protect plasma proteins from oxidation under conditions of oxidative stress *in vitro*. *Journal of the Brazilian Chemical Society*, *21*, 909–913.

Pandey, K. B., & Rizvi, S. I. (2011). Anti-oxidative action of resveratrol: Implications for human health. *Arabian Journal of Chemistry*, *4*, 293–298.

Papadopoulos, A. G., Nenadis, N., & Sigalas, M. P. (2016). DFT study of radical scavenging activity of sesame oil lignans and selected *in vivo* metabolites of sesamin. *Computational and Theoretical Chemistry*, *1077*, 125–132.

Papoutsis, K., Pristijono, P., Golding, J. B., Stathopoulos, C. E., Bowyer, M. C., Scarlett, C. J., et al. (2017). Enhancement of the total phenolic compounds and antioxidant activity of aqueous *Citrus limon* L. pomace extract using microwave pretreatment on the dry powder. *Journal of Food Processing and Preservation*, *41*(5), e13152.

Pari, L., & Jalaludeen, M. A. (2011). Protective role of sinapic acid against arsenic: Induced toxicity in rats. *Chemico-Biological Interactions*, *194*, 40–47.

Pasqualone, A., Laddomada, B., Spina, A., Todaro, A., Guzmàn, C., Summo, C., et al. (2018). Almond by-products: Extraction and characterization of phenolic compounds and evaluation of their potential use in composite dough with wheat flour. *LWT-Food Science and Technology*, *89*, 299–306.
Patel, M., & Naik, S. N. (2004). Gamma-oryzanol from rice bran oil—A review. *Journal of Scientific and Industrial Research*, *63*, 569–578.
Patras, A., Brunton, N. P., da Pieve, S., & Butler, F. (2009). Impact of high pressure processing on total antioxidant activity, phenolic, ascorbic acid, anthocyanin content and colour of strawberry and blackberry purées. *Innovative Food Science & Emerging Technologies*, *10*, 308–313.
Pawlus, A., Waffo-Téguo, P., Shaver, J., & Mérillon, J.-M. (2012). Stilbenoid chemistry from wine and the genus *Vitis*, a review. *Journal International des Sciences de la Vigne et du Vin*, *46*, 57–111.
Pegg, R. B., Amarowicz, R., & Naczk, M. (2005). Antioxidant activity of polyphenolics from bearberry-leaf (*Arctostaphylos uva-ursi* L.) Sprengel extract in met systems. *ACS Symposium Series*, *909*, 67–82.
Pękal, A., & Pyrzyńska, K. (2014). Evaluation of aluminum complexation reaction for flavonoid content assay. *Food Analytical Methods*, *7*, 1776–1782.
Peñarrieta, J. M., Alvarado, J. A., Åkesson, B., & Bergenståhl, B. (2008). Total antioxidant capacity and content of flavonoids and other phenolic compounds in canihua (*Chenopodium quinoa* Willd.) an Andean pseudocereal. *Molecular Nitrition and Food Research*, *52*, 708–717.
Pereira de Abreu, D. A., Losada, P., Maroto, J., & Cruz, J. M. (2010). Lipid damage during frozen storage of Atlantic halibut (*Hippoglossus hippoglossus*) in active packaging film containing antioxidants. *Food Chemistry*, *126*, 315–320.
Pereira de Abreu, D. A., Paseiro Losada, P., Maroto, J., & Cruz, J. M. (2010). Evaluation of the effectiveness of a new active packaging film containing natural antioxidants (from barley husks) that retard lipid damage in frozen Atlantic salmon (*Salmo salar* L.). *Food Research International*, *43*, 1277–1282.
Pereira de Abreu, D. A., Paseiro Losada, P., Maroto, J., & Cruz, J. M. (2011). Natural antioxidant active packaging film and its effect on lipid damage in frozen blue shark (*Prionace glauca*). *Innovative Food Science & Emerging Technologies*, *12*, 50–55.
Pihlava, J.-M., Nordlund, E., Heiniö, R.-L., Hietaniemi, V., Lehtinen, P., & Poutanen, K. (2015). Phenolic compounds in wholegrain rye and its fractions. *Journal of Food Composition and Analysis*, *38*, 89–97.
Popov, I. N., & Lewin, G. (1994). Photochemiluminescent detection of antiradical activity: II. Testing of nonenzymic water-soluble antioxidants. *Free Radicals in Biology and Medicine*, *17*, 267–271.
Popov, I. N., & Lewin, G. (1996). Photochemiluminescent detection of antiradical activity: IV: Testing of lipid-soluble antioxidants. *Journal of Biochemical and Biophysical Methods*, *31*, 1–8.
Prajapati, M. R., Patel, V., Parekh, T., & Subhash, R. (2013). Effect of in bio-processing on antioxidant activity of selected cereals. *Plant Science and Asian Journal of Research*, *3*, 66–72.
Prasad, K. (1997). Hydroxyl radical-scavenging property of secoisolariciresinol diglucoside (SDG) isolated from flax-seed. *Molecular and Cell Biochemistry*, *168*, 117–123.
Pulido, R., Bravo, L., & Saura-Calixto, F. (2000). Antioxidant activity of dietary polyphenols as determined by a modified ferric reducing/antioxidant power assay. *Journal of Agriculture and Food Chemistry*, *48*, 3396–33,402.
Qi, S., Huang, H., Huang, J., Wang, Q., & Wei, Q. (2015). Lychee (*Litchi chinensis* Sonn.) seed water extract as potential antioxidant and anti-obese natural additive in meat products. *Food Control*, *50*, 195–201.

Qiu, D., Guo, J., Yu, H., Yan, J., Yang, S., Li, X., et al. (2018). Antioxidant phenolic compounds isolated from wild *Pyrus ussuriensis* Maxim. Fruit peels and leaves. *Food Chemistry, 241*, 182–187.

Rababah, M., KEreifej, K. I., Alhamad, M. N., Al-Qudah, K. M., Rousan, L. M., Al-Mahasneh, M. A., et al. (2011). Effects of green tea and grape seed and TBHQ on physicochemical properties of Baladi goat meats. *International Journal of Food Properties, 14*, 1208–1216.

Rafiq, S., Kaul, R., Sofi, S. A., Bashir, N., Nazir, F., & Ahmad Nayik, G. (2018). Citrus peel as a source of functional ingredient: A review. *Journal of the Saudi Society of Agricultural Sciences, 17*, 351–358.

Ramalho, V. C., & Jorge, N. (2008). Antioxidant action of rosemary extract in soybean oil submitted to thermoxidation. *Grasas y Aceites, 59*, 128–131.

Raudoniute, I., Rovira, J., Venskutonis, P. R., Damasius, J., Rivero-Perez, M. D., & González-SanJose, M. L. (2011). Antioxidant properties of garden strawberry leaf extract and its effect on fish oil oxidation. *International Journal of Food Science and Technology, 46*, 935–943.

Re, R., Pellegrini, N., Proteggente, A., Pannala, A., Yang, M., & Rice-Evans, C. (1999). Antioxidant activity applying an improved ABTS radical cation decolorization assay. *Free Radical Biology and Medicine, 26*, 1231–1237.

Reddy, G. V. B., Sen, A. R., Nair, P. N., Reddy, K. S., Reddy, K. K., & Kondaiah, N. (2013). Effects of grape seed extract on the oxidative and microbial stability of restructured mutton slices. *Meat Science, 95*, 288–294.

Repo-Carrasco-Valencia, R., Hellström, J. K., Pihlava, J.-M., & Mattila, P. H. (2010). Flavonoids and other phenolic compounds in Andean indigenous grains: Quinoa (*Chenopodium quinoa*), kañiwa (*Chenopodium pallidicuale*) and kiwicha (*Amaranthus caudatus*). *Food Chemistry, 120*, 128–133.

Robards, K. (2003). Strategies for the determination of bioactive phenols in plants, fruit and vegetables. *Journal of Chromatography A, 1000*, 657–691.

Robbins, K. S., Greenspan, P., & Pegg, R. B. (2016). Effect of pecan phenolics on the release of nitric oxide from murine RAW 264.7 macrophage cells. *Food Chemistry, 212*, 681–687.

Rocha-Guzmán, N. E., González-Laredo, R. F., Ibarra-Pérez, F. J., Nava-Berúmen, C. A., & Gallegos-Infante, J.-A. (2007). Effect of pressure cooking on the antioxidant activity of extracts from three common bean (*Phaseolus vulgaris* L.) cultivars. *Food Chemistry, 100*, 31–35.

Rodríguez-Carpena, J. G., Morcuende, D., Andrade, M.-J., Kylli, P., & Estévez, M. (2011). Avocado (*Persea americana* Mill.) phenolics, *in vitro* antioxidant and antimicrobial activities, and inhibition of lipid and protein oxidation in porcine patties. *Journal of Agricultural and Food Chemistry, 59*, 5625–5635.

Rodríguez-Carpena, J.-G., Morcuende, D., & Estévez, M. (2011). Avocado by-products as inhibitors of color deterioration and lipid and protein oxidation in raw porcine patties subjected to chilled storage. *Meat Science, 89*, 166–173.

Rojas, M. C., & Brewer, M. S. (2007). Effect of natural antioxidants on oxidative stability of cooked, refrigerated beef and pork. *Journal of Food Science, 72*, S282–S288.

Roldan-Martin, E., Sanchez-Moreno, C., Lioria, R., de Ancos, B., & Cano, M. P. (2009). Onion high-pressure processing: Flavonol content and antioxidant activity. *LWT-Food Sciences and Technology, 42*, 835–841.

Ross, A. B., Shepherd, M. J., Schüpphaus, M., Sinclair, V., Alfaro, B., Kamal-Eldin, A., et al. (2003). Alkylresorcinols in cereals and cereal products. *Journal of Agricultural and Food Chemistry, 51*, 4111–4118.

Rostagno, M. A., Palma, M., & Barroso, C. G. (2004). Pressurized liquid extraction of isoflavones from soybeans. *Analytica Chimica Acta, 522*, 169–177.

Rusu, M. E., Gheldiu, A.-M., Mocan, A., Moldovan, C., Popa, D.-S., Tomuta, I., et al. (2018). Process optimization for improved phenolic compounds recovery from walnut (*Juglans regia* L.) septum: Phytochemical profile and biological activities. *Molecules, 23*(11), 2814.

Rybiński, W., Karamać, M., Sulewska, K., Börner, A., & Amarowicz, R. (2018). Antioxidant potential of grass pea seeds from European countries. *Foods*, 7(9), 142.

Saarinen, N. M., Penttinen, P. E., Smeds, A. I., Hurmerinta, T. T., & Mäkelä, S. I. (2005). Structural determinants of plant lignans for growth of mammary tumors and hormonal responses *in vivo*. *Journal of Steroid Biochemistry and Molecular Biology, 2005*(93), 209–219.

Saba, M. K., & Sogvar, O. B. (2016). Combination of carboxymethyl cellulose-based coatings with calcium and ascorbic acid impacts in browning and quality of fresh-cut apples. *LWT-Food Science and Technology, 66*, 165–171.

Sánchez-Moreno, C., Plaza, L., De Ancos, B., & Cano, M. P. (2006). Impact of high-pressure and traditional thermal processing of tomato purée on carotenoids, vitamin C and antioxidant activity. *Journal of the Science of Food and Agriculture, 86*, 171–179.

Sánchez-Moreno, C., Plaza, L., De Ancos, B., Martin, B. O., & Cano, M. P. (2005). Impact of high pressure and pulsed electric fields on bioactive compounds and antioxidant activity of orange juice in comparison with traditional thermal processing. *Journal of Agriculture and Food Chemistry, 53*, 4403–4409.

Santos, R. D., Shetty, K., & Da Silva Miglioranza, L. H. (2014). Oxidative stability of butter with added phenolics from Lamiaceae herbs and *in vitro* evaluation of potential cytotoxicity of rosemary (*Rosmarinus officinalis* L.) extract. *International Journal of Food Science and Technology, 49*, 768–775.

Santos-Zea, L., Antunes-Ricardo, M., Gutierrez-Uribe, J. A., García-Pérez, J. V., & Benedito, J. (2018). Effect of ultrasound transducer design on the acoustically-assisted supercritical fluid extraction of antioxidants from oregano. *Ultrasonics Sonochemistry, 47*, 47–56.

Schaich, K. M. (2006). Developing a rational basis for selection of antioxidant screening and testing methods. *Acta Horticulture, 709*, 79–94.

Seçmeler, Ö., Güçlü Üstündağ, Ö., Fernández-Bolaños, J., & Rodríguez-Gutiérrez, G. (2018). Effect of subcritical water and steam explosion pretreatments on the recovery of sterols, phenols and oil from olive pomace. *Food Chemistry, 265*, 298–307.

Sekhon-Loodu, S., Sumudu, N., Warnakulasuriyaa, H. P., Rupasinghe, V., & Shahidi, F. (2013). Antioxidant ability of fractionated apple peel phenolics to inhibit fish oil oxidation. *Food Chemistry, 140*, 189–196.

Serpen, A., Pelvan, E., Alasalvar, C., Mogol, B. A., Yavuz, H. T., Gökmen, V., et al. (2012). Nutritional and functional characteristics of seven grades of black tea produced in Turkey. *Journal of Agricultural and Food Chemistry, 60*, 7682–7689.

Sert, D., Arslan, D., Ayar, A., & Özcan, M. M. (2015). The effect of some spices extracts on oxidative stability of Yayik butter. *Archiv für Lebensmittelhygiene, 66*, 10–17.

Shah, M. A., Bosco, S. J. D., & Mir, S. A. (2014). Plant extracts as natural antioxidants in meat and meat products. *Meat Science, 98*, 21–33.

Shahidi, D., & Wanasundara, P. K. J. P. D. (1992). Phenolic antioxidants. *Critical Review of Food Science and Nutrition, 32*, 67–103.

Shahwar, D., Raza, M. A., Mughal, M. A. S., Abbasi, M. A., & Ahmad, V. U. (2010). Comparative study on the antioxidant and antimicrobial activities of stem-bark extract of *Litchi chinensis* and its organic fractions. *Journal of the Chemical Society of Pakistan, 32*, 357–362.

Shan, B., Cai, Y. Z., Brooks, J. D., & Corke, H. (2009). Antibacterial and antioxidant effects of five spice and herb extracts as natural preservatives of raw pork. *Journal of the Science of Food and Agriculture, 89*, 1879–1885.

Shan, B., Cai, Y. Z., Sun, M., & Corke, H. (2005). Antioxidant capacity of 26 spice extracts and characterization of their phenolic constituents. *Journal of Agricultural and Food Chemistry, 53*, 7749–7759.

Sheng, R., Lin, X., Zhang, J., Chol, K. S., Huang, W., Yang, B., et al. (2009). Design, synthesis and evaluation of flavonoid derivatives as potent AChE inhibitors. *Bioorganic and Medicinal Chemistry, 17*, 6692–6698.

Shrestha, K., Stevens, C. V., & De Meulenaer, B. (2012). Isolation and identification of a potent radical scavenger (canolol) from roasted high erucic mustard seed oil from Nepal and its formation during roasting. *Journal of Agricultural and Food Chemistry, 60*, 7506–7512.

Siddhuraju, P. (2006). The antioxidant activity and free radical-scavenging capacity of phenolics of raw and dry heated moth bean (*Vigna aconitifolia*) (Jacq.) Marechal seed extracts. *Food Chemistry, 99*, 149–157.

Siger, A., Czubinski, J., Dwiecki, K., Kachlicki, P., & Nogala-Kalucka, M. (2013). Identification and antioxidant activity of sinapic acid derivatives in *Brassica napus* L. seed meal extracts. *European Journal of Lipid Science and Technology, 115*, 1120–1138.

Singleton, V. L., Orthofer, R., & Lamuela-Raventhós, R. M. (1999). Analysis of total phenols and other oxidation substrates and antioxidants by means of Folin-Ciocalteu reagent. *Methods in Enzymology, 299*, 152–178.

Skrovankova, S., Sumczynski, D., Mlcek, J., Jurikova, T., & Sochor, J. (2015). Bioactive compounds and antioxidant activity in different types of berries. *International Journal of Molecular Sciences, 16*, 24673–24706.

Slavova-Kazakova, A., Karamać, M., Kancheva, V., & Amarowicz, R. (2016). Antioxidant activity of flaxseed extracts in lipid systems. *Molecules, 21*(1), 21010017.

Smuda, S. S., Mohsen, S. M., Olsen, K., & Hassan, A. M. (2018). Bioactive compounds and antioxidant activities of some cereal milling by-products. *Food Sciences and Technology, 55*, 1134–1142.

Sonmezdag, A. S., Kelebek, H., & Selli, S. (2018). Characterization of bioactive and volatile profiles of thyme (*Thymus vulgaris* L.) teas as affected by infusion times. *Journal of Food Measurement and Characterization, 12*, 2570–2580.

Stanger, M. C., Steffens, C. A., Soethe, C., Moreira, M. A., Both, V., & Brackmann, A. (2018). Phenolic compounds content and antioxidant activity of 'Galaxy' apples stored in dynamic controlled atmosphere and ultralow oxygen conditions. *Postharvest Biology and Technology, 144*, 70–76.

Starzyńska-Janiszewska, A., Stodolak, B., & Mickowska, B. (2014). Effect of controlled lactic fermentation on selected bioactive and nutritional parameters of tempeh obtained from unhulled common bean (*Phaseolus vulgaris* L.). *Journal of the Food Science and Agriculture, 94*, 359–366.

Strandås, C., Kamal-Eldin, A., Andersson, R., & Åman, P. (2008). Composition and properties of flaxseed phenolic oligomers. *Food Chemistry, 110*, 106–112.

Struijs, K., Vincken, J.-P., Doeswijk, T. G., Voragen, A. G. J., & Gruppen, H. (2009). The chain length of lignan macromolecule from flaxseed hulls is determined by the incorporation of coumaric acid and glucosides and ferulic acid glucosides. *Phytochemistry, 70*, 262–269.

Struijs, K., Vincken, J.-P., Verhoef, R., Voragen, A. G. J., & Gruppen, H. (2008). Hydroxycinnamic acids are ester-linked directly to glucosyl moieties within the lignan macromolecule from flaxseed hulls. *Phytochemistry, 69*, 1250–1260.

Suja, K. P., Abraham, J. T., Thamizh, S. N., Jayalekshmy, A., & Arumughan, C. (2004). Antioxidant efficacy of sesame cake extract in vegetable oil protection. *Food Chemistry, 84*, 393–400.

Suriano, S., Iannucci, A., Codianni, P., Fares, C., Russo, M., Pecchioni, N., et al. (2018). Phenolic acids profile, nutritional and phytochemical compounds, antioxidant properties in colored barley grown in southern Italy. *Food Research International, 113*, 221–233.

Szydłowska-Czerniak, A. (2013). Rapeseed and its products-sources of bioactive compounds: A review of their characteristics and analysis. *Critical Reviews in Food Science and Nutrition*, *53*, 307–330.

Takemoto, M., & Takemoto, H. (2018). Synthesis of theaflavins and their functions. *Molecules*, *23*(4), 918.

Tao, Y., Sun, D.-W., Górecki, A., Błaszczak, W., Lamparski, G., Amarowicz, R., et al. (2016). A preliminary study about the influence of high hydrostatic pressure processing in parallel with oak chip maceration on the physicochemical and sensory properties of a young red wine. *Food Chemistry*, *194*, 545–554.

Teixeira, A., Eiras-Dias, J., Castellarin, S. D., & Gerós, H. (2013). Berry phenolics of grapevine under challenging environments. *International Journal of Molecular Sciences*, *14*, 18,711–18,739.

Terán-Hilares, R., Chirinos, R., Pedreschi, R., & Campos, D. (2018). Enhanced antioxidant properties of tara (*Caesalpinia spinosa*) gallotannins by thermal hydrolysis and its synergistic effects with α-tocopherol, ascorbyl palmitate, and citric acid on sacha inchi (*Plukenetia volubilis*) oil. *Journal of Food Process Engineering*, *41*(1), e12613.

Terao, J. (2009). Dietary flavonoids as antioxidants. *Forum of Nutrition*, *61*, 87–94.

Thompsom, L. U., Boucher, B. A., Liu, Z., Cotterchio, M., & Kreiger, N. (2006). Phytoestrogen content of foods consumed in Canada, including isoflavones, lignans, and coumestan. *Nutrition and Cancer*, *54*, 184–201.

Tian, F., Li, B., Ji, B., Zhang, G., & Luo, Y. (2009). Identification and structure-activity relationship of gallotannins separated from Galla chinensis. *LWT-Food Science and Technology*, *42*, 1289–1295.

Timothy, C. N., Vishnu Priya, V., & Gayathri, R. (2018). Phytochemical analysis and total phenolic content of *Origanum vulgare* (oregano). *Drug Invention Today*, *10*, 1903–1905.

Tohidi, B., Rahimmalek, M., & Arzani, A. (2017). Essential oil composition, total phenolic, flavonoid contents, and antioxidant activity of Thymus species collected from different regions of Iran. *Food Chemistry*, *220*, 153–161.

Touré, A., & Xueming, X. (2010). Flaxseed lignans: source, biosynthesis, metabolism, antioxidant activity, bio-active components, and health benefits. *Comprehensive Review in Food Science and Food Safety*, *9*, 261–269.

Tsao, R., Yang, R., Xie, S., Sockovie, S., & Khanizadeh, S. (2005). Which polyphenolic compounds contribute to the total antioxidant activities of apple? *Journal of Agricultural and Food Chemistry*, *53*, 4989–4995.

Tudela, J. A., Cantos, E., Espin, J. C., Tomas-Barberan, F. A., & Gil, M. I. (2002). Induction of antioxidant flavonol biosynthesis in fresh-cut potatoes. Effect of domestic cooking. *Journal of Agricultural and Food Chemistry*, *50*, 5925–5931.

Turkmen, N., Sari, F., & Velioglu, S. (2005). The effect of cooking methods on total phenolics and antioxidant activity of selected green vegetables. *Food Chemistry*, *93*, 713–718.

Vashisth, T., Singh, R. K., & Pegg, R. B. (2011). Effects of drying on the phenolics content and antioxidant activity of muscadine pomace. *LWT-Food Science and Technology*, *44*, 1649–1657.

Vaz Patto, M. C., Amarowicz, R., Aryee, A. N. A., Boye, J. I., Chung, H.-J., Martín-Cabrejas, M. A., et al. (2015). Achievements and challenges in improving the nutritional quality of food legumes. *Critical Reviews in Plant Sciences*, *34*, 105–143.

Verma, B., Hucl, P., & Chibbar, R. N. (2008). Phenolic content and antioxidant properties of bran in 51 wheat cultivars. *Cereal Chemistry*, *85*, 544–549.

Vrhovsek, U., Rigo, A., Tonon, D., & Mattivi, F. (2004). Quantitation of polyphenols in different apple varieties. *Journal of Agricultural and Food Chemistry*, *52*, 6532–6538.

Vuorela, S., Meyer, A. S., & Heinonen, M. (2004). Impact of isolation method on the antioxidant activity of rapeseed meal phenolics. *Journal of Agricultural and Food Chemistry*, *52*, 8202–8207.

Waisundara, V. Y., Hsu, A., Tan, B. K., & Huang, D. (2009). Baicalin reduces mitochondrial damage in streptozotocin-induced diabetic Wistar rats. *Diabetes/Metabolism Research and Reviews*, *25*, 671–677.

Wakamatsu, D., Morimura, S., Sawa, T., Kida, K., Nakai, C., & Maeda, H. (2005). Isolation, identification, and structure of potent alkyl-peroxyl radical scavenger in crude canola oil, canolol. *Bioscience, Biotechnology, Biochemistry*, *69*, 1568–1574.

Waller, S. B., Madrid, I. M., Hoffmann, J. F., Picoli, T., Cleff, M. B., Chaves, F. C., et al. (2017). Chemical composition and cytotoxicity of extracts of marjoram and rosemary and their activity against *Sporothrix brasiliensis*. *Journal of Medical Microbiology*, *66*, 1076–1083.

Weidner, S., Amarowicz, R., Karamać, M., & Dąbrowski, G. (1999). Phenolic acids in caryopses of two cultivars of wheat, rye and triticale that display different resistance to preharvest sprouting. *European Food Research and Technology*, *210*, 109–113.

Weisz, G. M., Kammerer, D. R., & Carle, R. (2009). Identification and quantification of phenolic compounds from sunflower (*Helianthus annuus* L.) kernels and shells by HPLC-DAD/ESI-MS^n. *Food Chemistry*, *115*, 758–765.

Wijeratne, S. S. K., Amarowicz, R., & Shahidi, F. (2006). Antioxidant activity of almonds and their by-products in food model. *Journal of the American Oil Chemists' Society*, *83*, 223–230.

Wijesundera, C., Ceccato, C., Fagan, P., & Shen, Z. (2008). Seed roasting improves the oxidative stability of canola (*B. napus*) and mustard (*B. juncea*) seed oils. *European Journal of Lipid Science and Technology*, *110*, 360–367.

Wisam, H. M., Nahla, T. K., & Tariq, N. M. (2017). Antioxidant activities of thyme extracts. *Pakistan Journal of Nutrition*, *17*, 46–50.

Wright, J. S., Johnson, E. R., & Di Labio, G. A. (2001). Predicting the activity of phenolic antioxidants: theoretical method, analysis of substituent effects, and application to major families of antioxidants. *Journal of the American Chemical Society*, *123*, 1173–1183.

Wu, X., Boecher, G. R., Holden, J. M., & Haytowitz, D. B. (2004). Lipophylic and hydrophilic antioxidant capacity of common foods in the United States. *Journal of Agriculture and Food Chemistry*, *52*, 4026–4037.

Xu, B. J., & Chang, S. K. C. (2008). Total phenolic content and antioxidant properties of Elipse black beans (*Phaseolus vulgaris* L.). *Journal of Food Science*, *73*, H19–H27.

Xu, Z., Godber, J. S., & Xu, Z. (2001). Antioxidant activities of major components of gamma-oryzanol from rice bran using a linolenic acid model. *Journal of the American Oil Chemists' Society*, *78*, 465–469.

Xu, H., Wang, W., Liu, X., Yuan, F., & Gao, Y. (2015). Antioxidative phenolics obtained from spent coffee grounds (*Coffea arabica* L.) by subcritical water extraction. *Industrial Crops and Products*, *76*, 946–954.

Xu, B. J., Yuan, S. H., & Chang, S. K. C. (2007). Comparative analyses of phenolic composition, antioxidant capacity, and color of cool season legumes and other selected food legumes. *Journal of Food Science*, *72*, S167–S177.

Xue, J., Davidson, P. M., & Zhong, Q. (2013). Thymol nanoemulsified by whey protein-maltodextrin conjugates: the enhanced emulsifying capacity and antilisterial properties in milk by propylene glycol. *Journal of Agricultural and Food Chemistry*, *61*, 12720–12726.

Yan, L., Cao, Y., & Zheng, G. (2017). Optimization of subcritical water extraction of phenolic antioxidants from pomegranate (*Punica granatum* L.) peel by response surface methodology. *Analytical Methods*, *9*, 4647–4656.

Yang, X.-J., Dang, B., & Fan, M.-T. (2018). Free and bound phenolic compound content and antioxidant activity of different cultivated blue highland barley varieties from the Qinghai—Tibet plateau. *Molecules*, *23*(4), 879.

Yao, Y., Cheng, X., Wang, L., Wang, S., & Ren, G. (2011). Biological potential of sixteen legumes in China. *International Journal of Molecular Sciences*, *12*, 7048–7058.

Yashin, A., Yashin, Y., Wang, J. Y., & Nemzer, B. (2013). Antioxidant and antiradical activity of coffee. *Antioxidants, 2*, 230–245.
Yogesh, K., & Ali, J. (2014). Antioxidant potential of thuja (*Thuja occidentalis*) cones and peach (*Prunus persia*) seeds in raw chicken ground meat during refrigerated (4 ± 1 °C) storage. *Journal of Food Science and Technology, 51*, 1547–1553.
Yogesh, K., Jha, S. N., & Ahmad, T. (2014). Antioxidant potential of aqueous extract of some food grain powder in meat model system. *Journal of Food Science and Technology, 51*, 3446–3451.
Yu, J., Vasathan, T., & Temelli, F. (2001). Analysis of phenolic acids in barley by high-performance liquid chromatography. *Journal of Agricultural and Food Chemistry, 49*, 4352–4358.
Yu, H., Yang, G., Sato, M., Yamaguchi, T., Nakano, T., & Xi, Y. (2017). Antioxidant activities of aqueous extract from *Stevia rebaudiana* stem waste to inhibit fish oil oxidation and identification of its phenolic compounds. *Food Chemistry, 232*, 379–386.
Zadernowski, R., Naczk, M., & Nesterowicz, J. (2005). Phenolic acid profiles in some small berries. *Journal of Agricultural and Food, 53*, 2118–2124.
Zafrilla, P., Ferreres, F., & Tomás-Barberán, F. A. (2001). Effect of processing and storage on the antioxidant ellagic acid derivatives and flavonoids of red raspberry (*Rubus ideas*) jams. *Journal of Agricultural and Food Chemistry, 49*, 3652–3655.
Zago, E., Lecomte, J., Barouh, N., Aouf, C., Carré, P., Fine, F., et al. (2015). Influence of rapeseed meal treatments on its total phenolic content and composition in sinapine, sinapic acid and canolol. *Industrial Crops and Products, 76*, 1061–1070.
Żegarska, Z., Rafałowski, R., Amarowicz, R., Karamać, M., & Shahidi, F. (1998). Stabilization of butter with deodorized rosemary extract. *Zeitschrift für Lebensmitteluntersuchung und -Forschung A, 206*, 99–102.
Zhang, Y. J., de Witt, D. L., Murugesan, S., & Nair, M. G. (2004). Novel lipid peroxidation and cyclooxygenase inhibitory tannins from *Picrorhiza kurrora* seeds. *Chemistry & Biodiversity, 1*, 426–441.
Zieliński, H., Michalska, A., Amigo-Benavent, M., Del Castillo, M. D., & Piskula, M. K. (2009). Changes in protein quality and antioxidant properties of buckwheat seeds and groats induced by roasting. *Journal of Agricultural and Food Chemistry, 57*, 4771–4776.
Zupfer, J. M., Churchill, K. E., Rasmusson, D. C., & Fulcher, R. G. (1998). Variation in ferulic acid concentration among diverse barley cultivars measured by HPLC and microspectrophotometry. *Journal of Agricultural and Food Chemistry, 46*, 13,550–1354.

Further reading

Karamać, M., Kosińska, A., & Pegg, R. B. (2005). Comparison of radical-scavenging activities for selected phenolic acids. *Polish Journal of Food and Nutrition Sciences, 55*, 165–170.

CHAPTER TWO

Dietary fiber sources and human benefits: The case study of cereal and pseudocereals

María Ciudad-Mulero, Virginia Fernández-Ruiz, Mª Cruz Matallana-González, Patricia Morales*

Department of Nutrition and Food Science, Faculty of Pharmacy, Complutense University of Madrid, Madrid, Spain

*Corresponding author: e-mail address: patricia.morales@farm.ucm.es

Contents

Abstract

Dietary fiber (DF) includes the remnants of the edible part of plants and analogous carbohydrates that are resistant to digestion and absorption in the human small intestine with complete or partial fermentation in the human large intestine. DF can be classified into two main groups according to its solubility, namely insoluble dietary fiber (IDF) that mainly consists on cell wall components, including cellulose, some hemicelluloses, lignin and resistant starch, and soluble dietary fiber (SDF) that consists of non-cellulosic polysaccharides as non-digestible oligosaccharides, arabinoxylans (AX), β-glucans, some hemicelluloses, pectins, gums, mucilages and inulin. The intake of DF is associated with health benefits. IDF can contribute to the normal function of the intestinal tract and it has an important role in the prevention of colonic diverticulosis and constipation. SDF is extensively fermented by gut microbiota and it is associated with carbohydrate and lipid metabolism, with important health benefits due to its hypocholesterolemic

Advances in Food and Nutrition Research, Volume 90
ISSN 1043-4526
https://doi.org/10.1016/bs.afnr.2019.02.002

properties. Due to these nutritional and health properties, DF is widely used as functional ingredients in food industry, being whole grain cereals, pulses, fruits and vegetables the main sources of DF. Also some synthetic sources are employed, namely polydextrose, hydroxypropyl methylcellulose or cyclodextrins. The DF content of cereals varies depending on cultivars, their botanical components (pericarp, emdosperm and germ) and the processing conditions they have undergone (baking, extrusion, etc.). In cereal grains, AX are the predominant non-cellulose DF polysaccharides followed by cellulose and β-glucans, while in pseudocereals, pectins are quantitatively predominant.

1. Dietary fiber concept

Over the years, the definition of dietary fiber (DF) has been a topic of discussion. In Hipsley (1953) first introduced the term "dietary fiber" and defined it as "non-digestible constituents of the plant cell wall". In the 70s it was established that DF consists of the remnants of edible plant cells, polysaccharides, lignin, and associated substances resistant to digestion by the alimentary enzymes of humans. In particular, the constituents of DF included cellulose, hemicelluloses, lignin, gums, mucilage, oligosaccharides, pectin, and other associated minor substances as waxes, cutin or suberin. This definition prevailed for many years and led to the development of analytical methods for DF that complied with this definition (Dai & Chau, 2017; Macagnan, Da Silva, & Hecktheuer, 2016). The first set of AOAC 985.29/AACC 32-05.01 standard method was officially adopted in 1985, and several modifications were introduced in 1986 and 1988, which is primarily limited to total DF analysis (Li & Komarek, 2017). Until the 90s, the definition of DF was based primarily on analytical criteria, but physiological properties of DF determine its importance in the human health and its requirement in the human diet, so most scientists agree that the definition of DF should be physiologically based (Gray, 2006). Taking these considerations into account, in 2001, The American Association of Cereal Chemists (AACC) defined DF as "the remnants of the edible part of plants and analogous carbohydrates that are resistant to digestion and absorption in the human small intestine with complete or partial fermentation in the human large intestine." It includes polysaccharides, oligosaccharides, lignin and associated plant substances. DF exhibits one or more of either laxation (fecal bulking and softening; increased frequency; and/or regularity), blood cholesterol attenuation, and/or blood glucose decrease (AACC Report, 2001). In order to harmonize the concept of DF, in 2009 CODEX published its DF definition that includes carbohydrate polymers with 10 or more monomeric units (decision on whether to include carbohydrates of three to nine monomeric units should be left up to national authorities), which are not hydrolyzed by the endogenous enzymes in the small intestine of humans and belong to the following categories (Jones, 2014; Stephen et al., 2017):

- Edible carbohydrate polymers naturally occurring in the food as consumed.
- Carbohydrate polymers, which have been obtained from food raw material by physical, enzymatic or chemical means and which have been shown to have a physiological effect of benefit to health as demonstrated by generally accepted scientific evidence to competent authorities.
- Synthetic carbohydrate polymers, which have been shown to have a physiological effect of benefit to health as demonstrated by generally accepted scientific evidence to competent authorities.

CODEX indicates that when carbohydrate polymers derive from a plant origin, dietary fiber may include fractions of lignin and/or other compounds associated with polysaccharides in the plant cell walls. These compounds also may be measured by certain analytical methods for DF (Jones, 2014; Stephen et al., 2017).

In CODEX definition, it is admitted that there are three categories of DF, which are not necessarily equivalent. In general, this definition comprises all carbohydrate polymers that are not digested and none absorbed in the human small intestine. The first category includes intrinsic carbohydrates of the plant cell wall, characteristic of healthy diets, as the major form of fiber. The second and third categories describe extracted and synthetic carbohydrate polymers and clearly state that to include these categories as DF, it is necessary that the competent authorities confirm that its potential health benefits have been demonstrated by generally accepted scientific evidence (Macagnan et al., 2016).

In the opinion of European Food Safety Authority (EFSA), DF is defined as non-digestible carbohydrates plus lignin, including non-starch polysaccharides (NSP), resistant oligosaccharides, resistant starch and lignin associated with the DF polysaccharides. Among NSP, cellulose, hemicelluloses, pectins, hydrocolloids (i.e., gums, mucilages, glucans) are found. Resistant oligosaccharides include fructo-oligosaccharides (FOS) and galacto-oligosaccharides (GOS), among others. Resistant starch consists of physically enclosed starch, some types of raw starch granules, retrograded amylase chemically and/or physically modified starches (EFSA, 2010).

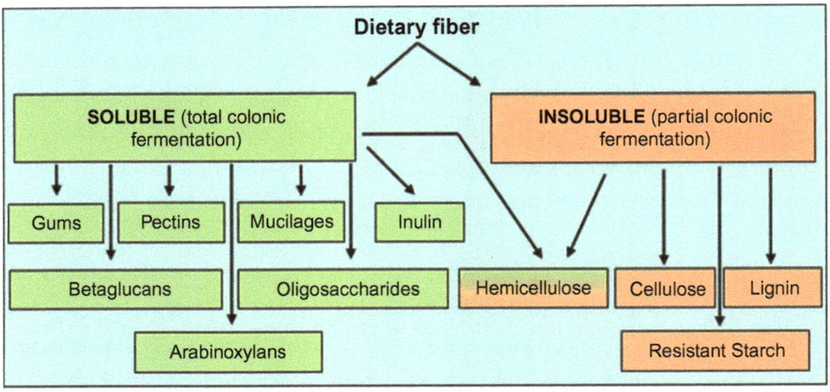

Fig. 1 Dietary fiber components. *Adapted from García Peris, P., & Velasco Gimeno, C. (2007). Evolución en el conocimiento de la fibra.* Nutrición Hospitalaria, *22(2), 20–25.*

2. Main dietary fiber constituents with health beneficial effects

Dietary fiber (DF) can be classified into two large groups according to its solubility (Fig. 1): insoluble dietary fiber (IDF) and soluble dietary fiber (SDF). IDF consists mainly of cell wall components, including cellulose, some hemicelluloses, lignin and resistant starch, while SDF consists of non-cellulosic polysaccharides as non-digestible oligosaccharides, arabinoxylans, β-glucans, some hemicelluloses, pectins, gums, mucilages and inulin (Dai & Chau, 2017; Dhingra, Michael, Rajput, & Patil, 2012; EFSA, 2010; Gray, 2006; Li & Komarek, 2017).

2.1 Insoluble dietary fiber (IDF)

IDF can contribute to the normal function of the intestinal tract. Its consumption is related with an increase of stool weight and decrease of colonic transit time. It has an important role in the prevention of colonic diverticulosis and constipation. Insoluble dietary fiber has an antioxidant potential that comes from phenolic compounds, and enhances certain health benefits (Tomic et al., 2017).

2.1.1 Cellulose

Cellulose is the main load-bearing constituent of the plant cell walls and it is located within a matrix of hemicelluloses, pectin, and also lignin. It is one of the most abundant natural biopolymers available which consists of linear

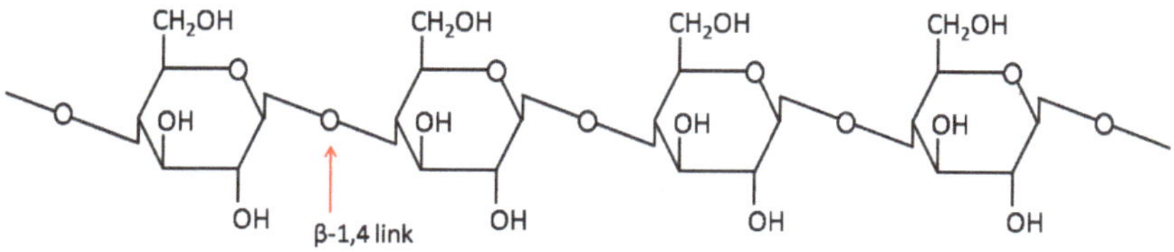

Fig. 2 Chemical structure of cellulose.

chains of β-(1 → 4) linked glucose monomers (Fig. 2) that are synthesized at the plasma membrane and are believed to aggregate into highly insoluble net that are often viewed as reinforcing rods in the cell wall composite (Burton & Fincher, 2014; Lattimer & Haub, 2010; Padayachee, Day, Howell, & Gidley, 2017).

Cellulose is water insoluble and resistant to digestive enzymes in the small intestine. However, it can be partial fermented by microbiota in the large intestine in turn producing short chain fatty acids (SCFA) (Lattimer & Haub, 2010). Moreover, cellulose has a key role on colon health by increasing the number of apoptotic epithelial cells in the large intestine, playing a protective role in the development of colon cancer. Due to its ability to capture water, it makes the stool bulky, improving the elimination of possible carcinogens and shortening bowel transit time (Dodevska et al., 2013; Dodevska, Šobajić, & Djordjević, 2015).

Cellulose is commonly present in cereals, legumes, fruits and vegetables, constituting about one quarter of the DF in grains and fruits and one third in vegetables and nuts (Gray, 2006; Mudgil & Barak, 2013; Yangilar, 2013).

2.1.2 Hemicellulose

Hemicellulose is a non-cellulosic component of both primary and secondary cell walls and it follows cellulose in abundance. Whereas cellulose is formed from units of glucose, different monomer units constitute hemicellulose. Hemicellulose consists in a heterogeneous group of polysaccharides made up of pyranoses and furanoses sugar units, including xylose, mannose, arabinose, glucose and galacturonic acid. Xylose and glucose are often the most abundant monomers found in hemicelluloses (Farhat et al., 2017; Mudgil & Barak, 2013; Ozyurt & Ötles, 2016; Padayachee et al., 2017).

Chemically, hemicelluloses can be grouped into four classes: xylans, xyloglucans, glucomannans and mixed linkage β-glucans. Xylans are composed of a backbone of β-(1 → 4)-D-xylose units with side chains that contain different sugars and sugar acid residues. These side chains include arabinose, glucose, galactose and in lower amounts, rhamnose, glucuronic acid and galacturonic acid. Xyloglucans are similar to the backbone of cellulose, consisting of β-(1→4)-linked D-glucopyranose units, but with frequent branching of α-D-xylose residues. Glucomannans consist of a branched backbone of β-(1→4)-linked D-mannose and D-glucose units. Mixed linkage (1→3, 1→4) β-glucans are other type of hemicelluloses that are restricted to grass species and some pteridophytes (Ozyurt & Ötles, 2016).

Hemicelluloses promote regular bowel movements by increasing hydration of the stool. These compounds bind cholesterol in the gut, preventing cholesterol absorption. Hemicelluloses are digested by microbiota increasing the number of beneficial bacteria in the gut and producing SCFA, which are used by colon cells as energetic substrate (Mudgil & Barak, 2013).

Hemicelluloses are principally present in cereal grains and about one third of the DF in vegetables, fruits, legumes and nuts consists of hemicelluloses (Dhingra et al., 2012; Mudgil & Barak, 2013).

2.1.3 Lignin

Lignin is not a polysaccharide but it is a complex random polymer containing about 40 oxygenated phenylpropane units including coniferyl, sinapyl and *p*-coumaryl alcohols that have undergone a complex dehydrogenative polymerization. Molecules of lignin vary in molecular weight and methoxyl content (Dhingra et al., 2012; Fuller, Beck, Salman, & Tapsell, 2016).

Lignin is one of the most chemically active components of the cell walls, being responsible for interactions with other dietary components and for decreasing bioavailability of nutrients. It also influences gastrointestinal physiology due to its water-holding capacity, increasing fecal bulk and stimulating the intestinal transit (Mudgil & Barak, 2013; Žilic et al., 2011).

Lignin is commonly found in foods with a woody component, as celery, and it is also present in the outer layer of cereal grains (Fuller et al., 2016; Mudgil & Barak, 2013).

2.1.4 Resistant starch

Starch is classified into three general types based on its rate of digestion: rapidly digestible, slowly digestible and resistant starch (Mohebbi, Homayouni, Azizi, & Hosseini, 2018). Resistant starch is defined as a portion of starch that resists digestion by human pancreatic amylase and brush border glycosidases in the small intestine of healthy humans and reaches the colon becoming available for fermentation by the microbiota (Chen, Bergman, McClung, Everette, & Tabien, 2017). Chemically, resistant starch is a linear polysaccharide of $(1 \rightarrow 4)$ α-D-glucan, essentially derived from the retrograded amylose fraction, and it has a relatively low molecular weight (1.2×10^5 Da) (Mohebbi et al., 2018). It is classified into five subtypes based on their mechanism of resistance to enzymatic digestion: I (encapsulated and physically inaccessible starch), II (resistant granules), III (retrograded amylose), IV (chemically modified starch) and V (amylase-lipid complex) (Kumar et al., 2018; Zhao et al., 2018).

Resistant starch type I is physically inaccessible to amylolytic and digestive enzymes and it passes the small intestine as such. It is present in whole kernel grain products (e.g., bread, seeds, pasta and legumes) (Fuentes-Zaragoza, Riquelme-Navarrete, Sánchez-Zapata, & Pérez-Álvarez, 2010; Lockyer & Nugent, 2017; Raigond, Ezekiel, & Raigond, 2015).

Resistant starch type II is found in raw starch granules, which are relatively dehydrated and have a compact structure that limits digestive enzymes ability to access it. It is present in raw potatoes, green bananas, high-amylose maize, ginkgo starch and some legumes (Chen et al., 2017; Fuentes-Zaragoza et al., 2010; Lockyer & Nugent, 2017).

Resistant starch type III is retrograded starch, primarily formed from amylose that has leached from starch granules after hydration. It is found in cooked potatoes, bread, corn flakes and food products with prolonged and/or repeated moist heat treatment (Chen et al., 2017; Fuentes-Zaragoza et al., 2010).

Type IV resistant starch is a group of chemically modified starches with similarity to resistant oligosaccharides and polydextrose, resistant to enzymatic hydrolysis. It is a constituent of some drinks and some foods in which modified starches have been used (certain breads and cakes) (Chen et al., 2017; Fuentes-Zaragoza et al., 2010; Raigond et al., 2015).

Finally, type V resistant starch is a kind of resistant starch arising from the formation of amylose–lipid complexes that can be formed during food processing and can also be prepared under controlled conditions. It comprises polysaccharides of water insoluble linear poly-α-$(1 \rightarrow 4)$-glucans and it is resistant to degradation by α-amylase. These polysaccharides promote the formation of SCFA, particularly butyrate. It is found in foods that contain naturally occurring amylose–lipid complexes, such as bread containing fat as an ingredient, or foods containing artificially made amylose–lipid complexes (Lockyer & Nugent, 2017; Raigond et al., 2015).

Due to its prebiotic effect, resistant starch contributes to maintenance of colonic health. During fermentation of resistant starch, it is produced high amount of butyrate, which is the principal nutrient of colonocytes and for this reason, resistant starch can reduce the risk of some colonic diseases, including colon cancer (Lockyer & Nugent, 2017). Resistant starch also presents hypoglycemic and hypocholesterolemic effects. It is not accessible to digestive enzymes, such as α-amylase and isoamylase and reduces postprandial blood glucose and insulin response, reducing glycemic and insulinemic responses to food. Due to hypocholesterolemic properties, resistant starch can improve cardiovascular health. For these reasons, the consumption of resistant starch improves gut health and can reduce the risk of several diseases, including colon cancer, diabetes and cardiovascular diseases (Chen et al., 2017; Raigond et al., 2015).

According to European Commission (2012), resistant starch has approved the following health claim: "Replacing digestible starches with resistant starch in a meal contributes to a reduction in the rise of blood glucose after that meal." This claim may be used in the label only for foods in which digestible starch has been replaced by resistant starch so that the final content of resistant starch is at least 14% of total starch.

2.2 Soluble dietary fiber (SDF)

SDF is hydrophilic, non-crystalline, and easily wetted by the aqueous gastrointestinal fluid, forming viscous colloidal dispersions or gels when hydrated. It is extensively fermented by gut microflora and it is associated with carbohydrate and lipid metabolism, showing hypocholesterolemic properties (Nair, Kharb, & Thompkinson, 2010). Due to their properties SDF are widely used in food industry to modify texture and rheology and to influence the colligative properties of food systems, thus improving the market-ability of the food product as health promoting or functional foods (Li, Liu, Wu, & Zhang, 2017).

2.2.1 Oligosaccharides

Recent definitions of dietary fiber have included oligosaccharides, such as fructo-oligosaccharides (FOS) and galacto-oligosaccharides (GOS) (Fig. 3), as sources of DF based on their physiological effects (Shortt et al., 2018).

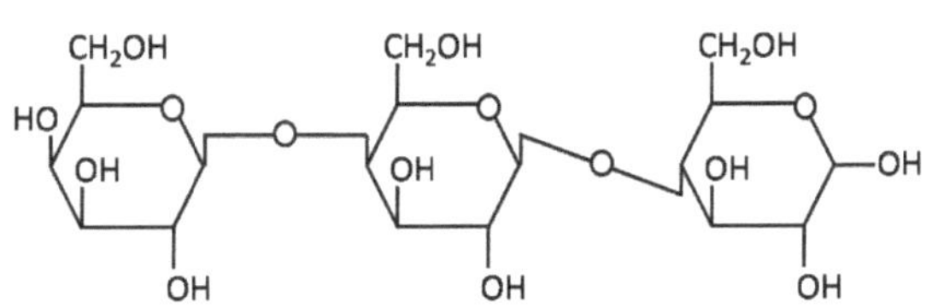

Fig. 3 Chemical structure of galacto-oligosaccharides.

Oligosaccharides are low molecular weight carbohydrates containing between 3 and 10 sugar units depending on degree of polymerization (Kothari, Patel, & Goyal, 2014).

Non-digestible oligosaccharides are natural constituents of many foods and often referred to as DF that resists digestion in the human small intestine, such as xylo-oligosaccharides (XOS).

They are associated with many health benefits, including positive effects on fermentation, mineral absorption, barrier function, fat metabolism, as well as, glycemic and insulin responses (Nauta & Garssen, 2013; Ou et al., 2016; Tanabe, Nakamura, & Oku, 2014). In particular, oligosaccharides positively affect to colon health by increasing bifidobacteria and lactic acid bacteria (Rainakari, Rita, Putkonen, & Pastell, 2016). Non-digestible oligosaccharides act as dietary prebiotics that are selectively fermented ingredients that result in specific changes, in the composition and/or activity of the gastrointestinal microbiota, thus conferring benefits upon consumer's health. Moreover, functional oligosaccharides, as feruloylated oligosaccharides, promote normal flora proliferation and pathogen suppression in the gastrointestinal tract (Fan, Zang, & Xing, 2015; Ou et al., 2016; Singh, Singh Jadaun, Narnoliya, & Pandey, 2017).

FOS are naturally present in asparagus (*Asparagus* L.), sugar beet (*Beta vulgaris* L.), garlic (*Allium sativum* L.), chicory (*Cichorium intybus* L.), onion (*Allium cepa* L.), Jerusalem artichoke (*Helianthus tuberosus* L.), wheat (*Triticum* L.), honey, banana (*Musa* L.), barley (*Hordeum vulgare* L.), tomato (*Solanum lycopersicum* L.) and rye (*Secale cereale* L.), whereas, milk (specially breast milk) is the main source of GOS (Singh et al., 2017).

2.2.2 Arabinoxylans

The arabinoxylans (AX) highlight within the dietary fiber components for its functional effect, both technological and nutritional, providing beneficial effects for the health of consumers.

These compounds are the main non cellulosic polysaccharides in cereals being part of the soluble fraction of the DF (Mendis & Simsek, 2014) and they are made up of a backbone of a linear chain of β-D-(1 → 4)-xylopyranose. This chain is substituted on the hydroxyl groups (–OH) of the 2- and 3-positions by L-arabinofuranosyl residues linked by β-(1 → 4) glycosidic bonds. Position 5 is commonly replaced with ferulic acid residues (Fig. 4), allowing cross-link bond formation by the oxidation of ferulic acid present in adjacent AX chains (Belitz & Grosch, 1997; Broekaert et al., 2011; Ciudad-Mulero et al., 2018; Lafiandra, Riccardi, & Shewry, 2014).

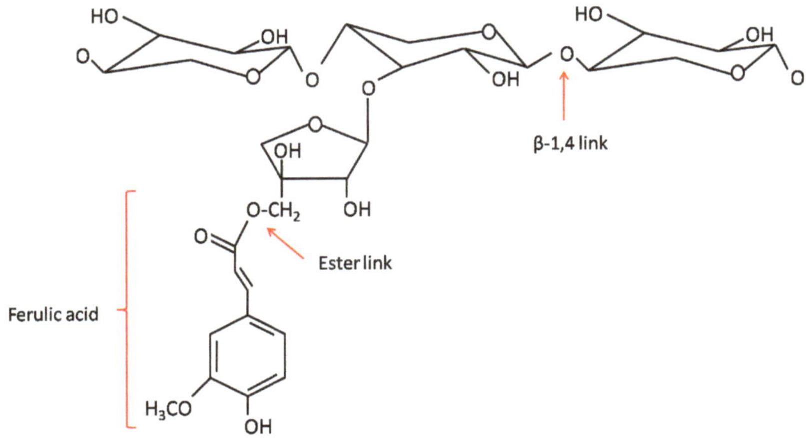

Fig. 4 Chemical structure of ferulated arabinoxylan.

The enzymatic hydrolysis of AX by xylanases and arabinofuranosidases produces arabinoxylan oligosaccharides (AXOS) and xylooligosaccharides (XOS) (Adams, Kroon, Williamson, Gilbert, & Morris, 2004), which are also considered dietary fiber and have several health effects, including immunomodulatory effect, hypocholesterolemic effect, control of type 2 diabetes, greater absorption of certain minerals, prebiotic effect, among others (Mendis & Simsek, 2014).

AX can be classified according to their physical properties of solubility, such as extractable in water (WEAX) or non-extractable in water (WUAX). The molecular weight of these polysaccharides varies from 10 to 10,000 kDa in the case of WEAX and exceeds 10,000 kDa in the case of WUAX (Niño-Medina et al., 2009). In the case of cereals, the WUAX are kept in the cell wall joined to other AX and other constituents of the cell through non-covalent interactions (hydrogen bonding) and covalent interactions (ester type bond). On the other hand, WEAX are weakly bound to the surface of the cell wall through incomplete cross-linking with other components or they may have undergone an initial enzymatic degradation in the grain (Van Craeyveld, 2009).

AX and their metabolites have important physiological and metabolic functions and improve health status. These compounds have protective effects against diseases with high prevalence in developed societies such as cardiovascular diseases, diabetes and certain types of cancer. Prebiotic effect of AX has been revealed. These compounds are resistant to gastric acidity, enzymatic hydrolysis and gastrointestinal absorption and they are also fermentable by gut microbiota. Moreover, AX can selectively stimulate growth and/or activity of intestinal bacteria, such as *Bifidobacterium* or *Lactobacillus*, which are beneficial to health (Broekaert et al., 2011; Gong et al., 2018; Grootaert et al., 2007; Neyrinck et al., 2011; Van Craeyveld, 2009). These compounds also influence the lipidic metabolism by reducing cholesterol and triglyceride levels, because AX promotes the excretion of lipids and regulates the activity of HMG-CoA reductase (Grootaert et al., 2007; Neyrinck et al., 2011; Saeed, Pasha, Anjum, & Sultan, 2011; Tong et al., 2014). Moreover, AX regulate glycemic metabolism improving blood glucose levels (Lu, Walker, Muir, & O′ Dea, 2004; Neyrinck et al., 2011). In this sense, according to European Commission (2012), particularly, AX obtained from wheat endosperm have approved the following health claim: "Consumption of AX as part of a meal contributes to a reduction of the blood glucose rise after that meal." This claim may be used only for food, which contains at least 8 g of AX-rich fiber produced from wheat endosperm (at least 60% AX by weight) per 100 g of available carbohydrates in a quantified portion as part of the meal. In addition to these health benefits, it is known that AX have antioxidant properties and immunomodulatory effects. This fact can be explained because AX are usually associated with ferulic acid, which is a polyphenol that has antioxidant activity. The increase of antioxidant activity is related with high immune cell functions (Akhtar et al., 2012; Ayala-Soto, Serna-Saldívar, García-Lara, & Pérez-Carrillo, 2014; Broekaert et al., 2011; Cao et al., 2011; Mendis & Simsek, 2014). Due to their prebiotic effect and their antioxidant and immunomodulatory activities, AX have also an important role in the prevention of colon cancer (Broekaert et al., 2011; Femia et al., 2010; Grootaert et al., 2007).

The main sources of AX are cereals, although they are also found in other foods as bamboo shoots (Qiu, Yadav, & Yin, 2017).

2.2.3 β-Glucans

β-Glucans are polysaccharides of D-glucose units connected through glycosidic linkages (Fig. 5). Their activity is influenced by differences in their structure, size of the polysaccharide chain, branches, and molecular weight. These compounds can be also classified according to its solubility, in soluble or insoluble β-glucans. Soluble viscous β-glucans fibers consist of β-(1 → 3/1 → 6)-D-linked glucose, whereas insoluble β-glucans fibers consist of β-(1 → 3/1 → 4)-D-linked glucose units (Baldassano, Accardi, & Vasto, 2017; Maheshwari, Sowrirajan, & Joseph, 2017; Sima, Vannucci, & Vetvicka, 2018).

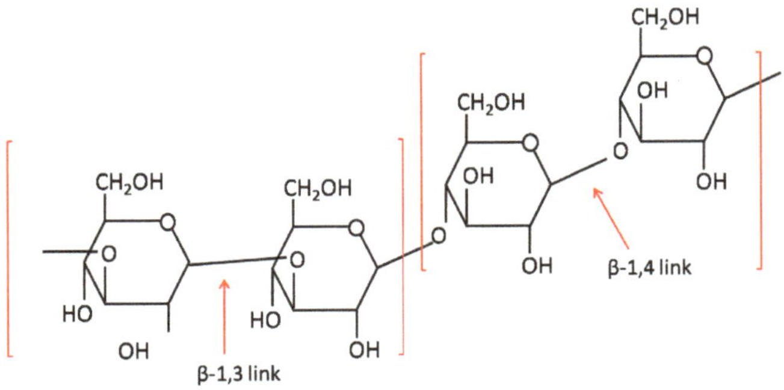

Fig. 5 Chemical structure of β-glucans.

During the later years, the β-glucans have gained much interest in the field of functional foods and actually these compounds are regarded as a potentially health promoting food ingredients (Baldassano et al., 2017). These compounds exhibit a broad spectrum of biological activities including anti-tumor, immune-modulating, anti-aging and anti-inflammatory properties (Zhu, Du, & Xu, 2016). Due to their functional effect and their benefits to human health, β-glucans have approved the following health claim according to European Commission (2012): "β-glucans contribute to the maintenance of normal blood cholesterol levels." This claim may be used only for food that contains at least 1 g of β-glucans from oats (*Avena sativa* L.), oat bran, barley (*Hordeum vulgare* L.), barley bran, or from mixtures of these sources per quantified portion.

β-Glucans are mainly present in endosperm cell walls of cereals, baker's yeast, certain mushrooms, algae and bacteria (Baldassano et al., 2017; Mohebbi et al., 2018).

2.2.4 Pectin

Pectin is a kind of water-soluble DF which is extensively used as a functional ingredient in food and beverage industries due to its thickening and gelling properties and as a colloidal stabilizer. Pectin is a complex group of polysaccharides present in plant cell walls, which act as intercellular cementing substance. It has a linear anionic region formed by D-galacturonic acid monomers, linked by α-(1 → 4) glycosidic bonds (Fig. 6), and branched regions primarily formed by various types of neutral monosaccharides (mainly rhamnose, xylose, mannose, and arabinose), linked together (Dhingra et al., 2012; Espinal-Ruiz, Parada-Alfonso, Restrepo-Sánchez, Narváez-Cuenca, & McClements, 2014).

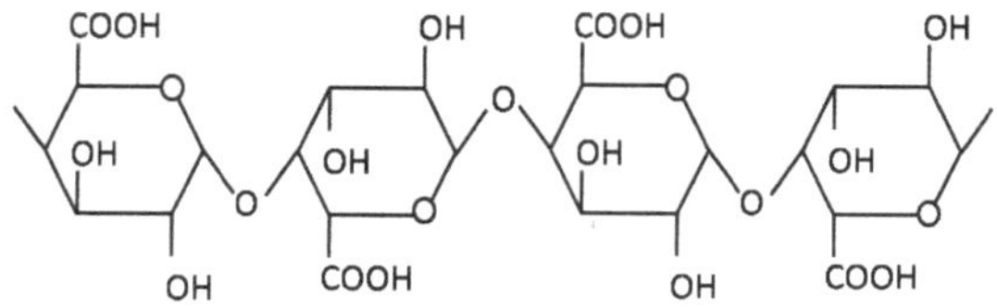

Fig. 6 Chemical structure of pectins.

These compounds are highly water-soluble and they are almost completely metabolized by colonic microbiote. Due to their gelling behavior, these soluble polysaccharides may decrease the rate of gastric emptying and influence small intestinal transit time. This explains their hypoglycemic properties. Pectin can contribute to reduce cholesterol levels because these compounds bind cholesterol, reducing its absorption and promoting their excretion (Dhingra et al., 2012; Espinal-Ruiz et al., 2014; Mudgil & Barak, 2013; Ozyurt & Ötles, 2016). Due to their hypoglycemic and hypocholesterolemic properties, pectins have approved the following health claims according to European Commission (2012): "Pectins contribute to the maintenance of normal blood cholesterol levels" (this claim may be used only for food which provides a daily intake of 6g of pectins) and "Consumption of pectins with a meal contributes to the reduction of the blood glucose rise after that meal" (this claim may be used only for food which contains 10g of pectins per quantified portion).

Pectins are found in plant cell walls as well as in the outer skin and rind of fruits and vegetables, e.g., the rind of an orange contains 30% pectin, an apple peel 15%, and onionskin 12% (Mudgil & Barak, 2013).

2.2.5 Gums

Gums are hydrocolloids derived from plant exudates, seeds and seaweed extracts (Fuller et al., 2016). These compounds are not digested in the upper intestinal tract and are resistant to the human digestive enzymes, being fermented in the large gut. This fermentation promotes the stimulation of the endogenous microbiota and the production of SCFA (Ozyurt & Ötles, 2016).

Therefore, gums are used in food production as a source of DF with prebiotic effects and are also used for their functional properties such as, improve food texture, retard starch retro-gradation, improve moisture retention and enhance the overall quality of the products during storage (Qasem et al., 2017).

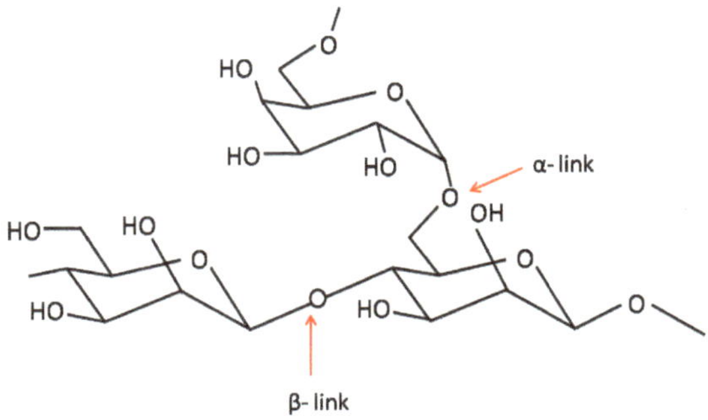

Fig. 7 Chemical structure of Guar gum.

Plant exudates are one of the main sources of gums, highlighting guar gum, gum arabic, gum tragacanth, karaya gum, etc.

Guar gum (Fig. 7) is a galactomannan isolated from the seed of *Cyamopsis tetragonolobus* (guar). Due to its thickener properties, it is used as food additive. Guar gum has prebiotic properties and it can improve bowel transit. It also shows hypoglycemic and hypolipidemic effects (Tungland & Meyer, 2002). In this sense, according to European Commission (2012), Guar gum has approved the following health claim: "Guar gum contributes to the maintenance of normal blood cholesterol levels." This claim may be used only for food that provides a daily intake of 10 g of guar gum.

The exudate from the acacia tree (*Acacia* Mill.) is known as gum arabic. It is a mixture of a complex arabinogalactan polysaccharide with a glycoprotein. For its stabilizing and emulsifying properties, gum arabic is used by the food industry as additive. It has bifidogenic effect and hypolipidemic properties (Tungland & Meyer, 2002).

Generally, the plants rich in gums are not used as food, but they are used as food additives. The most important gums in food belongs to different genus of *Leguminosae* family (Dhingra et al., 2012; Mataix Verdú, 2009).

2.2.6 Mucilage

As gums, mucilages are SDF that are used as gelling agents, thickeners, stabilizers and emulsifying agents (Fuller et al., 2016). Mucilages are polysaccharides constituted by large molecules of sugars and uronic acids linked by glycosidic bonds. Plant mucilages can be extracted from a variety of plant parts, including rhizomes, roots and seed endosperms. Not-water soluble mucilages swell and absorb considerable quantities of water, but only water-soluble mucilages can form viscous solutions. These compounds are widely used in pharmaceutical, food and cosmetics industry, as well as, in agriculture, textile, paper-industries (Troncoso, Zamora, & Torres, 2017).

Mucilages show a hampering effect on the diffusion of glucose, and can contribute to postpone the absorption and digestion of carbohydrates, which results in lowered postprandial blood glucose levels (Ozyurt & Ötles, 2016).

Mucilages are present in cells of the outer layers of seeds of the plantain family, e.g., *Plantago psyllium* L. (Mudgil & Barak, 2013).

2.3 Other compounds associated to fiber fraction

DF is often intimately associated in the plant cell structure with other organic compounds, such as vitamins, tannins, cutins, phytosterols, phytochemicals, etc. In the case of cereals products, DF is characterized for being constituted with different compounds that may be co-responsible for many of its physiological effects. An important amount of phenolic compounds (500–1500 mg/kg), mainly ferulic acid, is linked to the DF and this may explain why cereal DF has a marked antioxidant activity (Costabile et al., 2008). In cereal grains, phenolic compounds are mainly found as insoluble or bound forms, being linked to different carbohydrate and other components such as cellulose, lignin, and protein through ester bonds (Costabile et al., 2008; Gong et al., 2018; Mudgil & Barak, 2013; Yu & Ahmedna, 2013).

Cereals contain derivatives of cinnamic acid (ferulic acid, caffeic acid, *p*-coumaric acid, and sinapic acid) and benzoic acid (protocatechuic acid, *p*-hydroxybenzoic acid, salicylic acid, vanillic acid, and syringic acid), most of which are bound to IDF polysaccharides (Knudsen, Nørskov, Bolvig, Hedemann, & Lærke, 2017). This thematic will be further discussed in this volume (see chapter "Impact of molecular interactions with phenolic compounds on food polysaccharides functionality" by Corrine C. Dobson et al.).

Phytic acid is also associated with fiber in some foods, especially in cereal grains. Its phosphate group is strongly bind with positively charged ions such as iron, zinc, calcium and magnesium and may influence mineral absorption from the gastrointestinal tract. Also, phytic acid has the ability to suppressing iron-catalyzed oxidative reactions and it has recently received attention as an anticancer compound (Mudgil & Barak, 2013; Sidhu, Kabir, & Huffman, 2007).

In the case of cereals, dietary fiber is mainly located in bran, being bran also rich in minerals. It is reported that some dietary fiber components as arabinoxylans or inulin improve mineral absorption contributing to human health (Lattimer & Haub, 2010).

3. Functional dietary fiber effect

Whereas dietary fiber (DF) consists of "non digestible carbohydrates and lignin that are intrinsic and intact in plants," it is reported that functional dietary fiber consists of "isolated, non-digestible carbohydrates that have beneficial physiological effects in humans" (FNB, 2001).

The diets with a high content of DF, such as those rich in cereals, fruits and vegetables, have a demonstrated positive effect on human health. It is known that DF plays an important role in preventing several chronic diseases like obesity, coronary heart disease, and diabetes, and it is also associated with decreasing the prevalence of certain cancers (Dhingra et al., 2012; Kurek, Wyrwisz, Karp, & Wierzbicka, 2018).

DF intake, especially intake of whole grains or cereal fiber, tends to delay gastric emptying and create a sense of fullness and increased fiber intake is associated with increases in satiating gut hormones (Anderson et al., 2009).

Soluble dietary fiber (SDF) stimulates postprandial satiety in healthy humans by increasing postprandial levels of gastrointestinal hormones related with satiety (glucagon-like peptide and peptide YY), decreasing postprandial levels of the hormone that stimulates hunger (ghrelin) and by delaying the gastric emptying rate. For these reasons, DF can be useful against *obesity* (Anderson et al., 2009; Giacco, Costabile, & Riccardi, 2016; Grooms, Ommerborn, Pham, Djoussé, & Clark, 2013; Shinozaki, Okuda, Sasaki, Kunitsugu, & Shigeta, 2015).

Moreover, DF, and particularly insoluble dietary fiber (IDF), play an important role in the *gastrointestinal function*. IDF is especially effective in increasing fecal mass, decreasing intestinal transit time and promoting intestinal regularity. IDF can accelerate colonic transit by the colonic mucosa with mechanical stimulation/irritation with the increasing of secretion and peristalsis process (Anderson et al., 2009; Davison & Temple, 2018; El-Salhy, Ystad, Mazzawi, & Gundersen, 2017). IDF is used in the management of intestinal disorders, such as constipation, or in the prevention of the development of diverticulosis and diverticulitis (Nandi & Ghosh, 2015). Most non-absorbed carbohydrates have laxative effects, both by increasing bacterial mass or osmotic effects, and by water binding to remaining unfermented fiber (Mudgil & Barak, 2013). In general, cereal fibers are the most effective ones by increasing stool weight and the laxative effect of wheat bran is higher than other food matrix fibers (Slavin, 2013).

As it was already mention, different components of DF, including inulin, oligosaccharides, AX and resistant starch, have been reported to have a *prebiotic* role. Prebiotics are defined as selectively fermented ingredients that allow specific changes, both in the composition and/or activity in the gastrointestinal microbiota, that confer health benefits. This typically turns the composition of intestinal microbiota toward a relative increase in *Bifidobacterum* and/or *Lactobacillus* species (Fuller et al., 2016; Neyrinck et al., 2011; Slavin, 2013). The following requirements must be scientifically demonstrated to consider an ingredient as a prebiotic (García-Amezquita, Tejada-Ortigoza, Serna-Saldivar, & Welti-Chanes, 2018; Slavin, 2013):

- resistance to gastric acidity, hydrolysis by mammalian enzymes, and gastrointestinal absorption;
- be fermented by the intestinal microbiota; and
- selectively stimulate the growth and/or activity of specific intestinal bacteria potentially associated with health benefits.

The intake of prebiotics modifies the intestinal microbiota, promoting growth of *Bifidobacterium* and *Lactobacillus* sp., which are the main bacteria responsible of gut carbohydrate fermentation. The major products from the microbial fermentative activity in the gut are SCFA (short chain fatty acids), in particular, acetate, propionate, and butyrate. These SCFA reduced intestinal pH, improving the bioaccessibility of some minerals, as calcium and magnesium, and increasing the absorption of iron, being useful for the prevention of certain diseases as anemia and osteoporosis (Koh, De Vadder, Kovatcheva-Datchary, & Bäckhed, 2016; MacFarlane, MacFarlane, & Cummings, 2006; Ötles & Ozgoz, 2014; Teitelbaum & Walker, 2002).

Several authors reported a protective correlation between DF intake and *colon cancer* incidence. Especially, fibers from cereals and fruits showed a notable association with decreased colon cancer risk (Ötles & Ozgoz, 2014; Tao, Li, Li, & Li, 2018). It is accepted that the beneficial effects of a diet rich in DF are derived from the products of their fermentation by colonic microbiota, particularly, the production of butyrate. Among the SCFA, butyrate has been investigated most extensively. It is present at high levels (mM) in the gut lumen, is the primary energy source for colonocytes and also protects against colorectal cancer and inflammation (Encarnação, Abrantes, Pires, & Botelho, 2015; Koh et al., 2016; Ötles & Ozgoz, 2014; Teitelbaum & Walker, 2002). This SCFA is selectively absorbed in the colonic epithelium and it contributes to colonic homeostasis, having important functions as anti-inflammatory, antioxidant, and anti-carcinogenic actions. The butyrate capacity to act as a chemopreventive agent in a primary phase of colorectal cancer progression is based on its importance in the colon homeostasis, in activation of drug-metabolizing enzymes and in its capability to modulate the inflammatory process. Butyrate is also able to inhibit the growth of tumor cells, increasing apoptosis in human colonic tumor cell lines (Encarnação et al., 2015, 2018; Ötles & Ozgoz, 2014). As consequence of prebiotic effect of fermentable DF, the amount of pathogenic bacteria decreases in colon and therefore the production of carcinogenic substances are reduced (Ötles & Ozgoz, 2014).

A diet based on carbohydrate-rich foods with high fiber content, particularly whole grain products, may also contribute to prevent the metabolic syndrome, type 2 diabetes and cardiovascular diseases (Giacco et al., 2016; Johns et al., 2015; McRae, 2018).

There is an inverse association between DF intake, especially from cereals fibers, and type 2 *diabetes* prevalence (McRae, 2018; Yao et al., 2014). This protective effect of cereal fibers may be explained by the modulating impact of gut microbiota in different ways: improving glucose tolerance by different energy metabolism pathways (colonic fermentation and generation of SCFA), reducing inflammation and altering the immune response (Davison & Temple, 2018), as it will be detailed below:

– SDF can reduce postprandial glucose levels and the average daily blood glucose profile. When SDF is hydrated forming gel can increase the viscosity of stomach content, reducing the postprandial glycemic response. This reduction in postprandial blood glucose is correlated with the viscosity of the meal and the gastric transit time. Therefore, the SDF ability to delay both digestion and absorption of carbohydrates in the small intestine can explain their beneficial effects on postprandial glucose levels. However, the benefits of fiber-rich foods on postprandial glucose response depend not only on their viscosity but also on their ability to reduce the accessibility of starch to digestive enzymes. It is usually that starch granules present in natural fiber-rich foods are enveloped in fiber in order to reduce their interaction with α-amylases, slowing carbohydrate digestion.

Cereal DF may induce a relatively fast modification of colonic microbiota that, in turn, increases fiber fermentation and SCFA production. SCFA, in particular, propionate may contribute to improve insulin sensitivity, reducing insulin concentrations (Giacco et al., 2016).

- Diets high in fiber (specifically from cereal or vegetable sources) are significantly associated with a lower risk of *cardiovascular disease* and *coronary heart disease* (Johns et al., 2015; Threapleton et al., 2013). It is known that SDF has hypocholesterolemic properties; because its intake is related to a decrease of serum cholesterol and LDL cholesterol concentrations and higher DF consumption is also associated with increased plasma HDL cholesterol, which may contribute to their protective role against coronary heart disease. The hypocholesterolemic mechanisms of SDF include binding bile acids during the formation of micelles in the intestinal lumen, enhancing bile acid excretion and the physiological effects of fermentation products of SDF, mainly propionate (McRae, 2017; Thompkinson, Bhavana, & Kanika, 2014; Zhang, Cao, Yin, & Wang, 2018; Zhou et al., 2015).
- DF plays a multifaceted role in modulating tissue *immune responses*, *inflammation* in the intestine, and systemic inflammation. It seems that there is a beneficial relation between ingestion of DF and inflammatory processes. The amount of fiber intake is inversely related to the secretion of IL-6 and C-reactive protein. Butyrate and propionate show anti-inflammatory properties by inhibiting TNF-α, IL-8, IL-10, and IL-12 cytokines in immune and colonic cells. Diet with high fiber content can increase the proportion of CD8+ T cells and CD4+ T cells and increase NK cell activity (Janakiram, Mohammed, Madka, Kumar, & Rao, 2016).

The *metabolic syndrome* is a growing epidemic worldwide characterized by obesity, hyperlipidemia, hypertension, and insulin resistance. DF exerts protective cardiovascular benefits on several aspects of the metabolic syndrome, including waist circumference, blood glucose, dyslipidemia, blood pressure, insulin control, and the regulation of certain inflammatory markers (Jakobsdottir, Nyman, & Fåk, 2014; Merriam et al., 2012).

Due to their functional effect and their benefits to human health, there are some components of dietary fiber with approved health claims that can be included on the food labels in Europe (European Commission, 2012; Regulation (EC) No 1924/2006; Regulation (EU) No 1169/2011). These components, their correspondent health claim and the conditions of use of the claim are summarized in Table 1.

Table 1 Approved health claims related to dietary fiber components (European Commission, 2012).

Nutrient, substance, food or food category	Claim	Conditions of use of the claim	Conditions and/or restrictions of use of the food and/or additional statement or warning
Arabinoxylan produced from wheat endosperm	Consumption of arabinoxylan as part of a meal contributes to a reduction of the blood glucose rise after that meal	The claim may be used only for food, which contains at least 8 g of arabinoxylan (AX)-rich fiber produced from wheat endosperm (at least 60% AX by weight) per 100 g of available carbohydrates in a quantified portion as part of the meal. In order to bear the claim information shall be given to the consumer that the beneficial effect is obtained by consuming the arabinoxylan (AX)-rich fiber produced from wheat endosperm as part of the meal	—
Barley grain fiber	Barley grain fiber contributes to an increase in fecal bulk	The claim may be used only for food which is high in that fiber as referred to in the claim "high fiber" as listed in the Annex to Regulation (EC) No 1924/2006	—
β-Glucans	β-Glucans contribute to the maintenance of normal blood cholesterol levels	The claim may be used only for food, which contains at least 1 g of β-glucans from oats, oat bran, barley, barley bran, or from mixtures of these sources per quantified portion. In order to bear the claim information shall be given to the consumer that the beneficial effect is obtained with a daily intake of 3 g of β-glucans from oats, oat bran, barley, barley bran, or from mixtures of these β-glucans	—
β-Glucans from oats and barley	Consumption of β-glucans from oats or barley as part of a meal contributes to the reduction of the blood glucose rise after that meal	The claim may be used only for food, which contains at least 4 g of β-glucans from oats or barley for each 30 g of available carbohydrates in a quantified portion as part of the meal. In order to bear the claim information shall be given to the consumer that the beneficial effect is obtained by consuming the β-glucans from oats or barley as part of the meal	—

Table 1 Approved health claims related to dietary fiber components (European Commission, 2012).—cont'd

Nutrient, substance, food or food category	Claim	Conditions of use of the claim	Conditions and/or restrictions of use of the food and/or additional statement or warning
Guar gum	Guar gum contributes to the maintenance of normal blood cholesterol levels	The claim may be used only for food, which provides a daily intake of 10 g of guar gum. In order to bear the claim, information shall be given to the consumer that the beneficial effect is obtained with a daily intake of 10 g of guar gum	Warning of choking to be given for people with swallowing difficulties or when ingesting with inadequate fluid intake (advice on taking with plenty of water to ensure substance reaches stomach)
Hydroxypropyl methylcellulose (HPMC)	Consumption of hydroxypropyl methylcellulose with a meal contributes to a reduction in the blood glucose rise after that meal	The claim may be used only for food, which contains 4 g of HPMC per quantified portion as part of the meal. In order to bear the claim information shall be given to the consumer that the beneficial effect is obtained by consuming 4 g of HPMC as part of the meal	Warning of choking to be given for people with swallowing difficulties or when ingesting with inadequate fluid intake (advice on taking with plenty of water to ensure substance reaches stomach)
Hydroxypropyl methylcellulose (HPMC)	Hydroxypropyl methylcellulose contributes to the maintenance of normal blood cholesterol levels	The claim may be used only for food, which provides a daily intake of 5 g of HPMC. In order to bear the claim information shall be given to the consumer that the beneficial effect is obtained with a daily intake of 5 g of HPMC	Warning of choking to be given for people with swallowing difficulties or when ingesting with inadequate fluid intake (advice on taking with plenty of water to ensure substance reaches stomach)
Oat grain fiber	Oat grain fiber contributes to an increase in fecal bulk	The claim may be used only for food which is high in that fiber as referred to in the claim "high fiber" as listed in the Annex to Regulation (EC) No 1924/2006	—
Pectins	– Pectins contribute to the maintenance of normal blood cholesterol levels	– The claim may be used only for food, which provides a daily intake of 6 g of pectins. In order to bear the claim information shall be given to the consumer that the beneficial effect is obtained with a daily intake of 6 g of pectins	Warning of choking to be given for people with swallowing difficulties or when ingesting with inadequate fluid intake (advice on taking with plenty of water to ensure substance reaches stomach)

Continued

Table 1 Approved health claims related to dietary fiber components (European Commission, 2012).—cont'd

Nutrient, substance, food or food category	Claim	Conditions of use of the claim	Conditions and/or restrictions of use of the food and/or additional statement or warning
	– Consumption of pectins with a meal contributes to the reduction of the blood glucose rise after that meal	– The claim may be used only for food, which contains 10 g of pectins per quantified portion. In order to bear the claim, information shall be given to the consumer that the beneficial effect is obtained by consuming 10 g of pectins as part of the meal	
Resistant starch	Replacing digestible starches with resistant starch in a meal contributes to a reduction in the blood glucose rise after that meal	The claim may be used only for food in which digestible starch has been replaced by resistant starch so that the final content of resistant starch is at least 14% of total starch	—
Rye fiber	Rye fiber contributes to normal bowel function	The claim may be used only for food which is high in that fiber as referred to in the claim "high fiber" as listed in the Annex to Regulation (EC) No 1924/2006	—
Wheat bran fiber	Wheat bran fiber contributes to an acceleration of intestinal transit	The claim may be used only for food which is high in that fiber as referred to in the claim "high fiber" as listed in the Annex to Regulation (EC) No 1924/2006. In order to bear the claim information shall be given to the consumer that the claimed effect is obtained with a daily intake of at least 10 g of wheat bran fiber	—
Wheat bran fiber	Wheat bran fiber contributes to an increase in fecal bulk	The claim may be used only for food which is high in that fiber as referred to in the claim "high fiber" as listed in the Annex to Regulation (EC) No 1924/2006	—

Table 1 Approved health claims related to dietary fiber components (European Commission, 2012).—cont'd

Nutrient, substance, food or food category	Claim	Conditions of use of the claim	Conditions and/or restrictions of use of the food and/or additional statement or warning
Sugar beet fiber	Sugar beet fiber contributes to an increase in fecal bulk	The claim may be used only for food which is high in that fiber as referred to in the claim "high fiber" as listed in the Annex to Regulation (EC) No 1924/2006	—
Native chicory inulin	Chicory inulin contributes to normal bowel function by increasing stool frequency	Information shall be provided to the consumer that the beneficial effect is obtained with a daily intake of 12 g chicory inulin. The claim can be used only for food, which provides at least a daily intake of 12 g of native chicory inulin, a non-fractionated mixture of monosaccharides (<10%), disaccharides, inulin-type fructans and inulin extracted from chicory, with a mean degree of polymerization ≥9	—
Non-fermentable carbohydrates	Consumption of foods/drinks containing non-fermentable carbohydrates instead of fermentable carbohydrates contributes to the maintenance of tooth mineralization	In order to bear the claim, fermentable carbohydrates (1) should be replaced in foods or drinks by non-fermentable carbohydrates (2) in such amounts that consumption of such foods or drinks does not lower plaque pH below 5.7 during and up to 30 min after consumption. (1) Fermentable carbohydrates are defined as carbohydrates or carbohydrate mixtures as consumed in foods or beverages that lower plaque pH below 5.7, as determined in vivo or in situ by plaque pH telemetry tests, by bacterial fermentation during and up to 30 min after consumption. (2) Non-fermentable carbohydrates are defined as carbohydrates or carbohydrate mixtures as consumed in foods or beverages that do not lower plaque pH, as determined in vivo or in situ by plaque pH telemetry tests, below a conservative value of 5.7 by bacterial fermentation during and up to 30 min after consumption	—

Continued

Table 1 Approved health claims related to dietary fiber components (European Commission, 2012).—cont'd

Nutrient, substance, food or food category	Claim	Conditions of use of the claim	Conditions and/or restrictions of use of the food and/or additional statement or warning
Non-digestible carbohydrates	Consumption of foods/drinks containing non-fermentable carbohydrates instead of sugars induces a lower blood glucose rise after their consumption compared to sugar-containing foods/drinks	In order to bear the claim, sugars should be replaced in foods or drinks by non-digestible carbohydrates, which are carbohydrates neither digested nor absorbed in the small intestine, so that foods or drinks contain reduced amounts of sugars by at least the amount referred to in the claim REDUCED [NAME OF NUTRIENT] as listed in the Annex to Regulation (EC) No 1924/2006	—

4. Dietary fiber as functional food ingredient: Natural vs synthetic sources

Recent advances in the food and nutrition sciences support the concept that the diet has a significant role in the modulation of various functions in the body. As a consequence of this new concept, appears the term of functional food, which is defined as "a natural or processed food that contains known biologically active compounds that when in defined quantitative and qualitative amounts provide a clinically proven and documented health benefit. The functionality of functional foods is based on bioactive components, which may be contained naturally or added externally in the product but usually require formulation by appropriate technologies in order to optimize the required beneficial properties" (Rawat & Indrani, 2015).

In recent years, the interest in the development of functional foods has emerged due to the physiological and nutritional benefits they can provide. In this sense, DF, as it has been previously explained, is an excellent ingredient because it has numerous and well-known health benefits associated to its consumption. Dietary fiber has not only physiological properties, but it also provides several technological possibilities and it has a high applicability within food formulations. For these reasons, the food industry is continuously looking for new sources of DF to use as a functional food ingredient (García-Amezquita et al., 2018).

Considering the technological aspect, the incorporation of DF in foods allows the development of fiber-enriched products with functional properties. Among the main functional properties of DF that have technological interest are noteworthy the cation exchange capacity, the increase in water holding capacity and oil holding capacity, the stabilization of high fat food and emulsions, the gel forming ability, organic compounds absorptive properties, flavor, color, and rheological properties. These properties improve texture and reduce syneresis of the food products and they also provide appropriate sensory characteristics (Dello Staffolo, Sato, & Cunha, 2017; Maphosa & Jideani, 2016; Requena et al., 2016).

Food ingredients consisting of DF must meet specific conditions, including high total DF content (above 50%), low moisture and lipid content, low caloric value, and bland flavor when applied as an ingredient (García-Amezquita et al., 2018).

Over the past few years, the innovations in fiber ingredients have highly grown and the development of products rich in fiber as functional ingredient has been increased. Actually, there are a wide range of fiber products including traditional low-moisture product like breads, snacks and cereals and also innovative products with high content of fiber such as dairy or meat products and beverages (Ciudad-Mulero et al., 2018; López-Marcos, Bailina, Viuda-Martos, Pérez-Alvarez, & Fernández-López, 2015; Morales et al., 2015).

Moreover, some products based on meat or fish incorporate DF as fat replacer, emulsion stabilizer, water binder and for reduce lipid oxidation, improve cooking yield and improve texture of products. In the case of bakery industry, DF is incorporated in order to modify texture, increase volume, increase shelf life, modify bread volume, improve firmness of loaf, modify springiness, increase softness of the crumb, replace wheat flour or improve nutritional quality of bread and baked products. Dairy industry also adds DF in products as ice creams, yogurt or cheese for improve consistency, reduce syneresis and improve mouthfeel. Beverage industry includes DF in juices and other drinks in order to improve viscosity and stability, as well as bulking agent. It is usually that breakfast cereals, sweets and chocolates were fortified with DF and/or use this ingredient in their formulation as sugar substitute. In extruded products as pasta, DF was added as fortifying agent, to improve pseudoplastic behavior, stability, among others. Some fruits products, including jam and marmalade, also incorporate DF as functional ingredient (Maphosa & Jideani, 2016).

The sources of DF typically used by food industry can be separated into three classes: (a) hydrocolloids (mostly soluble polysaccharides), (b) bioactive oligosaccharides and (c) whole plant cell wall materials derived from cereal grains, fruits, and vegetables (Redgwell & Fischer, 2005).

(a) The hydrocolloids include a wide range of mixed viscous polysaccharides. These compounds derived from plant exudates (gum arabica and tragacanth), seeds (guar and locust bean gum) and seaweed extracts (agar, carrageenan and alginates). Gums and mucilages are hydrocolloids used in small amounts as gelling, thickening, stabilizing and emulsifying agents in certain food products (Mudgil & Barak, 2013).

(b) Bioactive oligosaccharides are widely used in food industry. The prebiotic effect of oligosaccharides (FOS and GOS) is widely used by food industry, e.g., added to infant formulas with the aim of achieving a bifidogenic effect on the gastrointestinal microbiota of the host (Vandenplas, Zakharova, & Dimitrieva, 2015). Also fructo-oligosaccharides (FOS) are used by juice industry as sucrose substitute sucrose without juice quality modifications, as in the case of pineapple, mango, and orange juices (Bali, Panesar, Bera, & Panesar, 2015).

Isomalto-oligosaccharides (IMOS) have higher possibilities for incorporation into widely varying of common food stuffs, especially in liquid foods and beverages as functional food ingredients, sugar replacement, slow energy release and to provide organoleptic functionality (Sorndech, Nakorn, Tongta, & Blennow, 2018).

As consequence of the fast growing of food processing industry, especially in the developing world, high amount of by-products are generated. These by-products are known to be sources of several bioactive compounds including DF (Sharma et al., 2016). The use of these food processing by-products (pomace, peel and pulp refuse, seed, oilcakes, stems hulls, husk and pods and bran) as a source of DF allows waste reduction and generates indirect income to food industry (Sharma et al., 2016). In this sense the fruit juice industry produces significant amounts of fruits by-products, which can be used by the food industry to develop new natural ingredients (Göksel Saraç & Dogan, 2016).

Cereal and pseudocereal polysaccharides are the main source of DF used by the food industry, whereas fruits and vegetables residues are being used as non-conventional sources of DF (Tejada-Ortigoza, García-Amezquita, Serna-Saldívar, & Welti-Chanes, 2016).

(c) Some of the most common natural dietary fiber by-products sources are detailed below:

- *Cereals*: Hulls and pods are the main grain by-product uses as an IDF source, highlighting wheat (*Triticum aestivum* L., *Triticum turgidum* L., *Triticum durum* Desf.), oat (*Avena sativa* L.), corn (*Zea mays* L.) and rice (*Oryza sativa* L.) bran. Cereal bran (wheat, rice, and barley (*Hordeum vulgare* L.) and oat bran) can be used in fiber-enriched pasta (Masli, Rasco, & Ganjyal, 2018; Rawat & Indrani, 2015; Sharma et al., 2016). β-Glucans are associated with lower serum cholesterol levels in hypercholesterolemic subjects, thus lowering the risk of cardiovascular disease, also have strong colloidal properties and therefore can be considered as good functional ingredients. The highest amount of β-glucans occurs in oat (2.2–7.8%) and barley (2.5–11.3%) bran, which are an important source of the DF (Abuajah, Ogbonna, & Osuji, 2015; Sidhu et al., 2007).
- *Pseudocereals*: They also constitute a good source of DF. Linen (*Linum usitatissimum* L.) has been used as feeding and fiber sources from ancient times in Asia, North Africa, and Europe

(Bolaños, Marchevsky, & Camiña, 2016) and amaranth (*Amaranthus caudatus* L.), quinoa (*Chenopodium quinoa* Willd.), and buckwheat (*Fagopyrum esculentum* Moench.) seeds are also considered a good source of DF (Boukid, Folloni, Sforza, Vittadini, & Prandi, 2018; Li & Zhu, 2018; Shevkani, Singh, Kaur, & Rana, 2014; Valcárcel-Yamani & da Silva Lannes, 2012; Wefers & Bunzel, 2015).

- *Pulses*: The use of pulses as a fiber source has gained attention due to their physicochemical properties as a functional ingredient for food formulations. However, it is limited by antimetabolic/antiphysiological substances such as protease inhibitors, lectins, saponins. In recent years, DF from pulses as soybeans, black soybeans, lentils, and peas has been used as stabilizers in beverages or dairy products. Resistant starch from pulses has been also widely employed as fiber supplements in bread (Kan et al., 2018; Tejada-Ortigoza et al., 2016).
- *Fruit*: DF is associated with significant content of bioactive compounds (flavonoids, carotenoids, etc.) and SDF is the prevalent fraction. For these reasons, fruit DF concentrates have better nutritional quality than those found in cereals (Rawat & Indrani, 2015). There are many fruits, including apple (*Malus* Mill.), peach (*Prunus persica* L.) and olive (*Olea europaea* L.), which are used for juice extraction. From all of them we can obtain by-products from which fiber fraction can be recovered, presenting a great potential as functional food ingredients. Orange and lemon by-products also constitute an important source of fiber since they are very rich in pectins (Rodríguez, Jiménez, Fernández-Bolaños, Guillén, & Heredia, 2006). Tropical fruits co-products obtained from mango (*Mangifera indica* L.), pineapple (*Ananas* Mill.) and passion fruit (*Passiflora edulis* Sims) have potential use as functional ingredient, providing DF (for nutritional and technological purposes) and/or natural antioxidants to food products (Selani et al., 2016). Date (*Phoenix dactylifera* L.) seed DF represents a new source of DF that could be successfully used in bakery products due to its appropriate chemical, physical, sensory, and baking properties (Shokrollahi & Taghizadeh, 2016).
- *Vegetables*: There are several vegetables such as pepper, artichoke, onion and asparagus, which contain both soluble and insoluble DF compounds, which can be used for new functional foods design (Rodríguez et al., 2006). Between vegetables, SDF obtained from the cellulose fraction of Chinese cabbage, represents a potential

source of DF with prebiotic, hypoglycemic, and hypolipidemic effects, and can be used as a new resource of functional beverages and nutraceutical products in the food industry (Park & Yoon, 2015).

- *Algae and seaweeds*: They have good amounts of DF and can be used by food industry as source of DF (Sharma et al., 2016). Edible macroalgae contain unusually high amounts of DF, ranging from 23.5% (*Codium reediae* P.C. Silva) to 64.0% of dry weight in *Gracilaria* spp. Soluble fiber comprises 52–56% of total fiber in commonly used green and red macroalgae and 67–85% in brown macroalgae. The major polysaccharide of brown alga is alginate, which comprises 14–40% of its dry mass. Alginate was first isolated in 1881 from *Laminaria* sp. Dietary alginates provide a sense of satiety and so have been explored as a weight control measure. Apart from alginates, the main polysaccharides from brown algae are β-glucans, cellulose, and heteroglycans (Wells et al., 2017).

In addition to the natural sources of DF, there are also *synthetic sources* used by the food industry. Synthetic derivatives of cellulose are included among synthetic sources of DF. These synthetic carbohydrate compounds are soluble and non-digestible, but hardly fermented by microbiota. Among the most extensive used of synthetic sources of DF are found *polydextrose, methylcellulose, hydroxypropyl methylcellulose and cyclodextrins,* which are detailed in the following.

- *Polydextrose* is a polysaccharide with an average degree of polymerization of about 10 glucose residues obtained by thermal polymerization of D-glucose in the presence of sorbitol and phosphoric acid. It can be used in some reduced energy products as a bulking agent to replace sugars and to provide texture. Its contribution to energy is lower (1 kcal/g), is partially fermented in colon and it can provide physiological effects similar to those of natural dietary fibers (Buttriss & Stokes, 2008; Carvalho Lago & Zapata Noreña, 2016).
- *Methylcellulose* is one of the most important commercial cellulose ethers that have been used with many industrial applications. It is usually synthesized by etherification of cellulose (reaction between cellulose, alkali and chloromethane or iodomethane). Methylcellulose is accepted for food applications in many countries over the world; it is identified as E-461 in the European Community as an emulsifier preventing the separation of two mixed liquids and texturing agents and it is also used as a thickener and gelling additive (European Regulation (CE) 1129/2011).

Like cellulose, it is a non-digestible, non-toxic, and non-allergic compound. It is used in bakery products as a texturing agent with the aim of gain volume, texture, and improved freshness of pastes. Methylcellulose is also used in gluten free products. In deep-frozen products as ice-creams, it reduces the ice crystal growth during freezing and thawing and it also helps to retain the shape and the heat-gelling properties of frozen products when they are fried. Methylcellulose is also used in the elaboration of sauces and creams for control the viscosity and emulsion stability. Moreover, it allows the reduction of fat in these type products, constituting a good alternative in the production of dietetic products. Methylcellulose is also employed to stabilize foams in cold drinks or for maintaining homogeneous dispersion of different components in food products (Nasatto et al., 2015).

- *Hydroxypropyl methylcellulose* is a modified water-soluble cellulose derivative that is often used in many products to improve functional properties. It is utilized for improve creaminess and texture in sauces and dressings, and thickening or gelling in many foods. It is considered as a potential fat replacer in the low-fat food system (Shin, Wicker, & Kim, 2017). As previously reported in Table 1, this compound has approved the following health claims according to European Commission (2012): "Consumption of hydroxypropyl methylcellulose with a meal contributes to a reduction in the blood glucose rise after that meal" (this claim may be used only for food which contains 4 g of hydroxypropyl methylcellulose per quantified portion as part of the meal. In order to bear the claim information shall be given to the consumer that the beneficial effect is obtained by consuming 4 g of hydroxypropyl methylcellulose as part of the meal) and "Hydroxypropyl methylcellulose contributes to the maintenance of normal blood cholesterol levels" (the claim may be used only for food which provides a daily intake of 5 g of hydroxypropyl methylcellulose. In order to bear the claim information shall be given to the consumer that the beneficial effect is obtained with a daily intake of 5 g of hydroxypropyl methylcellulose). In both case it is necessary to warn of risk of choking to be given for people with swallowing difficulties or when ingesting with inadequate fluid intake and it is also necessary to advice on taking with plenty of water to ensure that the substance reaches stomach.
- *Cyclodextrins* are a family of cyclic oligosaccharides typically containing six (α-cyclodextrin), seven (β-cyclodextrin), or eight (γ-cyclodextrin) $1 \rightarrow 4$-linked D-glucose units. They have aroused great interest in a

variety of industries, including those related to food, pharmaceuticals, cosmetics, chemicals, and agriculture. Cyclodextrins are produced from starch or starch derivatives by means of an enzymatic conversion catalyzed by cyclodextrin glycosyltransferase. Due to its high water solubility, ability to form complexes, and relatively high resistance to enzymatic hydrolysis, α-cyclodextrin has several applications in many fields, especially in the food industry. For technological uses in food, α-cyclodextrin has been used as a carrier and stabilizer for flavors, colors, and sweeteners; as a water-solubilizer for fatty acids and certain vitamins; as a flavor modifier in soya milk; and as an absorbent in confectionery products. The most important application of α-cyclodextrin in the food industry is its use as added soluble fiber, this compound reduces the absorption and bioavailability of dietary fat, which makes it useful as a weight loss supplement. For obese patients with type 2 diabetes, α-cyclodextrin is also effective in reducing and/or maintaining body weight. Besides weight control, α-cyclodextrin also provides other health benefits, including blood lipid profile control and postprandial glycemic response reduction without affecting the insulin response (Li, Chen, Gu, Chen, & Wu, 2014).

5. Dietary fiber content in cereals and pseudocereals

Grains could be classified in two main groups: cereals (wheat, rice, corn, oat, barley, rye) and pseudocereals (quinoa, amaranth, chia, buckwheat). Botanically, cereals are monocotyledonous grains that belong to *Poaceae* family, while pseudocereals are dicotyledonous grains belonging to several families such as *Polygonaceae*, *Amaranthaceae,* and *Lamiaceae*. From the nutritional point of view, cereals (with the exception of rice and corn) contains gluten, however, pseudocereals are gluten-free and constitute an alternative to celiac people (Boukid et al., 2018).

The dietary fiber composition of the main consumed cereals and pseudocereals all over the world will be described.

5.1 Dietary fiber content in cereals

The dietary fiber (DF) content of cereals varies depending on cultivars, their botanical components (such as pericarp, endosperm, and germ) and the processing conditions they have undergone (Sidhu et al., 2007). It is mainly located in the cell walls of the grains, being the outer layers, the seed coat and the pericarp, which contribute significantly to the IDF content of the grain (Rasane, Jha, Sabikhi, Kumar, & Unnikrishnan, 2015).

The DF fraction of cereals consists of non-starch polysaccharides (mainly arabinoxylans and β-glucans), resistant starch, oligosaccharides (mostly fructans) and the non-carbohydrate polyphenolic ether lignin (Knudsen et al., 2017; Rainakari et al., 2016).

5.1.1 Wheat

Wheat (*Triticum aestivum* L., *Triticum turgidum* L., *Triticum durum* Desf.) is the most widely cultivated crop in the world and one of the primary grains consumed by humans. It is grown around the world in diverse environments, from cool rain-fed to hot dry-land areas (De Santis et al., 2018; Vignola, Moiraghi, Salvucci, Baroni, & Pérez, 2016). The content of DF in wheat ranges from 9.2% to 20.0%, being the IDF the highest fraction and its content varies between 5.4% and 18.1%, while the amount of SDF in wheat grains ranges from 1.4% to 4.4%. Several authors reported that the principal wheat dietary fiber fraction are non-starch polysaccharides (NSP), being mixed-linkage β-glucans and AX the major components in wheat grain (Table 2), representing around 20% and 70%, respectively, of the NSP in wheat starch. Particularly, β-glucans, which are mainly present in the inner aleurone cell walls and subaleurone endosperm cell walls, was found in a lower amount (0.4–0.8%) comparing with other cereals, as barley, while AX content were found in a relative high content in this cereal variety (0.5–8.8%). The content of cellulose in wheat ranges from 1.9% to 2.5% and this cereal presents 0.8–1.5% of lignin (Amalraj & Pius, 2015; Ciccoritti et al., 2011; De Santis et al., 2018; Dodevska et al., 2013; Escarnot et al., 2015; Faltermaier, Waters, Becker, Arendt, & Gastl, 2014; Frølich et al., 2013; Knudsen et al., 2017; Marotti et al., 2012; Messia et al., 2017; Rainakari et al., 2016; Vignola et al., 2016; Vitaglione et al., 2008).

As previously reported in Table 1, wheat bran fiber has approved the following health claims according to European Commission (2012): "Wheat bran fiber contributes to an acceleration of intestinal transit" (this claim may be used only for food which is high in that fiber as referred to in the claim "high fiber." In order to bear the claim information shall be given to the consumer that the claimed effect is obtained with a daily intake of at least 10 g of wheat bran fiber) and "Wheat bran fiber contributes to an increase in fecal bulk" (this claim may be used only for food which is high in that fiber as referred to in the claim "high fiber").

Table 2 Dietary fiber (total, insoluble and soluble), β-glucans and arabinoxylans content in cereals (g/100 g edible portion).

	TDF	IDF	SDF	BG	AX	References
Wheat (*Triticum aestivum* L., *Triticum durum* Desf.)	9.2	—	—	0.4	0.5	Dodevska et al. (2013)
	12.4	5.4	4.4	0.5	6.9	Amalraj and Pius (2015) and Escarnot, Dornez, Verspreet, Agneessens, and Courtin (2015)
	11.6–17.0	10.2–14.7	1.4–2.3	—	4.0	De Santis et al. (2018) and Vitaglione, Napolitano, and Fogliano (2008)
	10.2–15-7	7.2–11.4	1.9–2.9	0.4–0.8	5.1–8.8	Messia, Candigliota, De Arcangelis, and Marconi (2017) and Rainakari et al. (2016)
	—	—	—	—	4.6	Ciccoritti, Scalfati, Cammerata, and Sgrulletta (2011)
	12.7–20.0	10.2–18.1	1.8–3.7		2.7–3.6	Marotti et al. (2012)
	14.2	—	—	0.6	7.1	Knudsen et al. (2017)
	13.5	—	—	0.8	5.6	Frølich, Aman, and Tetens (2013)
Rice (*Oryza sativa* L.)	9.2	1.0–3.8	2.9–5.2	0.4	0.5	Cáceres, Martínez-Villaluenga, Amigo, and Frias (2014) and Dodevska et al. (2013)
	9.9	5.4	4.4	—	—	Amalraj and Pius (2015)
	2.5	—	—	0.1	0.4	Knudsen et al. (2017)
	2.7–4.9	1.9–4.2	0.6–1.1	—	—	Prasad, Hymavathi, Ravindra Babu, and Longvah (2018)
Corn (*Zea mays* L.)	9.2	—	—	nd	1.4	Dodevska et al. (2013)
	14.9	9.4	5.4	—	—	Amalraj and Pius (2015)
	13.1–19.6	11.6–16.0	1.5–3.6	—	—	Vitaglione et al. (2008)
	11.6	—	—	0.1	4.7	Knudsen et al. (2017)
	3.7–8.6	3.1–6.1	0.5–2.5	—	—	Prasanthi, Naveena, Vishnuvardhana Rao, and Bhaskarachary (2017)
	8.3–10.7	8.0–9.1	0.3–1.6	—	—	Srichuwong et al. (2017)

Continued

Table 2 Dietary fiber (total, insoluble and soluble), β-glucans and arabinoxylans content in cereals (g/100 g edible portion).—cont'd

	TDF	IDF	SDF	BG	AX	References
Oat (*Avena sativa* L.)	13.7–30.1	—	11.5–20.0	2.7–3.5	—	Sterna, Zute, and Brunava (2016)
	10.3	6.5	3.8	2.3–8.5	—	Dhingra et al. (2012) and Rasane et al. (2015)
	10.6	—	—	4.6–5.6	—	Khan et al. (2016) and Tang and Tsao (2017)
	11.5–37.7	8.6–33.9	2.9–3.8	—	—	Vitaglione et al. (2008)
	9.8	—	—	3.8	2.1	Knudsen et al. (2017)
	10.2–12.1	6.0–7.1	4.1–4.9	—	—	Manthey, Hareland, and Huseby (1999)
	10.2	—	—	5.0	2.0	Frølich et al. (2013)
Barley (*Hordeum vulgare* L.)	15.4–18.1	7.1–10	6.1–9.3	4.7–8.0	3.1–4.1	Honců et al. (2016)
	17.4	11.5	5.9	5.2	4.0–5.4	Collar and Angioloni (2014) and Saeed et al. (2011)
	18.0–24.1	—	1.7–3.3	2.3–3.9	8.4–11.4	Teixeira, Nyman, Andersson, and Alminger (2016)
	16.8–27.9	—	—	3.3–9.2	5.1–9.1	Djurle, Andersson, and Andersson (2016)
	10.1	—	—	3.9–9.5	4.3–9.8	Messia et al. (2017) and Tang and Tsao (2017)
	20.8	—	3.0	4.2	—	Šterna, Zute, Jansone, and Kantane (2017)
	14.6–27.1	12.0–22.1	2.6–5.0	—	—	Vitaglione et al. (2008)
Rye (*Secalecereale* L.)	19.9	—	—	1.5	8.9	Frølich et al. (2013)
	15.2–20.9	11.1–16.0	3.7–4.5	1.7–2	3.1–4.3	Vitaglione et al. (2008) and Nyström et al. (2008)
	20.5	—	—	2.0	9.6	Knudsen et al. (2017)
	14.7–20.9	10.8–15.9	3.4–6.6	1.3–2.2	8–12.1	Hansen, Rasmussen, Knudsen, and Hansen (2003)
	9.6	—	3.6	1.5	5.3	Bucsella, Molnar, Harasztos, and Tömösközi (2016)

TDF: total dietary fiber, IDF: insoluble dietary fiber, SDF: soluble dietary fiber, BG: β-glucans, AX: arabinoxylans; nd: non-detected.

5.1.2 Rice

Rice (*Oryza sativa* L.) is one of the most cultivated and consumed cereal, especially in Asia. The content of DF in brown rice grains is higher than those of milled grains (i.e., white rice) because it is mainly located in hull, bran, and germ (Ji, Shin, Cho, & Lee, 2013). Rice DF content is around 2.5–9.9%, however, the proportion of IDF and SDF depends on the different rice variety (Table 2). IDF fraction content ranges between 1.0% and 5.4%, while the amount of SDF represent 0.6–5.2% in this cereal. IDF is higher than SDF in brown, black and basmati rice varieties, while the white, Bario and glutinous rice have higher amounts of SDF. The major components of SDF in rice are AX and β-D-glucans, while; cellulose and hemicellulose make up the IDF. Different authors reported that AX content varies from 0.4% to 0.5% and rice grain usually contains 0.1–0.4% of β-glucans. The content of resistant starch and cellulose in this cereal is 0.5% and 1.6%, respectively (Amalraj & Pius, 2015; Cáceres et al., 2014; Dodevska et al., 2013; Fernando, 2013; Knudsen et al., 2017; Prasad et al., 2018; Thomas, Bhat, & Kuang, 2015).

5.1.3 Corn

Corn (*Zea mays* L.) is one of the most important cereals cultivated after rice and wheat. The content of DF in corn ranges from 3.7% to 19.6%, being the IDF the highest fraction and its content varies between 3.1% and 16.0%, while the amount of SDF in corn is 0.3–5.4% (Table 2) (Amalraj & Pius, 2015; Dodevska et al., 2013; Knudsen et al., 2017; Prasanthi et al., 2017; Srichuwong et al., 2017; Vitaglione et al., 2008). Cellulose and hemicellulose are the main NSP present in corn grains, particularly in corn bran, which is widely used in several food products, such as breakfast cereals, to increase the dietary fiber contents. Corn bran obtained from the dry-milling process consists of about 22% cellulose and about 70% hemicelluloses. Corn bran is also rich in AX and glucuronoxylans (Ai & Jane, 2016). Corn is traditionally used as a food source for human nutrition after suffering various industrial processing. Particularly, corn fiber gum can potentially replace gum arabic for beverage flavor emulsification and it could be used as food additive (Yadav, Johnston, Hotchkiss Jr, & Hicks, 2007). Cellulosic fiber gel from corn bran could be employed as fat mimetic and corn bran and fibers could be also used as substrates for xylitol production (Kaur, Jha, Sabikhi, & Singh, 2014).

5.1.4 Oat

Oat (*Avena sativa* L.) consumption in human diet has been increased because of health benefits associated with its well-balanced nutritional composition.

Whole oat contains significant amount of DF, especially water soluble (1 → 3) (1 → 4) β-glucans. The content of DF in oat ranges from 9.8% to 37.7% (Table 2). IDF is the highest fraction and its content varies between 6.0% and 33.9%. The amount of SDF in oat is 2.9–20.0%. As previously mentioned, β-glucans are the most important compounds of oat DF with values between 2.3% and 8.5%. The intake of soluble oat β-glucans can lower the risk of coronary heart disease because these compounds can reduce the blood cholesterol level and blood pressure. Oat β-glucans can also reduce compounds, which are causative agents of colon cancer showing potential anti-cancerous properties. Also, oat grain contains around 2% of AX, 1.3% of cellulose and 2% of lignin. About 60% of oat grain is constituted by starch, which is the mainly constituent of endosperm. Oat also contains significant amount of resistant starch (25%) and other starch fractions, including rapidly digestible starch (7%) and slowly digestible starch (22%). Regular consumption of oat can be used to supplement these starches in diet (Dhingra et al., 2012; Frølich et al., 2013; Khan et al., 2016; Knudsen et al., 2017; Manthey et al., 1999; Rasane et al., 2015; Singh, De, & Belkheir, 2013; Sterna et al., 2016; Tang & Tsao, 2017; Vitaglione et al., 2008).

As previously reported in Table 1, oat grain fiber has approved the following health claim according to European Commission (2012): "Oat grain fiber contributes to an increase in fecal bulk." This claim may be used only for food which is high in that fiber as referred to in the claim "high fiber" (a claim that a food is high in fiber, and any claim likely to have the same meaning for the consumer, may only be made where the product contains at least 6 g of fiber per 100 g or at least 3 g of fiber per 100/kcal, according to Regulation (EC) No 1924/2006). Specifically, β-glucans from oats have also permitted the following health claim: "Consumption of β-glucans from oats as part of a meal contributes to the reduction of the blood glucose rise after that meal." This claim may be used only for food, which contains at least 4 g of β-glucans from oats for each 30 g of available carbohydrates in a quantified portion as part of the meal. In order to bear the claim information shall be given to the consumer that the beneficial effect is obtained by consuming the β-glucans from oats as part of the meal (European Commission, 2012).

5.1.5 Barley

Barley (*Hordeum vulgare* L.) is an excellent source of DF and, in particular, β-glucans that are the most important component of DF in terms of human diet and health benefits. The content of DF in barley ranges from 10.0% to 27.9% (Table 2). IDF is the highest fraction and its content varies between

7.1% and 22.1%. The amount of SDF in this cereal is 1.7–9.3%. The major components of barley DF are NSP, mainly cellulose, AX, β-glucans and oligosaccharides. AX content ranges from 3% to 11% and barley grain usually contains 2–10% of β-glucans. The location and the content of β-glucans in barley grain are particularly important from a technological and nutritional point of view. The cellulose content in barley ranges from 1.1% to 4.5% and this cereal presents 0.7–4.8% of lignin and resistant starch, respectively (Charalampopoulos, Wang, Pandiella, & Webb, 2002; Collar & Angioloni, 2014; Djurle et al., 2016; Frølich et al., 2013; Honců et al., 2016; Messia et al., 2017; Saeed et al., 2011; Šterna et al., 2017; Tang & Tsao, 2017; Teixeira et al., 2016; Vitaglione et al., 2008).

As previously described in Table 1, barley grain fiber has approved the following health claim according to European Commission (2012): "Barley grain fiber contributes to an increase in fecal bulk." This claim may be used only for food which is high in that fiber as referred to in the claim "high fiber" according to Regulation (EC) No 1924/2006. Specifically, β-glucans from barley have also permitted the follow health claim: "Consumption of β-glucans from barley as part of a meal contributes to the reduction of the blood glucose rise after that meal." This claim may be used only for food, which contains at least 4 g of β-glucans from barley for each 30 g of available carbohydrates in a quantified portion as part of the meal. In order to bear the claim information shall be given to the consumer that the beneficial effect is obtained by consuming the β-glucans from barley as part of the meal (European Commission, 2012).

5.1.6 Rye

Rye (*Secale cereale* L.) is a widely grown cereal in northern, central and Eastern Europe. It is used in bread and other products for human consumption or animal feed. Among commonly grown cereals, whole grain rye has the highest DF content, ranging from 9.6% to 20.9% (Table 2). IDF is the highest fraction and its content varies between 10.8% and 16.0%. The amount of SDF in rye is 3.4–6.6%. DF in rye consists of AX, cellulose, β-glucans, fructans, and lignin. In this respect, rye is similar to wheat, but the fiber content and the solubility of AX are higher in rye than in wheat. AX are the most abundant DF compounds in this cereal (3.1–12.1%) and they are found in different amounts and proportions in the different grain tissues. β-Glucans and fructans content ranges from 1.3% to 2.2% and from 4.5% to 6.4%, respectively, in rye grain. It is reported that this cereal content 2.9% of cellulose and 1.1% of lignin. WEAX and soluble β-glucans are responsible for the viscous properties of

SDF in rye, which may contribute to the technological functionalities and the various health effects of this cereal. SDF and fructans provide the most readily fermentable substrate for the microbiota in the large intestine, resulting in beneficial effects for human health. Moreover, rye IDF affects fecal bulk and intestinal transit time, decreasing the risk, and relieving symptoms, of constipation (Bucsella et al., 2016; Frølich et al., 2013; Hansen et al., 2003; Jonsson et al., 2018; Knudsen et al., 2017; Nyström et al., 2008; Rakha, Åman, & Andersson, 2010; Vitaglione et al., 2008).

As previously stated (Table 1), rye fiber has approved the following health claim according to European Commission (2012): "Rye fiber contributes to normal bowel function." This claim may be used only for food which is high in that fiber as referred to in the claim "high fiber" according to Regulation (EC) No 1924/2006.

5.2 Dietary fiber content in pseudocereals

In monocotyledonous cereal grains, such as wheat, rye, and corn, AX are the dominating non-cellulose DF polysaccharides followed by cellulose and β-glucans, however, in dicotyledonous pseudocereals, pectins are quantitatively predominant, as recently demonstrated for amaranth and quinoa varieties (Knudsen et al., 2017; Wefers & Bunzel, 2015).

5.2.1 Quinoa

Quinoa (*Chenopodium quinoa* Willd.) is a pseudocereal, which belongs to the *Chenopodiaceae* family. It was a basic food of the ancient civilizations of the Andes in South America. Quinoa is an excellent source of DF (both soluble and insoluble) with total values between 7% and 21.6% (Table 3), being the embryo richer than perisperm. This DF content is in the same range as found in cereal grains, being the starch the main carbohydrate component of quinoa with values higher than 50% (Alvarez-Jubete et al., 2010; Boukid et al., 2018; Gewehr et al., 2017; González Martín, Wells Moncada, Fischer, & Escudero, 2014; Gorinstein et al., 2008; Lamothe, Srichuwong, Reuhs, & Hamaker, 2015; Li & Zhu, 2017; Maradini Filho et al., 2017; Miranda et al., 2013; Pulvento et al., 2012; Srichuwong et al., 2017). Quinoa IDF represents78% of total DF content while SDF fraction constitutes 22% of quinoa DF, being SDF content higher than other cereals, such as wheat or corn (Gorinstein et al., 2008; Graf et al., 2015). The main monomeric units that constitute the components of IDF are galacturonic acid, arabinose, galactose, xylose, and glucose, while of quinoa SDF components are mainly constituted of glucose, galacturonic acid, and arabinose units (Graf et al., 2015).

Table 3 Dietary fiber (total, insoluble and soluble) content in pseudocereals (g/100g edible portion).

	TDF	IDF	SDF	References
Quinoa (*Chenopodium quinoa* Willd.)	7	—	—	Boukid et al. (2018)
	7–9.5	4.9–5.6	2.1–3.9	Srichuwong et al. (2017)
	14.2	—	—	Alvarez-Jubete, Arendt, and Gallagher (2010)
	1.8	1.4	0.4	Gorinstein et al. (2008)
	9.8	4.4	5.5	Gewehr, Danelli, De Melo, Flöres, and De Jong (2017)
	16.2–21.6	—	—	Pulvento et al. (2012)
	7.7–15.0			Li and Zhu (2017)
	11.6–15.1	9.9–12.2	0.4–2.9	Miranda et al. (2013)
Amaranth (*Amaranthus caudatus*L.)	8.9–20.6	—	—	Alvarez-Jubete et al. (2010), Boukid et al. (2018), and Valcárcel-Yamani and da Silva Lannes (2012)
	7.3	5.5	1.8	Srichuwong et al. (2017)
	11.8	9.1	2.7	Robin, Théoduloz, and Srichuwong (2015)
Chia (*Salvia hispanica* L.)	8.9	—	—	Boukid et al. (2018)
	34.4	—	—	Srichuwong et al. (2017)
	47.1–59.8	—	—	De Falco, Amato, and Lanzotti (2017)
	37–40	33–35	6–7	Ertaş-Öztürk and Şanlier (2017)
Buckwheat (*Fagopyrumesculentum* Moench.)	10.0	—	—	Boukid et al. (2018)
	11.9	5.8	6.1	Mir, Riar, and Singh (2018)
	7.0	2.2	4.8	Steadman, Burgoon, Lewis, Edwardson, and Obendorf (2001)

TDF: total dietary fiber, IDF: insoluble dietary fiber, SDF: soluble dietary fiber.

5.2.2 Amaranth

Amaranth (*Amaranthus caudatus* L.) is a good source of DF. Its leaves present a TDF between 6.95% and 9.65%, while grain fiber content is much higher than its leaves but slightly lower than wheat, ranging DF content from 19.5–27.9%, 35.1–49.3% and 33–44% in *A. cruentus*, *A. hypocondriacus*, and *A. caudatus*, respectively (Rastogi & Shukla, 2013). IDF is the prevalent fraction of amaranth DF (Table 3), being this fraction 75% of TDF. While, SDF represents around 25% of DF in amaranth grain and it is predominately composed of branched xyloglucans with a majority of di- and trisaccharide side chains, as well as pectic polysaccharides (Lamothe et al., 2015; Robin et al., 2015; Srichuwong et al., 2017). Amaranth also contained more than 25% water-insoluble β-(1,3)-D-glucan, which was less than in oats but higher than in other cereals and pseudocereals (Venskutonis & Kraujalis, 2013).

5.2.3 Chia

Chia (*Salvia hispanica* L.) is a medicinal and edible plant species used since ancient times by Mayan and Aztec populations (De Falco et al., 2017). Comparing DF content of this pseudocereal to traditional cereals, chia seeds has more fiber per 100 g of an edible portion than does barley, wheat, oats, corn and rice (Inglett & Chen, 2014) and authors have reported values up to 59.8% of TDF in chia (De Falco et al., 2017) (Table 3). Chia seeds constitute a potential ingredient in food industry applications due to its DF content, which represents values around 37–40%. IDF is the predominant fraction (33–35%) while SDF is present in lower amount (6–7%). Most of the insoluble forms are cellulose, hemicelluloses and lignin whereas SDF is mostly represented by mucilages (Ertaş-Öztürk & Şanlier, 2017). Chia mucilage is constituted of neutral sugars, indicating the presence of diverse carbohydrates on its structure. This compound is part of soluble dietary fiber fraction and it is known to have excellent water holding properties. Chia mucilage provides hydration, viscosity development and conservation of freshness, especially for baked foods, and it has properties that convert it into a potential fat substitute. The functional properties of chia hydrocolloids allow their use as a food component due their potential applications as emulsifier and stabilizer (Felisberto et al., 2015; Segura-Campos, Acosta-Chi, Rosado-Rubio, Chel-Guerrero, & Betancur-Ancona, 2014).

5.2.4 Buckwheat

Buckwheat (*Fagopyrum esculentum* Moench.) is a pseudocereal, which has gained increasing interest on industry and consumers over the past decade.

It belongs to the *Polygonaceae* family and its dehulled seeds are used in many traditional foods in different countries. Dietary fiber constituents of buckwheat are manly located in the cell walls of starchy endosperm, aleurone, seed coats and hulls, being cellulose, non-starch polysaccharides and lignin the main components of dietary fiber fraction in buckwheat. TDF in buckwheat ranges from 7% to 11.9% (Table 3), being SDF the prevalent fraction, with values between 4.8% and 6.1%. The main hemicellulosic polysaccharides in buckwheat DF are xyloglucans. The NSP contain a high amount of pecticpolysaccharides, especially arabinans and smaller amounts of linear galactans and homogalacturonan are also part of the fiber (Ahmed et al., 2014; Boukid et al., 2018; Mir et al., 2018; Steadman et al., 2001; Wefers & Bunzel, 2015).

6. Conclusions and future perspectives

The scientific interest in dietary fiberis widely growing, not only by its several nutritional and health benefits, but also by its potential applications in food industry as a functional ingredient. Currently, the main sources of dietary fiber are whole grain cereals, pulses, fruits and vegetables. Moreover, other synthetic sources of dietary fiber, such as polydextrose, hydroxypropyl methylcellulose or cyclodextrins, are also highly used in many food products. Several dietary fiber components such as pectins, arabinoxylans and β-glucans, have approved health claims that justified the importance of investigation about the properties and applications of dietary fiber constituents.

Acknowledgment

The authors are grateful to the ALIMNOVA research group (UCM-252/2017) for financial support.

References

AACC Report. (2001). The definition of dietary Fiber. *Cereal Foods World*, *46*(3), 112–126.

Abuajah, C. I., Ogbonna, A. C., & Osuji, C. M. (2015). Functional components and medicinal properties of food: A review. *Journal of Food Science and Technology*, *52*(5), 2522–2529.

Adams, E. L., Kroon, P. A., Williamson, G., Gilbert, H. J., & Morris, V. J. (2004). Inactivated enzymes as probes of the structure of arabinoxylans as observed by atomic force microscopy. *Carbohydrate Research*, *339*, 579–590.

Ahmed, A., Khalid, N., Ahmad, A., Abbasi, N. A., Latif, M. S. Z., & Randhawa, M. A. (2014). Phytochemicals and biofunctional properties of buckwheat: A review. *Journal of Agricultural Science*, *152*, 349–369.

Ai, Y., & Jane, J. L. (2016). Macronutrients in corn and human nutrition. *Comprehensive Reviews in Food Science and Food Safety, 15*, 581–598.
Akhtar, M., Tarik, A. F., Awais, M. M., Iqbal, Z., Muhammad, F., Shahid, M., et al. (2012). Studies on wheat bran Arabinoxylan for its immunostimulatory and protective effects against avian coccidiosis. *Carbohydrate Polymers, 90*, 333–339.
Alvarez-Jubete, L., Arendt, E. K., & Gallagher, E. (2010). Nutritive value of pseudocereals and their increasing use as functional gluten-free ingredients. *Trends in Food Science & Technology, 21*, 106–113.
Amalraj, A., & Pius, A. (2015). Influence of oxalate, phytate, tannin, dietary fiber, and cooking on calcium bioavailability of commonly consumed cereals and millets in India. *Cereal Chemistry, 92*(4), 389–394.
Anderson, J. W., Baird, P., Davis, R. H., Jr., Ferreri, S., Knudtson, M., Koraym, A., et al. (2009). Health benefits of dietary fiber. *Nutrition Reviews, 67*(4), 188–205.
Ayala-Soto, F. E., Serna-Saldívar, S. O., García-Lara, S., & Pérez-Carrillo, E. (2014). Hydroxycinnamic acids, sugar composition and antioxidant capacity of arabinoxylans extracted from different maize fiber sources. *Food Hydrocolloids, 35*, 471–475.
Baldassano, S., Accardi, G., & Vasto, S. (2017). Beta-glucans and cáncer: The influence of inflammation and gut peptide. *European Journal of Medicinal Chemistry, 142*, 486–492.
Bali, V., Panesar, P. S., Bera, M. B., & Panesar, R. (2015). Fructo-oligosaccharides: Production, purification and potential applications. *Critical Reviews in Food Science and Nutrition, 55*, 1475–1490.
Belitz, H., & Grosch, W. (1997). *Química de los Alimentos* (2th ed.). Spain: ACRIBIA S.A.
Bolaños, D., Marchevsky, E. J., & Camiña, J. M. (2016). Elemental analysis of amaranth, chia, sesame, linen, and quinoa seeds by ICP-OES: Assessment of classification by chemometrics. *Food Analytical Methods, 9*, 477–484.
Boukid, F., Folloni, S., Sforza, S., Vittadini, E., & Prandi, B. (2018). Current trends in ancient grains-based foodstuffs: Insights into nutritional aspects and technological applications. *Comprehensive Reviews in Food Science and Food Safety, 17*, 123–136.
Broekaert, W. F., Courtin, C. M., Verbeke, K., Van de Wiele, T., Verstraete, W., & Delcour, J. A. (2011). Prebiotic and other health-related effects of cereal-derived arabinoxylans, arabinoxylan-oligosaccharides, and xylooligosaccharides. *Critical Reviews in Food Science and Nutrition, 51*, 178–194.
Bucsella, B., Molnar, D., Harasztos, A. H., & Tömösközi, S. (2016). Comparison of the rheological and end-product properties of an industrial aleurone-rich wheat flour, whole grain wheat and rye flour. *Journal of Cereal Science, 69*, 40–48.
Burton, R. A., & Fincher, G. B. (2014). Plant cell wall engineering: Applications in biofuel production and improved human health. *Current Opinion in Biotechnology, 26*, 79–84.
Buttriss, J. L., & Stokes, C. S. (2008). Dietary fibre and health: An overview. *Nutrition Bulletin, 33*, 186–200.
Cáceres, P. J., Martínez-Villaluenga, C., Amigo, L., & Frias, J. (2014). Assessment on proximate composition, dietary fiber, phytic acid and protein hydrolysis of germinated Ecuatorian brown rice. *Plant Foods for Human Nutrition, 69*, 261–267.
Cao, L., Liu, X., Qian, T., Sun, G., Guo, Y., Chang, F., et al. (2011). Antitumor and immunomodulatory activity of arabinoxylans: A major constituent of wheat bran. *International Journal of Biological Macromolecules, 48*, 160–164.
Carvalho Lago, C., & Zapata Noreña, C. P. (2016). Polydextrose as wall material for microencapsulation of Yacon juice by spray drying. *Food and Bioprocess Technology, 9*, 2103–2113.
Charalampopoulos, D., Wang, R., Pandiella, S. S., & Webb, C. (2002). Application of cereals and cereal components in functional foods: A review. *International Journal of Food Microbiology, 79*, 131–141.

Chen, M. H., Bergman, C. J., McClung, A. M., Everette, J. D., & Tabien, R. E. (2017). Resistant starch: Variation among high amylose rice varieties and its relationship with apparent amylose content, pasting properties and cooking methods. *Food Chemistry, 234*, 180–189.

Ciccoritti, R., Scalfati, G., Cammerata, A., & Sgrulletta, D. (2011). Variations in content and extractability of durum wheat (*Triticum turgidum L. var durum*) Arabinoxylans associated with genetic and environmental factors. *International Journal of Molecular Sciences, 12*, 4536–4549.

Ciudad-Mulero, M., Barros, L., Fernandes, A., Berrios, J. D. J., Cámara, M., Morales, P., et al. (2018). Bioactive compounds and antioxidant capacity of extruded snack-type products developed from novel formulations of lentil and nutritional yeast flours. *Food & Function, 9*, 819–829.

Collar, C., & Angioloni, A. (2014). Nutritional and functional performance of high β-glucan barley flours in breadmaking: Mixed breads versus wheat breads. *European Food Research and Technology, 238*, 459–469.

Costabile, A., Klinder, A., Fava, F., Napolitano, A., Fogliano, V., Leonard, C., et al. (2008). Whole-grain wheat breakfast cereal has a prebiotic effect on the human gut microbiota: A double-blind, placebo-controlled, crossover study. *British Journal of Nutrition, 99*, 110–120.

Dai, F. J., & Chau, C. F. (2017). Classification and regulatory perspectives of dietary fiber. *Journal of Food and Drug Analysis, 25*, 37–42.

Davison, K. M., & Temple, N. J. (2018). Cereal fiber, fruit fiber, and type 2 diabetes: Explaining the paradox. *Journal of Diabetes and its Complications, 32*, 240–245.

De Falco, B., Amato, M., & Lanzotti, V. (2017). Chia seeds products: An overview. *Phytochemistry Reviews, 16*, 745–760.

De Santis, M. A., Kosik, O., Passmore, D., Flagella, Z., Shewry, P. R., & Lovegrove, A. (2018). Comparison of the dietary fibre composition of old and modern durum wheat (*Triticum turgidum* spp. *durum*) genotypes. *Food Chemistry, 244*, 304–310.

Dello Staffolo, M., Sato, A. C. K., & Cunha, R. L. (2017). Utilization of plant dietary fibers to reinforce low-calorie dairy dessert structure. *Food and Bioprocess Technology, 10*, 914–925.

Dhingra, D., Michael, M., Rajput, H., & Patil, R. T. (2012). Dietary fibre in foods: A review. *Journal of Food Science and Technology, 49*(3), 255–266.

Djurle, S., Andersson, A. A. M., & Andersson, R. (2016). Milling and extrusion of six barley varieties, effects on dietary fibre and starch content and composition. *Journal of Cereal Science, 72*, 146–152.

Dodevska, M. S., Djordjevic, B. I., Sobajic, S. S., Miletic, I. D., Djordjevic, P. B., & Dimitrijevic-Sreckovic, V. S. (2013). Characterisation of dietary fibre components in cereals and legumes used in Serbian diet. *Food Chemistry, 141*, 1624–1629.

Dodevska, M., Šobajić, S., & Djordjević, B. (2015). Fibre and polyphenols of selected fruits, nuts and green leafy vegetables used in Serbian diet. *Journal of the Serbian Chemical Society, 80*(1), 21–33.

EFSA. (2010). Panel on dietetic products, nutrition, and allergies (NDA); scientific opinion on dietary reference values for carbohydrates and dietary fibre. *EFSA Journal, 8*(3), 1462.

El-Salhy, M., Ystad, S. O., Mazzawi, T., & Gundersen, D. (2017). Dietary fiber in irritable bowel syndrome (review). *International Journal of Molecular Medicine, 40*, 607–613.

Encarnação, J. C., Abrantes, A. M., Pires, A. S., & Botelho, M. F. (2015). Revisit dietary fiber on colorectal cancer: Butyrate and its role on prevention and treatment. *Cancer Metastasis Reviews, 34*, 465–478.

Encarnação, J. C., Pires, A. S., Amaral, R. A., Gonçalves, T. J., Laranjo, M., Casalta-Lopes, J. E., et al. (2018). Butyrate, a dietary fiber derivative that improves irinotecan effect in colon cancer cells. *Journal of Nutritional Biochemistry, 56*, 183–192.

Ertaş-Öztürk, Y., & Şanlier, N. (2017). Chia seed (*Salvia hispanica* L.): It's place in nutrition and relation with health—A review. *Carpathian Journal of Food Science and Technology, 9*(4), 101–111.

Escarnot, E., Dornez, E., Verspreet, J., Agneessens, R., & Courtin, C. M. (2015). Quantification and visualization of dietary fibre components in spelt and wheat kernels. *Journal of Cereal Science, 62*, 124–133.

Espinal-Ruiz, M., Parada-Alfonso, F., Restrepo-Sánchez, L. P., Narváez-Cuenca, C. E., & McClements, J. (2014). Interaction of a dietary fiber (pectin) with gastrointestinal components (bile salts, calcium, and lipase): A calorimetry, electrophoresis, and turbidity study. *Journal of Agricultural and Food Chemistry, 62*, 12620–12630.

European Commission. (2012). *Commission Regulation (EU) No 432/2012 of 16 May, Establishing a list of permitted health claims made on foods, other than those referring to the reduction of disease risk and to children's development and health, 2012*. Official Journal of European Union. L136/1:40 (25/5/2012).

European Parliament & Council of the European Union. (2006). *Regulation (EC) No 1924/2006 of the European Parliament and of the Council of 20 December 2006 on nutrition and health claims made on foods*. Official Journal of European Union. L12/3:18 (18/01/2007).

European Parliament & Council of the European Union. (2011). *Regulation (EU) No 1169/2011 of the European Parliament and of the Council of 25 October 2011 on the provision of food information to consumers*. Official Journal of European Union. L304/18:63 (22/11/2011).

Faltermaier, A., Waters, D., Becker, T., Arendt, E., & Gastl, M. (2014). Common wheat (*Triticum aestivum* L.) and its use as a brewing cereal—A review. *Journal of the Institute of Brewing, 120*, 1–15.

Fan, P. H., Zang, M. T., & Xing, J. (2015). Oligosaccharides composition in eight food legumes species as detected by high-resolution mass spectrometry. *Journal of the Science of Food and Agriculture, 95*, 2228–2236.

Farhat, W., Venditti, R. A., Hubbe, M., Taha, M., Becquart, F., & Ayoub, A. (2017). A review of water-resistant hemicellulose-based materials: Processing and applications. *ChemSusChem, 10*, 305–323.

Felisberto, M. H. F., Wahanik, A. L., Rodrigues Gomes-Ruffi, C., Pedrosa Silva Clerici, M. T., Chang, Y. K., & Steel, C. J. (2015). Use of chia (*Salvia hispanica* L.) mucilage gel to reduce fat in pound cakes. *LWT–Food Science and Technology, 63*, 1049–1055.

Femia, A. P., Salvadori, M., Broekaert, W. F., François, I. E. J. A., Delcour, J. A., Courtin, C. M., et al. (2010). Arabinoxylan-oligosaccharides (AXOS) reduce preneoplastic lesions in the colon of rats treated with 1,2-dimethylhydrazine (DMH). *European Journal of Nutrition, 49*, 127–132.

Fernando, B. (2013). Rice as a source of fibre. *Journal of Rice Research, 1*(2), 1000–1101.

FNB (Food and Nutrition Board). (2001). *Dietary reference intakes: Proposed definition of dietary fiber*. Washington, DC: The National Academies Press.

Frølich, W., Aman, P., & Tetens, I. (2013). Whole grain foods and health—A Scandinavian perspective. *Food & Nutrition Research, 57*, 18503.

Fuentes-Zaragoza, E., Riquelme-Navarrete, M. J., Sánchez-Zapata, E., & Pérez-Álvarez, J. A. (2010). Resistant starch as functional ingredient: A review. *Food Research International, 43*, 931–942.

Fuller, S., Beck, E., Salman, H., & Tapsell, L. (2016). New horizons for the study of dietary fiber and health: A review. *Plant Foods for Human Nutrition, 71*, 1–12.

García-Amezquita, L. E., Tejada-Ortigoza, V., Serna-Saldivar, S. O., & Welti-Chanes, J. (2018). Dietary fiber concentrates from fruit and vegetable by-products: Processing, modification, and application as functional ingredients. *Food and Bioprocess Technology, 11*, 1439–1463.

Gewehr, M. F., Danelli, D., De Melo, L. M., Flöres, S. H., & De Jong, E. V. (2017). Nutritional and technological evaluation of bread made with quinoa flakes (*Chenopodium Quinoa* Willd). *Journal of Food Processing and Preservation, 41*, 1–8.

Giacco, R., Costabile, G., & Riccardi, G. (2016). Metabolic effects of dietary carbohydrates: The importance of food digestion. *Food Research International, 88*, 336–341.

Göksel Saraç, M., & Dogan, M. (2016). Incorporation of dietary fiber concentrates from fruit and vegetable wastes in butter: Effects on physicochemical, textural, and sensory properties. *European Food Research and Technology, 242*, 1331–1342.

Gong, L., Cao, W., Chi, H., Wang, J., Zhang, H., Liu, J., et al. (2018). Whole cereal grains and potential health effects: Involvement of the gut microbiota. *Food Research International, 103*, 84–102.

González Martín, M. I., Wells Moncada, G., Fischer, S., & Escudero, O. (2014). Chemical characteristics and mineral composition of quinoa by near-infrared spectroscopy. *Journal of the Science of Food and Agriculture, 94*, 876–881.

Gorinstein, S., Lojek, A., Ciz, M., Pawelzik, E., Delgado-Licon, E., Medina, O. J., et al. (2008). Comparison of composition and antioxidant capacity of some cereals and pseudocereals. *International Journal of Food Science and Technology, 43*, 629–637.

Graf, B. L., Rojas-Silva, P., Rojo, L. E., Delatorre-Herrera, J., Baldeón, M. E., & Raskin, I. (2015). Innovations in health value and functional food development of quinoa (*Chenopodium quinoa* Willd.). *Comprehensive Reviews in Food Science and Food Safety, 14*, 431–445.

Gray, J. (2006). Dietary fibre. Definition, analysis, phisiology & health (1st ed.). *IlsI Europe concise monograph series*. Belgium: ILSI Europe a.i.s.b.l.

Grooms, K. N., Ommerborn, M. J., Pham, D. Q., Djoussé, L., & Clark, C. R. (2013). Dietary fiber intake and cardiometabolic risks among US adults, NHANES 1999–2010. *The American Journal of Medicine, 126*, 1059–1067.

Grootaert, C., Delcour, J. A., Courtin, C. M., Broekaert, W. F., Verstraete, W., & Van de Wiele, T. (2007). Microbial metabolism and prebiotic potency of arabinoxylan oligosaccharides in the human intestine. *Trends in Food Science & Technology, 18*, 64–71.

Hansen, H. B., Rasmussen, C. V., Knudsen, K. E. B., & Hansen, A. (2003). Effects of genotype and harvest year on content and composition of dietary fibre in rye (*Secale cereale* L) grain. *Journal of the Science of Food and Agriculture, 83*, 76–85.

Hipsley, E. H. (1953). Letter: Dietary "fibre" and pregnancy toxaemia. *Medical Journal of Australia, 2*(9), 341–342.

Honců, I., Sluková, M., Vaculová, K., Sedláčková, I., Wiege, B., & Fehling, E. (2016). The effects of extrusion on the content and properties of dietary fiber components in various barley cultivars. *Journal of Cereal Science, 68*, 132–139.

Inglett, G. E., & Chen, D. (2014). Processing and physical properties of chia-oat hydrocolloids. *Journal of Food Processing and Preservation, 38*, 2099–2107.

Jakobsdottir, G., Nyman, M., & Fåk, F. (2014). Designing future prebiotic fiber to target metabolic syndrome. *Nutrition, 30*, 497–502.

Janakiram, N. B., Mohammed, A., Madka, V., Kumar, G., & Rao, C. V. (2016). Prevention and treatment of cancers by immune modulating nutrients. *Molecular Nutrition & Food Research, 60*, 1275–1294.

Ji, C. M., Shin, J. A., Cho, J. W., & Lee, K. T. (2013). Nutritional evaluation of immature grains in two Korean Rice cultivars during maturation. *Food Science and Biotechnology, 22*(4), 903–908.

Johns, D. J., Lindroos, A. K., Jebb, S. A., Sjöström, L., Carlsson, L. M. S., & Ambrosini, G. L. (2015). Dietary patterns, cardiometabolic risk factors, and the incidence of cardiovascular disease in severe obesity. *Obesity, 23*, 1063–1070.

Jones, J. M. (2014). CODEX-aligned dietary fiber definitions help to bridge the 'fiber gap'. *Nutrition Journal, 13*(34), 1–10.

Jonsson, K., Andersson, R., Knudsen, K. E. B., Hallmans, G., Hanhineva, K., Katina, K., et al. (2018). Rye and health—Where do we stand and where do we go? *Trends in Food Science & Technology*, *79*, 78–87.

Kan, L., Nie, S., Hu, J., Wang, S., Bai, Z., Wang, J., et al. (2018). Comparative study on the chemical composition, anthocyanins, tocopherols and carotenoids of selected legumes. *Food Chemistry*, *260*, 317–326.

Kaur, K. D., Jha, A., Sabikhi, L., & Singh, A. K. (2014). Significance of coarse cereals in health and nutrition: A review. *Journal of Food Science and Technology*, *51*(8), 1429–1441.

Khan, M. A., Nadeem, M., Rakha, A., Shakoor, S., Shehzad, A., & Khan, M. R. (2016). Structural characterization of oat bran (1→3), (1→4) -β-D-glucans by Lichenase hydrolysis through high-performance anion exchange chromatography with pulsed amperometric detection. *International Journal of Food Properties*, *19*, 929–935.

Knudsen, K. E. B., Nørskov, N. P., Bolvig, A. K., Hedemann, M. S., & Lærke, H. N. (2017). Dietary fibers and associated phytochemicals in cereals. *Molecular Nutrition & Food Research*, *61*(7), 1–15.

Koh, A., De Vadder, F., Kovatcheva-Datchary, P., & Bäckhed, F. (2016). From dietary fiber to host physiology: Short-chain fatty acids as key bacterial metabolites. *Cell*, *165*, 1332–1345.

Kothari, D., Patel, S., & Goyal, A. (2014). Therapeutic spectrum of nondigestible oligosaccharides: Overview of current state and prospect. *Journal of Food Science*, *79*(8), 1491–1498.

Kumar, A., Sahoo, U., Baisakha, B., Okpani, O. A., Ngangkham, U., Parameswaran, C., et al. (2018). Resistant starch could be decisive in determining the glycemic index of rice cultivars. *Journal of Cereal Science*, *79*, 348–353.

Kurek, M. A., Wyrwisz, J., Karp, S., & Wierzbicka, A. (2018). Effect of fiber sources on fatty acids profile, glycemic index, and phenolic compound content of in vitro digested fortified wheat bread. *Journal of Food Science and Technology*, *55*(5), 1632–1640.

Lafiandra, D., Riccardi, G., & Shewry, P. R. (2014). Improving cereal grain carbohydrates for diet and health. *Journal of Cereal Science*, *59*, 312–326.

Lamothe, L., Srichuwong, S., Reuhs, B. L., & Hamaker, B. R. (2015). Quinoa (*Chenopodium quinoa* W.) and amaranth (*Amaranthus caudatus* L.) provide dietary fibres high in pectic substances and xyloglucans. *Food Chemistry*, *167*, 490–496.

Lattimer, J. M., & Haub, M. D. (2010). Effects of dietary fiber and its components on metabolic health. *Nutrients*, *2*, 1266–1289.

Li, Z., Chen, S., Gu, Z., Chen, J., & Wu, J. (2014). Alpha-cyclodextrin: Enzymatic production and food applications. *Trends in Food Science & Technology*, *35*, 151–160.

Li, Y. O., & Komarek, A. R. (2017). Dietary fibre basics: Health, nutrition, analysis, and applications. *Food Quality and Safety*, *1*, 47–59.

Li, Q., Liu, R., Wu, T., & Zhang, M. (2017). Aggregation and rheological behavior of soluble dietary fibers from wheat bran. *Food Research International*, *102*, 291–302.

Li, G., & Zhu, F. (2017). Physicochemical properties of quinoa flour as affected by starch interactions. *Food Chemistry*, *221*, 1560–1568.

Li, G., & Zhu, F. (2018). Quinoa starch: Structure, properties, and applications. *Carbohydrate Polymers*, *181*, 851–861.

Lockyer, S., & Nugent, A. P. (2017). Health effects of resistant starch. *Nutrition Bulletin*, *42*, 10–41.

López-Marcos, M. C., Bailina, C., Viuda-Martos, M., Pérez-Alvarez, J. A., & Fernández-López, J. (2015). Properties of dietary fibers from agroindustrial coproducts as source for fiber-enriched foods. *Food and Bioprocess Technology*, *8*, 2400–2408.

Lu, Z. X., Walker, K. Z., Muir, J. G., & O′ Dea, K. (2004). Arabinoxylan fibre improves metabolic control in people with type II diabetes. *European Journal of Clinical Nutrition*, *58*, 621–628.

Macagnan, F. T., Da Silva, L. P., & Hecktheuer, L. H. (2016). Dietary fibre: The scientific search for an ideal definition and methodology of analysis, and its physiological importance as a carrier of bioactive compounds. *Food Research International, 85*, 144–154.

MacFarlane, S., MacFarlane, G. T., & Cummings, J. H. (2006). Review article: Prebiotics in the gastrointestinal tract. *Alimentary Pharmacology & Therapeutics, 24*, 701–714.

Maheshwari, G., Sowrirajan, S., & Joseph, B. (2017). Extraction and isolation of β-glucan from grain sources—A review. *Journal of Food Science, 82*(7), 1535–1545.

Manthey, F. A., Hareland, G. A., & Huseby, D. J. (1999). Soluble and insoluble dietary fiber content and composition in oat. *Cereal Chemistry, 76*(3), 417–420.

Maphosa, Y., & Jideani, V. A. (2016). Dietary fiber extraction for human nutrition—A review. *Food Reviews International, 32*(1), 98–115.

Maradini Filho, A. M., Ribeiro Pirozi, M., Da Silva Borges, J. T., Pinheiro Sant'Ana, H. M., Paes Chaves, J. B., & Dos Reis Coimbra, J. S. (2017). Quinoa: Nutritional, functional, and antinutritional aspects. *Critical Reviews in Food Science and Nutrition, 57*(8), 1618–1630.

Marotti, I., Bregola, V., Aloisio, I., Di Gioia, D., Bosi, S., Di Silvestro, R., et al. (2012). Prebiotic effect of soluble fibres from modern and old durum-type wheat varieties on *Lactobacillus* and *Bifidobacterium*strains. *Journal of the Science of Food and Agriculture, 92*, 2133–2140.

Masli, M. D. P., Rasco, B. A., & Ganjyal, G. M. (2018). Composition and physicochemical characterization of fiber-rich food processing byproducts. *Journal of Food Science, 83*(4), 956–965.

Mataix Verdú, J. (2009). Nutrición y Alimentación Humana (2nd ed.). In *Nutrientes y Alimentos*, (Vol. I), Spain: Ergon. (Chapter 7).

McRae, M. P. (2017). Dietary fiber is beneficial for the prevention of cardiovascular disease: An umbrella review of meta-analyses. *Journal of Chiropractic Medicine, 16*(4), 289–299.

McRae, M. P. (2018). Dietary fiber intake and type 2 diabetes mellitus: An umbrella review of meta-analyses. *Journal of Chiropractic Medicine, 17*(1), 44–53.

Mendis, M., & Simsek, S. (2014). Arabinoxylans and human health. *Food Hydrocolloids, 42*, 239–243.

Merriam, P. A., Persuitte, G., Olendzki, B. C., Schneider, K., Pagoto, S. L., Palken, J. L., et al. (2012). Dietary intervention targeting increased fiber consumption for metabolic syndrome. *Journal of the Academy of Nutrition and Dietetics, 112*(5), 621–623.

Messia, M. C., Candigliota, T., De Arcangelis, E., & Marconi, E. (2017). Arabinoxylans and β-glucans assessment in cereals. *Italian Journal of Food Science, 29*, 112–122.

Mir, N. A., Riar, C. S., & Singh, S. (2018). Nutritional constituents of pseudo cereals and their potential use in food systems: A review. *Trends in Food Science & Technology, 75*, 170–180.

Miranda, M., Vega-Gálvez, A., Martínez, E. A., López, J., Marín, R., Aranda, M., et al. (2013). Influence of contrasting environments on seed composition of two quinoa genotypes: Nutritional and functional properties. *Chilean Journal of Agricultural Research, 73*(2), 108–116.

Mohebbi, Z., Homayouni, A., Azizi, M. H., & Hosseini, S. J. (2018). Effects of beta-glucan and resistant starch on wheat dough and prebiotic bread properties. *Journal of Food Science and Technology, 55*(1), 101–110.

Morales, P., Berrios, J. D. J., Varela, A., Burbano, C., Cuadrado, C., Muzquiz, M., et al. (2015). Novel fiber-rich lentil flours as snack-type functional foods: An extrusion cooking effect on bioactive compounds. *Food & Function, 6*, 3135–3143.

Mudgil, D., & Barak, S. (2013). Composition, properties and health benefits of indigestible carbohydrate polymers as dietary fiber: A review. *International Journal of Biological Macromolecules, 61*, 1–6.

Nair, K. K., Kharb, S., & Thompkinson, D. K. (2010). Inulin dietary fiber with functional and health attributes—A review. *Food Reviews International, 26*, 189–203.
Nandi, I., & Ghosh, M. (2015). Studies on functional and antioxidant property of dietary fibre extracted from defatted sesame husk, rice bran and flaxseed. *Bioactive Carbohydrates and Dietary Fibre, 5*, 129–136.
Nasatto, P. L., Pignon, F., Silveira, J. L. M., Duarte, M. E. R., Noseda, M. D., & Rinaudo, M. (2015). Methylcellulose, a cellulose derivative with original physical properties and extended applications. *Polymers*, 7, 777–803.
Nauta, A. J., & Garssen, J. (2013). Evidence-based benefits of specific mixtures of non-digestible oligosaccharides on the immune system. *Carbohydrate Polymers, 93*, 263–265.
Neyrinck, A. M., Possemiers, S., Druart, C., Van de Wiele, T., De Backer, F., Cani, P. D., et al. (2011). Prebiotic effects of wheat Arabinoxylan related to the increase in Bifidobacteria, Roseburia and Bacteroides/Prevotella in diet-induced obese mice. *PLoS One, 6*, 1–12.
Niño-Medina, G., Carvajal-Millán, E., Rascon-Chu, A., Marquez-Escalante, J. A., Guerrero, V., & Salas-Muñoz, E. (2009). Feruloylated arabinoxylans and arabinoxylans gels: Structure, sources and applications. *Phytochemistry Reviews, 9*, 111–120.
Nyström, L., Lampi, A. M., Andersson, A. A. M., Kamal-Eldin, A., Gebruers, K., Courtin, C. M., et al. (2008). Phytochemicals and dietary Fiber components in Rye varieties in the HEALTHGRAIN diversity screen. *Journal of Agricultural and Food Chemistry, 56*, 9758–9766.
Ötles, S., & Ozgoz, S. (2014). Health effects of dietary fiber. *Acta Scientiarum Polonorum. Technologia Alimentaria, 13*(2), 191–202.
Ou, J. Y., Huang, J. Q., Song, Y., Yao, S. W., Peng, X. C., Wang, M. F., et al. (2016). Feruloylated oligosaccharides from maize bran modulated the gut microbiota in rats. *Plant Foods for Human Nutrition, 71*, 123–128.
Ozyurt, V. H., & Ötles, S. (2016). Effect of food processing on the physicochemical properties of dietary fibre. *Acta Scientiarum Polonorum. Technologia Alimentaria, 15*(3), 233–245.
Padayachee, A., Day, L., Howell, K., & Gidley, M. J. (2017). Complexity and health functionality of plant cell wall fibers from fruits and vegetables. *Critical Reviews in Food Science and Nutrition, 57*(1), 59–81.
Park, S. Y., & Yoon, K. Y. (2015). Enzymatic production of soluble dietary fiber from the cellulose fraction of Chinese cabbage waste and potential use as a functional food source. *Food Science and Biotechnology, 24*(2), 529–535.
Prasad, V. S. S., Hymavathi, A., Ravindra Babu, V., & Longvah, T. (2018). Nutritional composition in relation to glycemic potential of popular Indian rice varieties. *Food Chemistry, 238*, 29–34.
Prasanthi, P. S., Naveena, N., Vishnuvardhana Rao, M., & Bhaskarachary, K. (2017). Compositional variability of nutrients and phytochemicals in corn after processing. *Journal of Food Science and Technology, 54*(5), 1080–1090.
Pulvento, C., Riccardi, M., Lavini, A., Iafelice, G., Marconi, E., & D'andria, R. (2012). Yield and quality characteristics of quinoa grown in open field under different saline and non-saline irrigation regimes. *Journal of Agronomy and Crop Science, 198*, 254–263.
Qasem, A. A. A., Alamri, M. S., Mohamed, A. A., Hussain, S., Mahmood, K., & Ibraheem, M. A. (2017). Soluble fiber-fortified sponge cakes: Formulation, quality and sensory evaluation. *Food Measure, 11*, 1516–1522.
Qiu, S., Yadav, M. P., & Yin, L. (2017). Characterization and functionalities study of hemicellulose and cellulose components isolated from sorghum bran, bagasse and biomass. *Food Chemistry, 230*, 225–233.
Raigond, P., Ezekiel, R., & Raigond, B. (2015). Resistant starch in food: A review. *Journal of the Science of Food and Agriculture, 95*, 1968–1978.

Rainakari, A. I., Rita, H., Putkonen, T., & Pastell, H. (2016). New dietary fibre content results for cereals in the Nordic countries using AOAC 2011.25 method. *Journal of Food Composition and Analysis*, *51*, 1–8.
Rakha, A., Åman, P., & Andersson, R. (2010). Characterisation of dietary fibre components in rye products. *Food Chemistry*, *119*, 859–867.
Rasane, P., Jha, A., Sabikhi, L., Kumar, A., & Unnikrishnan, V. S. (2015). Nutritional advantages of oats and opportunities for its processing as value added foods—A review. *Journal of Food Science and Technology*, *52*(2), 662–675.
Rastogi, A., & Shukla, S. (2013). Amaranth: A new millennium crop of nutraceutical values. *Critical Reviews in Food Science and Nutrition*, *53*, 109–125.
Rawat, N., & Indrani, D. (2015). Functional ingredients of wheat-based bakery, traditional, pasta, and other food products. *Food Reviews International*, *31*, 125–146.
Redgwell, R. J., & Fischer, M. (2005). Dietary fiber as a versatile food component: An industrial perspective. *Molecular Nutrition & Food Research*, *49*, 521–535.
Requena, M. C., Aguilar González, C. N., Prado Barragán, L. A., Correia, T., Contreras Esquivel, J. C., & Rodríguez Herrera, R. (2016). Functional and physico-chemical properties of six desert-sources of dietary fiber. *Food Bioscience*, *16*, 26–31.
Robin, F., Théoduloz, C., & Srichuwong, S. (2015). Properties of extruded whole grain cereals and pseudocereals flours. *International Journal of Food Science and Technology*, *50*, 2152–2159.
Rodríguez, R., Jiménez, A., Fernández-Bolaños, J., Guillén, R., & Heredia, A. (2006). Dietary fibre from vegetable products as source of functional ingredients. *Trends in Food Science & Technology*, *17*, 3–15.
Saeed, F., Pasha, I., Anjum, F. M., & Sultan, M. T. (2011). Arabinoxylans and arabinogalactans: A comprehensive treatise. *Critical Reviews in Food Science and Nutrition*, *51*, 467–476.
Segura-Campos, M., Acosta-Chi, Z., Rosado-Rubio, G., Chel-Guerrero, L., & Betancur-Ancona, D. (2014). Whole and crushed nutlets of chia (*Salvia hispanica*) from Mexico as a source of functional gums. *Food Science and Technology (Campinas)*, *34*(4), 701–709.
Selani, M. M., Bianchini, A., Ratnayake, W. S., Flores, R. A., Massarioli, A. P., de Alencar, S. M., et al. (2016). Physicochemical, functional and antioxidant properties of tropical fruits co-products. *Plant Foods for Human Nutrition*, *71*, 137–144.
Sharma, S. K., Bansal, S., Mangal, M., Dixit, A. K., Gupta, R. K., & Mangal, A. K. (2016). Utilization of food processing by-products as dietary, functional, and novel fiber: A review. *Critical Reviews in Food Science and Nutrition*, *56*, 1647–1661.
Shevkani, K., Singh, N., Kaur, A., & Rana, J. C. (2014). Physicochemical, pasting, and functional properties of Amaranth seed flours: Effects of lipids removal. *Journal of Food Science*, *79*(7), 1271–1277.
Shin, W. K., Wicker, L., & Kim, Y. (2017). HPMC (hydroxypropyl methylcellulose) as a fat replacer improves the physical properties of low-fat tofu. *Journal of the Science of Food and Agriculture*, *97*, 3720–3726.
Shinozaki, K., Okuda, M., Sasaki, S., Kunitsugu, I., & Shigeta, M. (2015). Dietary fiber consumption decreases the risks of overweight and hypercholesterolemia in Japanese children. *Annals of Nutrition & Metabolism*, *67*, 58–64.
Shokrollahi, F., & Taghizadeh, M. (2016). Date seed as a new source of dietary fiber: Physicochemical and baking properties. *International Food Research Journal*, *23*(6), 2419–2425.
Shortt, C., Hasselwander, O., Meynier, A., Nauta, A., Noriega Fernández, E., Putz, P., et al. (2018). Systematic review of the effects of the intestinal microbiota on selected nutrients and non-nutrients. *European Journal of Nutrition*, *57*, 25–49.
Sidhu, J. S., Kabir, Y., & Huffman (2007). Functional foods from cereal grains. *International Journal of Food Properties*, *10*, 231–244.

Sima, P., Vannucci, L., & Vetvicka, V. (2018). β-Glucans and cholesterol (review). *International Journal of Molecular Medicine, 41*, 1799–1808.
Singh, R., De, S., & Belkheir, A. (2013). *Avena sativa* (oat), a potential nutraceutical and therapeutic agent: An overview. *Critical Reviews in Food Science and Nutrition, 53*, 126–144.
Singh, S. P., Singh Jadaun, J. S., Narnoliya, L. K., & Pandey, A. (2017). Prebiotic oligosaccharides: Special focus on Fructooligosaccharides, its biosynthesis and bioactivity. *Applied Biochemistry and Biotechnology, 183*, 613–635.
Slavin, J. (2013). Fiber and prebiotics: Mechanisms and health benefits. *Nutrients, 5*, 1417–1435.
Sorndech, W., Nakorn, K. N., Tongta, S., & Blennow, A. (2018). Isomalto-oligosaccharides: Recent insights in production technology and their use for food and medical applications. *LWT- Food Science and Technology, 95*, 135–142.
Srichuwong, S., Curti, D., Austin, S., King, R., Lamothe, L., & Gloria-Hernandez, H. (2017). Physicochemical properties and starch digestibility of whole grain sorghums, millet, quinoa and amaranth flours, as affected by starch and non-starch constituents. *Food Chemistry, 233*, 1–10.
Steadman, K. J., Burgoon, M. S., Lewis, B. A., Edwardson, S. E., & Obendorf, R. L. (2001). Buckwheat seed milling fractions: Description, macronutrient composition and dietary fibre. *Journal of Cereal Science, 33*, 271–278.
Stephen, A. M., Champ, M. M. J., Cloran, S. J., Fleith, M., van Lieshout, L., Mejborn, H., et al. (2017). Dietary fibre in Europe: Current state of knowledge on definitions, sources, recommendations, intakes and relationships to health. *Nutrition Research Reviews, 30*, 149–190.
Sterna, V., Zute, S., & Brunava, L. (2016). Oat grain composition and its nutrition benefice. *Agriculture and Agricultural Science Procedia, 8*, 252–256.
Šterna, V., Zute, S., Jansone, I., & Kantane, I. (2017). Chemical composition of covered and naked spring barley varieties and their potential for food production. *Polish Journal of Food and Nutrition Sciences, 67*(2), 151–158.
Tanabe, K., Nakamura, S., & Oku, T. (2014). Inaccuracy of AOAC method 2009.01 with amyloglucosidase for measuring non-digestible oligosaccharides and proposal for an improvement of the method. *Food Chemistry, 151*, 539–546.
Tang, Y., & Tsao, R. (2017). Phytochemicals in quinoa and amaranth grains and their antioxidant, anti-inflammatory, and potential health beneficial effects: A review. *Molecular Nutrition & Food Research, 61*(7), 1–16.
Tao, J., Li, Y., Li, S., & Li, H. B. (2018). Plant foods for the prevention and management of colon cancer. *Journal of Functional Foods, 42*, 95–110.
Teitelbaum, J. E., & Walker, W. A. (2002). Nutritional impact of pre- and probiotics as protective gastrointestinal organisms. *Annual Review of Nutrition, 22*, 107–138.
Teixeira, C., Nyman, M., Andersson, R., & Alminger, M. (2016). Effects of variety and steeping conditions on some barley components associated with colonic health. *Journal of the Science of Food and Agriculture, 96*, 4821–4827.
Tejada-Ortigoza, V., García-Amezquita, L. E., Serna-Saldívar, S. O., & Welti-Chanes, J. W. (2016). Advances in the functional characterization and extraction processes of dietary fiber. *Food Engineering Reviews, 8*, 251–271.
Thomas, R., Bhat, R., & Kuang, Y. T. (2015). Composition of amino acids, fatty acids, minerals and dietary fiber in some of the local and import rice varieties of Malaysia. *International Food Research Journal, 22*(3), 1148–1155.
Thompkinson, D. K., Bhavana, V., & Kanika, P. (2014). Dietary approaches for management of cardio-vascular health- a review. *Journal of Food Science and Technology, 51*(10), 2318–2330.
Threapleton, D. E., Greenwood, D. C., Evans, C. E. L., Cleghorn, C. L., Nykjaer, C., Woodhead, C., et al. (2013). Dietary fibre intake and risk of cardiovascular disease: Systematic review and meta-analysis. *British Medical Journal, 347*, 1–12.

Tomic, N., Dojnov, B., Miocinovic, J., Tomasevic, I., Smigic, N., Djekic, I., et al. (2017). Enrichment of yoghurt with insoluble dietary fiber from triticale—A sensory perspective. *LWT- Food Science and Technology*, *80*, 59–66.

Tong, L. T., Zhong, K., Liu, L., Qiu, J., Guo, L., Zhou, X., et al. (2014). Effects of dietary wheat bran arabinoxylans on colesterol metabolism of hypercholesterolemic hamsters. *Carbohydrate Polymers*, *112*, 1–5.

Troncoso, O. P., Zamora, B., & Torres, F. G. (2017). Thermal and rheological properties of the mucilage from the fruit of *Cordia lutea*. *Polymers from Renewable Resources*, *8*(3), 79–90.

Tungland, B. C., & Meyer, D. (2002). Nondigestible oligo- and polysaccharides (dietary fiber): Their physiology and role in human health and food. *Comprehensive Reviews in Food Science and Food Safety*, (3), 90–109.

Valcárcel-Yamani, B., & da Silva Lannes, S. C. (2012). Applications of quinoa (*Chenopodium Quinoa* Willd.) and Amaranth (*Amaranthus* Spp.) and their influence in the nutritional value of cereal based foods. *Food and Public Health*, *2*(6), 265–275.

Van Craeyveld, V. (2009). *Production and functional characterisation of arabinoxylan-oligosaccharides from wheat (*Triticum aestivum *L.) bran and psyllium (*Plantago ovata *Forsk) seed husk*. Thesis. Belgium: Universidad K.U. Leuven.

Vandenplas, Y., Zakharova, I., & Dimitrieva, Y. (2015). Oligosaccharides in infant formula: More evidence to validate the role of prebiotics. *British Journal of Nutrition*, *113*, 1339–1344.

Venskutonis, P. R., & Kraujalis, P. (2013). Nutritional components of Amaranth seeds and vegetables: A review on composition, properties, and uses. *Comprehensive Reviews in Food Science and Food Safety*, *12*, 381–412.

Vignola, M. B., Moiraghi, M., Salvucci, E., Baroni, V., & Pérez, G. T. (2016). Whole meal and white flour from argentine wheat genotypes: Mineral and arabinoxylans differences. *Journal of Cereal Science*, *71*, 217–223.

Vitaglione, P., Napolitano, A., & Fogliano, V. (2008). Cereal dietary fibre: A natural functional ingredient to deliver phenolic compounds into the gut. *Trends in Food Science & Technology*, *19*, 451–463.

Wefers, D., & Bunzel, M. (2015). Characterization of dietary fiber polysaccharides from Dehulled common buckwheat (*Fagopyrum esculentum*) seeds. *Cereal Chemistry*, *92*(6), 598–603.

Wells, M. L., Potin, P., Craigie, J. S., Raven, J. A., Merchant, S. S., Helliwell, K. E., et al. (2017). Algae as nutritional and functional food sources: Revisiting our understanding. *Journal of Applied Phycology*, *29*, 949–982.

Yadav, M. P., Johnston, D. B., Hotchkiss, A. T., Jr., & Hicks, K. B. (2007). Corn fiber gum: A potential gum arabic replacer for beverage flavor emulsification. *Food Hydrocolloids*, *21*, 1022–1030.

Yangilar, F. (2013). The application of dietary fibre in food industry: Structural features, effects on health and definition. Obtaining and analysis of dietary fibre: A review. *Journal of Food and Nutrition Research*, *1*(3), 13–23.

Yao, B., Fang, H., Xu, W., Yan, Y., Xu, H., Liu, Y., et al. (2014). Dietary fiber intake and risk of type 2 diabetes: A dose–response analysis of prospective studies. *European Journal of Epidemiology*, *29*, 79–88.

Yu, J., & Ahmedna, M. (2013). Functional components of grape pomace: Their composition, biological properties and potential applications. *International Journal of Food Science and Technology*, *48*, 221–237.

Zhang, H., Cao, X. R., Yin, M., & Wang, J. (2018). Soluble dietary fiber from Qing Ke (highland barley) brewers spent grain could alter the intestinal cholesterol efflux in Caco-2 cells. *Journal of Functional Foods*, *47*, 100–106.

Zhao, G., Zhang, R., Dong, L., Huang, F., Tang, X., Wei, Z., et al. (2018). Particle size of insoluble dietary fiber from rice bran affects its phenolic profile, bioaccessibility and functional properties. *LWT- Food Science and Technology*, *87*, 450–456.

Zhou, Q., Wu, J., Tang, J., Wang, J. J., Lu, C. H., & Wang, P. X. (2015). Beneficial effect of higher dietary fiber intake on plasma HDL-C and TC/HDL-C ratio among Chinese rural-to-urban migrant workers. *International Journal of Environmental Research and Public Health*, *12*, 4726–4738.

Zhu, F., Du, B., & Xu, B. (2016). A critical review on production and industrial applications of beta-glucans. *Food Hydrocolloids*, *52*, 275–288.

Žilic, S., Dodig, D., Milašinovic Šeremešic, M., Kandic, V., Kostadinovic, M., Prodanovic, S., et al. (2011). Small grain cereals compared for dietary fibre and protein contents. *Genetika*, *43*(2), 381–395.

CHAPTER THREE

Impact of molecular interactions with phenolic compounds on food polysaccharides functionality

Corrine C. Dobson[a,b,†], **Walid Mottawea**[a,†], **Alexane Rodrigue**[a,b,†], **Bruna L. Buzati Pereira**[b,c,†], **Riadh Hammami**[a], **Krista A. Power**[a], **Nicolas Bordenave**[a,*]
[a]School of Nutrition Sciences, Faculty of Health Sciences, University of Ottawa, Ottawa, ON, Canada
[b]Interdisciplinary School of Health Sciences, Faculty of Health Sciences, University of Ottawa, Ottawa, ON, Canada
[c]Internal Medicine Department, Botucatu Medical School, UNESP—Univ Estadual Paulista, Botucatu, Brazil
*Corresponding author: e-mail address: nicolas.bordenave@uottawa.ca

Contents

† These authors contributed equally to this work

Advances in Food and Nutrition Research, Volume 90
ISSN 1043-4526
https://doi.org/10.1016/bs.afnr.2019.02.010

Abstract

Commercial trends based of the emergence of plant-based functional foods lead to investigate the structure-function relationship of their main bioactive constituents and their interactions in the food matrix and throughout the gastro-intestinal tract. Among these bioactive constituents, dietary polysaccharides and polyphenols have shown to interact at the molecular level and these interactions may have consequences on the polysaccharides physical and nutritional properties. The methods of investigation and mechanisms of interactions between polysaccharides and polyphenols are reviewed in light of their respective technological and nutritional functionalities. Finally, the potential impact of the co-occurrence or co-ingestion of polyphenols and polysaccharides on the technological and nutritional functionality of the polysaccharides are investigated.

1. Introduction

Polysaccharides and polyphenols are two classes of bioactive food compounds. The functionalities of both type of compounds are fairly well known. Food polysaccharides in particular are used in food technology applications as physical modifiers of foods. Along with polyphenolic compounds, they are also known for the bioactivity and their association with reduced risks of non-communicable chronic diseases. Moreover, polysaccharides and polyphenols are often found in the same matrices but their molecular interactions and the consequences of these interactions have received little attention to date. Whereas some research has been carried out on the impact of these interactions on the bioavailability of the polyphenols or the organoleptic properties of the foods containing these molecules. There is an important knowledge gap about the impact of these interactions on the functionality of polysaccharides, despite emerging data that convincingly show this impact may not be negligible. We will review the main classes of food polysaccharides, their health and technological functionalities in relation with their molecular structure, and their co-occurrences with phenolic compounds. Then, we will review the current state of knowledge about their interactions at the molecular level and the consequences of these interactions, at the molecular level again, and on the health and technological functionalities of the polysaccharides.

2. Functional food polysaccharides

2.1 Structure, composition and occurrence of digestible and indigestible food polysaccharides

2.1.1 Starch

Starch is a major dietary glycemic carbohydrate and the most common polysaccharide in plants, where it is used for energy storage. Plant sources that produce starch include cereals, legumes and tubers. Starch structure is comprised of two macromolecules; amylose, a linear polymer of α-(1 → 4) D-glucose, and amylopectin, which has a linear backbone similar to amylose, and α-(1 → 4) D-glucose branches linked in α-(1 → 6) among themselves and to the backbone. The molar mass of amylose ranges from 10^5 to 10^6 g/mol and amylopectin 10^7–10^9 g/mol. Starch properties such as crystallinity, branching angle and length, and thermal properties vary by plant source and fine molecular structure (BeMiller, 2007).

Starch digestibility rate and kinetics is determined by starch supramolecular structure and ultimately its accessibility to digestive enzymes (Magallanes-Cruz, Flores-Silva, & Bello-Perez, 2017). A less open starch structure, imparted by a higher degree of crystallinity, makes glycosidic linkages less accessible to enzymes, slowing digestion rate. Starch digestion kinetics modulate glucose absorption in the small intestine (Zhang & Hamaker, 2009). Englyst, Kingman, and Cummings (1992) classified starch that is digested within the first 20 min of digestion as Rapidly Digestible Starch (RDS) while the term Slowly Digestible Starch (SDS) is used to describe starch that undergoes hydrolysis between the 20 and 120 min period of digestion in the small intestine. Resistant starch (RS) was defined as starch that does not undergo enzyme hydrolysis within 120 min of digestion and passes through the small intestine intact, and therefore can be classified as a dietary fiber.

2.1.2 Non-starch polysaccharides

Cellulose is a linear and water-insoluble pure β-(1 → 4)-linked glucan. As such, it is an insoluble dietary fiber that is found as the main structural polysaccharide in fruits, vegetables, pulses and cereals (BeMiller, 2007; Wrolstad, 2012).

Pectin is a common component of most flowering plants and found in fruits and vegetables, in the middle lamella of the cell walls. All pectic substances contain D-galacturonic acid as their main constituent but their structure is heterogeneous with the presence of additional neutral sugar monomers such as galactose, rhamnose, or arabinose and the esterification or acetylation of the monomers. The molar mass of pectins ranges approximately between 6×10^4 and 2×10^5 g/mol (BeMiller, 2007; Wrolstad, 2012). Apple (*Malus* sp.) pomace and citrus (*Citrus* sp.) peel are the major sources of pectin for commercial use. Highly esterified pectins form gels by increasing the solid content and lowering the pH of the pectin solution, whereas low-ester pectins form gels with divalent cations (Thakur, Singh, Handa, & Rao, 1997). Pectins are mostly water soluble and classified as fermentescible crude dietary fibers (Slavin & Lloyd, 2012).

Hemicelluloses are indigestible mixed glycans comprising glucose, xylose, arabinose, galactose, mannose, and rhamnose residues, resulting in a structural diversity that makes hemicelluloses difficult to be defined precisely as a class of polysaccharides.

Hemicelluloses are found in the cell walls of many different plant sources, and like cellulose, they provide rigidity to the structure of the plant. The majority of hemicelluloses are insoluble but due to their diverse structure and side chains, some soluble exceptions exist. The main classes of hemicelluloses found in food sources are arabinoxylans, found mostly in cereals such as wheat (*Triticum* sp.), maize (*Zea mays*), oats (*Avena sativa*), barley (*Hordeum vulgare*), and xyloglucans, found mostly in fruits and vegetables such as apples (*Malus pumila*), grapes (*Vitis vinifera*), strawberries (*Fragaria sp.*), oranges (*Citrus x sinensis*), tomatoes (*Solanum lycopersicum*), onions (*Allium cepa*), carrot (*Daucus carota* subsp. *sativus*) or lettuce (*Lactuca sativa*). Cereal arabinoxylans are essentially linear chains of xylose linked in β-(1$\rightarrow$4) and substituted in position C2 or C3 by arabinose residues making α-(1$\rightarrow$2) or α-(1$\rightarrow$3) linkages. The arabinose side groups can themselves be substituted with ferulic acid. Fruits and vegetables xyloglucans are cellulose with about three out of four glucose residues substituted in C6 position with a xylose by an α-(1$\rightarrow$6) linkage. The xylose side groups can themselves be substituted with galactose and/or fucose residues (BeMiller, 2007; Wrolstad, 2012). One of the nutritionally relevant subfamilies of hemicelluloses is the mixed linkage glucans and most notably β-glucans. They are composed of D-glucopyranosyl units linked linearly by β-(1$\rightarrow$3) and β-(1$\rightarrow$4) linkages and their molar mass ranges generally from approximately 3×10^4 to 2×10^6 g/mol (Tosh, Wood, Wang, & Weisz, 2004). Beta-glucans are present in cereals, such as rye (*Secale cereale*), barley (*Hordeum vulgare*), and wheat (*Triticum* sp.), but are most commonly found in oats (*Avena sativa*), accounting for approximately 50% of NSPs in oats. Beta-glucans can be isolated from cereals through milling, sieving, or air classification. Beta-glucans are water soluble (at pH 7) and very viscous in solution, a property that can be imparted to foodstuffs and digesta, and that is thought to provide β-glucans their beneficial health effects including blood glucose and cholesterol management (Webster & Wood, 2011). They exhibit bulking effects on digesta due to their ability to bind water and swell when consumed. Viscosity and water retention properties of β-glucans are related to their molar mass, as well as gel formation as evidenced by their reduced efficacy upon hydrolysis (Daou & Zhang, 2012).

Fructans are a polysaccharide consisting of monomers of fructose linked by β-(2 → 1) or β-(2 → 6) fructose-fructosyl linkages to a backbone of sucrose. Many cereals such as wheat (*Triticum* sp.), rye (*Secale cereale)* or vegetables contain fructans are believed to act as an energy reserve for plants. Notable examples of plants containing fructans include dicotyledons such as onion (*Allium cepa*), chicory (*Cichorium intybus*), and artichoke (*Cynara cardunculus*). Five different types of fructans exist based on different mechanisms of synthesis: levan-type, neolevan-type, graminan-type, inulin-type and neoinulin type. Specifically, inulin and oligofructose fit the classification of dietary fiber in that they are not digested or absorbed in the gastrointestinal tract of humans. Inulin-type fructans are of interest for their role as a pre-biotic dietary fiber (Apolinário et al., 2014; Roberfroid, 1993, 2002).

Gums are another class of non-starch polysaccharides generally extracted from plants' sap, beans or roots. Practically, the term "gum" generally refers to soluble, high-molar mass polysaccharides used as texture modifiers or stabilizers in the food or cosmetic industry. Xanthan gum has a heterogeneous structure and is composed of a cellulose chain backbone with trisaccharide side chains (mannose-β-(1 → 4)-glucoronic acid-β-(1 → 2)-mannose) and alternating glucose units joined via α-(1 → 3) linkages. It has a molar mass of approximately 2×10^6 g/mol. Xanthan is a water-soluble and acidic natural exopolysaccharide produced by the bacterium *Xanthomonas campestris* and has been of great interest due to its commercial applicability (BeMiller, 2007). Guar gum is a linear mannan with β-(1 → 4) linkages and galactan side chains (Pegg, 2012). It has an approximate molar mass of $10–15 \times 10^3$ g/mol and but is partially hydrolyzed for commercial purposes. It is sourced from the seeds of the plant *Cyamopsis tetragonoloba* Taub. found in India, Pakistan, and the United States (Texas) where it serves as the main storage carbohydrate (Zhang & Hamaker, 2009). Guar gum is a dietary fiber that has high viscosity at low temperatures and can increase the gastric content's viscosity. Guar gum, like other soluble fibers and gums, can increase stool weight and improve digestion and bowel health (Slavin & Greenberg, 2003). Modified celluloses include mainly carboxymethylcellulose, methylcellulose and hydroxypropyl methylcellulose. They are water-soluble derivatives of cellulose (extracted from wood or plant wastes) that can vary by the amount, nature and distribution of their substituent groups on the cellulose backbone and their molar mass. Their physico-chemical properties have made them ingredients of choice to modify the properties of food products. They are pH stable, odorless, tasteless, clear in solution, and varied in grades and viscosities (BeMiller, 2007).

2.2 Functionality of food polysaccharides

2.2.1 Health and nutritional functionality

2.2.1.1 Gastrointestinal health

The intestinal tract is a multifunctional cellular and secretory barrier that separates the luminal contents, including the microbiota, from the human/host tissues (Kim, Kim, Um, & Hong, 2010; Scaldaferri, Pizzoferrato, Gerardi, Lopetuso, & Gasbarrini, 2012). This barrier is under constant stress due to the changing luminal contents caused by diet, shifts in microbial activity and structure, and presence of other environmental toxins and allergens.

Alterations in one or many of the microenvironment components can drive microbial dysbiosis, impair barrier integrity, and initiate colonic and systemic inflammation (Natividad & Verdu, 2012). Therefore, approaches to maintain intestinal homeostasis in healthy individuals, and to re-establish intestinal health in diseased individuals is of current interest with food and food components, receiving increasing attention by those in the scientific community and food industry.

There are several factors known to influence intestinal homeostasis including stress, pathogenic microorganisms, antibiotics, and diet (Louis & O'Byrne, 2010). Insoluble non-digestible polysaccharides (NDPs) including cellulose and hemicelluloses, and soluble NDPs (pectin, inulin, gums, resistant starch, and arabinoxylans) help maintain a symbiotic colonic microbial community, promote the integrity and function of the colonic epithelium, and modulate immune and inflammatory responses which can aid in reducing the risk and severity of colon-associated chronic diseases. Insoluble NDPs, play a role in colon health by promoting laxation and fecal bulking and reducing fecal transit time through the intestinal tract (Eswaran, Muir, & Chey, 2013). On the other hand, soluble NDP aid in slowing intestinal transit time and transport of macronutrients through the small intestine, and are consumed by the microbiota resulting in the production of short chain fatty acids (SCFAs), including acetate, propionate, and butyrate, which are produced at an approximate molar ratio of 60:20:20 (Cummings, Pomare, Branch, Naylor, & Macfarlane, 1987). The degree to which SCFA are produced by the microbiota is dependent on the structure and molar mass of the NDP (Eswaran et al., 2013), as well as the composition of the microbiota which expresses appropriate enzymes for carbohydrate fermentation (Knudsen et al., 2018). For example, non-digestible short chain oligosaccharides (fructo-oligosaccharides (FOS), galacto-oligosaccharides (GOS)) are fermented more rapidly in the cecum and proximal colon compared to resistant starch, pectin, and gums, which are fermented more gradually throughout the colon, and insoluble NDPs being minimally fermented (Eswaran et al., 2013). The majority of acetate and propionate are transported across the epithelium via apical and basolateral plasma membranes monocarboxylate transporters (MCT-1 and MCT-4) (Sivaprakasam, Bhutia, Yang, & Ganapathy, 2017), and utilized systemically in metabolic processes (Cummings et al., 1987; den Besten et al., 2013), while butyrate is preferentially transported and used by the colonocytes as an energy source, thus emphasizing the importance of this NDP-derived microbial metabolite in intestinal health.

Mechanisms associated with beneficial intestinal health effects of microbial-derived SCFAs include enhancing the structure and integrity of the mucus and epithelial barriers by increasing villus/crypt length and mucin secretion through enhanced proliferation of epithelial and goblet cells (Burger-van Paassen et al., 2009; Edwards, Wilson, Hanlon, & Eastwood, 1992; Hamer et al., 2008; Ito et al., 2009; Willemsen, Koetsier, van Deventer, & van Tol, 2003), enhancing the expression and localization of tight junctional protein complexes which control intestinal permeability (Feng, Wang, Wang, Huang, & Wang, 2018; Peng, Li, Green, Holzman, & Lin, 2009); enhancing immunological homeostasis through activation of epithelial and immune cell surface G protein-coupled receptors (e.g., GPRs 41, 43) (Smith et al., 2013) and aryl hydrocarbon receptor (AhR) signaling (Jin et al., 2017), and inhibiting histone deacetylases (HDAC) which results in reduced carcinogenesis (Hamer et al., 2008; Jin et al., 2017). Butyrate has also been shown enhance epithelial barrier integrity and function and promote wound healing by increasing the expression of tight junction proteins such as ZO-1 and occludin in an in vitro model of small intestinal barrier dysfunction using IPEC-J2 cells (Ma et al., 2012). Peng et al. (2009) found the assembly of tight junction proteins to be regulated by AMP-activated protein kinase (AMPK), and that butyrate is involved in the activation of AMPK, thus accelerating tight junction assembly in a colon barrier model using Caco-2 cells. The increased expression of tight junction proteins has been demonstrated to promote barrier integrity through measurements of permeability, such as transepithelial electrical resistance (TER), which butyrate has been shown to increase (Peng, He, Chen, Holzman, & Lin, 2007). Furthermore, butyrate has been shown to reduce inflammation in colon HT-29 cells through inhibition of NF-κB and TNF-α (Inan et al., 2000). A reduction in inflammation is beneficial to the gut barrier, since inflammation is associated with increased tight junction permeability and hyper-responsiveness of GALT, often resulting in inflammatory intestinal diseases (Liu, Li, & Neu, 2005).

Several studies have demonstrated beneficial effects of NDPs, such as inulin (Casellas et al., 2007), pectins (Popov et al., 2006), resistant starch (Bassaganya-Riera et al., 2011), and partially hydrolyzed guar gum (Naito et al., 2006), as well as the SCFA, butyrate (Scheppach et al., 1992; Vernia et al., 2003; Vieira et al., 2012), on symptoms and progression of colitis, in patients with Inflammatory Bowel Diseases (IBD) and experimental animal models. In addition to the barrier promoting effects of SCFAs described above, the anti-inflammatory effects of NDPs are primarily due to SCFA production, in particular butyrate, which decreases pro-inflammatory mediators, such as TNF-α, leukotrienes, nitric oxide, IL-1β, IL-17 and IL-12 (Nishimura et al., 2010; Rodriguez-Cabezas et al., 2003; Ye et al., 2011). Additionally, NDPs and SCFAs increase levels of the anti-inflammatory cytokine IL-10, modulate the activation and infiltration of various immune cells (macrophages, eosinophil, T cells, B cells, dendritic cells, neutrophils), reduced Heat Shock Protein (HSP) 70 expression, and inhibit NF-κB activation and interferon (IFN)-γ-mediated apoptosis (Naito et al., 2006; Venkatraman et al., 2003; Vieira et al., 2012). Furthermore, NDPs modulation of the gut microbiota structure, also known as prebiotic effect, may also promote intestinal health. The dysbiotic microbial ecosystem observed in IBD patients may therefore be beneficially modulated by NDPs potentially reducing symptoms and disease severity (Steed, Macfarlane, & Macfarlane, 2008).

2.2.1.2 Colonic fermentation and prebiotic effect

The term prebiotic was introduced since 1995 to indicate indigestible food ingredients that induce the growth of one or a group of gut microorganisms, which results in stimulation of specific metabolic activity in the colon (Gibson & Roberfroid, 1995). Gut bacteria have the metabolic capacity to process indigestible food particles, including a variety of oligo- and polysaccharides (Tremaroli & Bäckhed, 2012). Several dietary NSPs such as inulin, fructooligosaccharides (FOS), glucooligosaccharides (GuOS), galactooligosaccharides (GaOS), xylooligosaccharides (XOS), and maltooligosaccharides (MOS) are used as prebiotics to promote the growth of beneficial bacteria in the gut (Hutkins et al., 2016; Singh, Jadaun, Narnoliya, & Pandey, 2017). Inulin, xylan, and FOS are the most popular prebiotics in food products. Indeed, the concept of prebiotics has emerged mainly from observations of *Bifidobacterium* enrichment during inulin fermentation (Gibson, Probert, Loo, Rastall, & Roberfroid, 2004). Recently, a double-blinded, randomized crossover intervention study has illustrated that inulin supplementation selectively induces the growth of *Bifidobacterium*, *Bilophila*, and *Anaerostipes* among colon bacteria (Vandeputte et al., 2017). This selective taxa enrichment was associated with increased abundance of dodecanal among fecal metabolites (Vandeputte et al., 2017).

Xylans are utilized exclusively by *Roseburia intestinalis*, *Eubacterium rectal*, and *Bacteroides* species (Chassard, Goumy, Leclerc, Del'homme, & Bernalier-Donadille, 2007; Leth et al., 2018; Mirande et al., 2010). However, these taxa have shown different affinities toward different xylan fragments. *R. intestinalis*, for example, prefers oligomers of 4–5 xylosyl units, while *Bacteroides* species target larger oligomers (Leth et al., 2018). These differential affinities may be attributed to the different modular organization of xylan degrading enzymes, such as xylanases (Leth et al., 2018; Zhang et al., 2014).

FOS supplementation decreased the relative abundance of Firmicutes, increased Bacteroidetes and altered >100 taxa of bacteria in the microbiome of obese mice (Everard et al., 2011). In addition, administration of FOS has been reported to induce *Bifidobacterium* and *Lactobacilli* while reducing butyrate producing bacteria (Liu, Li, et al., 2017; Roberfroid et al., 2010).

2.2.1.3 Metabolic health

Metabolic dysfunction associated with obesity is characterized by (i) an expansion of a dysfunctional adipose tissue (characterized by increased infiltrating immune cells (e.g., macrophages) and pro-inflammatory cytokine production (resistin, leptin, TNF- α, IL-6), and reduced synthesis of anti-inflammatory mediators (e.g., adiponectin)), and (ii) altered regulation of energy metabolism (hyperglycemia, hyperinsulinemia, and dyslipidemia) which contribute to the development of metabolic dysfunction-driven co-morbidities (Hotamisligil, 2006; Kyrgiou et al., 2017). An underappreciated fact within the general public is that obesity is actually a gut-associated disease, wherein the intestinal microenvironment is dramatically altered in obese humans: e.g., microbial community structure and activity (Delzenne & Cani, 2011; Ley, Turnbaugh, Klein, & Gordon, 2006; Rahat-Rozenbloom, Fernandes, Gloor, & Wolever, 2014; Santacruz et al., 2010; Schwiertz et al., 2010; Zhang et al., 2009), and intestinal epithelial barrier function (Salden et al., 2018; Troseid et al., 2013) compared to healthy lean controls.

Several pre-clinical and clinical studies have demonstrated that one key strategy to beneficially alter the obese intestinal microenvironment and impact inflammatory and metabolic dysfunctions is through consumption of specific dietary components. NDPs have been shown to induce numerous beneficial effects in obesity and metabolic health such as improving glycemic and insulinemic responses, thermogenesis, and increasing satiety and reduction of energy intake (Mozaffarian, 2016). Dietary supplementation of wheat bran-derived arabinoxylans oligosaccharides (AXOS) in HFD-induced obesity has been shown to counteract HFD-induced microbial dysbiosis, improve intestinal epithelial barrier function by upregulating tight junction protein expression, and reduce metabolic endotoxemia in mice (Neyrinck et al., 2011, 2012). Moreover, multiple aspects of the obese phenotype were dramatically improved by HFD-AXOS supplementation including reduced adipose tissue size, adipose tissue macrophage infiltration, inflammatory mediator expression and insulin resistance (i.e., metabolic dysfunction) (Neyrinck et al., 2011, 2012). Additionally, in animal HFD-induced obesity models, dietary intervention with inulin-type fructans (soluble fiber) were shown to beneficially alter the intestinal microbiota composition, reduce both endotoxemia and systemic inflammation and improve the obese metabolic dysfunction by increasing glucose tolerance, and reducing insulin resistance and visceral fat mass (Cani et al., 2007, 2009; Cani & Delzenne, 2009; Roberfroid et al., 2010). These beneficial effects in rodent models are also being translated into human clinical trials, wherein a recent double-blind, placebo-controlled intervention study was conducted in obese women supplemented with inulin/oligofructose (ITF) for 3 months (Dewulf et al., 2013). Compared to the placebo control, ITF supplementation altered the microbiota composition, which was associated with improved epithelial barrier integrity (via reduced serum bacterial lipopolysaccharide levels), and reduced circulating levels of C-reactive protein, a critical inflammatory clinical biomarker, and beneficially impacted several key metabolites implicated in obesity and/or diabetes (Dewulf et al., 2013). In a 2017 study (Salden et al., 2018), overweight subjects given an arabinoxylan-rich supplement daily for 6 weeks, demonstrated beneficial effects in microbial activity (increased production of anti-inflammatory microbial fermentation products, short chain fatty acids (SCFAs: acetate, propionate, and butyrate), increased biomarkers of intestinal epithelial barrier integrity (increased colon biopsy expression of tight junction proteins), and reduced blood pro-inflammatory mediators (e.g., TNFα), compared to placebo controls.

Increasing NDPs consumption may also decrease dietary energy absorption by way of diluting a diet's energy availability while maintaining other important nutrients. NDPs contribute little to the total caloric content of a diet (2 kcal/g), since it is resistant to digestion in the small intestine and depending on the structure of the NDPs, may be resistant to microbial fermentation in the large intestine (Lattimer & Haub, 2010). However, SCFAs produced following microbial fermentation of some NDPs, stimulate the production of intestinal glucagon-like peptide (GLP-1) and peptide YY (PYY) (Psichas et al., 2015) and brain neuropeptides (GABA) (Frost et al., 2014), which are known to play a role in inducing satiety. SCFAs can also modulate endogenous metabolic processes including lipolysis, insulin sensitivity and secretion, gluconeogenesis, and lipid storage (Russell et al., 2016). Furthermore, NDPs may also increase satiety because of their water-holding properties which increases fecal bulking and ultimately fecal output (Eastwood, Robertson, Brydon, & MacDonald, 1983). This increased bulking leads to higher levels of fullness and consequently reduces energy intake and body weight.

NDP intake has been inversely correlated with low glucose and insulin responses throughout the day, indicative of the therapeutic potential of a high-fiber diet in the treatment of diabetes (Russell et al., 2016). Viscous NDPs (e.g., glucomannan, guar gum, psyllium, β-glucan) exert acute improvements in glucose and insulin responses in individuals with type 2 diabetes (Marciani et al., 2001), potentially by absorbing water and reducing gastric emptying (Marciani et al., 2001; Russell et al., 2016). Highly viscous NDPs reduce intestinal motility, slowing the diffusion rate of starch digestion products, and reduce α-amylase accessibility (Marciani et al., 2001). This delay in gastric emptying can alter the rate of nutrient absorption, thus slowing the rate of glucose absorption during digestion (Holt, Heading, Carter, Prescott, & Tothill, 1979). This reduction in glucose absorption by soluble NDPs leads to lower blood glucose levels and reduces the need for insulin (Hannan et al., 2007). Delayed gastric emptying due to intake of oat β-glucan supplemented drink for 6 days was associated with increased satiety in both healthy subjects and subjects with type 2 diabetes (Yu, Ke, Li, Zhang, & Fang, 2014).

2.2.2 Functionality in food technology applications

2.2.2.1 Use as thickener and stabilizing agents in food

Cellulose and water-soluble gums have been used extensively to modify the physico-chemical properties of foods. Food innovations have been enabled in particular by the functions of water-soluble gums as hydrocolloids (BeMiller, 2007; Dar & Light, 2014). A few examples follow.

In purified powdered form or in microcrystalline form, cellulose (E460) has been added to provide non-caloric bulk and retain moisture, acting as a foam or emulsion stabilizer or fat replacer, particularly in reduced-calorie baked goods or extruded snacks (BeMiller, 2007).

Because of their ability to gel, pectins (E440) are often used in the food industry to stabilize emulsions, suspensions and foams, for example, in fruit juice concentrates. Pectins are also used in jam making and in frozen foods to maintain firmness and prevent loss of liquids (Sakai, Sakamoto, Hallaert, & Vandamme, 1993).

Arabinogalactans (E409) are used in the food industry to impart emulsifying, stabilizing, and water-binding properties. Due to their insolubility, hemicelluloses from cereals such as rice bran's arabinoxylans have been used to add dietary fibers in bread to improve the quality of the bread and its health benefits (Spiridon & Popa, 2008).

Fructans and inulin are water-soluble polysaccharides but their water holding capacity is not substantial enough to impart a significant increase in viscosity. Therefore, they have been used as bulking agents in food substances (Schneeman, 1999).

The viscosity of xanthan gum (E415) makes it suitable as a thickening agent and it can be combined with other gums like guar to enhance this property even further. Xanthan has commercial value for its applicability as a thickener or emulsifier in both the food industry and non-food industries like agriculture or cosmetics. It is commonly used in flour-based mixes for baking to improve quality such as volume and water retention (BeMiller, 2007).

Like xanthan, guar gum (E412) can impart stabilizing properties to a variety of foods including beverages, sauces, and condiments and can also impart viscosity to change the texture of food products (BeMiller, 2007). The health applications of guar gum have been investigated in terms of being able to manage cholesterol, glycaemia, and obesity, which is a common application of viscous dietary fibers (Butt, Shahzadi, Sharif, & Nasir, 2007).

Among other uses, cellulose derivatives (E461–469) have been used as thickeners and texture modifiers, as humectant, to form and stabilize emulsions and foams, to bind and hold water, as fat replacers, etc. (BeMiller, 2007).

2.2.2.2 Encapsulation of probiotics and bioactive compounds

Encapsulation has emerged to improve targeted site delivery and physical and chemical stability of sensitive bioactive ingredients such as probiotics, vitamins, minerals, peptides, phenolic compounds, phytosterols, and carotenoids (Tolve et al., 2016). The choice of encapsulating materials depends on different factors such as the proprieties of the core substance and the method of encapsulation. However, as they should be Generally

Recognized as Safe (GRAS) for human (Nazzaro, Orlando, Fratianni, & Coppola, 2012), polysaccharides are particularly suitable materials. The most popular polysaccharides employed in encapsulation include alginates, chitosan, gum arabic and carboxymethylcellulose (CMC) (Wani et al., 2016). Polysaccharides could be used alone (e.g., Ca- or Ba-alginate beads) or in combination (e.g., alginate-chitosan, alginate-pectin, CMC-κ-carrageenan beads) to improve the encapsulation efficiency (Madziva, Kailasapathy, & Phillips, 2006; Muhamad, Fen, Hui, & Mustapha, 2011). Polysaccharide beads usually used for targeted drug delivery of the bioactive core material in the human body such as alginate-chitosan and CMC-κ-carrageenan microcapsules (Muhamad et al., 2011; Wani et al., 2016). When these microcapsules are administered, swelling occurs in vivo followed by rupture of the microcapsules under certain conditions releasing its entrapped material (Liu et al., 2004). Chitosan could be also used in combination with proteins (e.g., β-lactoglobulin) as double wall coating, to sustain the release of core materials in the gut (Lee, Yim, Choi, Van Anh Ha, & Ko, 2012). Furthermore, polysaccharide encapsulation may be used as heat resistant microcapsules such as gum arabic-gelatin combinations or aluminum CMC-rice bran composite microcapsules (Chitprasert, Sudsai, & Rodklongtan, 2012; Wani et al., 2016).

Probiotics are living microorganisms that confer health benefits to the host when administered and reach the gut in adequate amounts (Food and Agriculture Organization of the United Nations, World Health Organization, Joint FAO/WHO Expert Consultation on Evaluation of Health and Nutritional Properties of Probiotics in Food including Powder Milk with Live Lactic Acid Bacteria, Joint FAO/WHO Working Group on Drafting Guidelines for the Evaluation of Probiotics in Food, 2006), typically commensal *Lactobacillus* and *Bifidobacterium* species isolated from the human gastrointestinal tract (GIT). The capacity to survive the challenging passage through the GIT is an important trait of probiotic strains (Masco, Crockaert, Van Hoorde, Swings, & Huys, 2007) as only if the colon is reached as viable metabolically active cells, a probiotic bacterium can exert health benefit to the consumer. Using different microencapsulation techniques, polysaccharides have shown significant progress in protecting probiotics from the drastic gastric environment and increased the viable amount of probiotics that reach the colon. For example, pectinate-chitosan beads showed the capability of delivering a high level of viable *Lactobacillus casei* to the intestine (Bepeyeva et al., 2017). Similarly, encapsulation of *L. plantarum* with κ-Carrageenan increased its survival toward the acidic pH but not against bile salts (Tee, Nazaruddin, Tan, & Ayob, 2014). Still, the encapsulation efficiency varies from one organism to another within the same encapsulation matrix and from one encapsulation techniques to another (Gani, Shah, Ahmad, Ashwar, & Masoodi, 2018; Tee et al., 2014). *Bifidobacteria*, for example, exhibited lower survival rate in a β-D-glucan matrix as compared to *Lactobacilli* (Gani et al., 2018). Moreover, freeze-drying and extrusion methods resulted in higher encapsulation efficiency of *L. plantarum* compared to emulsification technique (Tee et al., 2014). The double coating of the polysaccharide encapsulated probiotics with protein such as whey protein concentrate was proven to increase the survival of probiotics under drastic pH of 3 as compared to single polysaccharide coating (Iqbal, Zahoor, Huma, Jamil, & Unlu, 2018). Similarly, whey protein isolate and pectin were proved effective coating materials for improving the dual-coated liposomes encapsulation of bacteriocin MccJ25 and its controlled release (Gomaa, Martinent, Hammami, Fliss, & Subirade, 2017).

Using polysaccharides as probiotics encapsulating matrices have the advantage of being a prebiotic-probiotic combination. The synergistic combination of prebiotics and probiotics is known as synbiotics (Gibson & Roberfroid, 1995). Synbiotics help increases the survival of probiotics during passage through the GIT (Rioux, Madsen, & Fedorak, 2005). Therefore, the appropriate combination of both in a single product should ensure a superior activity as compared to that of prebiotic or probiotic alone. For instance, administration of FOS and *Lactobacillus paracasei* to weanling piglets increased the abundance of *Lactobacillus* and *Bifidobacterium* in fecal microbiota as compared to the administration of the probiotic or prebiotic alone (Nemcova, Bomba, Gancarcikova, Herich, & Guba, 1999). Similarly, the addition of inulin to alginate beads, the most popular matrix for probiotic encapsulation, provided more protection for different probiotics against acidic pH and provided a colonic controlled release system (Atia et al., 2017).

3. Co-occurrence of polysaccharides and phenolic compounds

3.1 Overview of dietary phenolic compounds

Phenolic compounds are present in numerous food sources. Here, we will focus on plant sources, such as cereals, fruits, vegetables, pulses and herbs. Dietary phenolic compounds can be grouped in six distinct categories: phenolic acids, flavonoids, tannins, stilbenoids, coumarins and polymeric lignans.

Phenolic acids are low molecular weight compounds, containing a phenolic ring and a carboxylic acid moiety. Two families of phenolic acids can be distinguished: hydroxybenzoic acids, derived from benzoic acid (e.g., gallic acid, protocatechuic acid, vanillic acid, syringic acid), and hydroxycinnamic acids, derived from cinnamic acid (e.g., *p*-coumaric acid, caffeic acid, ferulic acid, sinapic acid). Phenolic acids are ubiquitous in plants. However, they are mostly covalently bound to other plant compounds (polysaccharides, lignins or flavonoids) through ester, ether or acetal bonds, and free phenolic acids are only found in small quantities (Robbins, 2003).

Flavonoids are the largest family of phenolic compounds in food. Structurally, they are composed of two phenyl rings and a heterocyclic ring. Several sub-classes of monomeric flavonoids can be distinguished based on their structural features: flavonols (e.g., quercetin, kaempferol, myricetin, galangin, fisetin), flavones (e.g., apigenin, chrysin, luteolin), flavanols or flavan-3-ols (e.g., catechin, epicatechin, epigallocatechin, epicatechin gallate, epigallocatechin gallate), flavanones (e.g., eridictyol, hesperitin, naringenin), anthocyanidins (e.g., cyaniding, pelargonidin, delphinidin, peonidin, malvidin), isoflavonoids (e.g., genistein, daidzein, glycitein, formononetin) and chalcones (Fig. 1). In food sources, flavonoids can be found in various derivative forms: monomeric or oligomeric forms (e.g., proanthocyanidins which are flavanol oligomers, see condensed tannins below), aglycone *O*-glycoside or *C*-glycoside forms with mono- to tetra-saccharides, and conjugated forms (acylated or sulfated) (Andersen & Markham, 2006).

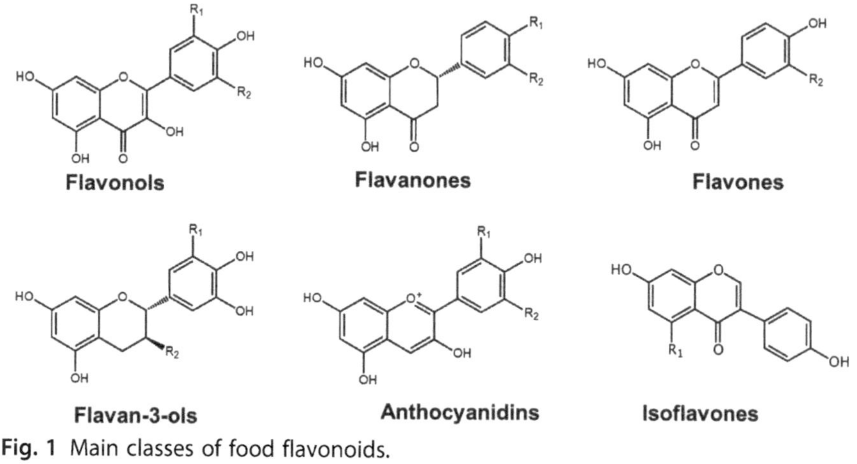

Fig. 1 Main classes of food flavonoids.

Tannins from plants are categorized as hydrolysable tannins (polymers of gallic acid in the form of ellagitannins or gallotannins, e.g., pentagalloylglucose, vescalagin, castalagin) and condensed tannins, also called proanthocyanidins (polymers of flavonoids which include procyanidin's, polymers of flavan-3-ols and their gallic acid esters, and prodelphidinins) (Serrano, Puupponen-Pimiä, Dauer, Aura, & Saura-Calixto, 2009).

Stilbenoids are hydroxylated derivatives of stilbene, a diarylethene. In dietary sources, stilbenoids can be found in aglycone (e.g., resveratrol, pterostilbene, piceatannol) or glycoside forms (e.g., piceid) (Dvorakova & Landa, 2017). Coumarins are derivatives of coumarin, a benzopyrone found in a number of plants (e.g., scopoletin, umbelliferone, aesculetin) (Kostova et al., 2011). Stilbenoids and coumarins are minor phenolic compounds in foods.

3.2 Naturally co-occurring polysaccharides and phenolic compounds

Phenolic compounds are found in numerous plant sources along with significant amounts of polysaccharides, which constitute the structure of the plants' cell walls.

Cereal grains mainly contain phenolic acids bound to cell wall materials, the free form concentration being generally $<100\,\mu g/g$, although it increases in germinated and malted grains. Cereal free phenolic acids are primarily gallic acid, vanillic acid, syringic acid, *p*-coumaric acid, *p*-hydroxybenzoic acid, protocatechuic acid, chlorogenic acid, and ferulic acid (Van Hung, 2016). Notably, sorghum grains contain high amounts of condensed tannins (Wu et al., 2012). These phenolic compounds are found in increasing concentration from the endosperm to the aleurone where most non-starch polysaccharides are found, including cellulose, hemicelluloses in the form of arabinoxylans. Notably, oats (*Avena sativa*), barley (*Hordeum vulgare*) and wheat (*Triticum* sp.) grains also contain mixed β-glucan gums, located throughout the grain, but in higher concentration in the bran layers whereas starch is mainly found in the grain's endosperm and represent generally $>50\%$ of the grain's dry weight (Thakur et al., 1997; Webster & Wood, 2011).

In fresh fruits, free phenolic compounds are very diverse. In fruits such as apples, pears and berries, the most abundant flavonoids are catechins (2–10 mg/100 g), flavonols (2–30 mg/100 g) and proanthocyanidins (1–15 mg/100 g). Generally, colored proanthocyanidins and flavonols such as quercetin are mainly present in the skin of fruits whereas catechins are mostly found in their flesh (Andersen & Markham, 2006). Fruits also contain significant amounts of phenolic acids (10–50 mg/100 g), principally caffeic acid, gallic acid, *p*-coumaric, *p*-hydroxybenzoic acid, and vanillic acid. The free form fraction is generally in the range of 30–50% with values as low as 10% and as high as 90% (Mattila, Hellström, & Törrönen, 2006). Notably, citrus (*Citrus* sp.) fruits are the only fruits to contain appreciable amounts of flavanones and flavones. Minor quantities of stilbenoids, such as resveratrol can also be found in dates (*Phoenix dactylifera*), strawberries (*Fragaria* sp.), or tomatoes (*Solanum lycopersicum*) (Sebastià, Montoro, León, & Soriano, 2017). Fresh vegetables mainly contain flavonols, essentially in the form of quercetin and kaempferol, and some flavones. Vegetables such as red cabbage (*Brassica oleracea*) and purple carrots (*Daucus carota* subsp. *sativus)* also contain proanthocyanidins (Mizgier et al., 2016). Fresh fruits and vegetables cell walls are structurally similar as they can be modeled as composite materials made of partially soluble polysaccharides reinforced by cellulose rigid rod. As described in Section 2.1.2, they differ in the composition of these polysaccharides: while they are mainly pectins in fruits, they are mainly xyloglucans in green vegetables; fructans and inulin can also be found in a number of vegetables such as artichokes (*Cynara cardunculus*), asparagus (*Asparagus officinalis*), onions (*Allium cepa*), garlic (*Allium sativa*), etc.

Pulses, such as dry beans (*Phaseolus vulgaris*), dry peas (*Pisum sativum*), lentils (*Lens culinaris*), and chickpeas (*Cicer arietinum*), can contain up to 100 mg/100 g of phenolic acids (particularly, chlorogenic acid), monomeric flavonoids (particularly catechin, epicatechin, as well as kaempferol and delphinidin glycosides) and procyanidin's (Giusti, Caprioli, Ricciutelli, Vittori, & Sagratini, 2017; Ramdath & Tsao, 2012). Dietary fibers of pulses are mainly composed of cellulose, as well as partially soluble high molecular weight pectins and hemicellulosic arabinoxylans (Brummer, Kaviani, & Tosh, 2015). Similar to cereals, pulses contain significant amount of starch mainly located in the cotyledons.

3.3 Co-occurrence in formulated food products

Numerous food formulations combine polysaccharide-rich ingredients with ingredients rich in free phenolic compounds: cereal bars, instant oatmeal with berries, wholemeal bread with fruit, vegetable or grain inclusions, etc. From a similar perspective, many starch or fiber-rich products are co-consumed with phenolic rich foods or beverages: whole grain cereals

Fig. 2 Oatmeal with berries (https://www.collegetimes.com/food-and-drink/7-porridge-variations-134552).

Fig. 3 Wholemeal pumpernickel bread sandwich (http://lavandulaofficinalis.blogspot.com/2010/12/pumpernickel-salmon-sandwich.html).

with fruits, coffee or tea in breakfast occasions (Fig. 2), whole grain cereal products (pasta, bread) with vegetables in sandwiches (Fig. 3), etc. This is also the case when starch or fiber-rich foods are co-consumed with taste enhancers: whereas herbs and spices are often used in small quantities making their fiber content irrelevant, they can contain significant amounts of free phenolic compounds (Andersen & Markham, 2006).

Therefore, starch, dietary fibers and free phenolic compounds are very often present simultaneously in a meal. Physical transformations such as cell wall disruption occurring during oral processing (mastication, in particular) and the stomach phase of digestion then favor their interactions in the food bolus and throughout the gastro-intestinal tract.

Additionally, in many food sources cited in the previous section, polysaccharides and phenolic compounds can be physically segregated (e.g., phytonutrients within cells and polysaccharides between cells and in cell walls). This is generally offset through thermomechanical processing, for example, during the making of fruit or vegetable purees, applesauce, whole grain-based beverages, etc.

Finally, many products containing fruits, vegetables or their extracts are combined with texture modifiers and improve their organoleptic properties. As a number of polysaccharides (xanthan, guar or locust bean gums, chemically modified starches) are used as texture agents, these types of formulation enhancements contribute to the *co*-occurrence of polysaccharides and phenolic compounds in the same matrix.

3.4 Co-occurrence in controlled delivery systems

Numerous encapsulation systems have been devised for the controlled and/or targeted delivery of bioactive compounds, especially phenolic compounds. Although proteins have been a material of choice for encapsulation due to their extensively documented ability to bind polyphenols, polysaccharide-based systems have been developed (Hu, Liu, Zhang, & Zeng, 2017): for example, maltodextrins encapsulating red wine polyphenols (Sanchez, Baeza, Galmarini, Zamora, & Chirife, 2013) or naringin (Pai, Vangala, Ng, Ng, & Tan, 2015), alginate and hydroxypropyl methylcellulose encapsulating green tea polyphenols (Belščak-Cvitanović et al., 2017), inulin and maltodextrin encapsulating coffee phenolic bioactives (Pettinato, Aliakbarian, Casazza, & Perego, 2017) or starch encapsulating catechin (Ahmad et al., 2019).

A recent study also reported the potential usefulness of amorphous solid dispersions (ASD) of cellulose derivatives for the delivery and bioavailability of quercetin. Indeed, in suspension conditions, cellulose derivatives ASD enhanced the solution concentration of quercetin, an otherwise poorly water-soluble flavonol, particularly an composite material made of cellulose acetate suberate and polyvinylpyrrolidone (Gilley et al., 2017). Although this work is not based on proper encapsulation, it highlights the potential of polysaccharides to act as non-covalent carriers for bioactive polyphenolic compounds.

4. Molecular interactions between polysaccharides and phenolic compounds

Historically, polysaccharide-polyphenol interactions seemed to arise from the study of the transformation of fruits such as apples (*Malus pumila*) and grapes (*Vitis vinifera*). As reviewed previously, binding between polyphenols (particularly tannins and phenolic acids) and polysaccharides (particularly pectins) has an impact on the extractability of the phenolic compounds and therefore the taste profile of wine (Bautista-Ortín, Ben Abdallah, Castro-López, Jiménez-Martínez, & Gómez-Plaza, 2016; Bindon, Bacic, & Kennedy, 2012; Bindon et al., 2016; Bindon & Kennedy, 2011; Bindon, Smith, Holt, & Kennedy, 2010; Bindon, Smith, & Kennedy, 2010; Busse-Valverde, Bautista-Ortín, Gómez-Plaza, Fernández-Fernández, & Gil-Muñoz, 2012; Busse-Valverde et al., 2010; Castro-López, Gómez-Plaza, Ortega-Regules, Lozada, & Bautista-Ortín, 2016; Cerpa-Calderón & Kennedy, 2008; Ducasse et al., 2010; Fournand et al., 2006; Gonçalves, Rocha, & Coimbra, 2012; Mekoue Nguela, Poncet-Legrand, Sieczkowski, & Vernhet, 2016; Riou, Vernhet, Doco, & Moutounet, 2002; Ruiz-Garcia, Smith, & Bindon, 2014; Springer & Sacks, 2014; Yacco, Watrelot, & Kennedy, 2016) and cider (Le Bourvellec, Bouchet, & Renard, 2005; Le Bourvellec et al., 2013; Le Bourvellec, Guyot, & Renard, 2004, 2009; Le Bourvellec, Le Quere, & Renard, 2007; Le Bourvellec & Renard, 2005; Le Bourvellec, Watrelot, Ginies, Imberty, & Renard, 2012; Renard et al., 2011; Renard, Baron, Guyot, & Drilleau, 2001; Watrelot, Le Bourvellec, Imberty, & Renard, 2013). Expanding this field of research, investigators have addressed to question as to whether the same interactions affect the phenolic compounds' bioavailability, as in purple carrots (*Daucus carota*) (Gómez-Mascaraque, Dhital, López-Rubio, & Gidley, 2017; Padayachee et al., 2013, 2012a, 2012b; Phan, D'Arcy, & Gidley, 2016; Phan, Flanagan, D'Arcy, & Gidley, 2017; Phan et al., 2015). The next two sections will review the general characteristics of polysaccharide-polyphenol binding that have been uncovered and the methods of investigation that have been refined through this body of research.

4.1 Molecular drivers of interaction and technical approach for their characterization

Polysaccharide-polyphenol binding has been characterized thermodynamically with adsorption or binding isotherms (Fig. 4). This technique consists essentially in quantifying the amount of bound and free phenolic compounds (concentration $[PP_b]$ and $[PP_f]$, respectively) in a solution containing a known amount of phenolic compounds and polysaccharides. Polyphenols-cell wall materials adsorption isotherms seemed to consistently follow type I Langmuir adsorption patterns, allowing the determination apparent affinity

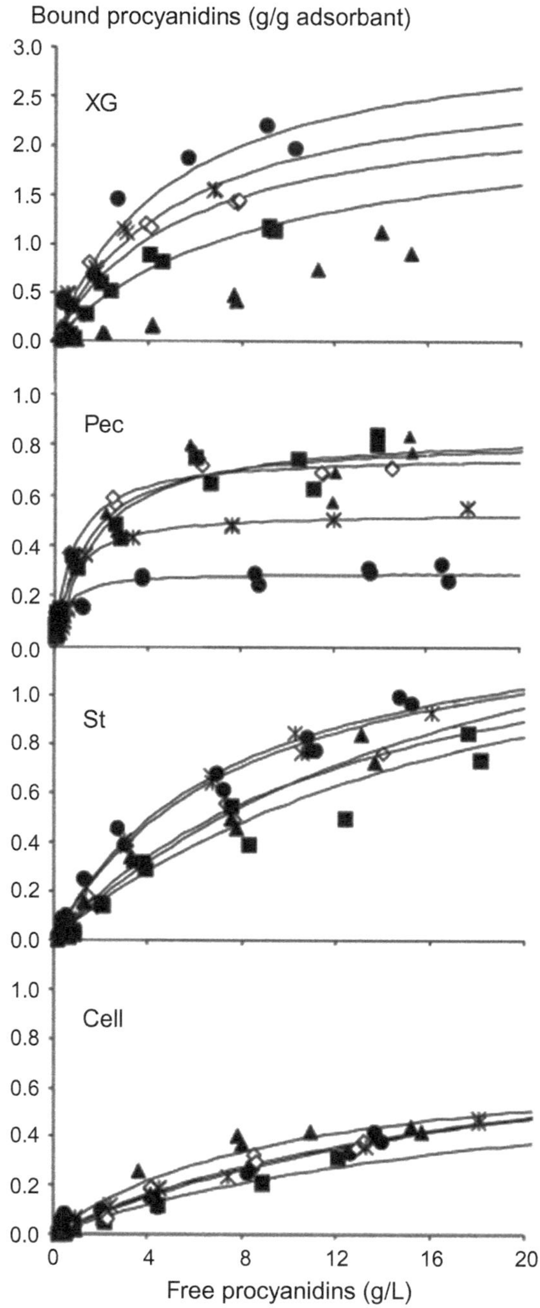

Fig. 4 Binding isotherms for procyanidin's at pH 3.8 and 25°C with cross-linked xyloglucan (XG), cross-linked pectin (Pec), starch (St), and cellulose (Cell), as function of the free procyanidin's concentration. The lines are the corresponding Langmuir adsorption isotherms. ● Adp 70. *Pdp 35. ■ Adp 10. ◇ Gdp 8 gall 22. ▲ Adp 3. Adp 3: purified apple polyphenol fraction of number average degree of polymerization (DPn) = 3; Adp 10: purified apple polyphenol fraction of DPn = 10; Adp 70: purified apple polyphenol fraction of DPn = 70; Pdp 35: purified pear polyphenol fraction of DPn = 35; Gdp 8 gall 22: purified grape seeds polyphenol fraction of DPn = 8 and % gall = 22. (Le Bourvellec & Renard, 2005).

between cell wall material and polyphenols (K_L) and number of binding sites (N_{max}) with the relationship:

$$[PP_b] = \frac{N_{max} \cdot K_L \cdot [PP_f]}{1 + K_L \cdot [PP_f]}$$

Nevertheless, binding isotherms require the physical separation of polysaccharides complexed with phenolic compounds from the rest of the solution, which is not always possible in the case of soluble polysaccharides. In these cases, Isothermal Titration Calorimetry (ITC) has been used to characterize polysaccharide-polyphenol binding events (Fig. 5). An ITC experiment consists in measuring the heat flow to or from the measurement cell maintained at constant temperature where a ligand (in this case, a solution of phenolic compounds) is injected into a solution of receptor

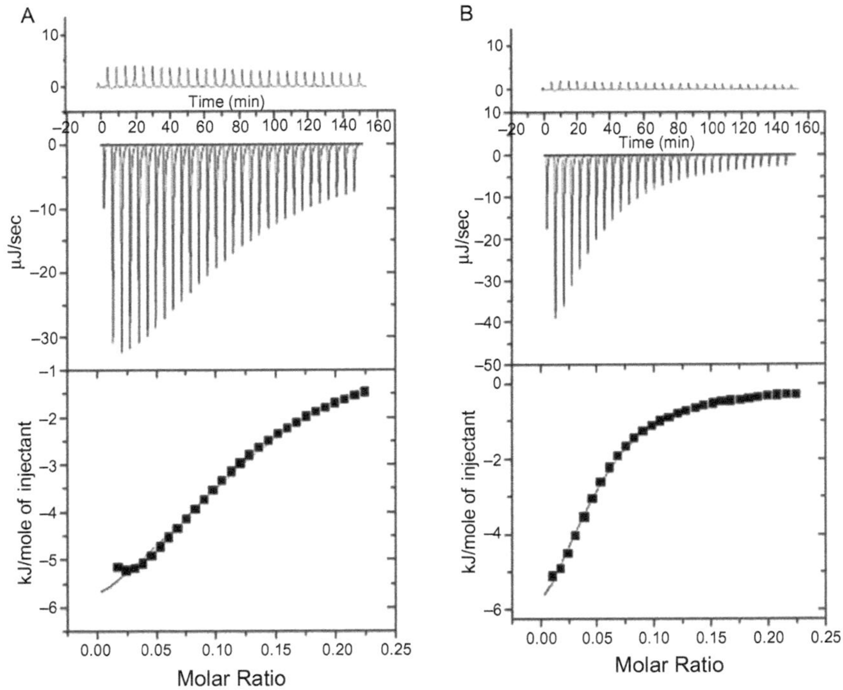

Fig. 5 Thermograms of titration of apple pectin (30 mM galacturonic acid equivalent) by (A) procyanidin's DP9 and (B) procyanidin's DP30 (30 mM (−)-epicatechin equivalent in both cases): (top) Control data obtained with procyanidin's in buffer; (middle) Measurement of heat release during the titration of apple pectins by procyanidin's; (bottom) Molar enthalpy change against a procyanidin's/apple pectin ratio after peak integration. The one-site fit curve is displayed as a thin line. (Watrelot et al., 2013).

(in this case, a solution or suspension of polysaccharide). Heat flow peaks are converted into enthalpy change per mole of ligand during injections, which is in turn plotted against the ligand/receptor molar ratio. This plot can be fitted to a theoretical titration curve in the form of:

$$Q = \frac{\left(1 + [M]_t.n.K_a + K_a.[L]_t\right) - \sqrt{\left(1 + [M]_t.n \times K_a + K_a.[L]_t\right)^2 - 4.[M]_t.n.K_a^2.[L]_t}}{2.\frac{K_a}{V}.\Delta H}$$

where Q is the cumulative heat, $[M]_t$ the total concentration of reactant in the sample cell, $[L]_t$ the total concentration of titrant added, V the volume of the sample cell, ΔH is the enthalpy change, K_a is the binding or association constant and n is the number of binding sites per molecule of receptor. Then, the van't Hoff equation $\Delta G = -RT \times \ln K_a = \Delta H - T\Delta S$ allows the determination of free energy (ΔG) and entropy (ΔS) of reaction/interaction.

ITC provides extensive thermodynamic characterization of the binding events, particularly about the fundamental drivers of these interactions. Indeed, in addition to the apparent affinity and number of binding sites, ΔH and ΔS are the enthalpy and entropy contributions to the interactions, reflective of the respective contributions of hydrogen-bonding and hydrophobic interactions in the observed binding.

Generally, polyphenols have been showed to bind polysaccharides non-selectively, partially reversibly and relatively quickly, in the range of minutes or tens of minutes.

In the case of apples (*Malus pumila*), co-binding of phenolic acids and proanthocyanidins could occur and lead to the formation of organized clusters onto the surface of cell wall materials. Binding was mainly dependent on polyphenol features such as the degree of polymerization and number of terminal hydroxyl groups (both promoting stronger and more extensive binding) (Le Bourvellec et al., 2005, 2007; Le Bourvellec & Renard, 2005; Watrelot et al., 2013). This was confirmed in a 2015 study where Phan et al. characterized the binding of ferulic acid, gallic acid, catechin, cyanidin-3-glucoside and chlorogenic acid to cellulose (Phan et al., 2015). However, binding did not follow the number of peripheral hydroxyl groups of the phenolic compounds. This may suggest that intrinsic factors other than the number of peripheral hydroxyl groups of the phenolic compounds might influence the characteristics of the binding events: for example, native charge of polyphenols, molecular flexibility or exposure of the polysaccharides binding sites.

From the polysaccharides perspective, binding of procyanidin's onto pectin's increases with the degree of methylation of pectin's, but decreased with its number of neutral sugar side groups (Le Bourvellec et al., 2012; Watrelot et al., 2013; Watrelot, Le Bourvellec, Imberty, & Renard, 2014). This suggests that polysaccharides favoring hydrophobic/H-bonding collaborative interactions and exhibiting limited steric hindrance of the binding site by side chains may be more prone to bind polyphenols effectively.

Table 1 also shows that apple procyanidin's affinity and bonding to polysaccharides are dependent on the nature and structure of the polysaccharides. Overall, affinity decreases from cross-linked pectin's, to starch, cross-linked xyloglucans and finally cellulose. These observations give indirect evidence for the impact of the supramolecular structure of the polysaccharides on their binding with phenolic compounds. Whereas pectin's and xyloglucans form gels that readily adsorb and bind phenolic compounds, the supramolecular organization of starch and cellulose based on inter-chain hydrogen bonding provides less binding sites for phenolic compounds, although starch granules allows more binding sites than cellulose due to the presence of pores.

Similarly, anthocyanin's and phenolic acids binding to materials mimicking purple carrot (*Daucus carota*) cell walls showed differences between cellulose alone and cellulose-pectin composite. In both cases, binding occurred in two-stages: an initial stage of direct phenolic-polysaccharide binding (over minutes-hours) followed by a second stage of phenolic stacking of clustering (over days-weeks). However, cellulose alone seemed to bind more phenolic compounds in the initial stage as compared to the cellulose-pectin composite, suggesting again the supramolecular arrangement impact binding events (Padayachee et al., 2012a, 2012b).

Cellulose's supramolecular structure also influenced the interactions. Liu, Martinez-Sanz, et al. (2017) found that cycles of freeze-drying and rehydration affected polyphenol binding. They found a positive correlation between cellulose's water content, its swelling ability and its binding capacity, suggesting again the crucial role of the availability of hydroxyl groups and therefore binding sites in the polysaccharide network.

(Li, Pernell, and Ferruzzi (2018) have provided further evidence of the role of starch granule pores on the absorption and adsorption of phenolic compounds by starch. In this work, starch granules were infused with phenolic acids and showed reduction of their degree of crystallinity and formation of V-type complexes. Retention of phenolic acids was observed to be up to 32 mg/g of starch.

Table 1 Apparent Langmuir parameters for binding isotherms of the different procyanidins/adsorbent combinations.

	Adsorbents											
	Cellulose			Starch			Cross-linked pectin			Cross-linked xyloglucan		
Fractions	K_L (L/g)	N_{max} (g/g)	r^2	K_L (L/g)	N_{max} (g/g)	r^2	K_L (L/g)	N_{max} (g/g)	r^2	K_L (L/g)	N_{max} (g/g)	r^2
Adp 3	0.10 ± 0.07	0.76 ± 0.35	0.96	0.05 ± 0.04	1.91 ± 0.99	0.96	0.58 ± 0.28	0.86 ± 0.12	0.92	0.003	–	–
Adp 10	0.05 ± 0.02	0.77 ± 0.24	0.97	0.05 ± 0.04	1.63 ± 0.85	0.93	0.70 ± 0.33	0.83 ± 0.09	0.94	0.13 ± 0.10	2.17 ± 0.94	0.92
Adp 70	0.04 ± 0.04	1.03 ± 0.73	0.94	0.13 ± 0.04	1.42 ± 0.20	0.98	2.20 ± 0.79	0.29 ± 0.02	0.93	0.20 ± 0.12	3.25 ± 1.04	0.95
Pdp 35	0.05 ± 0.02	0.95 ± 0.17	0.85	0.12 ± 0.02	1.41 ± 0.14	0.99	0.90 ± 0.20	0.57 ± 0.03	0.95	0.19 ± 0.12	2.83 ± 1.04	0.94
Gdp 8 gall 22	0.05 ± 0.03	0.92 ± 0.35	0.98	0.07 ± 0.04	1.50 ± 0.50	0.97	1.22 ± 0.38	0.76 ± 0.05	0.97	0.18 ± 0.13	2.49 ± 0.88	0.96

Adp 3: purified apple polyphenol fraction of number average degree of polymerization (DPn) = 3; Adp 10: purified apple polyphenol fraction of DPn = 10; Adp 70: purified apple polyphenol fraction of DPn = 70; Pdp 35: purified pear polyphenol fraction of DPn = 35; Gdp 8 gall 22: purified grape seeds polyphenol fraction of DPn = 8 and % gall = 22. (Le Bourvellec & Renard, 2005).

Other isolated studies have provided evidence for polyphenol non-covalent binding with polysaccharides in gel or dispersion form.

In a 2011 study, oats (*Avena sativa*) β-glucans showed they could adsorb green tea (*Camelia sinensis*) polyphenols through complexation, up to a level of 135 mg of phenolic compounds per g of β-glucans (Wu, Li, Ming, & Zhao, 2011). Similar observations with epigallocatechin gallate were made by Patel, Seijen-ten-Hoorn, and Velikov (2011). Complexes were formed with colloidal methylcellulose and characterized by ITC. Affinity was high ($K_a \geq 10^4 M^{-1}$) and binding stoichiometry was 21 mol of epigallocatechin gallate (EGCG) binding to 1 mol of a 30,000 g/mol methylcellulose, equivalent of 321 mg of bound EGCG per g of methylcellulose. This interesting work indirectly showed that cellulosic structures can bind effectively phenolic compounds when polymer-polymer interactions are limited, here due to peripheral methyl groups absent in cellulose. Binding stoichiometry was also shown to be high as compared to small phenolic acids, suggesting again that phenolic compounds with a higher number of H-bonding groups have a greater ability to bind polysaccharides.

Mixed glucans from corn silk also bound flavonoids (luteolin, apigenin and formononetin) (Guo, Ma, Xue, Gao, & Chen, 2018), in the range of 20–30 mg of flavonoids per g of glucans. Despite the size and number of hydroxyl groups of these flavonoids, this is similar to the levels observed by Li et al. (2018) with starch and phenolic acids. In this case, binding might be limited by the poor solubility of the flavonoids studied. Nevertheless, as with all the other examples cited here, binding was dominated by hydrogen bonding ($\Delta H = -3.6$ kJ/mol $\gg \Delta S = -22$ J/mol K).

4.2 Influence of chemical environment on polysaccharides-phenolic compounds interactions

A number of weak non-covalent interactions drive polysaccharide-polyphenol binding and make it strongly dependent on the molecular features of the polyphenols. Although Van der Waals, ionic and hydrophobic interactions play a role, these interactions are mainly driven by hydrogen-bonds. Therefore, the chemical environment does affect binding events.

In apples (*Malus pumila*), binding has been shown to be independent of ionic strength under normal food conditions, strongly pH-dependent for phenolic acids and pH-independent for higher molecular weight polyphenols (Le Bourvellec et al., 2004). Similar observations were made on polyphenol-cellulose interactions where pH was determined to be the most influential factor and ionic strength to play a negligible role

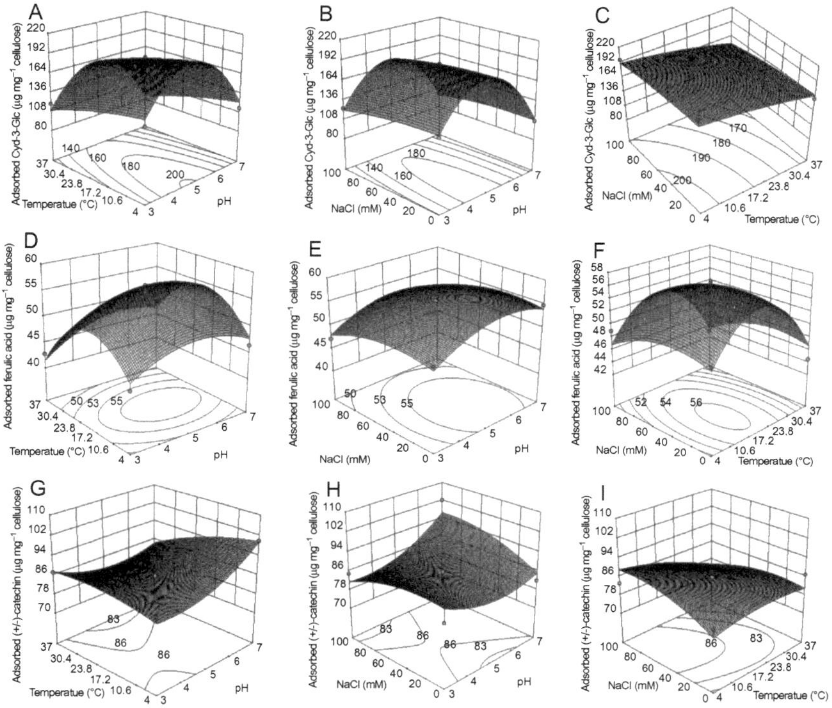

Fig. 6 Contour-surface plots of the combined effects of binding factors on the adsorption of Cya-3-glc (A, B, C), ferulic acid (D, E, F) and (+/−)-catechin (G, H, I) onto cellulose (Phan et al., 2016). All plots show adsorption in μg mg^{-1} of cellulose, as a function of temperature and pH (plots A, D, and G), as function of NaCl concentration and pH (plots B, E, and H), and as a function of NaCl concentration and temperature (plots C, F, and I).

(Phan et al., 2016), thus in both cases, temperature also played a role. Surprisingly, temperature did not seem to play a role only on the kinetics of binding, but also on the extent of binding. For example, Fig. 6 shows that binding of ferulic acid and cyanidin-3-glucoside (Cya-3-glc) exhibit a maximum around 15–20 °C. At elevated temperatures, this is consistent with hydrogen-bonding playing a major role in these interactions: molecular thermal agitation may prevent binding. However, it is difficult to explain the phenomenon at lower temperatures: reduced molecular mobility and flexibility of the polymers may play a role in less extensive binding.

In Wu et al.'s study on the adsorption of mixed green tea (*Camelia sinensis*) polyphenols onto oats (*Avena sativa*) β-glucans (Wu et al., 2011), the use of a Response Surface Methodology to optimize binding gave further evidence about the role of H-bonding in complexation through the role of pH and ionic strength into binding capacity, with pH tending toward 7 and increased ionic buffer concentrations were not favorable to binding.

5. Impact of polysaccharides-polyphenols interactions on the functionality of polysaccharides

5.1 Impact on polysaccharides health and nutritional properties

Numerous studies have looked into the impact of phenolic compounds on the nutritional properties of food and starch in particular. However, in most of these studies, the well-documented inhibition of digestive enzymes activity by phenolic compounds (Cao & Chen, 2012) is an important confounding factor making difficult to obtain evidence of the direct effect of polyphenol-polysaccharide complexation. Here, we will focus exclusively on how polyphenols can alter directly the functionality of polysaccharides.

5.1.1 Impact on colonic fermentation and prebiotic effect

A small fraction of polyphenols is absorbed in the upper intestinal tract, while as much as 90% of the polyphenolic molecules is bio-transformed in the colon (Etxeberria et al., 2013). Many recent studies have shed light on the impact of polyphenolic compounds on the composition and metabolism of gut microbiota (Mayta-Apaza et al., 2018; Mills et al., 2015; Tuohy, Conterno, Gasperotti, & Viola, 2012). For instance, polyphenols rich chocolate, coffee, and some fruits have been shown to increase fecal *Bifidobacteria* (Mills et al., 2015; Tuohy et al., 2012). Similarly, apple consumption, that is rich in polyphenols and fibers, has increased the relative abundance of Actinobacteria, *Bifidobacteria*, and SCFAs, while decreased the relative abundance of Bacteroidetes (Koutsos et al., 2017). The increased abundance of *Bifidobacterium* along with the enrichment of *Bacteroides* has been linked recently to tart cherry consumption, which is rich in anthocyanin's, flavonoids, chlorogenic, and neochlorogenic acids (Mayta-Apaza et al., 2018).

The changes in gut microbiota composition due to polyphenols consumption may influence the polysaccharides degradation pathways and energy metabolism (Xue et al., 2016). For instance, in vitro fermentation of three plant polyphenols (catechin, quercetin, and puerarin) by fecal microbiota have been shown to reduce the growth of Firmicutes and Bacteroidetes and decreased energy metabolism; however, the degradation ability of FOS by the fecal microorganisms was maintained (Xue et al., 2016).

Different studies have postulated that SCFAs production is enhanced upon administration of dietary fibers along with phenolic compounds. Thus, this hypothesis remains controversial. Four weeks of rats fed with a combination of cellulose and ellagitannin-rich raspberry and strawberry seeds increased their gut production of SCFAs and the proportion of isobutyric acid (Kosmala et al., 2015). Similarly, a combination of apple-pomace extract and FOS or cellulose has resulted in greater SCFAs production compared to rats consumption of fibers alone (Juskiewicz, Milala, Jurgonski, Krol, & Zdunczyk, 2011). Nevertheless, some other studies have not shown this synergism in SCFAs generation (Kosmala, Kolodziejczyk, Zdunczyk, Juskiewicz, & Boros, 2011; Wallace et al., 2015; Zdunczyk, Juskiewicz, & Estrella, 2006). The conjugation between polyphenolic compounds and polysaccharides reduces the bioavailability of both components in the small intestine leading to an increase in their availability for colonic fermentation by gut microbiota. This has been confirmed by studying the binding between cellulose-pectin composites and phenolic acids including caffeic, chlorogenic and ferulic acids (Padayachee et al., 2012b).

5.1.2 Impact on metabolic health

Diet-induced modulation of intestinal health and chronic disease may be mediated by mechanisms involving an interplay between NDP and phenolic compounds. Food-derived polyphenols, such as lignans, flavonoids and phenolic acids (e.g., hydroxycinnamic acids such as caffeic, ferulic or *p*-coumaric acids, and hydroxybenzoic acids such as gallic acid or ellagic acid)), and their microbial-derived secondary metabolites (e.g., enterolignans) can modulate a diverse range of biological processes including the microbial community structure (Duenas et al., 2015), enhancing intestinal barrier integrity (Suzuki & Hara, 2011; Ulluwishewa et al., 2011), attenuating oxidative stress (Giusti et al., 2017; Palla, Iqbal, Minhas, & Gilani, 2016; Sung & Park, 2013; Xu, Yuan, & Chang, 2007; Yao et al., 2010), modulating AhR signaling (Tokunaga, Woodin, & Stegeman, 2016; Zhang, Qin, & Safe, 2003), and reducing inflammation (Palla et al., 2016). Importantly, many phenolic compounds (70–95%) are covalently bound to cell wall NSPs and largely unavailable in upper GI until released by microbial-derived enzymes (e.g., esterases and hydroxylases) in the large intestines (Kroon, Faulds, Ryden, Robertson, & Williamson, 1997; Vitaglione et al., 2015). The subsequent breakdown of these polyphenols into low molecular weight phenolic acids enhances their absorption, bioavailability and bioactivity, contributing to their direct beneficial effects (Koppel, Maini Rekdal, & Balskus, 2017).

Therefore, two functional components of plant foods are consumed together in the food matrix and rarely in isolation; therefore, it is important to understand their combined effects and possible interactions.

Studies demonstrate that certain NDP and their microbial-derived SCFAs, can modulate the bioavailability and metabolism of phenolic compounds. For example, rodents fed diets supplemented with 0.5% (wt/wt) rutin, a flavonol glycoside enriched in foods such as asparagus, had altered microbial metabolism and bioavailability of its aglycone, quercetin, depending on the type of NSP in the diet (Tamura, Nakagawa, Tsushida, Hirayama, & Itoh, 2007). When consumed with pectin, more quercetin was produced and bioavailable, compared to when consumed with cellulose, which was attributed to the ability of pectin to enhance microbial activity and metabolism of rutin, thereby enhancing intestinal quercetin concentrations. Similar findings were observed in mice fed asparagus- (NDP- and rutin-rich) supplemented diet or diet containing purified rutin and cellulose, wherein more quercetin was produced and bioavailable in the asparagus-fed mice (Power et al., 2016). Improved health outcomes have also been demonstrated with combined NDPs and phenolic compounds. In one study, mice were fed a high-fat diet (HFD) supplemented with inulin, isoquercetin, or inulin and isoquercetin over a 12-week period (Tan et al., 2018). Mice fed the combination of NDP and flavonoids had attenuated weight gain, improved glucose tolerance, reduced hepatic lipid content, reduced serum acetate concentrations, and reduced adipocyte hypertrophy, compared to inulin or isoquercetin alone. These effects were thought to be mediated in part through enhanced metabolism and bioavailability of isoquercetin when combined with inulin, however future studies are required to confirm this hypothesis.

Mechanistically, the interaction between NDPs and phenolic compounds may occur through the interplay between their microbial metabolite. For example, butyrate and propionate can enhance colonocyte uptake of the phenolic acid, ferulic acid, and flavonoid, hesperetin, and enhance their phase II metabolism potentially through upregulation of apical MCT-1 expression and enhancing the activity of glucuronidase and sulfatase enzymes (Van Rymenant et al., 2017). Furthermore, butyrate has been shown to modulate the AhR-activation potential of flavonoids, in particular quercetin in Caco-2 cells, which may have downstream impact on inflammation and inflammation-associated diseases (Jin et al., 2018). On the other hand, phenolic compounds have been shown to differently modify the production of SCFA from NDP, with some studies showing enhancement, while other showing inhibitory effects on SCFA production (Fotschki et al., 2014, 2016; Mosele, Macia, & Motilva, 2015), which may be in part mediated through the anti-microbial effects of some phenolic compounds (Jurgonski et al., 2017; Puupponen-Pimia et al., 2001). Overall, although NDPs and phenolic compounds are naturally present in plant foods, mechanistic evidence demonstrating their interactions on human health and chronic diseases are limited, and may differ depending on the phenolic compound composition and NDP structure.

5.2 Impact on polysaccharides functionality in food technology applications

5.2.1.1 Impact of texture and stability of food systems

Among the effects of polyphenol-polysaccharide interactions on polysaccharides, the impact on the polysaccharides physical properties may be the least documented. However, several direct or indirect observation can suggest these effects may be important in food systems.

Nitta, Fang, Takemasa, and Nishinari (2004), Nishinari, Kim, Fang, Nitta, and Takemasa (2006) observed that the gelling behavior of tamarind seed xyloglucan were affected by the addition of epigallocateching gallate (EGCG), in particular, the storage modulus of tamarind seed xyloglucan exhibited a reversed U-shape relationship with the concentration of EGCG (Fig. 7), suggesting that EGCG may assist the formation of xyloglucan gel before turning into a plasticizer at high concentration.

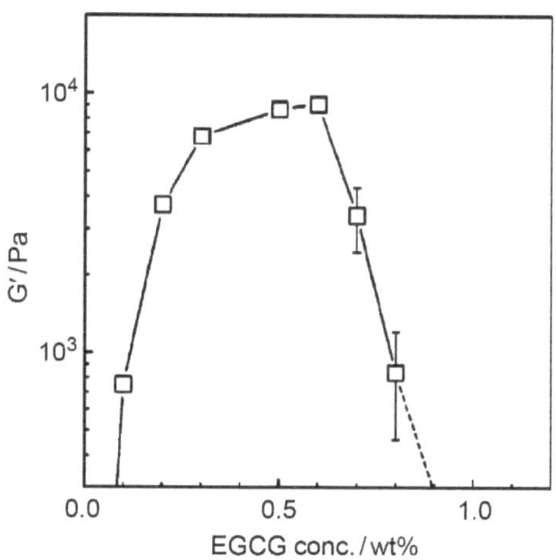

Fig. 7 G′ of 1 wt% tamarind seed xyloglucan gels as a function of epigallocatechin gallate concentration at 10 °C and 1 rad/s. The error bars represent standard deviations (Nitta, Fang, Takemasa, & Nishinari, 2004).

Table 2 Viscoelastic behavior of CMC film forming solutions with and without phenolic extract of murta leaves at 25 °C (Silva-Weiss et al., 2014).

	Crossover point		
	G′ = G″ (Pa)	**ω (rad/s)**	**η* (Pa.s)**
CMC	38	19	2.80
CMC + PP	53	86	0.87

Conversely, phenolic extracts from murta (*Myrtus* L.) leaves seemed to decrease the viscosity of polysaccharide-based film forming solutions (Silva-Weiss, Bifani, Ihl, Sobral, & Gómez-Guillén, 2014). In particular, at the crossover point in dynamic rheological testing (the frequency at which storage and loss moduli are equal), the phenolic extract decreased the apparent viscosity as well as the storage and loss moduli of the solution, and increased its gelling time (Table 2).

A similar observation was made on cassava-soy composite porridge in presence of phenolic-rich grape pomace. Again, the addition of a phenolic-rich materials contributed to a significant decrease of the viscosity of the extruded composite porridge (Oladiran & Emmambux, 2018). Interestingly, in these two latter studies, the impact on the physical properties of polysaccharides upon addition of phenolics seemed to be a minor observation point not related to polyphenol-polysaccharide interactions. Nevertheless, they support the idea that these interactions not only modify the function of polyphenols as observed before but also the properties of the polysaccharides themselves.

This consequence on polysaccharides function was directly observed in the interaction between methylcellulose and tannic acid (Patel, Seijen Ten-Hoorn, Hazekamp, Blijdenstein, & Velikov, 2013). After characterization of the non-covalent interaction by ITC, it was determined that tannic acid reduces the gelling temperature and enhances the foam and emulsion stabilizing properties of methylcellulose.

Finally, phenolic acid could also demonstrated ability to affect polysaccharides physical properties in the case of starch (Gilley et al., 2017). Caffeic acid and ferulic acid (but not gallic acid) association with potato starch significantly decreased starch peak viscosity, hot paste viscosity and cool paste viscosity through Rapid ViscoAnalyzer analysis. In the same study, amylopectin showed no changes in these pasting parameters with any of the phenolic acids. This would suggest that amylopectin is less prone to binding than amylose (in potato starch), probably due to steric hindrance created by amylopectin's dense branching, as already mentioned in Section 4.

5.2.1.2 Impact on probiotics and bioactive compounds encapsulation

The interactions between polyphenols, such as anthocyanin's, phenolic acids, and procyanidin's, and plant cell wall polysaccharides, like cellulose and pectin's have been employed to improve the microencapsulation efficiency of polysaccharides (Kardum & Glibetic, 2018). However, more studies are required to test the impact of polyphenols on the structural stability and delivery characteristics of polysaccharide-encapsulated bioactive compounds. Plant-rich polyphenols can act as cross-linkers that strengthen microencapsulation matrices and in the meantime as antioxidants that increase the stability of encapsulated bioactive materials. Sage polyphenols strengthened the shell matrix, improved encapsulation efficiency, reduced the oil oxidation and controlled the release of fish oil from gum arabic microencapsulates (Binsi et al., 2017). In addition, quercetin and green tea flavan-3-ols have enhanced the gelatinization of the starch (Wu, Lin, Chen, & Xiao, 2011). The polysaccharide-polyphenols interaction may enhance the encapsulation efficiency of polyphenols themselves. For instance, green tea polyphenols encapsulated in maltodextrin have exhibited a superior efficiency in the prevention of cardiovascular diseases as compared to non-capsulated extracts (Jung, Seong, Kim, Myong, & Chang, 2013).

6. Perspectives and conclusions

Whereas non-extractable phenolic compounds have been seen as co-passengers of dietary fibers for a long time, it is apparent that extractable polyphenols can be as well, whether it is due to intentional formulation or natural co-occurrence, and that they are able to change the polysaccharides properties. Polyphenols and polysaccharides are indeed very commonly present in the same matrix, but there is an important knowledge gap about their interactions and the consequences of these interactions on the polysaccharides themselves.

Firstly, green tea catechins, simple flavan-3-ols polymers and phenolic acids are often used as models, but data on the interactions of polysaccharides with other forms of phenolic compounds such as tannins, stilbenoids, and other flavonoids are scarce or completely inexistent although they are relevant in food systems or in gut health applications.

Secondly, some observations suggest strongly that polyphenols can alter the functions of polysaccharides through their molecular interactions. It is the case of encapsulation of polyphenolic extracts with polysaccharides, for the utilization of polysaccharides by the colonic microbiota in presence of polyphenols or for the film forming properties of polysaccharides in association with phenolic compounds. However, this review shows that very few studies hypothesize that polyphenol-polysaccharides interactions may actually be an important factor for the modified properties of the polysaccharides.

Therefore, more work is needed to fully understand the non-covalent interactions between phenolic compounds and polysaccharides, which could be used to tailor the functionality of the polysaccharides, whether it is for food technology applications or health benefits related to dietary fibers. This research has the potential to have impactful consequences as most health properties and products' health claims related to dietary fibers are based on their physical properties.

References

Ahmad, M., Mudgil, P., Gani, A., Hamed, F., Masoodi, F. A., & Maqsood, S. (2019). Nano-encapsulation of catechin in starch nanoparticles: Characterization, release behavior and bioactivity retention during simulated in-vitro digestion. *Food Chemistry*, *270*, 95–104.

Andersen, Ø. M., & Markham, K. R. (2006). *Flavonoids: Chemistry, biochemistry, and applications*. Boca Raton, FL: CRC, Taylor & Francis.

Apolinário, A. C., De Lima Damasceno, B. P. G., De Macêdo Beltrão, N. E., Pessoa, A., Converti, A., & Da Silva, J. A. (2014). Inulin-type fructans: A review on different aspects of biochemical and pharmaceutical technology. *Carbohydrate Polymers*, *101*(1), 368–378.

Atia, A., Gomma, A. I., Fliss, I., Beyssac, E., Garrait, G., & Subirade, M. (2017). Molecular and biopharmaceutical investigation of alginate-inulin synbiotic coencapsulation of probiotic to target the colon. *Journal of Microencapsulation*, *34*(2), 171–184.

Bassaganya-Riera, J., DiGuardo, M., Viladomiu, M., de Horna, A., Sanchez, S., Einerhand, A. W., et al. (2011). Soluble fibers and resistant starch ameliorate disease activity in interleukin-10-deficient mice with inflammatory bowel disease. *The Journal of Nutrition*, *141*(7), 1318–1325.

Bautista-Ortín, A., Ben Abdallah, R., Castro-López, L., Jiménez-Martínez, M., & Gómez-Plaza, E. (2016). Technological implications of modifying the extent of cell wall-proanthocyanidin interactions using enzymes. *International Journal of Molecular Sciences*, *17*(1), 123.

Belščak-Cvitanović, A., Jurić, S., Đorđević, V., Barišić, L., Komes, D., Ježek, D., et al. (2017). Chemometric evaluation of binary mixtures of alginate and polysaccharide biopolymers as carriers for microencapsulation of green tea polyphenols. *International Journal of Food Properties*, *20*(9), 1971–1986.

BeMiller, J. N. (2007). *Carbohydrate chemistry for food scientists* (2nd ed.). St. Paul, Minn: AACC International.

Bepeyeva, A., de Barros, J. M. S., Albadran, H., Kakimov, A. K., Kakimova, Z. K., Charalampopoulos, D., et al. (2017). Encapsulation of lactobacillus casei into calcium pectinate-chitosan beads for enteric delivery. *Journal of Food Science*, *82*(12), 2954–2959.

Bindon, K. A., Bacic, A., & Kennedy, J. A. (2012). Tissue-specific and developmental modifications of grape cell walls influence the adsorption of proanthocyanidins. *Journal of Agricultural and Food Chemistry*, *60*(36), 9249–9260.

Bindon, K. A., Carew, A. L., Mierczynska-Vasilev, A., Kassara, S., Kerslake, F., & Smith, P. A. (2016). Characterization of macromolecular complexes in red wine: Composition, molecular mass distribution and particle size. *Food Chemistry*, *199*, 838–846.

Bindon, K. A., & Kennedy, J. A. (2011). Ripening-induced changes in grape skin proanthocyanidins modify their interaction with cell walls. *Journal of Agricultural and Food Chemistry*, *59*(6), 2696–2707.

Bindon, K. A., Smith, P. A., Holt, H., & Kennedy, J. A. (2010). Interaction between grape-derived proanthocyanidins and cell wall material. 2. Implications for vinification. *Journal of Agricultural and Food Chemistry*, *58*(19), 10736–10746.

Bindon, K. A., Smith, P. A., & Kennedy, J. A. (2010). Interaction between grape-derived proanthocyanidins and cell wall material. 1. Effect on proanthocyanidin composition and molecular mass. *Journal of Agricultural and Food Chemistry*, *58*(4), 2520–2528.

Binsi, P. K., Nayak, N., Sarkar, P. C., Jeyakumari, A., Muhamed Ashraf, P., Ninan, G., et al. (2017). Structural and oxidative stabilization of spray dried fish oil microencapsulates with gum arabic and sage polyphenols: Characterization and release kinetics. *Food Chemistry*, *219*, 158–168.

Brummer, Y., Kaviani, M., & Tosh, S. M. (2015). Structural and functional characteristics of dietary fibre in beans, lentils, peas and chickpeas. *Food Research International*, *67*, 117–125.

Burger-van Paassen, N., Vincent, A., Puiman, P. J., van der Sluis, M., Bouma, J., Boehm, G., et al. (2009). The regulation of intestinal mucin MUC2 expression by short-chain fatty acids: Implications for epithelial protection. *Biochemistry Journal*, *420*(2), 211–219.

Busse-Valverde, N., Bautista-Ortín, A. B., Gómez-Plaza, E., Fernández-Fernández, J. I., & Gil-Muñoz, R. (2012). Influence of skin maceration time on the proanthocyanidin content of red wines. *European Food Research and Technology*, *235*(6), 1117–1123.

Busse-Valverde, N., Gómez-Plaza, E., López-Roca, J. M., Gil-Muñoz, R., Fernández-Fernández, J. I., & Bautista-Ortín, A. B. (2010). Effect of different enological practices on skin and seed proanthocyanidins in three varietal wines. *Journal of Agricultural and Food Chemistry*, *58*(21), 11333–11339.

Butt, M. S., Shahzadi, N., Sharif, M. K., & Nasir, M. (2007). Guar gum: A miracle therapy for hypercholesterolemia, hyperglycemia and obesity. *Critical Reviews in Food Science and Nutrition*, *47*(4), 389–396.

Cani, P. D., Amar, J., Iglesias, M. A., Poggi, M., Knauf, C., Bastelica, D., et al. (2007). Metabolic endotoxemia initiates obesity and insulin resistance. *Diabetes*, *56*(7), 1761–1772.

Cani, P. D., & Delzenne, N. M. (2009). Interplay between obesity and associated metabolic disorders: New insights into the gut microbiota. [review]. *Current Opinion in Pharmacology*, *9*(6), 737–743.

Cani, P. D., Possemiers, S., Van de Wiele, T., Guiot, Y., Everard, A., Rottier, O., et al. (2009). Changes in gut microbiota control inflammation in obese mice through a mechanism involving GLP-2-driven improvement of gut permeability. *Gut*, *58*(8), 1091–1103.

Cao, H., & Chen, X. (2012). Structures required of flavonoids for inhibiting digestive enzymes. *Anti-Cancer Agents in Medicinal Chemistry*, *12*(8), 929–939.

Casellas, F., Borruel, N., Torrejon, A., Varela, E., Antolin, M., Guarner, F., et al. (2007). Oral oligofructose-enriched inulin supplementation in acute ulcerative colitis is well tolerated and associated with lowered faecal calprotectin. *Alimentary Pharmacology & Therapeutics*, *25*(9), 1061–1067.

Castro-López, L. D. R., Gómez-Plaza, E., Ortega-Regules, A., Lozada, D., & Bautista-Ortín, A. B. (2016). Role of cell wall deconstructing enzymes in the proanthocyanidin–cell wall adsorption–desorption phenomena. *Food Chemistry*, *196*, 526–532.

Cerpa-Calderón, F. K., & Kennedy, J. A. (2008). Berry integrity and extraction of skin and seed proanthocyanidins during red wine fermentation. *Journal of Agricultural and Food Chemistry*, *56*(19), 9006–9014.

Chassard, C., Goumy, V., Leclerc, M., Del'homme, C., & Bernalier-Donadille, A. (2007). Characterization of the xylan-degrading microbial community from human faeces. *FEMS Microbiological Ecology*, *61*(1), 121–131.

Chitprasert, P., Sudsai, P., & Rodklongtan, A. (2012). Aluminum carboxymethyl cellulose-rice bran microcapsules: Enhancing survival of lactobacillus reuteri KUB-AC5. *Carbohydrate Polymers*, *90*(1), 78–86.

Cummings, J. H., Pomare, E. W., Branch, W. J., Naylor, C. P., & Macfarlane, G. T. (1987). Short chain fatty acids in human large intestine, portal, hepatic and venous blood. *Gut*, *28*(10), 1221–1227.

Daou, C., & Zhang, H. (2012). Oat Beta-glucan: Its role in health promotion and prevention of diseases. *Comprehensive Reviews in Food Science and Food Safety*, *11*(4), 355–365.

Dar, Y. L., & Light, J. M. (2014). *Food texture design and optimization*. Oxford, UK: Wiley-Blackwell.

Delzenne, N. M., & Cani, P. D. (2011). Interaction between obesity and the gut microbiota: Relevance in nutrition. [review]. *Annual Review of Nutrition*, *31*, 15–31.

den Besten, G., van Eunen, K., Groen, A. K., Venema, K., Reijngoud, D. J., & Bakker, B. M. (2013). The role of short-chain fatty acids in the interplay between diet, gut microbiota, and host energy metabolism. *Journal of Lipid Research*, *54*(9), 2325–2340.

Dewulf, E. M., Cani, P. D., Claus, S. P., Fuentes, S., Puylaert, P. G., Neyrinck, A. M., et al. (2013). Insight into the prebiotic concept: Lessons from an exploratory, double blind intervention study with inulin-type fructans in obese women. *Gut*, *62*(8), 1112–1121.

Ducasse, M.-A., Canal-Llauberes, R.-M., de Lumley, M., Williams, P., Souquet, J.-M., Fulcrand, H., et al. (2010). Effect of macerating enzyme treatment on the polyphenol and polysaccharide composition of red wines. *Food Chemistry*, *118*(2), 369–376.

Duenas, M., Munoz-Gonzalez, I., Cueva, C., Jimenez-Giron, A., Sanchez-Patan, F., Santos-Buelga, C., et al. (2015). A survey of modulation of gut microbiota by dietary polyphenols. *Biomedical Research International*, *2015*, 850902.

Dvorakova, M., & Landa, P. (2017). Anti-inflammatory activity of natural stilbenoids: A review. *Pharmacological Research*, *124*, 126–145.

Eastwood, M. A., Robertson, J. A., Brydon, W. G., & MacDonald, D. (1983). Measurement of water-holding properties of fibre and their faecal bulking ability in man. *British Journal of Nutrition*, *50*(3), 539–547.

Edwards, C. A., Wilson, R. G., Hanlon, L., & Eastwood, M. A. (1992). Effect of the dietary fibre content of lifelong diet on colonic cellular proliferation in the rat. *Gut*, *33*(8), 1076–1079.

Englyst, H. N., Kingman, S. M., & Cummings, J. H. (1992). Classification and measurement of nutritionally important starch fractions. *European Journal of Clinical Nutrition*, *46*(Suppl. 2), S33–S50.

Eswaran, S., Muir, J., & Chey, W. D. (2013). Fiber and functional gastrointestinal disorders. *American Journal of Gastroenterology*, *108*(5), 718–727.

Etxeberria, U., Fernandez-Quintela, A., Milagro, F. I., Aguirre, L., Martinez, J. A., & Portillo, M. P. (2013). Impact of polyphenols and polyphenol-rich dietary sources on gut microbiota composition. *Journal of Agricultural and Food Chemistry*, *61*(40), 9517–9533.

Everard, A., Lazarevic, V., Derrien, M., Girard, M., Muccioli, G. G., Neyrinck, A. M., et al. (2011). Responses of gut microbiota and glucose and lipid metabolism to prebiotics in genetic obese and diet-induced leptin-resistant mice. *Diabetes*, *60*(11), 2775–2786.

Feng, Y., Wang, Y., Wang, P., Huang, Y., & Wang, F. (2018). Short-chain fatty acids manifest stimulative and protective effects on intestinal barrier function through the inhibition of NLRP3 inflammasome and autophagy. *Cellular Physiology and Biochemistry*, *49*(1), 190–205.

Food and Agriculture Organization of the United Nations, World Health Organization, Joint FAO/WHO Expert Consultation on Evaluation of Health and Nutritional Properties of Probiotics in Food including Powder Milk with Live Lactic Acid Bacteria, Joint FAO/WHO Working Group on Drafting Guidelines for the Evaluation of Probiotics in Food, Probiotics in food : health and nutritional properties and guidelines for evaluation: Report of a Joint FAO/WHO Expert Consultation on Evaluation of Health and Nutritional Properties of Probiotics in Food including Powder Milk with Live Lactic Acid Bacteria, Cordoba, Argentina, 1–4 October 2001 [and] Report of a Joint FAO/WHO Working Group on Drafting Guidelines for the Evaluation of Probiotics in Food, London, Ontario, Canada, 30 April-1 May 2002, 2006, Food and Agriculture Organization of the United Nations, World Health Organization, Rome [Italy].

Fotschki, B., Juskiewicz, J., Jurgonski, A., Kolodziejczyk, K., Milala, J., Kosmala, M., et al. (2016). Anthocyanins in strawberry polyphenolic extract enhance the beneficial effects of diets with fructooligosaccharides in the rat cecal environment. *PLoS One*, *11*(2), e0149081.

Fotschki, B., Milala, J., Jurgonski, A., Karlinska, E., Zdunczyk, Z., & Juskiewicz, J. (2014). Strawberry ellagitannins thwarted the positive effects of dietary fructooligosaccharides in rat cecum. *Journal of Agricultural and Food Chemistry*, *62*(25), 5871–5880.

Fournand, D., Vicens, A., Sidhoum, L., Souquet, J.-M., Moutounet, M., & Cheynier, V. (2006). Accumulation and extractability of grape skin tannins and anthocyanins at different advanced physiological stages. *Journal of Agricultural and Food Chemistry*, *54*(19), 7331–7338.

Frost, G., Sleeth, M. L., Sahuri-Arisoylu, M., Lizarbe, B., Cerdan, S., Brody, L., et al. (2014). The short-chain fatty acid acetate reduces appetite via a central homeostatic mechanism. *Nature Communications*, *5*, 3611.

Gani, A., Shah, A., Ahmad, M., Ashwar, B. A., & Masoodi, F. A. (2018). Beta-d-glucan as an enteric delivery vehicle for probiotics. *International Journal of Biological Macromolecules*, *106*, 864–869.

Gibson, G. R., Probert, H. M., Loo, J. V., Rastall, R. A., & Roberfroid, M. B. (2004). Dietary modulation of the human colonic microbiota: Updating the concept of prebiotics. *Nutrition Research Reviews*, *17*(2), 259–275.

Gibson, G. R., & Roberfroid, M. B. (1995). Dietary modulation of the human colonic microbiota: Introducing the concept of prebiotics. *Journal of Nutrition*, *125*(6), 1401–1412.

Gilley, A. D., Arca, H. C., Nichols, B. L. B., Mosquera-Giraldo, L. I., Taylor, L. S., Edgar, K. J., et al. (2017). Novel cellulose-based amorphous solid dispersions enhance quercetin solution concentrations in vitro. *Carbohydrate Polymers*, *157*, 86–93.

Giusti, F., Caprioli, G., Ricciutelli, M., Vittori, S., & Sagratini, G. (2017). Determination of fourteen polyphenols in pulses by high performance liquid chromatography-diode array detection (HPLC-DAD) and correlation study with antioxidant activity and colour. *Food Chemistry*, *221*, 689–697.

Gomaa, A. I., Martinent, C., Hammami, R., Fliss, I., & Subirade, M. (2017). Dual coating of liposomes as encapsulating matrix of antimicrobial peptides: Development and characterization. *Frontiers in Chemistry*, *5*, 103.

Gómez-Mascaraque, G. L., Dhital, S., López-Rubio, A., & Gidley, M. J. (2017). Dietary polyphenols bind to potato cells and cellular components. *Journal of Functional Foods*, *37*, 283–292.

Gonçalves, F. J., Rocha, S. M., & Coimbra, M. A. (2012). Study of the retention capacity of anthocyanins by wine polymeric material. *Food Chemistry*, *134*(2), 957–963.

Guo, Q., Ma, Q., Xue, Z., Gao, X., & Chen, H. (2018). Studies on the binding characteristics of three polysaccharides with different molecular weight and flavonoids from corn silk (Maydis stigma). *Carbohydrate Polymers*, *198*, 581–588.

Hamer, H. M., Jonkers, D., Venema, K., Vanhoutvin, S., Troost, F. J., & Brummer, R. J. (2008). Review article: The role of butyrate on colonic function. *Alimentary Pharmacology & Therapeutics*, *27*(2), 104–119.

Hannan, J. M., Ali, L., Rokeya, B., Khaleque, J., Akhter, M., Flatt, P. R., et al. (2007). Soluble dietary fibre fraction of Trigonella foenum-graecum (fenugreek) seed improves glucose homeostasis in animal models of type 1 and type 2 diabetes by delaying carbohydrate digestion and absorption, and enhancing insulin action. *British Journal of Nutrition*, *97*(3), 514–521.

Holt, S., Heading, R. C., Carter, D. C., Prescott, L. F., & Tothill, P. (1979). Effect of gel fibre on gastric emptying and absorption of glucose and paracetamol. *Lancet*, *1*(8117), 636–639.

Hotamisligil, G. S. (2006). Inflammation and metabolic disorders. *Nature*, *444*(7121), 860–867.

Hu, B., Liu, X., Zhang, C., & Zeng, X. (2017). Food macromolecule based nanodelivery systems for enhancing the bioavailability of polyphenols. *Journal of Food and Drug Analysis*, *25*(1), 3–15.

Hutkins, R. W., Krumbeck, J. A., Bindels, L. B., Cani, P. D., Fahey, G., Jr., Goh, Y. J., et al. (2016). Prebiotics: Why definitions matter. *Current Opinion in Biotechnology*, *37*, 1–7.

Inan, M. S., Rasoulpour, R. J., Yin, L., Hubbard, A. K., Rosenberg, D. W., & Giardina, C. (2000). The luminal short-chain fatty acid butyrate modulates NF-kappaB activity in a human colonic epithelial cell line. *Gastroenterology*, *118*(4), 724–734.

Iqbal, R., Zahoor, T., Huma, N., Jamil, A., & Unlu, G. (2018). In-vitro GIT tolerance of microencapsulated Bifidobacterium bifidum ATCC 35914 using polysaccharide-protein matrix. *Probiotics and Antimicrobial Proteins*.

Ito, H., Satsukawa, M., Arai, E., Sugiyama, K., Sonoyama, K., Kiriyama, S., et al. (2009). Soluble fiber viscosity affects both goblet cell number and small intestine mucin secretion in rats. *Journal of Nutrition*, *139*(9), 1640–1647.

Jin, U. H., Cheng, Y., Park, H., Davidson, L. A., Callaway, E. S., Chapkin, R. S., et al. (2017). Short chain fatty acids enhance aryl hydrocarbon (ah) responsiveness in mouse colonocytes and Caco-2 human colon cancer cells. *Scientific Reports*, 7(1), 10163.

Jin, U. H., Park, H., Li, X., Davidson, L. A., Allred, C., Patil, B., et al. (2018). Structure-dependent modulation of aryl hydrocarbon receptor-mediated activities by flavonoids. *Toxicological Sciences*, *164*(1), 205–217.

Jung, M. H., Seong, P. N., Kim, M. H., Myong, N. H., & Chang, M. J. (2013). Effect of green tea extract microencapsulation on hypertriglyceridemia and cardiovascular tissues in high fructose-fed rats. *Nutrition Research and Practice*, 7(5), 366–372.

Jurgonski, A., Juskiewicz, J., Fotschki, B., Kolodziejczyk, K., Milala, J., Kosmala, M., et al. (2017). Metabolism of strawberry mono- and dimeric ellagitannins in rats fed a diet containing fructo-oligosaccharides. *European Journal of Nutrition*, *56*(2), 853–864.

Juskiewicz, J., Milala, J., Jurgonski, A., Krol, B., & Zdunczyk, Z. (2011). Consumption of polyphenol concentrate with dietary fructo-oligosaccharides enhances cecal metabolism of quercetin glycosides in rats. *Nutrition*, *27*(3), 351–357.

Kardum, N., & Glibetic, M. (2018). Polyphenols and their interactions with other dietary compounds: Implications for human health. *Advances in Food and Nutrition Research*, *84*, 103–144.

Kim, S. J., Kim, M. C., Um, J. Y., & Hong, S. H. (2010). The beneficial effect of vanillic acid on ulcerative colitis. *Molecules*, *15*(10), 7208–7217.

Knudsen, K. E. B., Lærke, H. N., Hedemann, M. S., Nielsen, T. S., Ingerslev, A. K., Nielsen, D. S. G., et al. (2018). Impact of diet-modulated butyrate production on intestinal barrier function and inflammation. *Nutrients*, *10*(10), Art. no. 1499.

Koppel, N., Maini Rekdal, V., & Balskus, E. P. (2017). Chemical transformation of xenobiotics by the human gut microbiota. *Science*, *356*(6344).

Kosmala, M., Kolodziejczyk, K., Zdunczyk, Z., Juskiewicz, J., & Boros, D. (2011). Chemical composition of natural and polyphenol-free apple pomace and the effect of this dietary ingredient on intestinal fermentation and serum lipid parameters in rats. *Journal of Agricultural and Food Chemistry, 59*(17), 9177–9185.

Kosmala, M., Zdunczyk, Z., Juskiewicz, J., Jurgonski, A., Karlinska, E., Macierzynski, J., et al. (2015). Chemical composition of defatted strawberry and raspberry seeds and the effect of these dietary ingredients on polyphenol metabolites, intestinal function, and selected serum parameters in rats. *Journal of Agricultural and Food Chemistry, 63*(11), 2989–2996.

Kostova, I., Bhatia, S., Grigorov, P., Balkansky, S., Parmar, V. K., Prasad, A., et al. (2011). Coumarins as antioxidants. *Current Medicinal Chemistry, 18*(25), 3929–3951.

Koutsos, A., Lima, M., Conterno, L., Gasperotti, M., Bianchi, M., Fava, F., et al. (2017). Effects of commercial apple varieties on human gut microbiota composition and metabolic output using an in vitro colonic model. *Nutrients, 9*(6), Art. no. 533.

Kroon, P. A., Faulds, C. B., Ryden, P., Robertson, J. A., & Williamson, G. (1997). Release of covalently bound ferulic acid from fiber in the human colon. *Journal of Agricultural and Food Chemistry, 45*(3), 661–667.

Kyrgiou, M., Kalliala, I., Markozannes, G., Gunter, M. J., Paraskevaidis, E., Gabra, H., et al. (2017). Adiposity and cancer at major anatomical sites: Umbrella review of the literature. *British Medical Journal, 356*, j477.

Lattimer, J. M., & Haub, M. D. (2010). Effects of dietary fiber and its components on metabolic health. *Nutrients, 2*(12), 1266–1289.

Le Bourvellec, C., Bouchet, B., & Renard, C. M. G. C. (2005). Non-covalent interaction between procyanidins and apple cell wall material. Part III: Study on model polysaccharides. *Biochimica et Biophysica Acta (BBA)—General Subjects, 1725*(1), 10–18.

Le Bourvellec, C., Gouble, B., Bureau, S., Loonis, M., Plé, Y., & Renard, C. M. G. C. (2013). Pink discoloration of canned pears: Role of procyanidin chemical depolymerization and procyanidin/cell wall interactions. *Journal of Agricultural and Food Chemistry, 61*(27), 6679–6692.

Le Bourvellec, C., Guyot, S., & Renard, C. M. G. C. (2004). Non-covalent interaction between procyanidins and apple cell wall material. *Biochimica et Biophysica Acta (BBA) - General Subjects, 1672*(3), 192–202.

Le Bourvellec, C., Guyot, S., & Renard, C. M. G. C. (2009). Interactions between apple (Malus x domestica Borkh.) polyphenols and cell walls modulate the extractability of polysaccharides. *Carbohydrate Polymers, 75*(2), 251–261.

Le Bourvellec, C., Le Quere, J.-M., & Renard, C. M. G. C. (2007). Impact of noncovalent interactions between apple condensed tannins and cell walls on their transfer from fruit to juice: Studies in model suspensions and application. *Journal of Agricultural and Food Chemistry, 55*(19), 7896–7904.

Le Bourvellec, C., & Renard, C. M. G. C. (2005). Non-covalent interaction between procyanidins and apple cell wall material. Part II: Quantification and impact of cell wall drying. *Biochimica et Biophysica Acta (BBA)—General Subjects, 1725*(1), 1–9.

Le Bourvellec, C., Watrelot, A. A., Ginies, C., Imberty, A., & Renard, C. M. G. C. (2012). Impact of processing on the noncovalent interactions between procyanidin and apple cell wall. *Journal of Agricultural and Food Chemistry, 60*(37), 9484–9494.

Lee, P. S., Yim, S. G., Choi, Y., Van Anh Ha, T., & Ko, S. (2012). Physiochemical properties and prolonged release behaviours of chitosan-denatured beta-lactoglobulin microcapsules for potential food applications. *Food Chemistry, 134*(2), 992–998.

Leth, M. L., Ejby, M., Workman, C., Ewald, D. A., Pedersen, S. S., Sternberg, C., et al. (2018). Differential bacterial capture and transport preferences facilitate co-growth on dietary xylan in the human gut. *Nature Microbiology, 3*(5), 570–580.

Ley, R. E., Turnbaugh, P. J., Klein, S., & Gordon, J. I. (2006). Microbial ecology: Human gut microbes associated with obesity. [clinical trial]. *Nature*, *444*(7122), 1022–1023.

Li, M., Pernell, C., & Ferruzzi, M. G. (2018). Complexation with phenolic acids affect rheological properties and digestibility of potato starch and maize amylopectin. *Food Hydrocolloids*, 77, 843–852.

Liu, F., Li, P., Chen, M., Luo, Y., Prabhakar, M., Zheng, H., et al. (2017). Fructooligosaccharide (FOS) and Galactooligosaccharide (GOS) increase Bifidobacterium but reduce butyrate producing bacteria with adverse glycemic metabolism in healthy young population. *Scientific Reports*, 7(1), 11789.

Liu, Z., Li, N., & Neu, J. (2005). Tight junctions, leaky intestines, and pediatric diseases. *Acta Paediatrica*, *94*(4), 386–393.

Liu, D., Martinez-Sanz, M., Lopez-Sanchez, P., Gilbert, E. P., & Gidley, M. J. (2017). Adsorption behaviour of polyphenols on cellulose is affected by processing history. *Food Hydrocolloids*, *63*, 496–507.

Liu, X. D., Xue, W. M., Liu, Q., Yu, W. T., Fu, Y. L., Xiong, X., et al. (2004). Swelling behaviour of alginate-chitosan microcapsules prepared by external gelation or internal gelation technology. *Carbohydrate Polymers*, *56*(4), 459–464.

Louis, P., & O'Byrne, C. P. (2010). Life in the gut: Microbial responses to stress in the gastrointestinal tract. *Science Progress*, *93*(Pt 1), 7–36.

Ma, X., Fan, P. X., Li, L. S., Qiao, S. Y., Zhang, G. L., & Li, D. F. (2012). Butyrate promotes the recovering of intestinal wound healing through its positive effect on the tight junctions. *Journal of Animal Science*, *90*(Suppl. 4), 266–268.

Madziva, H., Kailasapathy, K., & Phillips, M. (2006). Evaluation of alginate-pectin capsules in cheddar cheese as a food carrier for the delivery of folic acid. *LWT- Food Science and Technology*, *39*(2), 146–151.

Magallanes-Cruz, P. A., Flores-Silva, P. C., & Bello-Perez, L. A. (2017). Starch structure influences its digestibility: A review. *Journal of Food Science*, *82*(9), 2016–2023.

Marciani, L., Gowland, P. A., Spiller, R. C., Manoj, P., Moore, R. J., Young, P., et al. (2001). Effect of meal viscosity and nutrients on satiety, intragastric dilution, and emptying assessed by MRI. *American Journal of Physiology. Gastrointestinal and Liver Physiology*, *280*(6), G1227–G1233.

Masco, L., Crockaert, C., Van Hoorde, K., Swings, J., & Huys, G. (2007). *In Vitro* assessment of the gastrointestinal transit tolerance of taxonomic reference strains from human origin and probiotic product isolates of Bifidobacterium. *Journal of Dairy Science*, *90*(8), 3572–3578.

Mattila, P., Hellström, J., & Törrönen, R. (2006). Phenolic acids in berries, fruits, and beverages. *Journal of Agricultural and Food Chemistry*, *54*(19), 7193–7199.

Mayta-Apaza, A. C., Pottgen, E., De Bodt, J., Papp, N., Marasini, D., Howard, L., et al. (2018). Impact of tart cherries polyphenols on the human gut microbiota and phenolic metabolites in vitro and in vivo. *The Journal of Nutritional Biochemistry*, *59*, 160–172.

Mekoue Nguela, J., Poncet-Legrand, C., Sieczkowski, N., & Vernhet, A. (2016). Interactions of grape tannins and wine polyphenols with a yeast protein extract, mannoproteins and β-glucan. *Food Chemistry*, *210*, 671–682.

Mills, C. E., Tzounis, X., Oruna-Concha, M. J., Mottram, D. S., Gibson, G. R., & Spencer, J. P. (2015). In vitro colonic metabolism of coffee and chlorogenic acid results in selective changes in human faecal microbiota growth. *British Journal of Nutrition*, *113*(8), 1220–1227.

Mirande, C., Kadlecikova, E., Matulova, M., Capek, P., Bernalier-Donadille, A., Forano, E., et al. (2010). Dietary fibre degradation and fermentation by two xylanolytic bacteria Bacteroides xylanisolvens XB1AT and Roseburia intestinalis XB6B4 from the human intestine. *Journal of Applied Microbiology*, *109*(2), 451–460.

Mizgier, P., Kucharska, A. Z., Sokół-Łętowska, A., Kolniak-Ostek, J., Kidoń, M., & Fecka, I. (2016). Characterization of phenolic compounds and antioxidant and anti-inflammatory properties of red cabbage and purple carrot extracts. *Journal of Functional Foods*, *21*, 133–146.

Mosele, J. I., Macia, A., & Motilva, M. J. (2015). Metabolic and microbial modulation of the large intestine ecosystem by non-absorbed diet phenolic compounds: A review. *Molecules*, *20*(9), 17429–17468.

Mozaffarian, D. (2016). Dietary and policy priorities for cardiovascular disease, diabetes, and obesity: A comprehensive review. *Circulation*, *133*(2), 187–225.

Muhamad, I. I., Fen, L. S., Hui, N. H., & Mustapha, N. A. (2011). Genipin-cross-linked kappa-carrageenan/carboxymethyl cellulose beads and effects on beta-carotene release. *Carbohydrate Polymers*, *83*(3), 1207–1212.

Naito, Y., Takagi, T., Katada, K., Uchiyama, K., Kuroda, M., Kokura, S., et al. (2006). Partially hydrolyzed guar gum down-regulates colonic inflammatory response in dextran sulfate sodium-induced colitis in mice. *The Journal of Nutritional Biochemistry*, *17*(6), 402–409.

Natividad, J. M., & Verdu, E. F. (2012). Modulation of intestinal barrier by intestinal microbiota: Pathological and therapeutic implications. *Pharmacological Research: The Official Journal of the Italian Pharmacological Society*.

Nazzaro, F., Orlando, P., Fratianni, F., & Coppola, R. (2012). Microencapsulation in food science and biotechnology. *Current Opinion in Biotechnology*, *23*(2), 182–186.

Nemcova, R., Bomba, A., Gancarcikova, S., Herich, R., & Guba, P. (1999). Study of the effect of Lactobacillus paracasei and fructooligosaccharides on the faecal microflora in weanling piglets. *Berliner Und Munchener Tierarztliche Wochenschrift*, *112*(6–7), 225–228.

Neyrinck, A. M., Possemiers, S., Druart, C., Van de Wiele, T., De Backer, F., Cani, P. D., et al. (2011). Prebiotic effects of wheat arabinoxylan related to the increase in bifidobacteria, Roseburia and Bacteroides/Prevotella in diet-induced obese mice. *PLoS One*, *6*(6), e20944.

Neyrinck, A. M., Van Hee, V. F., Piront, N., De Backer, F., Toussaint, O., Cani, P. D., et al. (2012). Wheat-derived arabinoxylan oligosaccharides with prebiotic effect increase satietogenic gut peptides and reduce metabolic endotoxemia in diet-induced obese mice. *Nutrition & Diabetes*, *2*, e28.

Nishimura, T., Andoh, A., Hashimoto, T., Kobori, A., Tsujikawa, T., & Fujiyama, Y. (2010). Cellobiose prevents the development of dextran sulfate sodium (DSS)-induced experimental colitis. *Journal of Clinical Biochemistry and Nutrition*, *46*(2), 105–110.

Nishinari, K., Kim, B., Fang, Y., Nitta, Y., & Takemasa, M. (2006). Rheological and related study of gelation of xyloglucan in the presence of small molecules and other polysaccharides. *Cellulose*, *13*(4), 365–374.

Nitta, Y., Fang, Y., Takemasa, M., & Nishinari, K. (2004). Gelation of xyloglucan by addition of epigallocatechin Gallate as studied by rheology and differential scanning calorimetry. *Biomacromolecules*, *5*(4), 1206–1213.

Oladiran, D. A., & Emmambux, N. M. (2018). Nutritional and functional properties of extruded cassava-soy composite with grape pomace. *Starch—Stärke*, *70*(7–8), 1700298.

Padayachee, A., Netzel, G., Netzel, M., Day, L., Mikkelsen, D., & Gidley, M. J. (2013). Lack of release of bound anthocyanins and phenolic acids from carrot plant cell walls and model composites during simulated gastric and small intestinal digestion. *Food & Function*, *4*(6), 906.

Padayachee, A., Netzel, G., Netzel, M., Day, L., Zabaras, D., Mikkelsen, D., et al. (2012a). Binding of polyphenols to plant cell wall analogues—Part 1: Anthocyanins. *Food Chemistry*, *134*(1), 155–161.

Padayachee, A., Netzel, G., Netzel, M., Day, L., Zabaras, D., Mikkelsen, D., et al. (2012b). Binding of polyphenols to plant cell wall analogues—Part 2: Phenolic acids. *Food Chemistry, 135*(4), 2287–2292.

Pai, D. A., Vangala, V. R., Ng, J. W., Ng, W. K., & Tan, R. B. H. (2015). Resistant maltodextrin as a shell material for encapsulation of naringin: Production and physicochemical characterization. *Journal of Food Engineering, 161*, 68–74.

Palla, A. H., Iqbal, N. T., Minhas, K., & Gilani, A. H. (2016). Flaxseed extract exhibits mucosal protective effect in acetic acid induced colitis in mice by modulating cytokines, antioxidant and antiinflammatory mechanisms. *International Immunopharmacology, 38*, 153–166.

Patel, A. R., Seijen Ten-Hoorn, J., Hazekamp, J., Blijdenstein, T. B. J., & Velikov, K. P. (2013). Colloidal complexation of a macromolecule with a small molecular weight natural polyphenol: Implications in modulating polymer functionalities. *Soft Matter, 9*(5), 1428–1436.

Patel, A. R., Seijen-ten-Hoorn, J., & Velikov, K. P. (2011). Colloidal complexes from associated water soluble cellulose derivative (methylcellulose) and green tea polyphenol (epigallocatechin gallate). *Journal of Colloid and Interface Science, 364*(2), 317–323.

Pegg, A. (2012). *The application of natural hydrocolloids to foods and beverages* (pp. 175–196). Ingredients and Flavourings: Natural Food Additives.

Peng, L., He, Z., Chen, W., Holzman, I. R., & Lin, J. (2007). Effects of butyrate on intestinal barrier function in a Caco-2 cell monolayer model of intestinal barrier. *Pediatric Research, 61*(1), 37–41.

Peng, L., Li, Z. R., Green, R. S., Holzman, I. R., & Lin, J. (2009). Butyrate enhances the intestinal barrier by facilitating tight junction assembly via activation of AMP-activated protein kinase in Caco-2 cell monolayers. *Journal of Nutrition, 139*(9), 1619–1625.

Pettinato, M., Aliakbarian, B., Casazza, A. A., & Perego, P. (2017). Encapsulation of antioxidants from spent coffee ground extracts by spray drying. *Chemical Engineering Transactions, 57*, 1219–1224.

Phan, A. D. T., D'Arcy, B. R., & Gidley, M. J. (2016). Polyphenol-cellulose interactions: Effects of pH, temperature and salt. *International Journal of Food Science & Technology, 51*(1), 203–211.

Phan, A. D. T., Flanagan, B. M., D'Arcy, B. R., & Gidley, M. J. (2017). Binding selectivity of dietary polyphenols to different plant cell wall components: Quantification and mechanism. *Food Chemistry, 233*, 216–227.

Phan, A. D. T., Netzel, G., Wang, D., Flanagan, B. M., D'Arcy, B. R., & Gidley, M. J. (2015). Binding of dietary polyphenols to cellulose: Structural and nutritional aspects. *Food Chemistry, 171*, 388–396.

Popov, S. V., Markov, P. A., Nikitina, I. R., Petrishev, S., Smirnov, V., & Ovodov, Y. S. (2006). Preventive effect of a pectic polysaccharide of the common cranberry Vaccinium oxycoccos L. on acetic acid-induced colitis in mice. *World journal of gastroenterology: WJG, 12*(41), 6646–6651.

Power, K. A., Lu, J. T., Monk, J. M., Lepp, D., Wu, W., Zhang, C., et al. (2016). Purified rutin and rutin-rich asparagus attenuates disease severity and tissue damage following dextran sodium sulfate-induced colitis. *Molecular Nutrition & Food Research, 60*(11), 2396–2412.

Psichas, A., Sleeth, M. L., Murphy, K. G., Brooks, L., Bewick, G. A., Hanyaloglu, A. C., et al. (2015). The short chain fatty acid propionate stimulates GLP-1 and PYY secretion via free fatty acid receptor 2 in rodents. *International Journal of Obesity, 39*(3), 424–429.

Puupponen-Pimia, R., Nohynek, L., Meier, C., Kahkonen, M., Heinonen, M., Hopia, A., et al. (2001). Antimicrobial properties of phenolic compounds from berries. *Journal of Applied Microbiology, 90*(4), 494–507.

Rahat-Rozenbloom, S., Fernandes, J., Gloor, G. B., & Wolever, T. M. (2014). Evidence for greater production of colonic short-chain fatty acids in overweight than lean humans. *International Journal of Obesity*, *38*(12), 1525–1531.

Ramdath, D. D., & Tsao, R. (2012). Bioactives and health benefits of lentils (Lens culinaris L.). *Cereals and pulses: nutraceutical properties and health benefits*, 217–228.

Renard, C. M. G. C., Baron, A., Guyot, S., & Drilleau, J. F. (2001). Interactions between apple cell walls and native apple polyphenols: Quantification and some consequences. *International Journal of Biological Macromolecules*, *29*(2), 115–125.

Renard, C. M. G. C., Le Quéré, J. M., Bauduin, R., Symoneaux, R., Le Bourvellec, C., & Baron, A. (2011). Modulating polyphenolic composition and organoleptic properties of apple juices by manipulating the pressing conditions. *Food Chemistry*, *124*(1), 117–125.

Riou, V., Vernhet, A., Doco, T., & Moutounet, M. (2002). Aggregation of grape seed tannins in model wine—Effect of wine polysaccharides. *Food Hydrocolloids*, *16*(1), 17–23.

Rioux, K. P., Madsen, K. L., & Fedorak, R. N. (2005). The role of enteric microflora in inflammatory bowel disease: Human and animal studies with probiotics and prebiotics. *Gastroenterology Clinics of North America*, *34*(3), 465 —+.

Robbins, R. J. (2003). Phenolic acids in foods: An overview of analytical methodology. *Journal of Agricultural and Food Chemistry*, *51*(10), 2866–2887.

Roberfroid, M. (1993). Dietary fiber, inulin, and oligofructose: A review comparing their physiological effects. *Critical Reviews in Food Science and Nutrition*, *33*(2), 103–148.

Roberfroid, M. B. (2002). Functional foods: Concepts and application to inulin and oligofructose. *British Journal of Nutrition*, *87*(Suppl. 2), S139–S143.

Roberfroid, M., Gibson, G. R., Hoyles, L., McCartney, A. L., Rastall, R., Rowland, I., et al. (2010). Prebiotic effects: Metabolic and health benefits. *British Journal of Nutrition*, *104*(Suppl 2), S1–63.

Rodriguez-Cabezas, M. E., Galvez, J., Camuesco, D., Lorente, M. D., Concha, A., Martinez-Augustin, O., et al. (2003). Intestinal anti-inflammatory activity of dietary fiber (Plantago ovata seeds) in HLA-B27 transgenic rats. *Clinical Nutrition*, *22*(5), 463–471.

Ruiz-Garcia, Y., Smith, P. A., & Bindon, K. A. (2014). Selective extraction of polysaccharide affects the adsorption of proanthocyanidin by grape cell walls. *Carbohydrate Polymers*, *114*, 102–114.

Russell, W. R., Baka, A., Bjorck, I., Delzenne, N., Gao, D., Griffiths, H. R., et al. (2016). Impact of diet composition on blood glucose regulation. *Critical Reviews in Food Science and Nutrition*, *56*(4), 541–590.

Sakai, T., Sakamoto, T., Hallaert, J., & Vandamme, E. J. (1993). Pectin, pectinase, and protopectinase: Production, properties, and applications. *Advances in Applied Microbiology*, *39*(C), 213–294.

Salden, B. N., Troost, F. J., Wilms, E., Truchado, P., Vilchez-Vargas, R., Pieper, D. H., et al. (2018). Reinforcement of intestinal epithelial barrier by arabinoxylans in overweight and obese subjects: A randomized controlled trial: Arabinoxylans in gut barrier. *Clinical Nutrition*, *37*(2), 471–480.

Sanchez, V., Baeza, R., Galmarini, M. V., Zamora, M. C., & Chirife, J. (2013). Freeze-drying encapsulation of red wine polyphenols in an amorphous matrix of maltodextrin. *Food and Bioprocess Technology*, *6*(5), 1350–1354.

Santacruz, A., Collado, M. C., Garcia-Valdes, L., Segura, M. T., Martin-Lagos, J. A., Anjos, T., et al. (2010). Gut microbiota composition is associated with body weight, weight gain and biochemical parameters in pregnant women. *British Journal of Nutrition*, *104*(1), 83–92.

Scaldaferri, F., Pizzoferrato, M., Gerardi, V., Lopetuso, L., & Gasbarrini, A. (2012). The gut barrier: New acquisitions and therapeutic approaches. *Journal of Clinical Gastroenterology*, *46*(Suppl), S12–S17.

Scheppach, W., Sommer, H., Kirchner, T., Paganelli, G. M., Bartram, P., Christl, S., et al. (1992). Effect of butyrate enemas on the colonic mucosa in distal ulcerative colitis. [Clinical trial randomized controlled trial]. *Gastroenterology*, *103*(1), 51–56.
Schneeman, B. O. (1999). Fiber, inulin and oligofructose: Similarities and differences. *Journal of Nutrition*, *129*(7 Suppl), 1424S–1427S.
Schwiertz, A., Taras, D., Schafer, K., Beijer, S., Bos, N. A., Donus, C., et al. (2010). Microbiota and SCFA in lean and overweight healthy subjects. *Obesity*, *18*(1), 190–195.
Sebastià, N., Montoro, A., León, Z., & Soriano, J. M. (2017). Searching trans-resveratrol in fruits and vegetables: A preliminary screening. *Journal of Food Science and Technology*, *54*(3), 842–845.
Serrano, J., Puupponen-Pimiä, R., Dauer, A., Aura, A.-M., & Saura-Calixto, F. (2009). Tannins: Current knowledge of food sources, intake, bioavailability and biological effects. *Molecular Nutrition & Food Research*, *53*(S2), S310–S329.
Silva-Weiss, A., Bifani, V., Ihl, M., Sobral, P. J. A., & Gómez-Guillén, M. C. (2014). Polyphenol-rich extract from murta leaves on rheological properties of film-forming solutions based on different hydrocolloid blends. *Journal of Food Engineering*, *140*, 28–38.
Singh, S. P., Jadaun, J. S., Narnoliya, L. K., & Pandey, A. (2017). Prebiotic oligosaccharides: Special focus on Fructooligosaccharides, its biosynthesis and bioactivity. *Applied Biochemistry and Biotechnology*, *183*(2), 613–635.
Sivaprakasam, S., Bhutia, Y. D., Yang, S., & Ganapathy, V. (2017). Short-chain fatty acid transporters: Role in colonic homeostasis. *Comprehensive Physiology*, *8*(1), 299–314.
Slavin, J. L., & Greenberg, N. A. (2003). Partially hydrolyzed guar gum: Clinical nutrition uses. *Nutrition*, *19*(6), 549–552.
Slavin, J. L., & Lloyd, B. (2012). Health benefits of fruits and vegetables. *Advances in Nutrition (Bethesda, Md)*, *3*(4), 506–516.
Smith, P. M., Howitt, M. R., Panikov, N., Michaud, M., Gallini, C. A., Bohlooly, Y. M., et al. (2013). The microbial metabolites, short-chain fatty acids, regulate colonic Treg cell homeostasis. *Science*, *341*(6145), 569–573.
Spiridon, I., & Popa, V. I. (2008). Hemicelluloses: Major sources, properties and applications. *Monomers, Polymers and Composites From Renewable Resources*, 289–304.
Springer, L. F., & Sacks, G. L. (2014). Protein-precipitable tannin in wines from Vitis vinifera and interspecific hybrid grapes Vitis ssp.: Differences in concentration, extractability, and cell wall binding. *Journal of Agricultural and Food Chemistry*, *62*(30), 7515–7523.
Steed, H., Macfarlane, G. T., & Macfarlane, S. (2008). Prebiotics, synbiotics and inflammatory bowel disease. *Molecular Nutrition & Food Research*, *52*(8), 898–905.
Sung, M. K., & Park, M. Y. (2013). Nutritional modulators of ulcerative colitis: Clinical efficacies and mechanistic view. *World journal of gastroenterology: WJG*, *19*(7), 994–1004.
Suzuki, T., & Hara, H. (2011). Role of flavonoids in intestinal tight junction regulation. *Journal of Nutritional Biochemistry*, *22*(5), 401–408.
Tamura, M., Nakagawa, H., Tsushida, T., Hirayama, K., & Itoh, K. (2007). Effect of pectin enhancement on plasma quercetin and fecal flora in rutin-supplemented mice. *Journal of Food Science*, *72*(9), S648–S651.
Tan, S., Caparros-Martin, J. A., Matthews, V. B., Koch, H., O'Gara, F., Croft, K. D., et al. (2018). Isoquercetin and inulin synergistically modulate the gut microbiome to prevent development of the metabolic syndrome in mice fed a high fat diet. *Scientific Reports*, *8*(1), 10100.
Tee, W. F., Nazaruddin, R., Tan, Y. N., & Ayob, M. K. (2014). Effects of encapsulation on the viability of potential probiotic lactobacillus plantarum exposed to high acidity condition and presence of bile salts. *Food Science and Technology International*, *20*(6), 399–404.
Thakur, B. R., Singh, R. K., Handa, A. K., & Rao, M. A. (1997). Chemistry and uses of pectin—A review. *Critical Reviews in Food Science and Nutrition*, *37*(1), 47–73.

Tokunaga, S., Woodin, B. R., & Stegeman, J. J. (2016). Plant lignan secoisolariciresinol suppresses pericardial edema caused by dioxin-like compounds in developing zebrafish: Implications for suppression of morphological abnormalities. *Food and Chemical Toxicology*, *96*, 160–166.

Tolve, R., Galgano, F., Caruso, M. C., Tchuenbou-Magaia, F. L., Condelli, N., Favati, F., et al. (2016). Encapsulation of health-promoting ingredients: Applications in foodstuffs. *International Journal of Food Sciences and Nutrition*, *67*(8), 888–918.

Tosh, S. M., Wood, P. J., Wang, Q., & Weisz, J. (2004). Structural characteristics and rheological properties of partially hydrolyzed oat β-glucan: The effects of molecular weight and hydrolysis method. *Carbohydrate Polymers*, *55*(4), 425–436.

Tremaroli, V., & Bäckhed, F. (2012). Functional interactions between the gut microbiota and host metabolism. *Nature*, *489*, 242–249.

Troseid, M., Nestvold, T. K., Rudi, K., Thoresen, H., Nielsen, E. W., & Lappegard, K. T. (2013). Plasma lipopolysaccharide is closely associated with glycemic control and abdominal obesity: Evidence from bariatric surgery. *Diabetes Care*, *36*(11), 3627–3632.

Tuohy, K. M., Conterno, L., Gasperotti, M., & Viola, R. (2012). Up-regulating the human intestinal microbiome using whole plant foods, polyphenols, and/or fiber. *Journal of Agricultural and Food Chemistry*, *60*(36), 8776–8782.

Ulluwishewa, D., Anderson, R. C., McNabb, W. C., Moughan, P. J., Wells, J. M., & Roy, N. C. (2011). Regulation of tight junction permeability by intestinal bacteria and dietary components. *The Journal of Nutrition*, *141*(5), 769–776.

Van Hung, P. (2016). Phenolic compounds of cereals and their antioxidant capacity. *Critical Reviews in Food Science and Nutrition*, *56*(1), 25–35.

Van Rymenant, E., Abranko, L., Tumova, S., Grootaert, C., Van Camp, J., Williamson, G., et al. (2017). Chronic exposure to short-chain fatty acids modulates transport and metabolism of microbiome-derived phenolics in human intestinal cells. *The Journal of Nutritional Biochemistry*, *39*, 156–168.

Vandeputte, D., Falony, G., Vieira-Silva, S., Wang, J., Sailer, M., Theis, S., et al. (2017). Prebiotic inulin-type fructans induce specific changes in the human gut microbiota. *Gut*, *66*(11), 1968–1974.

Venkatraman, A., Ramakrishna, B. S., Shaji, R. V., Kumar, N. S., Pulimood, A., & Patra, S. (2003). Amelioration of dextran sulfate colitis by butyrate: Role of heat shock protein 70 and NF-kappaB. *American Journal of Physiology. Gastrointestinal and Liver Physiology*, *285*(1), G177–G184.

Vernia, P., Annese, V., Bresci, G., d'Albasio, G., D'Inca, R., Giaccari, S., et al. (2003). Topical butyrate improves efficacy of 5-ASA in refractory distal ulcerative colitis: Results of a multicentre trial. [Clinical trial multicenter study randomized controlled trial]. *European Journal of Clinical Investigation*, *33*(3), 244–248.

Vieira, E. L., Leonel, A. J., Sad, A. P., Beltrao, N. R., Costa, T. F., Ferreira, T. M., et al. (2012). Oral administration of sodium butyrate attenuates inflammation and mucosal lesion in experimental acute ulcerative colitis. *The Journal of Nutritional Biochemistry*, *23*(5), 430–436.

Vitaglione, P., Mennella, I., Ferracane, R., Rivellese, A. A., Giacco, R., Ercolini, D., et al. (2015). Whole-grain wheat consumption reduces inflammation in a randomized controlled trial on overweight and obese subjects with unhealthy dietary and lifestyle behaviors: Role of polyphenols bound to cereal dietary fiber. *The American Journal of Clinical Nutrition*, *101*(2), 251–261.

Wallace, A. J., Eady, S. L., Hunter, D. C., Skinner, M. A., Huffman, L., Ansell, J., et al. (2015). No difference in fecal levels of bacteria or short chain fatty acids in humans, when consuming fruit juice beverages containing fruit fiber, fruit polyphenols, and their combination. *Nutrition Research*, *35*(1), 23–34.

Wani, T. A., Shah, A. G., Wani, S. M., Wani, I. A., Masoodi, F. A., Nissar, N., et al. (2016). Suitability of different food grade materials for the encapsulation of some functional foods well reported for their advantages and susceptibility. *Critical Reviews in Food Science and Nutrition, 56*(15), 2431–2454.

Watrelot, A. A., Le Bourvellec, C., Imberty, A., & Renard, C. M. G. C. (2013). Interactions between Pectic compounds and Procyanidins are influenced by methylation degree and chain length. *Biomacromolecules, 14*(3), 709–718.

Watrelot, A. A., Le Bourvellec, C., Imberty, A., & Renard, C. M. G. C. (2014). Neutral sugar side chains of pectins limit interactions with procyanidins. *Carbohydrate Polymers, 99*, 527–536.

Webster, F. H., & Wood, P. J. (2011). *Oats: Chemistry and technology* (2nd ed.). St Paul, MN: AACC International Press.

Willemsen, L. E., Koetsier, M. A., van Deventer, S. J., & van Tol, E. A. (2003). Short chain fatty acids stimulate epithelial mucin 2 expression through differential effects on prostaglandin E(1) and E(2) production by intestinal myofibroblasts. *Gut, 52*(10), 1442–1447.

Wrolstad, R. E. (2012). *Food carbohydrate chemistry*. Hoboken, N.J.: Wiley-Blackwell.

Wu, Z., Li, H., Ming, J., & Zhao, G. (2011). Optimization of adsorption of tea polyphenols into oat β-glucan using response surface methodology. *Journal of Agricultural and Food Chemistry, 59*(1), 378–385.

Wu, Y., Li, X., Xiang, W., Zhu, C., Lin, Z., Li, J., et al. (2012). Presence of tannins in sorghum grains is conditioned by different natural alleles of Tannin1. *Proceedings of the National Academy of Sciences, 109*(26), 10281–10286.

Wu, Y., Lin, Q. L., Chen, Z. X., & Xiao, H. X. (2011). The interaction between tea polyphenols and rice starch during gelatinization. *Food Science and Technology International, 17*(6), 569–577.

Xu, B. J., Yuan, S. H., & Chang, S. K. (2007). Comparative analyses of phenolic composition, antioxidant capacity, and color of cool season legumes and other selected food legumes. *Journal of Food Science, 72*(2), S167–S177.

Xue, B., Xie, J., Huang, J., Chen, L., Gao, L., Ou, S., et al. (2016). Plant polyphenols alter a pathway of energy metabolism by inhibiting fecal Bacteroidetes and Firmicutes in vitro. *Food & Function*, 7(3), 1501–1507.

Yacco, R. S., Watrelot, A. A., & Kennedy, J. A. (2016). Red wine tannin structure–activity relationships during fermentation and maceration. *Journal of Agricultural and Food Chemistry, 64*(4), 860–869.

Yao, J., Wang, J. Y., Liu, L., Li, Y. X., Xun, A. Y., Zeng, W. S., et al. (2010). Anti-oxidant effects of resveratrol on mice with DSS-induced ulcerative colitis. *Archives of Medical Research, 41*(4), 288–294.

Ye, M. B., Bak, J. P., An, C. S., Jin, H. L., Kim, J. M., Kweon, H. J., et al. (2011). Dietary beta-glucan regulates the levels of inflammatory factors, inflammatory cytokines, and immunoglobulins in interleukin-10 knockout mice. *Journal of Medicinal Food, 14*(5), 468–474.

Yu, K., Ke, M. Y., Li, W. H., Zhang, S. Q., & Fang, X. C. (2014). The impact of soluble dietary fibre on gastric emptying, postprandial blood glucose and insulin in patients with type 2 diabetes. *Asia Pacific Journal of Clinical Nutrition, 23*(2), 210–218.

Zdunczyk, Z., Juskiewicz, J., & Estrella, I. (2006). Cecal parameters of rats fed diets containing grapefruit polyphenols and inulin as single supplements or in a combination. *Nutrition, 22*(9), 898–904.

Zhang, M. L., Chekan, J. R., Dodd, D., Hong, P. Y., Radlinski, L., Revindran, V., et al. (2014). Xylan utilization in human gut commensal bacteria is orchestrated by unique modular organization of polysaccharide-degrading enzymes. *Proceedings of the National Academy of Sciences of the United States of America, 111*(35), E3708–E3717.

Zhang, H., DiBaise, J. K., Zuccolo, A., Kudrna, D., Braidotti, M., Yu, Y., et al. (2009). Human gut microbiota in obesity and after gastric bypass. *Proceedings of the National Academy of Sciences of the United States of America*, *106*(7), 2365–2370.

Zhang, G., & Hamaker, B. R. (2009). Slowly digestible starch: Concept, mechanism, and proposed extended glycemic index. *Critical Reviews in Food Science and Nutrition*, *49*(10), 852–867.

Zhang, S., Qin, C., & Safe, S. H. (2003). Flavonoids as aryl hydrocarbon receptor agonists/antagonists: Effects of structure and cell context. *Environmental Health Perspectives*, *111*(16), 1877–1882.

CHAPTER FOUR

Plant phenolics as functional food ingredients

Celestino Santos-Buelga[a,*], Ana M. González-Paramás[a], Taofiq Oludemi[b], Begoña Ayuda-Durán[a], Susana González-Manzano[a]

[a]Grupo de Investigación en Polifenoles (GIP-USAL), Universidad de Salamanca, Salamanca, Spain
[b]Mountain Research Center (CIMO), Polytechnic Institute of Bragança, Bragança, Portugal
*Corresponding author: e-mail address: csb@usal.es

Contents

Advances in Food and Nutrition Research, Volume 90
ISSN 1043-4526
https://doi.org/10.1016/bs.afnr.2019.02.012

Abstract

Phenolic compounds have attracted much attention in recent times as their dietary intake has been associated with the prevention of some chronic and degenerative diseases that constitute major causes of death and incapacity in developed countries, such as cardiovascular diseases, type II diabetes, some types of cancers or neurodegenerative disorders like Alzheimer's and Parkinson's diseases. Nowadays it is considered that these compounds contribute, at least in part, for the protective effects of fruit and vegetable-rich diets, so that the study of their role in human nutrition has become a central issue in food research. This chapter reviews the current knowledge on the phenolic compounds as food components, namely their occurrence in the diet, bioavailability and metabolism, biological activities and mechanisms of action. Besides, the approaches for their extraction from plant matrices and technological improvements regarding their preparation, stability and bioavailability in order to be used as functional food ingredients are also reviewed, as well as their legal situation regarding the possibility of making "health claims" based on their presence in food and beverages.

1. Introduction

The surge in aged population, the busy lifestyle and the lack of time, together with the increase in adoption of healthy lifestyle has stimulated industry to research and develop healthier and more nutritious foods. These foods are frequently referred to as "functional foods," "nutraceuticals" or "(dietary) supplements," terms that although have different meaning are frequently used interchangeably. In general, these terms do not have legal/regulatory definition, although some proposals have been made. The term "nutraceutical" (a combination of the words "nutrient" and "pharmaceutical") was coined in 1989 by De Felice (1995), who defined them as "foods (or part of a food) that provide medical or health benefits, including the prevention and/or treatment of a disease." The concept of "functional food" was first introduced in Japan in the mid-1980s for foods containing ingredients with functions for health, and more recently, the Academy of Nutrition and Dietetics in United States has defined them as "whole foods along with fortified, enriched or enhanced foods that have a potentially beneficial effect on health when consumed as part of a varied diet on a regular basis at effective levels" (Crowe & Francis, 2013). To distinguish both concepts, Kalra (2003) proposed to refer as functional when they comprise nutritional components required for human's healthy survival, and nutraceuticals when the aim is to treat/prevent a disease or disorder. Nutraceuticals can be considered dietary supplements (i.e., sold in discrete presentations similar to drugs: pills, extracts, tablets, etc.) that deliver a concentrated form of a presumed bioactive agent, presented in a non-food matrix, and used with the purpose of enhancing health in dosages that exceed those that could be obtained from normal foods (Espín, García-Conesa, & Tomás-Barberán, 2007).

Currently, >80% of the food active compounds are obtained from natural sources (Qilong et al., 2013), and among them polyphenols, some of the most common phytochemicals found in the nutraceutical market. In this sense, this chapter aims to overview the most recent advances in polyphenols as food ingredients, including a description on polyphenols structure and their presence in food, discussion about their bioavailability and metabolism, their putative mechanisms of action and the health benefits attributed to this class of phytochemicals. In addition, a state of the art about the most recent techniques for their extraction from plant matrices and the technological improvements in their stability and bioavailability are reviewed. Finally, a discussion on the legal situation for the use of these ingredients and the possibility to include "health claims" based on their presence in food products is made.

2. Description

Phenolic compounds are a large group of plant secondary metabolites that constitute a heterogeneous group of molecules with a diversity of chemical structures. They are widespread in higher plants, where they play relevant roles, being involved in the mechanisms of natural resistance against biotic and abiotic stresses. They contribute to plant structural integrity, UV photoprotection, reproduction, or internal regulation of plant cell signaling, and act as chemotactic factors, as chemical modulators of plant communication with insects and microbes, and as phytoalexins against pathogens and herbivores (Lattanzio, Kroon, Quideau, & Treutter, 2008). These metabolites are uncommon in algae and fungi, being limited to a few classes of phenolics, with flavonoids almost completely absent (Lattanzio et al., 2008). Phenolic compounds are also abundant in many plant foods and derived products, where they contribute to sensory, technological and health properties.

Plant phenolics derive from the shikimate/phenylpropanoid pathway, the acetate/malonate polyketide pathway or the combination of both (Fig. 1), being commonly classified in two major classes: flavonoids (flavan-3-ols, flavones, flavonols, flavanones, dihydroflavonols, anthocyanins, isoflavones and chalcones) and non-flavonoids, including phenolic alcohols, phenolic

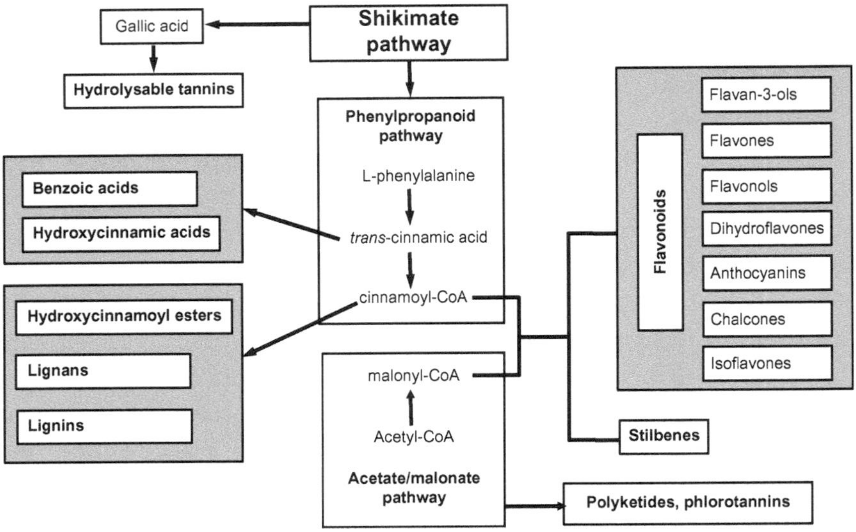

Fig. 1 Biosynthetic pathways of the main classes of plant phenolic compounds.

acids and derivatives (e.g., hydroxybenzoic and hydroxycinnamic acids and their esters), stilbenes and lignans (Fig. 2). Other phenolic groups, such as naturally occurring quinones (benzo-, naphto- and anthraquinones), xanthones, aurones, lignins or phlorotannins are not going to be dealt with in this chapter.

In their natural media, the phenolic compounds may occur in free form, as glycosylated, prenylated or acylated derivatives. In plant tissues and foods, phenolic acids are often found in combinations with polyols like glucose or quinic acid, while most flavonoids (except flavan-3-ols, which are mainly found as aglycones), lignans and stilbenes are usually present as glycosides. Some phenolic compounds also occur as polymerized structures, such as the so-called tannins, from which two classes are distinguished: condensed and hydrolysable tannins. Condensed tannins (also known as proanthocyanidins) are polymers of flavan-3-ol units linked through C—C bonds established between the C4 of one flavan-3-ol unit and the C8 or C6 of another unit (B-type linkage); occasionally they may also contain an additional ether linkage between the C2 of the upper unit and the oxygen-bearing C7 or C5 of the lower unit (A-type linkage). Hydrolysable tannins are composed of polyols linked to at least one gallic acid (gallotannins) or one hexahydroxydiphenic acid (ellagitannins) (Fig. 3).

In addition, some phenolic derived products that are formed during food and beverage processing and storage might also be added as new phenolic classes, owing to their structural relationship and contribution to sensory

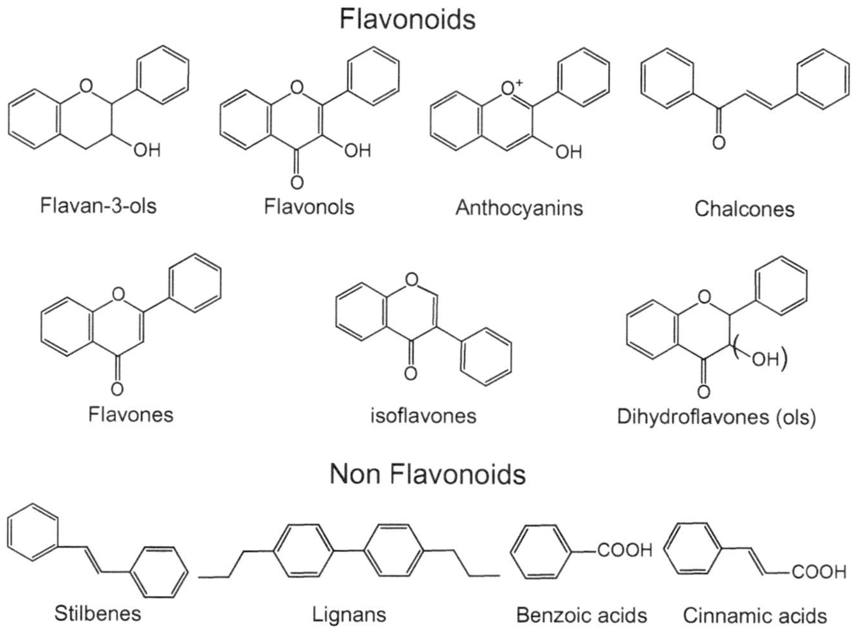

Fig. 2 Basic structures of main classes of phenolic compounds.

and functional properties of phenolic compounds in food. These are, for instance, the cases of the flavanol-derived thearubigins and theaflavins (Fig. 3) present in black tea leaves (Santos-Buelga & Scalbert, 2000), or the anthocyanin-derived pigments, such as pyranoanthocyanins (Fig. 3) and flavanol-anthocyanin condensed pigments formed in fruit derivatives like red wine or jams (Santos-Buelga & González-Paramás, 2019).

Quite commonly plant phenolic compounds are also referred to as "polyphenols," although they may not contain various phenolic hydroxyls in their structure. According to Quideau, Deffieux, Douat-Casassus, and Pouysegu (2011), the term "polyphenols" should be reserved to design "plant secondary metabolites derived from the shikimate-derived phenylpropanoid and/or the polyketide pathway(s), featuring more than one phenolic ring and being devoid of any nitrogen-based functional group in their most basic structural expression." Even more restrictive was Edwin Haslam, who defined plant polyphenols as water-soluble substances able to precipitate proteins that have molecular masses between 500 and 3000 and possess 12–16 phenolic hydroxy groups and 5–7 aromatic rings per 1000 Da of relative molecular mass (Haslam, 1998). Any of those definitions would exclude a great deal of chemically and biosynthetically-related phenolic

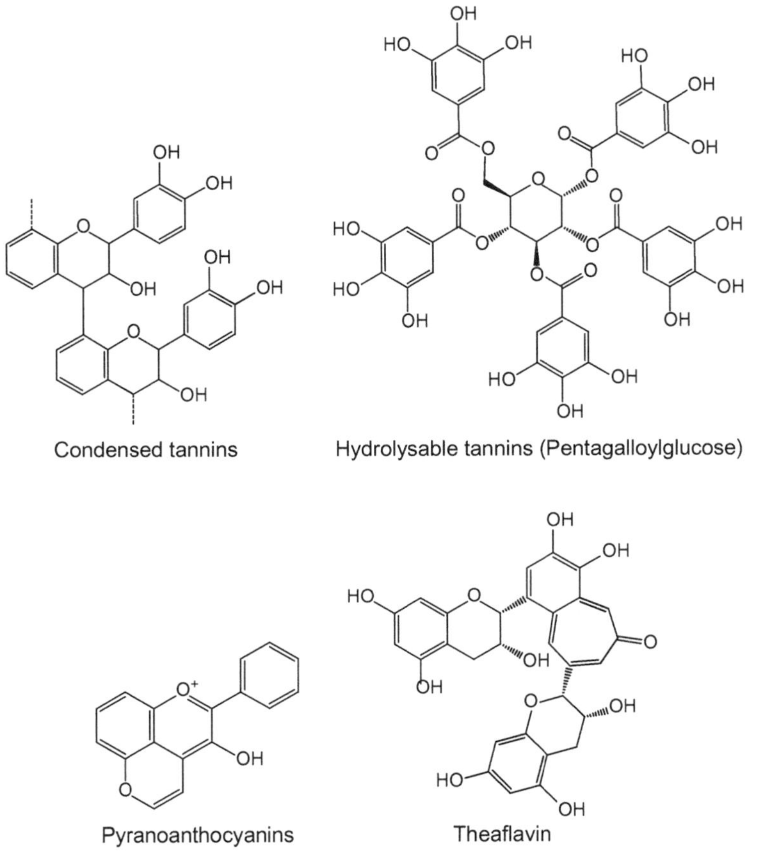

Fig. 3 Structures of some complex polyphenols.

compounds, such as lignin polymers or compounds consisting of only one aromatic ring, whatever the number on substituting hydroxyl groups that they may have (e.g., many phenolic acids and derivatives). Nevertheless, due to its usual employment among both scientists and public, in this chapter, the term "polyphenols" will be used as synonym of "phenolic compounds," even when we may not refer to "true polyphenols" according to the definitions of Quideau et al. (2011) or Haslam (1998).

3. Polyphenols as food components

3.1 Occurrence in food

Phenolic acids and flavonoids are the polyphenol classes most commonly found in foods. Two types of phenolic acids are distinguished derived from hydroxybenzoic (HBA) and hydroxycinnamic acids (HCA), these latter being most abundant in plants and food.

The contents of hydroxybenzoic acids in food are usually low, with the exception of certain Rosaceae fruits and green and black tea. Gallic and ellagic acids are usually the commonest phenolic acids, although they occur in good extent as a part of the hydrolysable tannins. Contents of gallic acid up to 3.5 g/kg have been reported in black tea leaves, and concentrations of 20–50 mg/L of brew have been estimated for a typically prepared tea infusion (Tomás-Barberán & Clifford, 2000). The highest levels of ellagic acid (EA) are found in berries, although a relevant part of it is present as ellagitannins (ETs). Contents of ETs from 1 to 400 mg/100 g fresh weight have been reported in berries from different authors, as reviewed by Landete (2011), with the highest concentrations found in raspberries, arctic bramble and cloudberries (Kähkönen, Hopia, & Heinonen, 2001; Koponen, Happonen, Mattila, & Törrönen, 2007; Määttä-Riihinen, Kamal-Eldin, Mattila, González-Paramás, & Törrönen, 2004). Actually, ETs are the main phenolics found in berries of the genus Rubus, while they represent the second largest group in genus Fragaria (strawberry) after anthocyanins (Kähkönen et al., 2001). Other relevant sources of ETs are pomegranate and walnut. Contents of ETs as high as 2000 mg/L have been determined in pomegranate juice (Fischer, Carle, & Kammerer, 2011; Gil, Tomás-Barberán, Hess-Pierce, Holcroft, & Kader, 2000), while 59 mg of total EAs/100 g dry weight have been reported in walnut (Daniel et al., 1989).

The most frequent hydroxycinnamic acids are caffeic acid and ferulic acid. These acids are rarely found in free form, but they commonly occur in foods and beverages conjugated with sugars or organic acids, especially quinic or tartaric acids. Chlorogenic acids (CGA), a family of esters formed between hydroxycinnamic acids and quinic acid, are the most ubiquitous, especially the caffeoylquinic acid isomers (Clifford, 2000a). Coffee is one of the richest dietary sources of CGA, and for many consumers must be the major dietary source. It has been estimated that a 200 mL-cup of coffee may supply 70–350 mg CGA (Clifford, 1999). Other important sources for some populations are some fruits, such as blueberries, kiwis, plums, cherries or apples, with hydroxycinnamate contents in the range of 0.5–2 g/kg (Manach, Scalbert, Morand, Rémésy, & Jiménez, 2004), aubergines (600 mg/kg CGA), Asteraceae vegetables like lettuce, endive and artichoke (50–500 mg/kg total cinnamates) or green mate (107–133 mg CGA per approx. 200 mL brew) (Clifford, 1999).

Cinnamoyl-tartaric acid esters, such as caffeoyltartaric acid (caftaric acid), are especially abundant in grapes, with concentrations in grape juice that may reach up to 600 mg/L (Clifford, 1999). Cereals are the main sources of ferulic acid, although it is mostly present in bound form in the outer parts of the grain, with very low amounts in the endosperm (Manach et al., 2004); while whole wheat contains some 20–30 mg/kg cinnamic acids esterified to polysaccharides, wheat bran may reach some 4–7 g/kg and maize bran as much as 30 g/kg (Clifford, 1999).

Flavonoids are the largest class of polyphenols in plants and food, with >8000 naturally occurring compounds documented (Andersen & Markham, 2006), although the number of newly reported structures is continually growing.

Flavan-3-ols are majority flavonoids in many foods, being present in many types of fruits and vegetables, tea, cocoa and red wine; green tea and dark chocolate are considered by far the richest sources (Manach et al., 2004). Flavan-3-ols occur as monomeric (catechins) and oligo/polymeric forms (proanthocyanidins or condensed tannins). Catechin and epicatechin are the main monomers in fruits, whereas gallocatechin, epigallocatechin, and epigallocatechin gallate are mainly found in tea. Tea is probably the most important source of catechins in many countries (Hollman & Arts, 2000). Black tea contains lower catechin levels, as they are oxidized during the processing of tea leaves to complex forms, i.e., theaflavins (dimers) and thearubigins (polymers) (Santos-Buelga & Scalbert, 2000). An average serving of black tea (235 mL) may supply about 140 mg of flavonoids, from which >70% are thearubigins (around 100 mg per serving), 10% theaflavins and 8% catechins (Lakenbrink, Lapczynski, Maiwald, & Engelhardt, 2000). Proanthocyanidins are mainly found in berries, cocoa, some fruits like apple and plum, nuts, beans and some cereals (sorghum). Due to their difficult analysis and the fact that they are in part linked to matrix structures in insoluble forms, the contents of proanthocyanidins in plants and foods are not well known and usually underestimated. In a study on Spanish foods, de Pascual-Teresa, Santos-Buelga, and Rivas-Gonzalo (2000) found highest contents of flavan-3-ols (monomers to trimers) in broad beans (up to 184 mg/100 g fresh weight), followed by apples and plums (up to 50 mg/100 g) and chocolate (60% cocoa; up to 20.9 mg/100 g). Greater concentrations were determined by Gu et al. (2004) for total flavanols in U.S. commodities, including proanthocyanidin polymers in their estimation. Relevant levels were found in whole sorghum grain (>1900 mg/100 g), chocolate (>1600 mg/100 g), pinto beans (800 mg/100 g), some berries (>650 mg/100 g) or hazelnuts (around 500 mg/100 g) (Gu et al., 2004).

Around 30 anthocyanidins (i.e., anthocyanin aglycones) have been identified in nature, but only six of them are widespread: cyanidin, delphinidin, petunidin, peonidin, pelargonidin, and malvidin, being cyanidin glycosides the most common anthocyanins in foods (Santos-Buelga & González-Paramás, 2019). The most important anthocyanin food sources belong to Rosaceae fruits (berries, cherries, plums, apples), with contents that range from a few milligrams to >1000 mg per 100 g fw, reaching the highest levels in berries like blackcurrants, blackberry, blueberries or chokeberry (Andersen & Jordheim, 2013; Clifford, 2000b). Anthocyanins are also abundant in certain cereals and leafy and root vegetables, such as pigmented potatoes, eggplant, cabbage, or red onion, with values as high as 1400 mg/100 g found in purple corn and purple sweet potato (Andersen & Jordheim, 2013; Clifford, 2000b). Young red wines are also a relevant source of anthocyanins, with concentrations that may reach >500 mg/L (Santos-Buelga & González-Paramás, 2019).

Quercetin glycosides are the most ubiquitous flavonols in food, with kaempferol myricetin and isorhamnetin derivatives also well represented. They are found in many fruits and vegetables, although concentrations are usually below 10 mg/kg (Hertog, Hollman, & Katan, 1992), except for some products like onions, with contents of quercetin that may reach >600 mg/kg fw in some varieties (Crozier, Lean, McDonald, & Black, 1997), kale (around 110 mg quercetin/kg and up to 470 mg kaempferol/kg) and broccoli (30–37 mg quercetin/kg and 60–72 mg kaempferol/kg) (Hollman & Arts, 2000). Broad beans are a relevant source of myricetin (26 mg/kg) (Hertog et al., 1992).

Flavones (luteolin and apigenin glycosides) are mostly present in herbs and some vegetables, being parsley and celery the most important edible sources. Contents up to 40 mg/kg of luteolin and 191 mg/kg of apigenin have been reported in celery stalks (Crozier et al., 1997; Hertog et al., 1992), whereas celery leaves contain as much as 200 and 750 mg/kg of luteolin and apigenin, respectively (Hollman & Arts, 2000).

Flavanones are only found in significant concentrations in citrus fruits. Orange juice contains between 200 and 600 mg hesperidin/L, and the whole fruit may contain up to five times more. Contents of naringin ranging 73–481 mg/L have been reported in grapefruit juice, and 150–249 mg/L of narirutin in mandarin juice (Tomás-Barberán & Clifford, 2000).

Despite their restricted distribution, in regions with high consumption of citrus fruits, such as Mediterranean countries, the intake of flavanones might even exceed that other more widespread flavonoids like flavonols. Apple and derived juice and cider are the only relevant source of dihydrochalcones (namely phloretin derivatives) with concentrations that may reach >200 mg/L in freshly-prepared apple juices (Suárez-Valles, Santamaria-Victorero, Mangas, & Blanco, 1999).

Isoflavones, lignans and stilbenes are classified as phytoestrogens. Isoflavones have quite limited distribution in the plant kingdom being restricted to leguminous species. Soybeans and their processed products are by far their main dietary sources. Concentration ranges of total isoflavones in soybeans from 18 to 562 mg/100 g, and from 60 to 265 mg/100 g in soy flour have been reported, whereas soy-derived products, like tofu, miso or soy milk usually present contents below 100 mg/kg (Cassidy, Hanley, & Lamuela-Raventos, 2000; Mortensen et al., 2009). Their consumption is undoubtedly different among Western and Asian countries, where fermented soy products are part of the traditional diet; although the intake by Western vegetarians and soy-consumers is higher than for the rest of population, it is still low compared to intakes in Asian populations (Jaganath & Crozier, 2010).

Lignans and stilbenoids are non-flavonoid phytoestrogens present in several foods but usually as minor constituents. The only richest dietary source of lignans is linseed (flaxseed), that mostly contains secoisolariciresinol, with concentrations as high as 527 and 675 mg/kg being reported in flaxseed flour and meals, respectively (Cassidy et al., 2000). Other oleaginous seeds (soybean), algae, some legumes (lentils), cereal brans, and certain vegetables (garlic, asparagus, carrots, broccoli) and fruits (pears, plums) have been identified as minor lignan sources (Cassidy et al., 2000; Thompson, Robb, Serraino, & Cheung, 1991). In the human organism, plant lignans are metabolized by the gut microflora to the so-called mammalian lignans or enterolignans, enterodiol and enterolactone (Thompson et al., 1991), which would be the actual compounds responsible for the beneficial effects on human health that have been associated to lignan consumption, such as cancer protective effects (Adlercreutz, 2007).

Stilbenes are widely distributed in liverworts and higher plants. However, they are commonly found in the roots, barks, rhizomes and leaves, while their concentrations in edible parts are low, so that they are incorporated in very small amounts in the human diet (Cassidy et al., 2000). The major dietary sources of stilbenes are grapes, grape juices and wine, and peanuts and peanut butter (Cassidy et al., 2000). Contents in the range 0.3–15 mg/L have been reported in red wines (Manach et al., 2004). Resveratrol (3,5,4′-trihydroxystilbene) is one of the most studied phytochemicals regarding its biological activity and putative health benefits on human health (Rauf et al., 2017). However, owing to it is an extremely minor component in the human diet, its beneficial effects seem unlikely at normal food intakes (Manach et al., 2004), although it can be explored as a possible therapeutic agent (Rauf et al., 2017).

Phenolic alcohols, such as tyrosol and hydroxytyrosol and derived compounds like their esters with elenoic acid, such as oleuropein, present in the olive tree have also given relevance in recent years by their putative healthy effects against some types of cancer (breast, prostate and colon cancer). High concentrations of oleuropein are found in olive leaves (60–90 mg/g dry weight) (Soler-Rivas, Espín, & Wichers, 2000). García, Romero, and Brenes (2018) reported oleuropein contents up to 1411.0 ± 452.7 mg/kg, and of hydroxytyrosol up to 1133.1 ± 110.6 mg/kg in Spanish olives preserved in acidified brine of the Hojiblanca and Manzanilla cultivars. However, the levels of these compounds are dramatically reduced during processing to obtain table olives and olive oil, as they have to be removed due to the bitter taste that they impart. Concentrations of hydroxytyrosol in the range 9.4 ± 2.4 to 40.9 ± 6.3 mg/kg were determined by García et al. (2018) in American and Spanish of commercial black ripe olives, while no oleuropein was detected. Contents of hydroxytyrosol + tyrosol from 100 to 400 mg/kg oil were found in a screening on Spanish virgin olive oils from different varieties (Romero & Brenes, 2012).

3.2 Dietary intake of polyphenols

The interest in the associations between polyphenols consumption and health promotion has made the estimation of their dietary intake a point of interest. However, accurate data on phenolic composition in foods and beverages are not readily available and easy to obtain. On the one hand, due to their structural diversity there are no single and standardized analytical methods that allow the analysis of all polyphenol classes or compounds, so that the results obtained for a given food may vary depending on the methods employed. On the other hand, important variations in the qualitative and quantitative phenolic composition may exist within a particular plant product, as influenced by varietal, agronomic and environmental conditions, as well as growth or maturation stages. In addition, food processing and storage may involve processes of degradation and structural transformations, thus changing contents and composition profiles (Santos-Buelga & González-Paramás, 2014).

A first gross estimation of the human intake of phenolic compounds was made by Kühnau (1976), that estimated an average daily consumption of flavonoids in the United States diet of around 1 g, with catechins and biflavans (i.e., proanthocyanidins) as major contributors (more than two-thirds of flavonoid intake). According to that author, beverages and drinks (tea, coffee, cocoa, wine, beer) would account for around 40% of total flavonoid intake, and fruits, berries and fruit juices around 30%. More recent estimations of the total and individual polyphenol intake have been made taking advantage of data on polyphenol composition in foods, collected in databases compiled by different organisms over the last 20 years, especially the Phenol-Explorer and USDA databases (see Section 3.4). The calculated values for the dietary intake of total polyphenols show important variations, between a range from a few hundred mg/day to >1800 mg/day, as also do the types of phenolic classes consumed, depending on the region and target population, as well as on the methodology used for the assessment.

A daily polyphenol intake of 863 mg was calculated for Finnish adults (Ovaskainen et al., 2008), using data on food consumption obtained from a 48-h dietary interview and data on phenolic contents incorporated in the Finnish National Food Composition database. Phenolic acids derivatives were the main group of consumed polyphenols (75% of total intake), followed by proanthocyanidins (14%), anthocyanins and other flavonoids (10%), with coffee, cereals and berries and other fruits as the main dietary sources. Pérez-Jiménez et al. (2011) estimated a mean polyphenol intake of 1193 ± 510 mg/day in the French diet from subjects of the SU.VI.MAX (SUpplémentation en VItamines et Minéraux AntioXydants) cohort, using 24-h dietary records and data on polyphenol composition obtained from the Phenol-Explorer database. Hydroxycinnamic acid esters and proanthocyanidins were the most largely consumed polyphenols, being non-alcoholic beverages and fruits the main contributors to polyphenol intake. A food frequency questionnaire and the Phenol-Explorer database were also employed by Grosso, Stepaniak, Topor-Mądry, Szafraniec, and Pająk (2014) for the estimation of the intake and dietary sources of polyphenols in a Polish cohort from the HAPIEE (Health, Alcohol and Psychosocial Factors In Eastern Europe) study. They calculated a mean total polyphenol intake of 1756.5 ± 695.8 mg/day (median 1662.5 mg/day) with similar contribution of flavonoids (897 mg/day) and phenolic acids (800 mg/day), being newly caffeoylquinic acids (mostly originated from coffee) and catechin-related compounds (mostly from tea and cocoa derivatives) the most consumed individual types of compounds.

A comprehensive study on dietary polyphenol consumption was performed by Zamora-Ros et al. (2016) within the frame of the EPIC study (European Prospective Investigation into Cancer and Nutrition) conducted in 10 European countries (Denmark, France, Germany, Greece, Italy, Norway, Spain, Sweden, the Netherlands, and the United Kingdom). Estimations were made based on the information collected using a standardized 24-h dietary recall software linked with Phenol-Explorer database. They calculated mean total polyphenol intakes in the range 584–744 mg/day in Greece (the lowest consumption) to 1626–1786 mg/day in Denmark (the highest). The study found a large heterogeneity in both the nature of polyphenols and levels of intake across countries, although the main food sources for individual polyphenols were similar among regions, with coffee, tea and fruits as major contributors. Phenolic acid derivatives, namely caffeoylquinic acids, were the best represented phenolic class (52.5–56.9%), except in men from Mediterranean countries and in United Kingdom health-conscious consumers, where they were flavonoids (49.1–61.7%), mostly proanthocyanidin oligomers and polymers.

Using a food frequency questionnaire and the Phenol-Explorer database, Godos, Marventano, Mistretta, Galvano, and Grosso (2017) estimated a mean intake of polyphenols of 663.7 mg/day in adult subjects from Catania (Sicily, Italy), mostly phenolic acids (362.7 mg/day) and flavonoids (258.7 mg/day). Nuts were the main dietary sources of polyphenols, whereas tea and coffee were major contributors for flavanols and hydroxycinnamic acids, respectively, fruits for anthocyanins, citrus for flavanones, and vegetables for flavones and flavonols. Similar intakes of total polyphenols (683.3 ± 5.8 mg/day) were obtained by Vitale et al. (2018) for an Italian cohort of people with type 2 diabetes, using data of food consumption obtained with the EPIC food frequency questionnaire and the Phenol-Explorer and USDA databases. Equal contribution was found for flavonoids (47.5%) and phenolic acids (47.4%), with non-alcoholic beverages as the main food source (35.5% of polyphenol intake), followed by fruits (23.0%), alcoholic beverages (14.0%) and vegetables (12.4%).

A total polyphenol consumption of 820 ± 323 mg/day (443 ± 218 mg/day of flavonoids and 304 ± 156 mg/day of phenolic acids) was calculated by Tresserra-Rimbau et al. (2014) for Spanish adults included in the PREDIMED study, a 5-year feeding trial aimed at assessing the effects of the

Mediterranean diet on the prevention of cardiovascular disease. Compared with other countries, olives and olive oil represented an important differential factor to the profile of phenolic compounds consumed by this Spanish population. Lower intakes of polyphenols (332.7 ± 237.9 mg/day; median 299 mg/day) were, however, estimated by Karam, Bibiloni, and Tur (2018) for older adults from the Mediterranean Island of Mallorca (Spain), with flavonoids as the most consumed phenolic class (170.3 mg/day) and among them flavan-3-ols; alcoholic drinks, namely red wine, were the main contributors (118.3 mg/day) to polyphenol intake, followed by fruits (98.6 mg/day). Much greater daily intakes of polyphenols in the Spanish diet, ranging between 2590 and 3016 mg, were estimated by Saura-Calixto, Serrano, and Goñi (2007), which included in their calculation non-extractable polyphenols that contributed almost double amount than extractable ones.

Miranda, Steluti, Fisberg, and Marchioni (2016), using 24-h dietary recalls and the Phenol-Explorer database, calculated a mean total intake of polyphenols of 377.5 ± 15.3 mg/day in adults from São Paulo (Brazil). Phenolic acids (284.8 ± 15.9 mg/day) were the main polyphenol class, with coffee as their major source, whereas flavonoids were much lower (54.6 ± 3.5 mg/day). Higher average intakes of polyphenols (1198.6 mg/day) were estimated by Nascimento-Souza, de Paiva, Pérez-Jiménez, Castro Franceschini, and Ribeiro (2018) on an elderly population from another Brazilian region (Viçosa), also using a recall of habitual consumption and the Phenol-Explorer database. Newly, caffeoylquinic acids, largely originated from coffee, were the main dietary contributors to polyphenol intake. An average total flavonoid intake of 626 mg/day was recently estimated for Australian adults, with flavan-3-ols being the major contributors, and especially thearubigins from tea (Murphy, Walker, Dyer, & Bryan, 2019).

All in all, despite the broad distribution of phenolic compounds and their large content variations across plant-derived foods, on a global scale, the most important commodities that are relevant contributors to their dietary intake are usually associated to coffee, tea, red wine and cocoa, with fruits and vegetables generally in a second level (Crozier, Jaganath, & Clifford, 2009). As for compound classes, phenolic acid derivatives and flavonoids are the most abundant polyphenols in the diet, with predominance of one or another depending on the dietary habits. Hydroxycinnamoyl derivatives, and especially chlorogenic acids, would be the main phenolic acids consumed by most populations, whereas among flavonoids, flavan-3-ols, followed by anthocyanins and flavonols, are prominent.

3.3 Health implications of dietary polyphenols

Early observations on the beneficial effects of phenolic compounds from food were made in the 1930s by Szent-Györgyi and collaborators (Bentsáth, Rusznyak, & Szent-György, 1936; Bruckner & Szent-Györgyi, 1936), who found that flavonoid extracts obtained from lemon juice and paprika were able to counteract the vascular symptoms associated to the deficiency of ascorbic acid in man and guinea pigs. Based on it, they proposed a vitamin nature for flavonoids and called them "vitamin P" (Benthsáth, Rusznyák, & Szent-Györgyi, 1937; Rusznyak & Szent-Gyorgyi, 1936), a term that was dropped in 1950 once it was demonstrated that they were not indispensable (Anonymous, 1950). In recent years, the interest on flavonoids and other phenolic compounds from food has renewed owing to the accumulated epidemiological evidences that point to the existence of inverse correlations between their dietary intake and reduced incidence and mortality from several degenerative diseases. First observations referred to a reduction in the risk of coronary heart disease as related with flavone and flavonol intake (Hertog, Feskens, Hollman, Katan, & Kromhout, 1993; Hertog et al., 1995), but further associations have also been established for other chronic conditions and distinct polyphenol classes, especially different flavonoids groups and lignans. A great deal of studies concerns cardiovascular diseases (e.g., Grosso et al., 2017; Knekt, Jarvinen, Reunanen, & Maatela, 1996; McCullough et al., 2012; Wang, Ouyang, Liu, & Zhao, 2014; Wang et al., 2014), but also type II diabetes (Jacques et al., 2013; Liu et al., 2014; Tresserra-Rimbau et al., 2016; Xiao & Hogger, 2015; Xie, Huang, & Su, 2016; Zamora-Ros et al., 2014), some types of cancers (Boffetta et al., 2010; Hirvonen, Virtamo, Korhonen, Albanes, & Pietinen, 2001; Hui et al., 2013; Zamora-Ros et al., 2012), or cognitive decline and neurodegenerative disorders like Alzheimer's and Parkinson's diseases (Commenges et al., 2000; Devore, Kang, Breteler, & Grodstein, 2012; Letenneur, Proust-Lima, Le, Dartigues, & Barberger-Gateau, 2007).

Nowadays, phenolic compounds are considered, at least in part, responsible for the health protective effects of fruit and vegetable-rich diets. Nonetheless, evidences contributed by the epidemiological studies are still insufficient to claim undisputed positive health effects relating to polyphenol consumption, particularly with regard to long-term dietary ingestion (Vauzour, Rodriguez-Mateos, Corona, Oruna-Concha, & Spencer, 2010). Most of the available information on the biological activity and effects of the phenolic compounds have been obtained from in vitro, ex vivo and animal studies, whereas data directly obtained in humans are still scarce

(Ferreira, Martins, & Barros, 2017). Human intervention studies are usually restricted to short-term trials on a reduced number of people, often using supplementation with polyphenol preparations or pure compounds. Indeed, assessing the effects of dietary polyphenols is tricky and the conclusions may be biased by the fact that their sources (fruits and vegetables) are also rich in other components with putative healthy effects, such as vitamins, minerals, dietary fiber or antioxidants, while little dense in caloric nutrients. Actually, the lack of appropriate control study populations together with the insufficient knowledge on the phytochemical contents in food and beverages are common limitations in epidemiological and human intervention studies. Long-term, randomized, controlled, dietary intervention trials with appropriate controls are required in order to assess the unequivocal role that polyphenols play in preventing human disease (Vauzour et al., 2010). On the other hand, there is still insufficient knowledge on how age, genetics or gut microbiota influence polyphenol bioavailability. Furthermore, polyphenol bioaccessibility is highly dependent upon the food matrix and the manner in which the food is prepared. For instance, it is known that polyphenols can bind onto dietary fibers (e.g., hemicelluloses), which decreases their accessibility for absorption after ingestion in the upper digestive tract, thus increasing the fraction that reaches the colon, where polyphenols might be released by the action of bacteria (Jakobek & Matić, 2019). Other combinations that may affect polyphenol bioaccessibility can also take place with divalent metals or proteins. Further knowledge on all these aspects is required in order to establish the compounds and metabolites that are ultimately responsible for the in vivo activity of polyphenols, as well as to help define adequate biomarkers of their intake (Vauzour et al., 2010).

3.4 Databases and biomarkers

A limitation of most observational studies investigating the relation between polyphenols and health conducted so far is the difficulty to make accurate estimations of the phenolic consumption by individuals. On the one hand, dietary assessments are typically based on self-reported dietary recalls, food frequency questionnaires or diet diaries, thus relying on the participants' ability to report their own food intake (Zamora-Ros, Touillaud, Rothwell, Romieu, & Scalbert, 2014). On the other hand, reliable data on phenolic composition in food are scarce, a shortcoming that is being overcome as long as more complete databases are available.

Food databases including data on polyphenols and other phytochemicals have started to be compiled over the last 20 years. A database on isoflavones was firstly released by the Nutrient Data Laboratory (NDL) of the United States Department of Agriculture (USDA) in 1999 (updated in 2008) (U.S. Department of Agriculture, 2008). Data on flavonoids and proanthocyanidins were published in 2003 and 2004, respectively, and further merged and expanded to build a unique database in 2007 containing entries for some 50 polyphenols, namely flavonoids, phenolic acids, lignans and stilbenes (U.S. Department of Agriculture, 2011). A limitation in this database was that it only contains data for aglycones, whose chemical and biological properties may greatly differ from those of their glycosides, which are the compounds commonly present in foods.

In Europe, the eBASIS database (Bioactive Substances in Food Information Systems; http://ebasis.eurofir.org/Default.asp) was developed as a part or the EuroFIR initiative (European Food Information Resource; http://www.eurofir.org/), aimed at standardizing and harmonizing food composition data in Europe (Unwin et al., 2016). This database compiles data on 17 classes of plant bioactive compounds, and among them flavonoids, isoflavones, phenolic acids and lignans, in major European plant foods. It includes information not only on food composition but also on physiological effects, in vitro or in vivo biological activity, food processing, or biomarkers (Gry et al., 2007).

Phenol-Explorer (http://phenol-explorer.eu/) is a web-based database that contains representative mean content values for >500 polyphenols (glycosides, esters and aglycones) and 450 foods (Neveu et al., 2010). The values are expressed in standard units (mg/100 g of fresh weight and mg/100 mL for beverages) after conversion of the units found in the original publications. The web interface allows making queries on the data to identify foods containing a given polyphenol or polyphenols present in a given food. Further, this database has been enriched with data on human metabolites, as well as with information on the influence of food processing and preparation on polyphenol composition, thus allowing obtaining information on the intake of these compounds as they are consumed (Rothwell et al., 2013).

FooDB (http://foodb.ca/), supported by The Metabolomics Innovation Centre (TMIC) of Canada, is probably the most comprehensive database on food composition. It contains information regarding both nutrient and non-nutrient components, with detailed compositional, biochemical, chemical and physico-chemical data on the compounds, their food sources and concentrations, and putative physiological and health effects. Information can be browsed by food (listing foods by their chemical composition) or compound (listing chemicals by their food sources).

PhytoHub (http://phytohub.eu/) is another online database dedicated to dietary phytochemicals. Around 1000 compounds representing all polyphenol classes, terpenoids, or alkaloids are included. It provides information on their main dietary sources (extracted from the literature and online databases such as FooDB and Phenol-Explorer, with a direct link to FooDB food cards), physico-chemical characteristics and mass and spectral data, as well as about known human metabolites, and potential metabolites predicted through in silico expert systems.

Regarding metabolites, although not dealing strictly with polyphenols, it is also worth mentioning the Human Metabolome Database (http://www.hmdb.ca), a freely available electronic database containing about 114,100 entries on metabolites found in the human body, with many data fields hyperlinked to other databases (KEGG, PubChem, MetaCyc, ChEBI, PDB, UniProt, and GenBank) (Wishart et al., 2013). Another interesting web resource is the FOODBALL Portal (http://foodmetabolome.org/foodball) developed within the Food Biomarkers Alliance (FoodBAll), a project funded by the European Commission under the Joint Programming Initiative "A Healthy Diet for a Healthy Life," aimed at identifying and quantifying dietary biomarkers to be used for nutritional assessment and research. Besides, the Exposome-Explorer database (http://exposome-explorer.iarc.fr/) has started to be developed at the International Agency for Research on Cancer (IARC) in collaboration with the University of Alberta (Canada), with the aim of collecting biomarkers of dietary exposure that can be used for biomonitoring or disease etiology studies (Neveu et al., 2017).

Despite the increasing availability of data on polyphenol contents in food and metabolites, the accurate measurement of the polyphenol intake is still challenging. Firstly, there is a large number of existing compounds, distributed across a wide range of foods, which levels and profiles are strongly influenced by agronomic and environmental factors, as well as by the changes that may take place during food processing, storage and cooking. Moreover, data reported in the literature used to feed databases are sometimes of low quality due to insufficient food description, badly detailed sample collection and/or the use of non-validated methods. Another limitation of databases is that they only include contents of extractable polyphenols, soluble in the aqueous or organic solvents usually employed for their extraction from foods. However, a variable fraction of non-extractable polyphenols also exists in fruits and vegetables, accounting in many cases for >50% of the total polyphenol content, which is usually overlooked (Pérez-Jiménez & Saura-Calixto, 2015). Non-extractable polyphenols make part of the dietary fibers and may be degraded by the colonic microbiota releasing products that could contribute to the physiological effects of dietary phenolics.

The use of biomarkers is a promising alternative to overcome some of the indicated limitations, as they may better reflect exposure to polyphenols than intake measurements, as well as reduce biases associated with self-reporting diet assessment (Zamora-Ros, Touillaud, et al., 2014). However, the number of robust biomarkers for either individual or total polyphenol intake is yet very limited. The level of total phenolics in urine, as determined by the Folin-Ciocalteau reagent has been suggested as a biomarker for evaluating the dietary intake of polyphenols (Roura, Andres-Lacueva, Estruch, & Lamuela-Raventos, 2006). Nevertheless, the measurement of total polyphenols as a biomarker does not consider their large diversity in terms of structure, physicochemical properties, bioavailability and biological effects. Some metabolites have been proposed for the assessment of the intake of particular types of polyphenols, such as *S*-equol for soy isoflavones (Setchell, Brown, & Lydeking-Olsen, 2002), ellagic acid and urolithins for ellagitannins (Cerdá, Tomás-Barberán, & Espín, 2005), or enterodiol and enterolactone for lignans (Adlercreutz, 2007). However, the formation of these metabolites is dependent on the intestinal microbiota that may differ among individuals, thus limiting their reliability as biomarkers of the polyphenol intake for the whole of a population, although they could serve as a metabolic signature reflecting the catabolic capacity of the microbiome of each individual, and therefore indirectly be considered a marker of the individual gut microbiota composition, richness, diversity, and functionality (Tomás-Barberán, Selma, & Espin, 2018).

Indeed, defining adequate biomarkers for polyphenol intake is a tricky question, as there are marked differences in their metabolism and kinetics of appearance in systemic circulation. Previous enzymatic processes (deglycosylation, deesterification, depolymerization, etc.) may be required for the absorption of compounds, which are in part produced by the gut microbiota, so that the compounds may be absorbed in the large intestine, which takes longer times (6–8 h) than for those taken up in the small intestine (1–2 h). That means that the time of collection of samples after ingestion of a food needs to be long enough to cover full absorption (Ulaszewska et al., 2019). Further, in intestinal epithelial cells and liver, the compounds undergo xenobiotic phase I and/or phase II transformation. Therefore, for most polyphenols only metabolites are found in human plasma that in most cases are rapidly cleared with hardly retention in kidney, so that their total levels in plasma are very low, usually in the nanomolar range, which poses challenges for their analysis. Genetic polymorphisms and the diversity in the composition of the gut microbiota also cause great differences between individuals regarding their ability to metabolize polyphenols, adding an extra layer of complexity to metabolite analysis (Ulaszewska et al., 2019). The production of *S*-equol may serve as an example, with only 25–30% of the adult population in Western countries being able to produce it from soy products containing isoflavones (Setchell & Clerici, 2010).

4. Activity and mechanisms of action

4.1 Antioxidant activity

For years most of the beneficial health effects of polyphenols have been associated to their ability to act as effective scavengers of most types of oxidizing species, such as reactive oxygen and nitrogen species (RONS), through mechanisms that involve the transfer of an H atom or of a single electron to the radical stabilizing it (Procházková, Boušová, & Wilhelmová, 2011). In the case of flavonoids, the presence of a catechol group in the B-ring is the most significant determinant for scavenging of RONS, owing to its ability to donate hydrogen. Further structural criteria for optimal scavenging activity are the presence of a 2,3-double bond conjugated with a 4-oxo function in the C-ring and a 3- (and 5-) hydroxy group, as they provide extensive electron delocalization over the three-ring system and confer higher stability to the derived aroxyl radical (Bors, Heller, Michel, & Saran, 1990). Some authors, however, have questioned the stability of the formed flavonoid aroxyl radicals and have described their conversion into more reactive secondary radicals, such as quinones or semiquinones, that may give rise to pro-oxidant or potentially cytotoxic effects (Metodiewa, Jaiswal, Cenas, Dickancaité, & Segura-Aguilar, 1999).

The ability of polyphenols to prevent the toxicity of redox active metal ions, such as iron or copper has been less considered than their scavenging capacity. These cations are believed to catalyze the production of oxidant species leading to oxidation at different cellular levels (lipids, DNA or proteins). In the presence of hydrogen peroxide, Fe(II) catalyzes the formation of hydroxyl radicals ($OH^•$) by the Fenton reaction, whereas the reaction of Cu(II) with H_2O_2 leads to the formation of both $OH^•$ and superoxide

($O_2^{\bullet -}$) radicals. Polyphenols and in particular flavonoids can form stable metal complexes through their multiple OH groups and the carbonyl moiety, whenever present, removing a causal factor for the development of free radicals (Leopoldini, Russo, & Toscano, 2011).

Polyphenols can also regulate the oxidative status of the cell by inhibiting oxidative enzymes responsible for superoxide production, such as xanthine oxidase and protein kinase C (Ferriola, Cody, & Middleton, 1989). The interference with nitric oxide-synthase (NOS) activity is another potential mechanism to decrease oxidative damage in the cell. NO, produced by the oxidation of L-arginine catalyzed by NO synthases (NOS), interacts with free radicals, especially $O_2^{\bullet -}$, producing peroxynitrites. Although it is not clearly understood how polyphenols inhibit induction of NOS and NO production, they would possess ability to directly scavenge molecules of both NO and peroxynitrite once produced (Choi et al., 2002).

Despite the abundant literature about the antioxidant capacity of polyphenols, it is necessary to consider that these compounds are, in general, little bioavailable and largely biotransformed in the organism, so that their levels as such in body fluids, tissues and cells are usually very low and well below those of other physiological antioxidants, like urate, α-tocopherol, ascorbate or glutathione (Hollman, 2014). All in all, what seems clear is that the notion of these compounds acting as "systemic" antioxidants is unlikely to be the (sole) explanation for their putative health effects. Nowadays, other hypotheses are emerging to explain the in vivo activity of polyphenols, such as the possibility that they could act as modulators of gene expression and intracellular signaling cascades vital to cellular function (Williams, Spencer, & Rice-Evans, 2004).

4.2 Polyphenol–protein interactions

Another mechanism classically associated to some biological activities of polyphenols is their ability to bind a variety of proteins, including different enzymes. Main driving forces in these interactions are hydrogen bonding and hydrophobic effects (Hagerman, Rice, & Ritchard, 1998; Oh, Hoff, Armstrong, & Haff, 1980). Hydrogen bonding can be established between electronegative nitrogen or oxygen atoms from the amino and phenolic hydroxyl groups and positively charged hydrogen atoms from neighboring hydroxyl or amino groups of another polyphenol or protein molecules (Haslam, 1998). The keto group on the C-ring existing in some flavonoids, such as flavones and flavonols, could also participate in hydrogen bonding, as well as the glycosyl residues (Dangles & Dufour, 2008).

Hydrophobic interactions can take place between the benzenic ring of phenolic compounds and the apolar side chains of amino acids such as leucine, lysine or proline in proteins (Oh et al., 1980). The presence of proline is apparently a common characteristic of proteins with high binding affinities toward polyphenols (Hagerman & Butler, 1981). Proline residues possess a flat, rigid and hydrophobic surface, which favors the interactions with other planar hydrophobic surfaces such as benzenic rings (Murray, Williamson, Lilley, & Haslam, 1994). Furthermore, proline residues contribute to maintain the peptide in an extended conformation, thereby providing a bigger surface of protein to binding (Baxter, Lilley, Haslam, & Williamson, 1997).

Condensed and hydrolysable tannins are the classes of polyphenols more usually involved in the interactions with proteins. Tannins can act as multidentate ligands, so that one tannin molecule is able to bind to more than one protein at one time or to bind to more than one point in the same protein (Charlton, Haslam, & Williamson, 2002). The interactions are strongly influenced by the pH value, being higher at pH values close to the isoelectric point (pI) of the protein (Yan & Bennick, 1995). The ability to complex with proteins increases with tannin size and degree of galloylation probably because they have more interaction sites, although highly polymerized structures have more difficulty to bind proteins due to their lower flexibility and solubility in aqueous media (de Freitas & Mateus, 2001).

Protein–polyphenol interactions have been associated to anti-nutritional effects as they may lead to the inhibition of digestive enzymes decreasing the efficiencies of proteins and nutrient utilization (Butler, 1992). On the other hand, binding to enzymes involved in oxidative stress, such as xanthine oxidase or lipoxygenase, might also contribute to the antioxidant effects of polyphenols as it leads to enzyme inhibition and subsequent decrease in ROS production. Similarly, the interactions with specific proteins, such as protein kinases, phase I and phase II metabolism enzymes or transcription factors, could also play a determining role in the biological effects of polyphenols (Dangles & Dufour, 2008).

4.3 Pleiotropic effects of polyphenols

Polyphenols and their metabolites are increasingly recognized to exhibit a pleiotropic character, affecting multiple molecular targets, such as the modulation of signaling, energy-sensitive, oxidative stress and inflammation-related pathways, mitochondrial function or epigenetic modifications, most of them interconnected (Barrajón-Catalán et al., 2014).

Phenolic compounds have been shown to be able to modulate cell oxidative stress through the regulation of oxidative stress-sensitive pathways, such as the antioxidant response element (ARE) regulatory system (Chen, Yu, Owuor, & Kong, 2000). This activation would be related to the intrinsic ability of certain polyphenols to form potentially toxic quinones in cellular media, thus boosting the expression of enzymes for their own detoxication (Lee-Hilz et al., 2006), such as phase II detoxifying enzymes (e.g., NAD(P)H-quinone oxidoreductase, glutathione *S*-transferase, and UDP-glucuronosyl transferase) and antioxidant enzymes (e.g., glutathione peroxidase, catalase or superoxide dismutase) (Masella, Di Benedetto, Varì, Filesi, & Giovannini, 2005; Nagata, Takekoshi, Takagi, Honma, & Watanabe, 1999). The up-regulation of gene expression through induction of the ARE is triggered by the activation of Nrf2 (nuclear factor-erythroid 2 p45-related factor 2), a transcription factor that has been shown to be activated by different flavonoids, such as quercetin (Granado-Serrano, Martín, Bravo, Goya, & Ramos, 2012), epigallocatechin-gallate (Tsai et al., 2011), or resveratrol (Samsami-Kor, Daryani, Asl, & Hekmatdoost, 2015).

Polyphenol activity has also been associated to the ability to modulate energy metabolism and energy-sensing pathways. Leptin and adiponectin are adipokines involved in the glycemic control and energy homeostasis. Adiponectin increases glucose uptake in muscles and insulin sensitivity, suppresses gluconeogenesis in hepatocytes and increases fatty acid oxidation, while leptin is related to insulin resistance, increases energy expenditure and reduces food intake (Eseberri, Lasa, Churruca, & Portillo, 2013). Some polyphenols like resveratrol and its metabolites have been shown to be able to decrease leptin expression and secretion while increasing adiponectin's (Eseberri et al., 2013; Szkudelska, Nogowski, & Szkudelski, 2009).

The effects of polyphenols on energy homeostasis and inflammatory processes have been linked to the activation of AMPK (AMP-activated protein kinase) and subsequent inhibition of the mTOR (mammalian target of rapamycin) signaling pathway (Barrajón-Catalán et al., 2014). This pathway is involved in the regulation of adipose tissue functions such as adipogenesis, thermogenesis or lipid metabolism, and also modulates processes like mitochondrial biogenesis, hypoxia signaling, autophagy and cell cycle progression (Cai, Dong, & Liu, 2016). Some polyphenols like virgin olive oil secoiridoids (e.g., oleuropein and decarboxymethyl-oleuropein) were shown to be able to activate AMPK, suggesting them as gerosuppressant agents with potential application in the prevention and treatment of aging-related diseases, like cancer or diabetes (Menéndez et al., 2013). The inhibition of the mTOR gerogene has also been suggested to be related with the ability of phenolic compounds to mimic caloric restriction (Menéndez et al., 2013), a factor that is known to prolong lifespan in distinct organisms, including mammals.

Caloric restriction mimetic effects and lifespan extension have been reported for compounds like quercetin and the stilbenes resveratrol and piceatannol in evolutionarily distant species, such as *Saccharomyces cerevisiae*, the nematode *Caenorhabditis elegans*, *Drosophila melanogaster*, Zebra fish or mice (Baur et al., 2006; Howitz et al., 2003; Wood et al., 2004), and attributed to the activation of sirtuins, a family of highly conserved NAD^+-dependent protein deacetylases that modulate longevity and other age-related events.

Antiproliferative effects of polyphenols such as resveratrol (Yan et al., 2010) and virgin olive oil secoiridoids (Menéndez et al., 2013) have been related to the up-regulation of several heat shock proteins (HSPs) during endoplasmic reticulum stress, leading to the activation of unfolded protein response (UPR) and subsequent cell cycle arrest (Barrajón-Catalán et al., 2014).

The ability to reverse adverse epigenetic regulation involved in pathological conditions through the modulation of microRNA (miRNA) expression, histone acetylation, or DNA methylation has also been proposed as a mechanism to explain the effects of different polyphenols, such as anthocyanins, catechins, soy isoflavones or phenolic-rich extracts. Understanding how polyphenols can control small non-coding RNAs and regulate physiological mechanisms related to different pathological conditions, such as inflammation or obesity would allow for the development of dietary approaches to prevent metabolic complications (Correa & Rogero, 2019). Besides, dietary polyphenol-targeted epigenetics might become an attractive approach for disease prevention and intervention (Joven, Micol, Segura-Carretero, Alonso-Villaverde, & Menéndez, 2014; Pan, Lai, Wu, & Ho, 2013; Russo et al., 2017).

Most of the discussed activities and mechanisms have been shown in studies performed in vitro and cell or animal models and with isolated phenolic compounds or purified extracts. However, it is unclear whether they could explain the in vivo effects that have been associated to dietary polyphenols. Aspects like bioavailability, interactions with the gut microbiota, types of metabolites and their distribution and activity, molecular targets or toxicity must be still resolved.

4.4 Harmful effects

Despite their benefits, polyphenols may also cause adverse effects, especially in vulnerable populations, such as those with genetic polymorphisms in genes related to the polyphenol metabolic pathways. In general, when consumed as food components, polyphenols usually show low toxicity; however, adverse effects might take place for highly fortified foods or when ingested as supplements (Correa & Rogero, 2019).

The biological effects of many polyphenols have been described to follow a hormetic behavior, so that while they induce beneficial effects at low doses they act as toxic agents at higher levels. As previously indicated, in cell and tissue media polyphenols may behave as pro-oxidants. At low concentrations this activity has been associated to promotion of antioxidant defenses resulting in overall cell protection, but above certain pro-oxidant level the antioxidant cell response is overcome leading to oxidative stress (Tang & Halliwell, 2010).

The pro-oxidant activity of polyphenols might lead to carcinogenic or genotoxic effects. The production of forestomach and kidney tumors has been observed in rodents fed caffeic acid at high concentrations (Hagiwara et al., 1991). Carcinogenic effects on kidney were also observed for long-term dietary administration of quercetin (40–1900 mg/kg/day) to rats (Dunnick & Halley, 1992), whereas treatment with green tea catechins enhanced chemically-induced colon carcinogenesis in rats (Hirose et al., 2001). The consumption of green tea dietary supplements has also been associated to hepatotoxicity in several observational studies and related to liver oxidative stress probably induced by epigallocatechin-3-gallate (EGCG) or its metabolites (Mazzanti, Di Sotto, & Vitalone, 2015; Mazzanti et al., 2009). Treatment with tea polyphenols, and especially EGCG, has been shown to be cytotoxic in rat hepatocytes by producing an increase in ROS production and collapse of the mitochondrial membrane potential (Galati, Lin, Sultan, & O'Brien, 2006). Whereas traditional tea infusion is considered safe, for food supplements, experts from the European Food Safety Authority (EFSA) concluded that doses of EGCG at 800 mg/day may be associated with initial signs of liver damage (EFSA Panel on Food Additives and Nutrient Sources Added to Food, 2018). A level of around 300 mg/day has been estimated as a conservative limit for the consumption of EGCG delivered in solid dosage in adult individuals (Hu, Webster, Cao, & Shao, 2018).

The induction of estrogenic effects has also been described for some polyphenols. It has been suggested that endocrine-disrupting properties of isoflavones (or their metabolites) may compromise the growth and pubertal development of children fed soy-based formulas (Kim et al., 2011), as well as adversely affect women at-risk for estrogen-sensitive breast cancer and endometrial cancer (Zhong et al., 2016). Nevertheless, a recent report by the EFSA found no risk of taking isoflavone-containing food supplements for peri- and post-menopausal women (EFSA Panel on Food Additives and Nutrient Sources Added to Food, 2015). Independently of their estrogenicity, soy isoflavones could also induce antithyroid effects by inhibiting thyroid peroxidase, which might increase the risk of goiter. This activity may also include additional soy components, and other factors could be required, such as iodine deficiency (Doerge & Sheehan, 2002).

High consumption of polyphenols may also have antinutritional effects due to their metal-chelating properties. In particular, different phenolic compounds, such as tea catechins, quercetin or hydrolysable tannins, have been shown to be able to reduce iron absorption. This inhibitory effect can add to that of phytic acid, especially in diets rich in cereals and legumes, increasing the risk of iron deficiency in individuals with marginal iron status (Hurrell & Egli, 2010; Petry, Egli, Zeder, Walczyk, & Hurrell, 2010). By contrast, it has also been suggested that diets rich in polyphenols might be beneficial for groups at risk of iron loading, such as subjects with hereditary hemochromatosis (Lesjak et al., 2014). Tannins may also behave as antinutritional compounds because of their ability to interact with proteins and inhibit digestive enzymes, leading to decreased feed efficiency and reduced growth rate in experimental animals (Butler, 1992).

Polyphenols also may also affect drug bioavailability and pharmacokinetics, owing to their capacity to modulate the expression of genes related with oxidative stress and xenobiotic metabolism, like cytochrome P450 monooxygenases and phase II conjugation enzymes, as well as interfere with membrane transporters involved in drug excretion. This could either result in induction or inhibition of the metabolism of chemotherapeutic drugs and nutrients like some vitamins (Galli, 2007; Moon, Wang, & Morris, 2006). However, most of these effects have been shown in in vitro or ex vivo studies, and it has not been proven that these effects also occur in human intakes from habitual diets, which are usually lower than the doses used in the studies (Correa & Rogero, 2019).

5. Bioavailability and metabolism of polyphenols

The physiological effects of food polyphenols not only depend on their intrinsic activities, but also they are strongly influenced by their bioavailability. Following consumption, polyphenols can be subject to modifications in the upper part of the gastrointestinal tract, be absorbed and biotransformed in the small gut or, more often, reach the gut, where they are going to interact with the colon microbiota being catabolized to a range of phenolic metabolites, which might be absorbed and distributed by systemic circulation to different biological targets. In the end, the actual compounds that can be present in human compartments may be different from and possess distinct bioactivity than the original polyphenols present in food. The high variety of phenolic structures, their bioavailabilities and the different molecular mechanisms of action involved, together with the interindividual variability in composition and activity of gut microbiota, and aspects such as diet composition, food matrix or gastrointestinal transit time, are all variables that influence the effects of polyphenols in the human organism (Williamson, Kay, & Crozier, 2018).

For a compound being bioavailable (that is, being absorbed and becoming available at the site of action) has to be bioaccessible (that is, released from the food matrix in the gastrointestinal tract and become accessible to absorption). Bioaccessibility depends on the physicochemical characteristics of the compound (e.g., structure, solubility, …) and is strongly influenced by the food matrix, i.e., interactions with other components such as fibers, lipids and proteins, and their capacity to inhibit digestive enzymes.

The bioavailability of polyphenols can greatly differ among compounds and compound classes, although in most cases is considered to be low (Thilakarathna & Rupasinghe, 2013). It has been estimated that <5–10% of the phenolics consumed might be taken up in the small intestine to be rapidly conjugated by phase II enzymes (Clifford, 2004), so that they are going to appear in plasma mostly as conjugated metabolites, which would be the forms able to reach the physiological targets (Kroon et al., 2004). The remaining non-absorbed compounds reach the large intestine where they interact with the colonic microflora yielding a variety of catabolites that may be bioactive and contribute to explain a relevant part of the effects of polyphenols in the human organism.

5.1 Absorption and metabolic transformations in the small intestine

In general, the absorption of polyphenols in the digestive tract starts in the ileum and may take place by transcellular (through the cell by passive diffusion or membrane transporters) or paracellular transport (through the intercellular spaces).

Flavonoids mostly occur in plants and foods as glycosides that cannot be generally absorbed as such. Some glycosides, however, can be hydrolyzed by the enzyme lactase phloridzin hydrolase (LPH) in the intestinal epithelium to release the aglycone, which can enter the enterocyte by passive diffusion because of its increased lipophilicity (Day, Gee, Dupont, Johnson, & Williamson, 2003). Saliva and oral microbiota also show β-glucosidase activity, so that deglycosylation might already start in the oral cavity (Requena et al., 2010). A few specific flavonoid glycosides might also go into intestinal cells using membrane transporters, such as sodium-dependent glucose transporter SGLT1 (Day et al., 2003). There are evidences that some anthocyanins might be absorbed intact in a small percentage in the stomach (Lila, Burton-Freeman, Grace, & Kalt, 2016; McGhie, Ainge, Barnett, Cooney, & Jensen, 2003). Oligomeric and polymeric proanthocyanidins because of their size are not absorbed and most pass unaltered to the gut (Rios et al., 2002). Some minor amounts of catechin monomers and dimers have been occasionally detected in plasma, suggesting that some degree of proanthocyanidin depolymerization may occur in the gastrointestinal tract (Zhang et al., 2016), although in vivo studies including feeds to ileostomists have not supported this conclusion (Del Rio et al., 2013).

Phenolic acids are mainly found in food as esters with carboxylic acids or sugars and, among them, chlorogenic acids (i.e., hydroxycinnamoylquinic acids) are particularly prominent. Esters of phenolic acids are basically stable to the acidic conditions of the stomach and go intact into the intestine. In assays carried out in healthy subjects and ileostomists, Stalmach, Steiling, Williamson, and Crozier (2010) concluded that approx. 30% of chlorogenic acids intake may be taken up in the small intestine, whereas the remaining 70% goes to the gut where it could be subject to degradation of the microbiota. In most cases, hydrolysis and release of the parent hydroxycinnamic acid seems required before absorption, which can take place in the small intestine by the action of intestinal of pancreatic esterases (Andreasen, Kroon, Williamson, & Garcia-Conesa, 2001; Kroon, Faulds, Ryden, Robertson, & Williamson, 1997). Nevertheless, some studies have also detected caffeoylquinic acids in urine (Gonthier, Verny, Besson, Remesy, & Scalbert, 2003) or plasma (Lafay et al., 2006) of rats, as well as in human plasma after coffee intake (Monteiro, Farah, Perrone, Trugo, & Donangelo, 2007), suggesting that a part of hydroxycinnamoyl esters are absorbed as such. Hydroxycinnamic acids, namely ferulic acid, are also found bound to polysaccharides in cereals. These forms could be hydrolyzed in some extent by xylanases in the upper part of the gastrointestinal tract, although it is supposed that most of them reach the colon intact where they could be degraded by the microbiota (Couteau, McCartney, Gibson, Williamson, & Faulds, 2001). The absorption of hydroxycinnamic acids, present in food in free form or released from their esters, would take place in both the small and large intestine (Zhao, Egashira, & Sanada, 2003).

Once absorbed, the polyphenols are subject to the action of phase II enzymes, catechol-*o*-methyltransferase (COMT), UDP-glucuronosyltransferase (UGT) and sulfotransferase, to yield conjugated metabolites (glucuronides, sulfates and methylated derivatives) that will be found circulating in plasma. Except for particular compounds, like some catechins or certain isoflavones, flavonoid aglycones are not found as such in the blood (Poquet, Clifford, & Williamson, 2009; Terao, 2009). Phenolic acids appear in plasma, bile, and urine as free and conjugated forms as glucuronides, sulfates, and sulfoglucuronides. Conjugation starts in the enterocyte, although it mostly occurs in the liver (Plumb et al., 1999; Zhao, Egashira, & Sanada, 2004). Free ferulic and caffeic acids have been found in rat plasma after oral administration of caffeic acid, together with their sulfate and glucuronide conjugates, suggesting that not only sulfate and glucuronide conjugation was produced but also methylation (Azuma et al., 2000). Actually, methylation of hydroxycinnamic acids has been shown to occur in cultures of Caco-2 cells (Kern et al., 2003).

After conversions in the liver, enterohepatic transport in the bile may occur and some metabolites are recycled back to the small intestine. Additionally, the conjugated metabolites, especially glucuronide derivatives, can be pumped out into the gut lumen by efflux transporters such as multidrug resistance-associated proteins (MRPs) and breast cancer resistance protein (BCRP), decreasing in this way even more the bioavailability of polyphenols and limiting their exposure to target organs (Zhang et al., 2015).

The final concentration of phenolic metabolites that can be found in human plasma is usually very low and situated in the nanomolar to low micromolar range (Clifford, 2004), concentrations that are far below those of other physiological antioxidants (Hollman, 2014), thereby a direct antioxidant effect might be only expected in tissues directly exposed to polyphenols, like the gastrointestinal tract.

Some authors have suggested that polyphenols may overcome the challenges posed by their poor bioavailability through their accumulation in some body compartments, in view of their capacity to establish H-bonds and hydrophobic interactions with biomembranes (Laranjinha, 2010). Despite the ability of the conjugated phenolic metabolites to cross membranes is greatly reduced and requires specific transporters (Williamson, Barron, Shimoi, & Terao, 2005), in some tissues or under some physiological situations, such as inflammation (O'Leary et al., 2001; Shimoi et al., 2001) deconjugation of the metabolites can occur, so that aglycones may be released contributing to the in vivo effects of dietary polyphenols (Menéndez et al., 2011; Terao, Murota, & Kawai, 2011).

5.2 Polyphenol metabolism by gut microbiota

As stated above, the large majority of the consumed polyphenols are not absorbed in the small intestine or are recycled after metabolism, reaching the large intestine where they can be metabolized by the colonic microflora. Bacterial enzymes catalyze a wide variety of reactions that occur under anaerobic conditions and are based mainly on processes of reduction and/or hydrolysis, such as dehydroxylation, demethylation, decarboxylation, reduction of double bonds, hydrolysis of glucuronides, sulfates and glycosides, or ring cleavage (Stevens & Maie, 2016).

According to Williamson and Clifford (2017), phenolic compounds can be classified into three categories regarding microbiota catabolism. The first one would include the compounds for which no information is available, such as phlorotannins, naphthoquinones and anthraquinones, coumarins or pyranoanthocyanins. The second group is compounds associated with unique catabolites, e.g., tyrosols from oleuropein and related compounds, *S*-equol and 5-hydroxy-equol from isoflavones, urolithins from ellagitannins, or diarylbutanes from lignans. The third and predominant group comprises those substrates that despite being structurally different give rise to similar catabolites.

The removal of glycosyl or ester residues by beta-glycosidases and esterases could take place as a first step of the gut metabolism in the case of flavonoids and chlorogenic acids, respectively. The released flavonoid aglycones might further suffer a breakdown of their C-ring, followed by dehydroxylation and/or methoxylation (Serra et al., 2012). The predominant catabolites would be aromatic and phenolic acids with zero to three aromatic hydroxyls, or their mono- or dimethoxy analogs. Among others, 3,4-dihydroxyphenylacetic acid, 3,4-dihydroxybenzoic acid (protocatechuic acid), homovanillic acid, 3-(3,4-dihydroxyphenyl) propionic acid, and 3-(3-hydroxyphenyl) propionic acid, have been described as products from the bacterial metabolism of different flavonoid classes, such as flavones, flavonols, flavanones or anthocyanins, but also theaflavins or thearubigins and other polyphenols like chlorogenic acids. Some of the produced phenolic acids can be further decarboxylated to the corresponding phenols or methyl-phenols. All these types of products can further appear in plasma and urine conjugated with glucuronic acid and/or sulfate (Aura et al., 2005, 2002; Selma, Espín, & Tomás-Barberán, 2009; Williamson & Clifford, 2017). Catechins and oligomeric proanthocyanidins also undergo the opening of the C-ring followed by different reactions, like lactonization, decarboxylation, dehydroxylation or oxidation. Phenyl-gamma-valerolactones and phenylvaleric acids have been described as exclusive microbial metabolites of flavan-3-ols. Phenylvaleric acids can be subsequently biotransformed by successive loss of carbon atoms to give rise to different phenylacetic, phenylpropionic and hydroxybenzoic acids, and in minor extension to hippuric, vanillic and homovanillic acids (Rechner et al., 2004; Williamson & Clifford, 2010).

As for other polyphenol classes, sulfates and glucuronides of the parent hydroxycinnamic acids and their dihydro derivatives have been identified as major metabolites from the metabolism of hydroxycinnamate derivatives (Stalmach et al., 2009). The catabolism of ellagitannins gives place to ellagic acid that is then sequentially metabolized by the intestinal microbiota to urolithins. Urolithin A and B glucuronides and dimethylellagic acid-glucuronide have been reported as major metabolites in urine of individuals consuming ellagitannin-rich pomegranate juice (Seeram et al., 2006). Plant lignans are metabolized by the human gut microbiota to enterodiol and enterolactone. The bacteria involved in enterodiol formation are part of the dominant intestinal microbiota, whereas those producing enterolactone are minor, so that much larger amounts of enterodiol are formed (Clavel et al., 2005). Bioavailability of *trans*-resveratrol is very low, with glucuronides and sulfates and dihydroresveratrol conjugates reported as main metabolites (Walle, 2011). Dihydroresveratrol is formed by the hydrogenation of the double bond of resveratrol by the intestinal microflora (Juan, Alfaras, & Planas, 2010).

5.3 Interactions polyphenols–microbiota

It is important to remark that there is a great interindividual variation both in the amount and the profile of phenolic metabolites produced after polyphenol intake. These differences can be attributed in part to the existence in the human gut microbiome of the so-called "enterotypes," which are defined as networks of co-abundant microbial populations dominated by the prominent presence of one of these three genera: *Ruminococcus*, *Bacteroides* and

Prevotella (Arumugam et al., 2011). Nevertheless, modifications in minority microbial groups also play a determining role in the catabolism of some polyphenol classes, as they may differ in their ability to produce specific metabolites. For instance, urolithin production has been associated to greater populations of bacteria of the genus *Gordonibacter* (Romo-Vaquero et al., 2015) and *Akkermansia muciniphila* (Li, Lin, Vanhoutte, Woo, & Xu, 2016). The stratification of individuals according to their gut metabolic phenotypes (metabotypes) is crucial to understand the health effects of some dietary polyphenols, as reported for isoflavones or ellagitannins. It is well known that there are individuals with equol-producer and non-producer metabotypes, and that this determines the estrogenic effects produced by soy isoflavones (Setchell et al., 2002). Also, different urolithin-producing groups of individuals have been described regarding ellagitannin catabolism: those where the production of urolithins takes already place at zero time, others where it can be induced after some time of exposure to ellagitannins, and a final one for which induction of urolithin production by dietary supplementation is not possible (Tomás-Barberán, García-Villalba, González-Sarrías, Selma, & Espín, 2014). Belonging to one or another urolithin metabotype could explain the interindividual variability in the lipid-lowering effects and influence on cardiovascular risk biomarkers induced by consumption of ellagitannin-containing foods (González-Sarrías et al., 2017).

Unabsorbed polyphenols and metabolites can also have an impact on the composition of the gut microbiota. For instance, they have been suggested to be able to modulate (decrease) the Firmicutes/Bacteroidetes ratio (Etxeberria et al., 2015; Jin et al., 2018), which has been linked to obesity trends in humans, with Firmicutes being better represented in obese individuals (Cox & Blaser, 2013). Distinct polyphenols have been related with prebiotic-like effects, leading to an increase in the abundance of beneficial bacterial species, including Bifidobacteria and Lactobacilli spp. (Cardona, Andrés-Lacueva, Tulipani, Tinahones, & Queipo-Ortuño, 2013; Etxeberria et al., 2015; Griffin et al., 2017; Hervert-Hernández, Pintado, Rotger, & Goñi, 2009; Pozuelo et al., 2012), while producing a reduction in the levels of Bacteroides, Streptococci, Enterobacteriacae or Clostridia (Fiesel, Gessner, Most, & Eder, 2014; Kafantaris et al., 2017; Viveros et al., 2011). All these modifications have also a significant influence on the bioavailability and metabolism of the phenolic compounds, leading to changes in the profile of produced metabolites, which together with the alterations in the functionality of the microbiota would have subsequent effects on host health (Cueva et al., 2017; Ozdal et al., 2016; Tomás-Barberán, Selma, & Espin, 2016).

The potential of polyphenols and metabolites to modulate the composition of the gut microbiota is not easy to assess. Most studies have been carried out in animal models, isolated bacteria cultures, or in vitro incubations with human fecal flora. However, despite they provide useful information, those studies do not reflect well what may happen in humans. On the one hand, the composition of gut microbiota is highly variable between animal species. On the other hand, different bacteria also possess different ability to metabolize polyphenols and/or show different sensitivity against them. For instance, whereas some *Lactobacillus* species are intolerant to catechins (e.g., *L. fermentum*, *L. acidophilus*, *L. vaginalis*), other (*L. plantarum*, *L. casei*, *L. bulgaricus*) grow best in the presence of oligomeric procyanidins (Tabasco et al., 2011). Incubations with fecal samples do not truly represent the microbiota composition or the metabolic competence of the human gastrointestinal tract, as some species are strongly bound to the gut surface and may not be voided, and others are sensitive to oxygen and may not survive transfer to the culture medium (Williamson & Clifford, 2017). Studies in human volunteers provide a more realistic situation, although they do not always lead to clear or concluding results. Many intervention studies with different polyphenol substrates (e.g., green tea, red wine, fruit or whole cereal preparations) have failed to show substantial changes in the gut microbiota, while in others just modest changes were detected and subjected to considerable interindividual variations. It has been suggested that possibly the changes in the composition of the gut microbiota cannot be detected in discrete observations, but they take several generations to develop, or that the persons must be exposed to the particular diet since an early age (Wu et al., 2016). These limitations notwithstanding, collectively studies demonstrate that the composition of the human gut microbiota can be modulated in vivo by supplementation with some polyphenol-rich commodities, but that modulation is not an inevitable consequence, depending at least in part on the individual metabotype (Williamson & Clifford, 2017).

6. Preparation of extracts and compounds

6.1 Extraction from natural sources

The emergence of phenolic compounds as added-value ingredients with potential application in the pharmaceutical, food and nutraceutical industries, has drawn significant attention in recent years. Considering all the health promoting benefits associated with these classes of compounds, their recovery from different natural matrices is becoming a very hot topic in the multidisciplinary area of applied chemistry, biology and biotechnology. Among several techniques of extraction, the conventional extraction methods such as maceration and heat-assisted extraction in their simplest form basically involve mixing the matrix and the solvent, allowing for interaction and subsequent release of bioactive compounds. These processes require long extraction times, high solvent and energy consumption and usually result in low extraction yields (Azmir et al., 2013). Bearing in mind these constraints, several non-conventional extraction techniques have been utilized to maximize extraction of phenolic compounds, such as microwave-assisted (MAE), hydrostatic pressure (HPAE), pulsed electric field (PEF), supercritical fluid (SFE) and ultrasound-assisted extraction (UAE), offering advantages of short extraction time, lower solvent usage and higher extraction yield (Oludemi et al., 2018).

UAE and MAE are effective techniques that maximize extraction yield in a short time, using moderate energy and low solvent consumption, while offering protection to thermo-labile bioactive compounds (Heleno et al., 2016; Medina-Torres, Ayora-Talavera, Espinosa-Andrews, Sánchez-Contreras, & Pacheco, 2017; Vieira et al., 2017). UAE is based on the principle of acoustic cavitation, which can disrupt the cell walls thereby allowing the release of bioactive compounds. MAE uses microwave energy to heat up samples, moisture present in the matrices is evaporated and further generates pressure on the cell wall that favors leaching out the target compounds. HPAE is a faster, effective, non-thermal and environment-friendly extraction process that operates under very high pressures ranging from 10 to 1000 MPa. In this technique, higher pressure favors solvent penetration into cells and hence high compounds release (Briones-Labarca, Giovagnoli-Vicuña, & Cañas-Sarazúa, 2018; Haining & Yongkun, 2017). SFE is a sustainable extraction process that utilizes supercritical fluids such as CO_2, ethane, butane, pentane, nitrous oxide, ammonia, trifluoromethane, or water. This technique requires high capital investment, but it is environment friendly, safe, non-flammable and non-toxic (Ghafoor, Al-Juhaimi, & Choi, 2012; Santos, Villaverde, Silva, Neto, & Silvestre, 2012). PEF is an emerging non-thermal technology that involves exposing solid samples to electric field pulses of high intensity (100–300 V/cm to 20–80 kV/cm) during short periods, which induce a transmembrane potential difference across the cell membrane. When the potential difference reaches a critical value, the cell membrane starts to collapse allowing for increased intracellular metabolite extraction (López, Puértolas, Condón, Álvarez, & Raso, 2008; López-Giral et al., 2015). An interesting feature is the possibility of combining two techniques to enhance extraction efficiency. Hybrid systems have been, for instance, applied to prepare phenolic compounds rich extracts from pomegranate peel (*Punica granatum* L.) (Rocchetti et al., 2018), blackberry (*Rubus* sp.) (Pasquel-Reátegui, Machado, Barbero, Rezende, & Martínez, 2014) or malagueta pepper (*Capsicum frutescens* L.) (Santos, Aguiar, Barbero, Rezende, & Martínez, 2015), among others.

In recent times, novel green extraction solutions have emerged as environmental-friendly alternatives to conventional extraction procedures, such as the use of the so-called deep eutectic solvents (DES). DES are composed of two non-toxic components, one with capacity to be a hydrogen-bond acceptor (HBA), namely quaternary ammonium, tetraalkylammonium or phosphonium salts, and the other possessing hydrogen-bond donor (HBD) properties, like acids, alcohols, amines or carbohydrates (Cunha & Fernandes, 2018). DES present the advantages of easy preparation, non-flammability, lower toxicity, adjustable viscosity, low volatility, solubility in water, high extraction yield and high solubilization strength for distinct compounds, and they have been used for the extraction of phenolics from various matrices (Chanioti & Tzia, 2018; Vieira et al., 2018).

Obtaining enhanced polyphenol-rich extracts from a given matrix depends on the extraction method used. Different variables, such as solvent type and proportion, temperature, time, molecular affinity between solvent and solute, solid–liquid ratio, and particle sizes have an influence in the extraction yield (Azmir et al., 2013). To effectively carry out an optimization, the influence of each defined variable should be independently assessed. Nevertheless, the application of mathematical models such as the response surface methodology (RSM) is gaining place among the scientific community. Through RSM design, the optimization of possible interactions among experimental variables is allowed simultaneously with the prediction of the most efficient conditions. RSM has been successfully applied to optimize the extraction of polyphenols from different plant materials (e.g., Campone et al., 2018; Heleno et al., 2016; Jiménez-López et al., 2018; Oludemi et al., 2018; Pinela et al., 2018; Vieira et al., 2017).

Beyond climate change, the most important challenges of our globalized world are the increased demand for energy, unsustainable consumption and production of food, and waste generation (King et al., 2017). A new approach to sustainability is the adoption of the circular economy model, where sustainable and resource-efficient policies are adopted, one of which involves the conversion of low-value side streams/residues/wastes into more valuable products (Zabaniotou & Kamaterou, 2018). As the agro-industry continues to generate wastes and by-products, their re-use has been emphasized over the last years owing to their richness in bioactive compounds. The valorization of these agro-industrial by-products has been widely reported due to its usefulness as ingredients in functional foods, supplements, cosmetics and nutraceutical products (Pinela et al., 2017).

Coffee is one of the most important food commodities all over the world, but this product carries a huge economic and environmental burden in the form waste streams with potential to be converted into various high value-added biomolecules (Janissen & Huynh, 2018). Some of its wastes, such as spent coffee grounds, coffee pulp and husk, coffee silver skin and coffee beans produced during treatment and processing of coffee cherries, milling of dried beans, roasting of green coffee beans and coffee brewing, represent excellent sources of polyphenols (tannins and chlorogenic acids) and caffeine (Janissen & Huynh, 2018; Mirón-Mérida, Yáñez-Fernández, Montáñez-Barragán, & Barragán, 2019; Pettinato, Casazza, Ferrari, Palombo, & Perego, 2019). Viticulture constitutes a relevant traditional activity in several countries in Southwestern Europe (i.e., France, Greece, Italy, Portugal, and Spain). This industry produces bio-residues either in the form of organic wastes, wastewater, emission of greenhouse gases and inorganic residues, which are discarded in open areas potentially causing environmental problem or in some cases used as organic fertilizer or intended for animal feed (Teixeira et al., 2014). Grape pomaces remaining after winemaking still contain large residual amounts of polyphenols, namely anthocyanins and flavan-3-ols (catechins and proanthocyanidins), in addition to other components such as dietary fiber, lipids, proteins and minerals of potential exploitation (García-Lomillo, González-San José, Pino-García, Rivero-Pérez, & Muñiz-Rodríguez, 2014). Much research has been devoted to the extraction of phenolic compounds from grape pomaces using different solvents and procedures (Barba, Zhu, Koubaa, Sant'Ana, & Orlien, 2016). As a result, a large number of grape phenolic extracts have been introduced onto the market and used in dietary supplements, as well as for cosmetic or pharmaceutical purposes (Fontana, Antoniolli, & Bottini, 2013). Lettuce being one of the most important fresh-cut vegetable in the world also generates wastes in the form of the external leaves removes and core. High-pressure homogenization pre-treatment and ultrasound-assisted extraction have been applied for enhanced recovery of polyphenols from lettuce (caffeoylquinic acid, caffeoyltartaric acid, isochlorogenic acid, chicoric acid, luteolin 7-O-glucuronide and quercetin-3-O-glucuronide) (Viacava, Roura, & Agüero, 2015). With global onion production predicted to increase significantly up to over 100 million in the coming years, onion skin bio-wastes have become an important source of phenolic compounds. Recently, Munir, Kheirkhah, Baroutian, Quek, and Young (2018) obtained quercetin and kaempferol-rich extracts from waste onion skins using subcritical water extraction in a high-pressure reactor.

A summary of recently published works exploring the use of some of the novel/green/sustainable/non-conventional techniques mentioned above to the extraction of phenolic compounds from different natural matrices is collected in Table 1. However, these technologies still present challenges, as the choice of extraction conditions varies depending on the plant material and compounds of interest, thus making it difficult to set an operating condition. In addition, most studies have been conducted in lab-based extractors, so that there is still a need for further developments to scale-up the procedures to industrial extractors with high loading capacity while maintaining an attractive cost.

6.2 Biotechnological production of polyphenols

A large number of polyphenols are synthesized by plants with a potential to be utilized as interesting health promoting biomolecules. Accessibility to each individual compound in a complex mixture of biomolecules vary among different matrices, seasonal and geographical variation, abundance of crop species and large capital investment to achieve optimum polyphenol extraction from these natural matrices. Large-scale chemical synthesis of many of these biomolecules is complicate, as it involves a series of chemical reactions, laborious purification, variable yields, expensive precursors and/or use of toxic chemicals and catalysts, making it less profitable and posing risks to human health (Chouhan, Sharma, Zha, Guleria, & Koffas, 2017). Plant cell cultures in bioreactor-based systems have been tried for large-scale recovery of polyphenols, although they present serious challenges, such as formation of aggregates, need of reducing exposure of cells to lighting, slow growth rates, susceptibility to stresses, difficulty of cultivation, or unavailability of convenient genetic manipulation techniques (Chouhan et al., 2017; Fowler & Koffas, 2009).

Nowadays, the pathways responsible for the biosynthesis of these biomolecules have been genetically engineered and integrated into microbial host strains, allowing for large-scale production of individual polyphenols using

Table 1 Recent developments in the preparation of phenolic compounds rich extract from natural sources (plant materials and by-products).

Matrices	Part used	Extraction conditions	Bioactive compounds	References
MAE				
Olea europaea L.	Olive pomace	*Solvents* (choline chloride–glycerol, choline chloride–maltose, choline chloride–citric acid and choline chloride–lactic acid), *Power* (200 W), *Temperatures* (40 or 60 °C), *Time* (30 min)	Oleuropein, hydroxytyrsol, caffeic acid, vanillin rutin, luteolin	Chanioti and Tzia (2018)
Vitis vinifera L.	Grape pomace	*Frequency* (2458 MHz), *Power* (1000 W)	Malvidin-3-*O*-glucoside, epicatechin, rutin, quercetin, catechin	Caldas et al. (2018)
Coffea arabica L.	Spent coffee grounds	*Solvent* (ethanol and deionized water), *Time* (10 min), *Power* (500 W), *Solid–liquid ratio* (100 g/L)	Chlorogenic acid and its derivatives	Pettinato et al. (2019)
Hibiscus sabdariffa L.	Calyces	*Solvents* (Water–ethanol 6–84%), *Power* (850–1500 W), *Temperature* (50–164 °C), *Time* (3–20 min), *Solid–liquid ratio* (100 g/L)	Gallic acid, chlorogenic acid, rutin, quercetin, quercetin-glucoside, *p*-coumaric acid, myricetin, quercitrin	Pimentel-Moral et al. (2018)
Origanum glandulosum L.	Aerial parts	*Solvents* (Water–ethanol 0%, 50% and 100%), *Temperature* (30, 90, 150 °C), *Time* (1, 5.5, 10 min), *Solid–liquid ratio* (50 g/L)	Caffeic acid hexoside, luteolin-*O*-diglucuronide, rosmarinic acid derivative, rosmarinic acid, isosalvianolic acid derivative	Nabet et al. (2019)

UAE				
Olea europaea L.	Olive pomace	*Solvents* (choline chloride–glycerol, choline chloride–maltose, choline chloride–citric acid, and choline chloride–lactic), *Frequency* (60 kHz), *Power* (280 W), *Temperatures* (40–60 °C), *Time* (30 min)	Oleuropein, hydroxytyrsol, rutin, caffeic acid, vanillin, luteolin	Chanioti and Tzia (2018)
Olea europaea L.	Fruit powder	*Solvent* (80% methanol), *Power* (240 W), *Temperature* (30–70 °C), Time (10–50 min), *Solid–liquid ratio* (10–50 g/L)	Gallic acid, hydroxytyrosol, rutin oleuropein, *p*-hyroxybenzoate salicylic, benzoic acid	Deng et al. (2017)
Solanum melongena L.	Fruits	*Solvent* (acidified water), *Frequency* (12 kHz), *Power* (400 W), *Temperature* (25 °C), *Time* (30–50 min)	Anthocyanins	Ferarsa et al. (2018)
Morus alba L.	Leaves	*Power* (600 W), *Frequency* (25 kHz), *Temperature* (40 °C), *Time* (30 min)	Phenolic acids, catechin, epicatechin, rutin, astragalin, quercetin	Zhou et al. (2018)
Solanum tuberosum L.	Potato peels	*Solvent* (Ethanol–water 50–100%), *Temperature* (20–50 °C), *Time* (15–45 min), *Solid–liquid ratio* (20–90 g/L)	1-*O*-caffeoylquinic acid, chlorogenic acid, 4-*O*-caffeoylquinic acid, caffeic acid	Riciputi et al. (2018)
Citrus reticulata B.	Mandarin peels	*Solvent* (80% acetone), *Frequency* (38.5 kHz), *Power* (30.34–59.36 W), *Temperature* (30–50 °C), *Time* (20–40 min), *Solid–liquid ratio* (8 g/L)	Hesperidin	Nipornram, Tochampa, Rattanatraiwong, and Singanusong (2018)

Continued

Table 1 Recent developments in the preparation of phenolic compounds rich extract from natural sources (plant materials and by-products).—cont'd

Matrices	Part used	Extraction conditions	Bioactive compounds	References
Vitis vinifera L.	Grape pomace	*Solvent* (44% ethanol), *Frequency* (20 kHz), *Power* (500 W), *Temperature* (50 °C), *Time* (3 min)	Hydroxycinamic acid derivatives, flavonols, tannins, catechins and anthocyanins	Poveda, Loarce, Alarcón, Díaz-maroto, and Alañón (2018)
Lactuca sativa L.	Lettuce waste	*Power* (400 W), *Frequency* (24 kHz), *Temperature* (50 °C)	*O*-caffeoylquinic acid, chicoric acid, caffeoyltartaric acid, isochlorogenic acid, luteolin 7-*O*-glucuronide, quercetin-3-*O*-glucuronide	Plazzotta and Manzocco (2018)
HPAE				
Olea europaea L.	Olive pomace	*Solvent* (choline chloride–glycerol, choline chloride–maltose, choline chloride–citric acid, and choline chloride–lactic), *Pressure* (300–600 MPa), *Time* (5–0 min)	Oleuropein, hydroxytyrosol, caffeic acid, vanillin, rutin, luteolin	Chanioti and Tzia (2018)
Vitis vinifera L.	Grape marc	*Solvent* (water–ethanol 50%), *Pressure* (10.0 ± 0.5 MPa), *Temperature* (40–100 °C), *Time* (220 min)	Anthocyanins	Tamires et al. (2019)
Citrus sinensis L.	Orange peel	*Solvent* (75%, 50% and 100% ethanol), *Pressure* (10 MPa), *Temperatures* (45, 55, and 65 °C), *Time* (40 min)	Hesperidin and narirutin	Barrales et al. (2018)
Nasturtium officinale W.T. Aiton	Whole plant	*Solvent* (0–100% ethanol), *Pressure* (0.1–600 MPa), *Time* (1.5–33.5 min), *Temperature* (20 °C), *Solid–liquid ratio* (30 g/L)	Quercetin glycoside derivative, kaempferol glycoside derivatives, phenolic acids	Pinela et al. (2018)

Lonicerae Japonicae flos Thunb.	Whole plant	*Solvent* (water–ethanol 90%), *Pressure* (0–2000 bar), *Time* (30–150 s)	Neochlorogenic acid, chlorogenic acid, 3,5-dicaffeoylquinic acid, 4,5-dicaffeoylquinic acid	Duan et al. (2018)
SFE				
Hibiscus sabdariffa L.	Dried calyces	*Solvent* (CO_2–ethanol, 7–15%), *Pressure* (150–350 bars), *Time* (90 min), *Temperature* (40–60 °C)	Chlorogenic acid, 5-*O*-caffeoylshikimic acid, methylepigallocatechin, myricetin	Pimentel-Moral et al. (2019)
Allium cepa L.	Onion skin	*Pressure* (100 bar), *Temperature* (40 °C), *Time* (120 min)	Protocatechuic acid, quercetin, quercetin-7,4′-diglycoside, quercetin 3,4′-diglycoside	Campone et al. (2018)
Arbutus unedo L.	Fruit	*Pressure* (100, 175 and 250 bar), *Temperature* (40, 55 and 70 °C)	Gallic acid, gallotannins protocatechuic acid, catechin, ellagic acid, ellagitannins, anthocyanins	Alexandre, Matias, Duarte, and Bronze (2018)
Acca sellowiana (O. Berg) Burret	Guava peel	*Pressure* (200 and 300 bar), *Temperature* (40 and 55 °C), *Solid–liquid ratio* (150 g/L)	Ferulic acid, ellagic acid, vanillic acid	Santos, Baggio Ribeiro, Micke, Vitali, and Hense (2019)

sustainable and environmentally friendly resources (Milke, Aschenbrenner, Marienhagen, & Kallscheuer, 2018). Biotechnological production of polyphenols like lignins, salicylates, coumarins, hydroxycinnamoyl derivatives, pigments, and flavonoids is possible because of the presence of similar metabolic pathways, enzymes, co-substrates and precursors in both plants and microorganisms (Fowler & Koffas, 2009). The most commonly used eukaryotic and prokaryotic organisms for metabolic engineering are *Saccharomyces cerevisiae* and the Gram-negative bacterium *Escherichia coli* (Chouhan et al., 2017). Further biotechnological advancements have seen the introduction of other species such as *Corynebacterium glutamicum*, *Lactococcus lactis*, or *Streptomyces venezuelae*, as viable alternative hosts for microbial production of polyphenols (Kallscheuer et al., 2016; Milke et al., 2018). Naringenin and pinocembrin are flavanones derived from the phenylpropanoids *p*-coumaric acid and cinnamic acid, respectively, which are among the first plant-derived polyphenols produced by metabolic pathway reconstruction in *E. coli* (Hwang, Masafumi, Yasuo, & Sueharu, 2003). The authors used three enzymes from different sources, namely phenylalanine ammonia lyase (PAL) from *Rhodotorula rubra*, coumarate:coenzyme A ligase (4CL) from *Streptomyces coelicolor*, and chalcone synthase (CHS) from *Glycyrrhiza echinata*. Katsuyama, Matsuzawa, Funa, and Horinouchi (2008) reported increase recovery of curcuminoids up to 100 mg/L, from an *E. coli* system carrying 4CL and curcuminoid synthase (CUS) from rice (*Oryza sativa* L.) in the presence of supplemented phenylpropanoid acids (*p*-coumaric acid, cinnamic acid and ferulic acid). Good yields of the anthocyanin cyanidin 3-*O*-glucoside (350 mg/L) were obtained in *E. coli* supplemented with (+)-catechin, after optimizing cultivation and induction parameters in order to improve the expression of anthocyanidin synthase (ANS) and 3-*O*-glycosyltransferase (Lim et al., 2015). *Saccharomyces cerevisiae* was also genetically engineered to produce naringenin up to 400 μM, solely from glucose in the presence of CS and tyrosine ammonia lyase (Koopman et al., 2012). Different flavonoids were also produced in *S. cerevisiae* including 4CL (*Petroselinum crispum*), CHS (*Petunia hybrida*) and other genes, with the most abundant titers recovered being kaempferol, 26.57 ± 2.66 mg/L and quercetin, 20.38 ± 2.57 mg/L (Rodriguez et al., 2017). Kallscheuer, Vogt, and Marienhagen (2017) were able to develop a strategy for enhanced resveratrol recovery using *C. glutamicum* by reversing the β-oxidative phenylpropanoid degradation pathway, thus avoiding any ammonia lyase activity.

Some challenges toward enhancing polyphenols recovery from these microbial hosts include insufficient supply of precursor molecules needed to increase polyphenol synthesis by the microbial metabolism, and low activity of plant-derived enzymes in heterologous hosts (Milke et al., 2018). The limiting step during microbial synthesis of polyphenols is the low activity of these heterologous enzymes. Even though researches from diverse background are optimistic about the great potential of phenolic compounds recovery using genetical engineering techniques, further studies are still needed to address the challenges toward achieving economically feasible microbial polyphenol production.

6.3 Emerging technologies to improve the bioavailability of phenolic compounds

As stated above, the biological activities displayed in vitro by the phenolic substances do not always translate when tested in in vivo models, mainly due to their low bioavailability at the target sites (Souza, Casanova, & Costa, 2015). These compounds face several environmental limitations, such as instability due to the interference with other compounds and sensitivity to heat, light, pH or temperature, metabolic transformations (methylation, glucuronidation, and sulfation), high rate of metabolism and rapid elimination from the human body, which will result in a loss of their efficiency and effectiveness (Massounga-Bora, Ma, Li, & Liu, 2018). These constraints should be taken into account for the design of supplements or functional foods fortified with phenolic compounds (Faridi, Assadpour, & Mahdi, 2018). With this aim, new strategies have been developed in the form of nano- and microtechnologies that employ wall/coating materials or capsule membranes to protect active principles, improve stability, and/or provide their gradual release over time, assuring correct delivery at the target sites (Dias, Ferreira, & Barreiro, 2015). The encapsulation can be performed using different methods, such as coacervation, liposomes, fluid bed coating, extrusion-, spray-, emulsion-, ultrasound-, or supercritical fluid-based processes, among others (Dias et al., 2015). The choice of the technique is driven by the desired application of the product, and selected methods should be capable to produce particles with homogeneous size distribution, with high encapsulation efficiency, high loading capacity, and able to ensure a sustained release (Vincekovi et al., 2017). Some of the used technologies are described below.

Spray-based processes. These systems are the most used encapsulation techniques. They consist of converting a solution, suspension or emulsion of the bioactive ingredient through the dispersion of the stream, by a process of atomization, into individual dried powder particles in a single step that allows the active principle to be entrapped (O'Sullivan, Norwood, O'Mahony, & Kelly, 2019). This technique offers economic and processing advantages over other techniques and is also suitable for formulating both heat-sensitive, heat-resistant and low solubility bioactive molecules (Vieira da Silva, Barreira, & Oliveira, 2016).

Coacervation. This methodology employs hydrocolloids, primarily proteins and polysaccharides, to induce the formation of a shell surrounding the active ingredient by modification of the media environment (pH, ionic strength, temperature, solubility). It is classified as simple, when only one hydrocolloid is used, or complex, when two oppositely charged hydrocolloids are used. Both approaches have been successfully used to encapsulate food ingredients, while providing a controlled release (Timilsena, Akanbi, Khalid, Adhikari, & Barrow, 2019).

Emulsion-based processes. Emulsion-based techniques allow for encapsulation of both water and oil soluble biomolecules. Emulsions can either be oil/water emulsion, when oil droplets are dispersed in an aqueous phase, or water/oil emulsion when the water droplets are dispersed in an oil phase (Vieira da Silva et al., 2016). Emulsification is also a critical step in producing microcapsules/microparticles in other encapsulation processes (Dias et al., 2015).

Fluidized-bed coating. This is one of the most efficient microencapsulation techniques, which is finding ever-growing applications in the food and pharmaceutical industries (Dewettinck & Huyghebaert, 1999). It involves suspending the core material by an air stream under controlled temperature and humidity, and then spraying the coating material around the bioactive compound (Bakry et al., 2016).

Liposomes. Liposomes are spherical vesicles (15–1000 nm), made up of a bilayer of phospholipids, with a hydrophobic tail and a hydrophilic head, that allow the encapsulation of hydrophilic compounds within the core of the liposome and the hydrophobic ones partitioned within the bilayer. They present advantages namely large-scale production, ability to carry both hydrophilic and lipophilic moieties and biocompatibility with a wide variety of bioactive compounds (Bonechi et al., 2019; Guldiken et al., 2019). Liposomes have been successfully used for the incorporation of phenolic compounds in food matrices (Rashidinejad, Birch, Sun-Waterhouse, & Everett, 2014).

One of the main goals of microencapsulation is the controlled release of the active ingredients, which can be addressed to delay compounds release until the right time or to provide a sustained delivery, when gradual release is desired (Aguiar, Estevinho, & Santos, 2016). The mechanism behind the controlled release depends on the physico-chemical properties of the wall materials and the type of substance microencapsulated. It can take place by diffusion, dissolution, erosion, digestion or mechanical disruption, and is triggered due to environmental changes (i.e., temperature, pH, pressure, and ionic force) (Vincekovi et al., 2017).

6.4 The use of extracts or pure compounds as functional food ingredients

Extracts may contain diverse bioactive compounds, not only distinct polyphenols, but also other hydrophilic or lipophilic substances, such as antioxidant carotenoids or tocopherols, which can have either similar, overlapping, different or complementary biological capacities. When they are consumed, the total biological capacity of the individual bioactive components may be modified via synergistic, additive, or antagonistic interactions among them, which may in turn alter their physiological impacts (Phan, Paterson, Bucknall, & Arcot, 2018; Wang, Meckling, Marcone, Kakuda, & Tsao, 2011). When an individual compound with a desired biological property is to be utilized as a food supplement, large doses are often needed. The use of extracts with the appropriate composition makes it possible to lower such doses due to the cumulative effect of the individual compounds present in the extract. This also helps prevent some side effects associated with the use of large amounts of the individual compound (Malongane, McGaw, & Mudau, 2017). Significant antioxidant synergism between β-carotene paired with puerarin, quercetin and rutin was reported with an increase antioxidant activity up to 50% (Han et al., 2011). Rosmarinic acid was shown to have a synergistic interaction with α-tocopherol that resulted in caffeic acid generation, increasing their antioxidant capacity (Panya et al., 2012). Higher antioxidant activity was also reported for three phenolic acids, found in coffee brew: caffeic, chlorogenic and ferulic acid, and its mixture with alpha-tocopherol (Neunert, Górnaś, Dwiecki, Siger, & Polewski, 2015). These examples show that synergistic antioxidant response in extract combinations may not necessarily result from interaction among polyphenols but also with other bioactive molecules.

As for phenolic mixtures, distinct observations have been made depending on the compounds and the model system employed. Heo, Kim, Chung, and Kim (2007) did not find synergistic antioxidant effects but only additive effects in different in vitro combinations of various polyphenols (chlorogenic acid, catechin and other flavonoids), using the ABTS radical-scavenging assay. By contrast, a synergistic interaction between chlorogenic, gallic, protocatechuic and vanillic acids was reported (Hugo et al., 2012). Oppositely, Pinelo, Manzocco, Nuñez, and Nicoli (2004) found an antagonistic effect when catechin, resveratrol and quercetin interacted at three different temperatures using the DPPH assay.

Antagonistic interactions were also described in myricetin/naringenin combinations as assessed by the oxygen radical absorbance capacity (Freeman, Eggett, & Parker, 2010). In studies in rat models, Arias, Macarulla, Aguirre, Milton, and Portillo (2016) found that while either resveratrol or quercetin did not show any significant reduction in adipose tissue weights, in a diet supplemented with a mixture of both compounds, body fat accumulation and triacylglycerol metabolism were remarkably reduced indicating in vivo synergy as anti-obesity ingredients. The interactions between resveratrol, quercetin, catechin and ethyl gallate, as antiproliferative agents against vascular smooth muscle cell (VSMC) proliferation were explored by Kurin et al. (2012). All four polyphenols inhibited serum-induced VSMC proliferation when applied individually, however, the inhibitory efficacy was significantly enhanced when they were combined, which promote the idea of a multi-bioactive component treatment approach for atherogenesis. The combinatorial interactions of curcumin and silymarin, both major components in spice turmeric and milk thistle, respectively, were evaluated on their action against cancer cell proliferation (Montgomery, Adeyeni, San, Heuertz, & Ezekiel, 2016). Single compound-treated cells showed less activity when compared to combination treated cells, with the latter exhibiting marked cell rounding and membrane blebbing of apoptotic cells. The mechanisms underlying the combined biological effects of these phytochemicals are still unknown and, as such, further studies are needed to assist in future design, standardization and optimization of mixtures of different natural extract based on their synergistic or antagonist interactions.

Despite the benefits of utilizing extracts as food ingredients, some limitations may also exist derived from the occasional presence of antinutritional or toxic compounds, such as alkaloids, tannins, saponins or mycotoxins (Malongane et al., 2017). Some of these naturally occurring compounds can affect nutrient utilization, especially proteins, vitamins, and minerals, by binding to them and hence reducing their absorption and bioavailability and inhibiting enzyme activities in the gastrointestinal tract (Nikmaram et al., 2017). The effectiveness of saponins as bioactive ingredients has been associated to their impact on biological cell membranes, however, they sometimes promote permeability of intestinal mucosal cells, thereby facilitating the uptake of substances that are normally not absorbed (Nikmaram et al., 2017). Protease inhibitors in legume extracts may have anti-nutritional effects on human, hampering the protein digestibility and growth impairment (Guillamón et al., 2008). Further studies have to be conducted to shed light on these possible adverse interactions and balance the intake of combined mixtures of compounds, so as to maximize the supplementation processes to improve the nutritional quality and health promoting benefits of the extracts.

7. Current situation and prospects

7.1 Legal requirements

Although according to the literature numerous beneficial properties have been associated to different polyphenols classes, not many products on the market containing these bioactives are labeled with a health claim message due to legal restrictions. Making health claims about foods is an area that gets more and more complicated over time, especially since there are different levels of qualified health claims. International legislative frameworks are now well developed in countries such as Japan, United Stated of America (USA) or the European Union (EU).

Japan was the first country to recognize functional foods as a separate category, when in 1991 they introduced the term FOSHU (Foods for Specific Health Use) referring to foods containing ingredients with functions for health, and officially approved to claim their physiological effects on the human body. However, claims for disease-risk reduction are only allowed under FOSHU for calcium and folic acid, but not for any polyphenol. Japan has a clear system to organize approved compounds, classifying them into three main categories:

- *Qualified FOSHU*: Food with health function which is not clearly substantiated on the scientific evidence, or food with certain effectiveness but without established mechanism of the effective element for the function. It allows health claims with some conditions.
- *Standardized FOSHU*: The product has enough FOSHU approvals and there is scientific evidence on its health effects.
- *Reduction of risk FOSHU*: The effectiveness of the product is clinically and nutritionally well established.

In order to sell a food as FOSHU, the Government evaluates the claimed effects and safety, and the Secretary-General of the Consumer Affairs Agency (CAA) gives approval for the labeling of each food product that satisfies the requirements. These requirements include: (a) to prove effectiveness based on scientific evidence including clinical studies, (b) absence of any safety issues as assessed from historical consumption pattern data and additional safety studies conducted in humans, and (c) determination of the functional component responsible for the beneficial physiological action and guarantee of compatibility with product specifications by the time of consumption (Yamada, Sato-Mito, Nagata, & Umegaki, 2008). In April 2015, the Japanese CAA started a new labeling system for food functional ingredients, called Foods with Function Claims (FFC). The system was introduced in order to make more products available clearly labeled with certain nutritional or health functions, as well as to enable consumers to make more informed choices. Unlike FOSHU, FFC is only a notification system; the government does not evaluate the safety and effectiveness of these foods and the product is not individually pre-approved, although information about its safety and effectiveness should be submitted to the CAA before the product is marketed. This system allows food business operators, under their own responsibility, to make claims that a food product is helpful for maintaining and promoting health based on scientific evidence. Such as scientific evidence must be obtained from intervention studies in human or systematic literature reviews (Consumer Affairs Agency, 2015). Table 2 summarizes the differences between FOSHU and FFC products.

In addition to these functional food categories, the CAA agency also establishes claims related with "Foods with Nutrient Function Claims (FNFC)" for products containing certain amounts of a nutrient whose function has already been substantiated by scientific evidence. These products

Table 2 Comparison of two main food labeling systems in Japan.

	Foods for specified health uses (FOSHU)	**Foods with function claims (FCC)**
Regulatory	Approval by the CAA agency	Only submit the required information (no review and no approval by the CAA)
Scientific evidence	Clinical trial	Clinical trial or systematic literature review
Responsibility	Agency	Company
Timeline	Evaluation period: >6 months	60 days before placed on the market
Cost	High	Low

can bear a nutrient function claim prescribed by the standards without submitting a notification to the government.

A few botanical-derived products containing polyphenols have been approved as FOSHU. Commercial teas containing polyphenols from leaves of guava (*Psidium guajava* L.) were approved in the category of "foods related to blood sugar levels" and recommended for subjects with pre-diabetes (Deguchi & Miyazak, 2010). The CAA also approved the marketing as FOSHU products of different catechin-rich tea beverages (green and oolong teas), containing amounts of EGCG from 10.2 to 41.9 mg/100 mL, due to the various health-promoting functions of catechins, especially those for mitigating triacylglycerol and body fat (Maeda-Yamamoto & Ohtani, 2018). However, excessive ingestion of EGCG may deleteriously affect liver function, so the consumption of green tea-based FOSHU beverages should be limited to one bottle per day (Maruyama et al., 2017). Similar claims have also been approved for chlorogenic acid, quercetin glycosides and apple procyanidins, whereas soybean isoflavones have a claim related to the promotion of osteogenesis (Maeda-Yamamoto, 2017).

In contrast to the FOSHU scheme, where only around 1100 products have been approved since 1991, >400 foods were labeled with function claims (FFC) in the first year of application of the new category of functional foods, and currently near 1000 foods with function claims have been notified. These FFC are usually present in the marked as processed foods and include numerous examples of products containing different phenolic compounds: isoflavones from kudzu flower to help reduce visceral fat and high body mass index; procyanidin B1, monoglycosyl hesperidin, gallic acid and polyphenols from *Terminalia bellerica* to decrease serum triglyceride and LDL cholesterol levels; cacao flavanols that help maintain normal blood pressure in moderately hypertensive individuals; lutein, cyanidin-3-glucoside or anthocyanins of blueberries to contribute to focus adjustment function, or flavonoid glycosides from Gingko leaf to increase memory accuracy as a component of cognitive function (Maeda-Yamamoto & Ohtani, 2018).

Although nutraceuticals and functional foods are food marketing concepts and there are no U.S. regulatory definitions to accommodate them separately from other foods, food label claims have been regulated by the Food and Drug Administration (FDA) since 1990 through the Nutrition Labeling and Education Act (NLEA) (González-Díaz, Gil-González, & Ávarez-Dardet, 2018). Within the context of these regulations, the labeling of food may not include any information about the usefulness of a food to cure, mitigate, treat, or prevent a disease, but food labels can present information about how a food may affect a structure or function of the body and claims that describe how a food or food component may affect disease risk (Hoadley & Rowlands, 2014). All FDA-approved health claims are generic and not for the exclusive use of the petitioner. The FDA conducts an evidence-based review to ascertain the scientific validity of the claim. It reviews and authorizes the health claims by three means (Agarwal, Hordvik, & Morar, 2014; Lalor & Wall, 2011):

- *Claims based on Significant Scientific Agreement (SSA)*: Claims under the NLEA amendments require an FDA assessment by qualified experts that the totality of the scientific evidence supports the dietary substance/disease relationship; this means that the validity of the relationship is not likely to be reversed by new and evolving science. Under this regulation, FDA has authorized general health claims like "fruits and vegetables and reduced risk of cancer" or "fruits, vegetables and grain products that contain fiber, particularly soluble fiber, and reduced risk of coronary heart disease."
- *Claims based on Authoritative statement*: Since 1997, the FDA Modernization Act (FDAMA) allows the use of health claims based on authoritative statements from a scientific body of the U.S. Government or the National Academy of Sciences. If in the period of 120 days after the companies' notification the FDA did not act to prohibit or modify the claim, the claim could be used. Only four claims have been authorized under this category.
- *Qualified health claims*: FDA permits the use of a health claim when there is emerging, but credible, scientific evidence for a relationship between a food and reduced risk of a disease or health-related condition. The FDA uses the term qualified health claim to refer to health claims for which the scientific evidence does not meet the SSA standard. These claims have to include qualifying language as part of the claim, indicating that the evidence supporting the claim is limited. Qualified health claims include some related to food rich in polyphenols, e.g., "green tea and risk of breast and prostate cancers," "tomatoes and prostate, ovarian, gastric, and pancreatic cancers," "nuts and coronary heart disease." Nevertheless, although they are permitted, in every case the FDA concludes that there is little scientific evidence supporting these claims. A listing of qualified health claim enforcement discretion decisions is posted on the FDA Website (https://www.fda.gov/Food/LabelingNutrition/ucm072756.htm).

In the European Union (EU), all foods making nutrition or health claims are subject to specific legislation through Regulation 1924/2006 that describes a health claim as "any claim that states, suggests or implies that a relationship exists between a food category, a food or one of its constituents and health." The regulation also includes reduction of disease risk claims defined as "claims that state, suggest, or imply that the consumption of a food category, a food, or one of its constituents significantly reduces a risk factor in the development of a human disease." The aim of this regulation is to ensure that any claim made on a food label in the EU is clear, accurate and substantiated to enable consumers make informed and meaningful choices when it comes to food and drinks. The regulation involves a pre-marketing approval system and scientific evidence-based assessment of nutrition and health claims (Khedkar, Ciliberti, & Bröring, 2016). Although the European Food Safety Authority (EFSA) evaluates if health claims are sufficiently scientifically substantiated to be included in the EU Register of Nutrition and Health Claims, it is the European Commission that decides whether or not any new claim will be approved. EFSA uses standardized protocols to elaborate opinions based on three questions: (1) the development of enough characterization of the food on which the claim is done; (2) the existence of enough data on the biological effects and physiological benefits, and (3) the existence of clinical trials with human subjects to support the claimed effect (Baenas et al., 2018).

The European regulations establish different types of health claims:

- *Function claims* (article 13), i.e., health claims other than those referring to the reduction of disease risk and to children's development and health. They include health claims describing or referring to growth, development and functions of the body, psychological and behavioral functions, slimming or weight-control, and satiety or reduction of available energy from diet. Health claims based on generally accepted scientific data (article 13.1) are only allowed when included on a list. The first list of permitted health claims according with this regulation was published in the Commission Regulation (EU) no. 432/2012 and amended with later regulations in 2013 and 2016. Any additions of claims to the list based on newly developed scientific data and/or that include a request for the protection of proprietary data shall be adopted after application for individual authorization. The updated list of evaluated health claims is on the webpage of the European Commission (http://ec.europa.eu/food/safety/labelling_nutrition/claims/register). According to this regulation, two health claims related polyphenols have been authorized: one referring to olive oil polyphenols and their contribution to the protection of blood lipids from oxidative stress (Commission Regulation (EU)

432/2012), and the other on cocoa flavanols to help maintain endothelium-dependent vasodilation, which contributes to normal blood flow. However, many other requested claims have not been authorized on the basis of the scientific evidence assessed for the claimed effect for the food is not sufficiently substantiated. Some examples are: natural berries for a heart-friendly diet; olive biophenols for combating bacterial infections; phenolic compounds from cranberry and lingonberry as health-promoting antioxidants; red wine polyphenols to help vascular functions that contribute to a healthy cardiovascular system; apple extract powder containing polyphenols to help decrease the blood glucose levels; cocoa flavanols help to promote healthy cells by neutralizing free radicals; cocoa flavanols for maintenance and promotion of a normal blood pressure, or flavonoids from green tea, apple and onion to reduce the absorption of carbohydrates and visceral fat.

- *Reduction of disease risk claims* (article 14). These are only allowed after submission of an application to EFSA and approval through the Standing Committee on the Food Chain and Animal Health. The principles for scientific assessment established by the EFSA are very strict and unlike FDA do not include evidence grading. The application shall include, among others, information about the characteristics of the nutrient or substance, or the food or the category of food, in respect of which the health claim is to be made, copies of the studies that have been carried out with regard to the health claim, and a proposal for the wording of the health claim for which authorization is sought including. Examples of proposed, but non-authorized, health claims related to phenolic compounds present in dietary supplements are: health claim application on CranMax® or Uroval® (products containing cranberry (*Vaccinium macrocarpon*) powder standardized for proanthocyanidins content) and reduction of the risk of urinary tract infection by inhibiting the adhesion of certain bacteria; and OPC Plus® or OPC Premium®, containing 40 mg oligomeric procyanidins and berry-blend to increase the microcirculation and to reduce blood cholesterol levels, thus reducing the risks of chronic venous insufficiency and cardiovascular disease.

Since its adoption in 2006, the implementation of the regulation remains incomplete since health claims on plants and their preparations used in foods are not yet fully regulated. For this reason, the European Commission is nowadays under a REFIT (Regulatory Fitness and Performance Programme) evaluation.

In conclusion, full regulatory approval for claims across the world requires the support of scientific evidence, but there are differences in the requirements and the level of scientific evidence required to approve a health claim. While in the United States and Japan a health claim that is suggested but not fully supported by scientific evidence is known as a qualified health claim and is permitted, it is not authorized in the EU. Since this causes consumer confusion and develops an uneven playing pitch for the industry, a consensus would be advisable as to the level of scientific evidence required to approve a health claim (Lalor & Wall, 2012).

7.2 Emerging trends

On the developed world many vegetables are widely used directly as food but also due to their healthy properties. Enriched extracts of polyphenols from herbs and vegetables have numerous applications in herbal medicine formulations, added to beverage or starch-based foods, used as condiments or infused into cosmetics. Polyphenols are one of the most researched bioactive compounds because of their wide distribution in nature and also due to their versatility as agents that can improve human health and enhance the shelf-life of foods (Adebooye, Alashi, & Aluko, 2018).

The U.S. and Europe polyphenol markets are projected to reach $584,907 million by 2025, and a volume of 17,892 tons, which represents 7.7% of increase from 2018 to 2025 (Allied Market Research, 2018). However, the market trend suggests that global polyphenol economy in 2024 would be led by Asia Pacific, with about 40% of the global demand, followed by Europe (Grand View Research, 2016). The single, most powerful trend in today's marketplace is consumers' desire for foods and ingredients that are "naturally functional," so developments of food and beverages from plant-based are rising (Mirosa & Mangan-Walker, 2018; Song & Im, 2018). Grapeseed segment dominated the U.S. and Europe polyphenol market, especially due to its antioxidant and antiaging properties along with the increase in demand from personal care and skin care market, although green tea, apple and maracuja/passion fruit also represent important sectors, and in minor extension other segments, like olives, cocoa, and pomegranate. Among the different categories of functional foods, functional beverages (polyphenol-rich beverages in the form of juices, energy drinks, and enhanced water) followed by functional foods, specially snacking products, were the segments that accounted for the highest contribution in the U.S.

and Europe polyphenol market in 2017. During the last 2 years, numerous bakery products have been formulated incorporating polyphenols from different matrices, e.g., pomegranate seeds in bread (Bustamante, Hinojosa, Robert, & Escalona, 2017), green tea polyphenols in bread (Ye, Georges, & Selomulya, 2018), apple pomace in biscuits (Alongi, Melchior, & Anese, 2018), or grape skin pomace in muffins (Bender et al., 2017).

One of the research focuses of the industry of polyphenols is to optimize their recovery during extraction, as well as to identify the bioactive compounds that constitute the polyphenol extract (Sulaiman, Sajak, Ooi, Supriatno, & Seow, 2011). The development of an efficient procedure for the extraction, proper analysis, and characterization of phenolic compounds from different sources is a challenging task, owing to their structural diversity, complex matrices, and interaction with other cellular components. The use of green and economically feasible modern extraction procedures, as reviewed in Section 6.1, represents a promising approach for overcoming current limitations to the exploitation of polyphenols as bioactive compounds, as well as to explore their wide-reaching applications on an industrial scale and in emerging global markets (Ameer, Shahbaz, & Kwon, 2017). Some recent patents have been developed in the field of polyphenols, both to innovate in the extraction process and in the formulation of food including the polyphenolic extracts. Lores-Aguin, Garcia Jares, Alvarez Casas, and Llompart (2014) patented a straightforward method with few steps for obtaining polyphenol-rich extracts with anti-oxidant and anti-bacterial properties from white-grape residues, which can be used on an industrial scale, essentially in the cosmetic, pharmaceutical and/or food industries. In the same way, a method to produce and antioxidant phenolic rich grape extract, exhibiting an ORAC value of at least 10,000 μmol Trolox Equivalent/g, was patented in the United States (Ying, Xiong, Chen, & Yang, 2013). Also, an innovative method for stably dispersing microparticulated water-insoluble bioactive polyphenols in a beverage was patented by Zhang and Mutilangi (2013).

Maybe the most important key that limits the authorization of health claims related to polyphenols present in functional beverages or food is their bioavailability and the incomplete elucidation of their mechanisms of action. For this reason, the interest in studies that can address the definition of good biomarkers of intake and/or effects have been increased nowadays. Metabolomics approaches are carrying out with the aim to detect and identify metabolites present in different body fluids or tissues that can afford the understanding of the in vivo transformation of polyphenols

(González-Paramás, Ayuda-Durán, Martínez, González-Manzano, & Santos-Buelga, 2018; Rienks, Barbaresko, & Nöthlings, 2017; Rothwell et al., 2016). On the other hand, efforts are necessary to conduct clinical trials that actually are limited because difficulty in obtaining funding, ethical considerations and stringent conditions by the safety agencies (double-blinded, randomized, placebo-controlled, wash-out periods, cross-over studies and complex inclusion and exclusion criteria) (Brown, Caligiuri, Brown, & Pierce, 2018). Attention must also be paid to the effective dosages used in the clinical trials, determining whether nutritional low and chronic administration of functional food can or not play a role in health, and whether an isolated substance has the same efficacy when ingested in a concentrated form, as when naturally present in a whole food (Pinto da Costa, 2017).

In order to optimize delivery of bioactives, the food industry is improving the formulation of functional foods containing polyphenols, designing new matrices to increase compound stability, bioactivity and bioavailability. Within each matrix, different aspects, such as interaction of polyphenols with other food components like proteins, fats, carbohydrates and minerals, have been shown to influence the release, stability, accessibility and digestibility of phenolic compounds (Crowe, 2013; Zhang et al., 2014). For example, protein-rich ingredients like soybean flour have been used to bind blueberry anthocyanins resulting in a stable ingredient capable of delivering more anthocyanins to the intestinal tract compared to an equal amount of blueberry juice (Ribnicky et al., 2014). Taking into account the advance in the knowledge of the active forms of polyphenols and the interest in their increased absorption, research is progressing in the encapsulation of the phenolic compounds not only to protect them from adverse conditions such as light, oxidation, temperature or hydrolysis, but also for delivery of the stable active form to the appropriate segment of the gastrointestinal tract for their release and uptake (Chen, Gnanaraj, Arulselvan, El-Seedi, & Teng, 2019; Dias et al., 2015; Oidtmann et al., 2012).

8. Concluding remarks

The putative benefits of the consumption of phenolic compounds on the prevention of major chronic diseases have attracted the interest of the consumers and food industry. However, there are still many gaps to fill in the knowledge of their actual effects on human health, which prevent doing recommendations about their dietary intake and limit their use as functional ingredients for foods. Further research must still be done on aspects such as bioavailability, pharmacokinetics, biological targets, mechanisms of action, actual bioactive compounds, active doses or possible adverse effects. Appropriate evaluation methods have also to be developed to adequately assess their health benefits, including the definition of robust biomarkers of their consumption and effects. All this knowledge is required not only to promote improved recommendations on the consumption of phenolic compounds, but also to get authorization for making health claims based on their use as nutraceuticals or functional food ingredients. As for the industry, the availability of suitable sources and techniques for their extraction, the definition of efficient and safe doses, and the development of adequate ways for their incorporation into food, so as to improve their stability, bioavailability and proper delivering at target sites, are technological key issues that require further consideration. No doubt that in the coming years, we are going to see notable advances in all these aspects and assist to an increasing presence in the market of phenolic based functional foods, nutraceuticals, cosmeceuticals and drugs.

References

Adebooye, O. C., Alashi, A. M., & Aluko, R. E. (2018). A brief review on emerging trends in global polyphenol research. *Journal of Food Biochemistry, 42*, e12519.

Adlercreutz, H. (2007). Lignans and human health. *Critical Reviews in Clinical Laboratory Sciences, 44*, 483–525.

Agarwal, S., Hordvik, S., & Morar, S. (2014). Nutrition and health-related labeling claims for functional foods and dietary supplements in the United States. In D. Bagchi (Ed.), *Nutraceutical and functional food regulations in the United States and around the world* (pp. 141–150). San Diego: Academic Press.

Aguiar, J., Estevinho, B. N., & Santos, L. (2016). Microencapsulation of natural antioxidants for food application—The specific case of coffee antioxidants—A review. *Trends in Food Science and Technology, 58*, 21–39.

Alexandre, A. M. R. C., Matias, A., Duarte, C. M. M., & Bronze, M. R. (2018). High-pressure CO_2 assisted extraction as a tool to increase phenolic content of strawberry-tree (Arbutus unedo) extracts. *Journal of CO□ Utilization, 27*, 73–80.

Allied Market Research. (2018). U.S. and Europe polyphenol market. In *Opportunities and forecast, 2018-2025*. Retrieved from https://www.alliedmarketresearch.com/u.s.-and-europe-polyphenol-market (January, 2019).

Alongi, M., Melchior, S., & Anese, M. (2018). Reducing the glycemic index of short dough biscuits by using apple pomace as a functional ingredient. *LWT—Food Science and Technology, 100*, 300–305.

Ameer, K., Shahbaz, H. M., & Kwon, J.-H. (2017). Green extraction methods for polyphenols from plant matrices and their byproducts: A review. *Comprehensive Reviews in Food Science and Food Safety, 16*, 295–315.

Andersen, Ø. M., & Jordheim, M. (2013). Basic anthocyanin chemistry and dietary sources. In T. C. Wallace & M. Giusti (Eds.), *Anthocyanins in health and disease* (pp. 13–90). Boca Raton: CRC Press.

Andersen, Ø. M., & Markham, K. R. (2006). *Flavonoids: Chemistry, biochemistry and applications*. Boca Raton: CRC Press.

Andreasen, M. F., Kroon, P. A., Williamson, G., & Garcia-Conesa, M. T. (2001). Esterase activity able to hydrolyze dietary antioxidant hydroxycinnamates is distributed along the intestine of mammals. *Journal of Agricultural and Food Chemistry, 49*, 5679–5684.

Anonymous. (1950). Use of the term vitamin P. *Nature, 166*, 543.

Arias, N., Macarulla, M. T., Aguirre, L., Milton, I., & Portillo, M. P. (2016). The combination of resveratrol and quercetin enhances the individual effects of these molecules on triacylglycerol metabolism in white adipose tissue. *European Journal of Nutrition, 55*, 341–348.

Arumugam, M., Raes, J., Pelletier, E., Le Paslier, D., Yamada, T., Mende, D. R., et al. (2011). Enterotypes of the human gut microbiome. *Nature, 473*, 174–180.

Aura, A. M., Martin-Lopez, P., O'Leary, K. A., Williamson, G., Oksman-Caldentey, K. M., Poutanen, K., et al. (2005). In vitro metabolism of anthocyanins by human gut microflora. *European Journal of Nutrition, 44*, 133–142.

Aura, A. M., O'Leary, K. A., Williamson, G., Ojala, M., Bailey, M., Puupponen-Pimia, R., et al. (2002). Quercetin derivatives are deconjugated and converted to hydroxyphenylacetic acids but not methylated by human fecal microflora in vitro. *Journal of Agricultural and Food Chemistry, 50*, 1725–1730.

Azmir, J., Zaidul, I. S. M., Rahman, M. M., Sharif, K. M., Mohamed, A., Sahena, F., et al. (2013). Techniques for extraction of bioactive compounds from plant materials: A review. *Journal of Food Engineering, 117*, 426–436.

Azuma, K., Ippoushi, K., Nakayama, M., Ito, H., Higashio, H., & Terao, J. (2000). Absorption of chlorogenic acid and caffeic acid in rats after oral administration. *Journal of Agricultural and Food Chemistry, 48*, 5496–5500.

Baenas, N., Abellán, A., Rivera, S., Moreno, D. A., García-Viguera, C., & Domínguez-Perles (2018). Foods and supplements. In C. H. Galanakis (Ed.), *Polyphenols: Properties, recovery, and applications*: Woodhead Publishing.

Bakry, A. M., Abbas, S., Ali, B., Majeed, H., Abouelwafa, M. Y., Mousa, A., et al. (2016). Microencapsulation of oils: A comprehensive review of benefits, techniques, and applications. *Comprehensive Reviews in Food Science and Food Safety, 15*, 143–182.

Barba, F. J., Zhu, Z., Koubaa, M., Sant'Ana, A. S., & Orlien, V. (2016). Green alternative methods for the extraction of antioxidant bioactive compounds from winery wastes and by-products: A review. *Trends in Food Science and Technology, 49*, 96–109.

Barrajón-Catalán, E., Herranz-López, M., Joven, J., Segura-Carretero, A., Alonso-Villaverde, C., Menéndez, J. A., et al. (2014). Molecular promiscuity of plant polyphenols in the management of age-related diseases: Far beyond their antioxidant properties. *Advances in Experimental Medicine and Biology, 824*, 141–159.

Barrales, F. M., Silveira, P., Barbosa, P. D. P. M., Ruviaro, A. R., Paulino, B. N., Pastore, G. M., et al. (2018). Recovery of phenolic compounds from citrus by-products using pressurized liquids—An application to orange peel. *Food and Bioproducts Processing, 112*, 9–21.

Baur, J. A., Pearson, K. J., Price, N. L., Jamieson, H. A., Lerin, C., Kalra, A., et al. (2006). Resveratrol improves health and survival of mice on a high-calorie diet. *Nature, 444*, 337–342.

Baxter, N. J., Lilley, T. H., Haslam, E., & Williamson, M. P. (1997). Multiple interactions between polyphenols and a salivary proline-rich protein repeat result in complexation and precipitation. *Biochemistry, 36*, 5566–5577.

Bender, A. B. B., Speroni, C. S., Salvador, P. R., Loureiro, B. B., Lovatto, N. M., Goulart, F. R., et al. (2017). Grape pomace skins and the effects of its inclusion in the technological properties of muffins. *Journal of Culinary Science & Technology, 15*, 143–157.

Benthsáth, A., Rusznyák, S. T., & Szent-Györgyi, A. (1937). Vitamin P. *Nature, 139*, 326–327.

Bentsáth, A., Rusznyak, S. T., & Szent-György, A. (1936). Vitamin nature of flavones. *Nature, 138*, 798.

Boffetta, P., Couto, E., Wichmann, J., Ferrari, P., Trichopoulos, D., Bueno-de-Mesquita, H. B., et al. (2010). Fruit and vegetable intake and overall cancer risk in the European prospective investigation into cancer and nutrition (EPIC). *Journal of the National Cancer Institute, 102*, 529–537.

Bonechi, C., Donati, A., Tamasi, G., Pardini, A., Rostom, H., Leone, G., et al. (2019). Chemical characterization of liposomes containing nutraceutical compounds: Tyrosol, hydroxytyrosol and oleuropein. *Biophysical Chemistry, 246*, 25–34.

Bors, W., Heller, W., Michel, C., & Saran, M. (1990). Flavonoids as antioxidants: Determination of radical scavenging efficiencies. *Methods in Enzymology, 186*, 343–355.

Briones-Labarca, V., Giovagnoli-Vicuña, C., & Cañas-Sarazúa, R. (2018). Optimization of extraction yield, flavonoids and lycopene from tomato pulp by high hydrostatic pressure-assisted extraction. *Food Chemistry, 278*, 751–759.

Brown, L., Caligiuri, S. P. B., Brown, D., & Pierce, G. N. (2018). Clinical trials using functional foods provide unique challenges. *Journal of Functional Foods, 45*, 233–238.

Bruckner, V., & Szent-Györgyi, A. (1936). Chemical nature of citrin. *Nature, 138*, 1057.

Bustamante, A., Hinojosa, A., Robert, P., & Escalona, V. (2017). Extraction and microencapsulation of bioactive compounds from pomegranate (*Punica granatum* var. Wonderful) residues. *International Journal of Food Science & Technology, 52*, 1452–1462.

Butler, L. G. (1992). Antinutritional effects of condensed and hydrolysable tannins. In R. W. Hemingway & P. E. Laks (Eds.), *Plant polyphenols: Synthesis, properties and significance* (pp. 693–698): Plenum Press.

Cai, H., Dong, L. Q., & Liu, F. (2016). Recent advances in adipose mTor signaling and function: Therapeutic prospects. *Trends in Pharmacological Sciences, 37*, 303–317.

Caldas, T. W., Mazza, K. E. L., Teles, A. S. C., Mattos, G. N., Iraidy, A., Brígida, S., et al. (2018). Phenolic compounds recovery from grape skin using conventional and non-conventional extraction methods. *Industrial Crops and Products, 111*, 86–91.

Campone, L., Celano, R., Lisa, A., Pagano, I., Carabetta, S., Di, R., et al. (2018). Response surface methodology to optimize supercritical carbon dioxide/co-solvent extraction of brown onion skin by-product as source of nutraceutical compounds. *Food Chemistry, 269*, 495–502.

Cardona, F., Andrés-Lacueva, C., Tulipani, S., Tinahones, F. J., & Queipo-Ortuño, M. I. (2013). Benefits of polyphenols on gut microbiota and implications in human health. *The Journal of Nutritional Biochemistry, 24*, 1415–1422.

Cassidy, A., Hanley, B., & Lamuela-Raventos, R. M. (2000). Isoflavones, lignans and stilbenes—Origins, metabolism and potential importance to human health. *Journal of the Science of Food and Agriculture, 80*, 1044–1062.

Cerdá, B., Tomás-Barberán, F. A., & Espín, J. C. (2005). Metabolism of antioxidant and chemopreventive ellagitannins from strawberries, raspberries, walnuts, and oak-aged wine in humans: Identification of biomarkers and individual variability. *Journal of Agricultural and Food Chemistry, 53*, 227–235.

Chanioti, S., & Tzia, C. (2018). Extraction of phenolic compounds from olive pomace by using natural deep eutectic solvents and innovative extraction techniques. *Innovative Food Science and Emerging Technologies, 48*, 228–239.

Charlton, A. J., Haslam, E., & Williamson, M. P. (2002). Multiple conformations of the proline-rich protein/epigallocatechin gallate complex determined by time-averaged nuclear overhauser effects. *Journal of the American Chemical Society, 124*, 9899–9905.

Chen, L., Gnanaraj, C., Arulselvan, P., El-Seedi, H., & Teng, H. (2019). A review on advanced microencapsulation technology to enhance bioavailability of phenolic compounds: Based on its activity in the treatment of type 2 diabetes. *Trends in Food Science & Technology, 85*, 149–162. https://doi.org/10.1016/j.tifs.2018.11.026.

Chen, C., Yu, R., Owuor, E. D., & Kong, A. N. (2000). Activation of antioxidant-response element (ARE), mitogen activated protein kinases (MAPKs) and caspases by major green tea polyphenol components during cell survival and death. *Archives of Pharmacal Research, 23*, 605–612.

Choi, J. S., Chung, H. Y., Kang, S. S., Jung, M. J., Kim, J. W., No, J. K., et al. (2002). The structure-activity relationship of flavonoids as scavengers of peroxynitrite. *Phytotherapy Research, 16*, 232–235.

Chouhan, S., Sharma, K., Zha, J., Guleria, S., & Koffas, M. A. G. (2017). Recent advances in the recombinant biosynthesis of polyphenols. *Frontiers in Microbiology, 8*, 1–16.

Clavel, T., Henderson, G., Alpert, C. A., Philippe, C., Rigottier-Gois, L., Dore, J., et al. (2005). Intestinal bacterial communities that produce active estrogen-like compounds enterodiol and enterolactone in humans. *Applied and Environmental Microbiology, 71*, 6077–6085.

Clifford, M. N. (1999). Chlorogenic acids and other cinnamates—Nature, occurrence and dietary burden. *Journal of the Science of Food and Agriculture, 79*, 362–372.

Clifford, M. N. (2000a). Chlorogenic acids and other cinnamates—Nature, occurrence and dietary burden. *Journal of the Science of Food and Agriculture, 80*, 1033–1043.

Clifford, M. N. (2000b). Anthocyanins—Nature, occurrence and dietary burden. *Journal of the Science of Food and Agriculture, 80*, 1063–1072.

Clifford, M. N. (2004). Diet-derived phenols in plasma and tissues and their implications for health. *Planta Medica, 70*, 1103–1114.

Commenges, D., Scotet, V., Renaud, S., Jacqmin-Gadda, H., Barberger-Gateau, P., & Dartigues, J. F. (2000). Intake of flavonoids and risk of dementia. *European Journal of Epidemiology, 16*, 357–363.

Consumer Affairs Agency. (2015). Food labelling division, Guidelines on notification of foods with function claims. Retrieved from http://www.caa.go.jp/foods/index18.html. (March, 2019).

Correa, T. A. F., & Rogero, M. M. (2019). Polyphenols regulating microRNAs and inflammation biomarkers in obesity. *Nutrition, 59*, 150–157.

Couteau, D., McCartney, A. L., Gibson, G. R., Williamson, G., & Faulds, C. B. (2001). Isolation and characterization of human colonic bacteria able to hydrolyse chlorogenic acid. *Journal of Applied Microbiology, 90*, 873–881.

Cox, M., & Blaser, M. J. (2013). Pathways in microbe-induced obesity. *Cell Metabolism, 17*, 883–894.

Crowe, K. M. (2013). Designing functional foods with bioactive polyphenols: Highlighting lessons learned from original plant matrices. *Journal of Human Nutrition & Food Science, 1*, 1018.

Crowe, K. M., & Francis, C. (2013). Position of the academy of nutrition and dietetics: Functional foods. *Journal of the Academy of Nutrition and Dietetics, 113*, 1096–1103.

Crozier, A., Jaganath, I. B., & Clifford, M. N. (2009). Dietary phenolics: Chemistry, bioavailability and effects on health. *Natural Product Reports, 26*, 1001–1043.

Crozier, A., Lean, M. E. J., McDonald, M. S., & Black, C. (1997). Quantitative analysis of the flavonoid content of commercial tomatoes, onions, lettuce, and celery. *Journal of Agricultural and Food Chemistry, 45*, 590–595.

Cueva, C., Gil-Sánchez, I., Ayuda-Durán, B., González-Manzano, S., González-Paramás, A. M., Santos-Buelga, C., et al. (2017). An integrated view of the effects of wine polyphenols and their relevant metabolites on gut and host health. *Molecules, 22*, 99.

Cunha, S. C., & Fernandes, J. O. (2018). Extraction techniques with deep eutectic solvents. *Trends in Analytical Chemistry, 105*, 225–239.

Dangles, O., & Dufour, C. (2008). Flavonoid-protein binding processes and their impact on human health. In F. Daayf & V. Lattanzio (Eds.), *Vol. 1. Recent advances in polyphenol research* (pp. 67–87). Oxford, UK: Wiley-Blackwell.

Daniel, E. M., Krupnick, A. S., Heur, Y. H., Blinzler, J. A., Nims, R. W., & Stoner, G. D. (1989). Extraction, stability, and quantitation of ellagic acid in various fruits and nuts. *Journal of Food Composition and Analysis, 2*, 338–349.

Day, A. J., Gee, J. M., Dupont, M. S., Johnson, I. T., & Williamson, G. (2003). Absorption of quercetin-3-glucoside and quercetin-4′-glucoside in the rat small intestine: The role of lactase phlorizin hydrolase and the sodium-dependent glucose transporter. *Biochemical Pharmacology, 65*, 1199–1206.

De Felice, S. L. (1995). The nutraceutical revolution: Its impact on food industry R&D. *Trends in Food Science & Technology, 6*, 59–61.

de Freitas, V. A. P., & Mateus, N. (2001). Structural features of procyanidin interactions with salivary proteins. *Journal of Agricultural and Food Chemistry, 49*, 940–945.

de Pascual-Teresa, S., Santos-Buelga, C., & Rivas-Gonzalo, J. C. (2000). Quantitative analysis of flavan-3-ols in Spanish foodstuffs and beverages. *Journal of Agricultural and Food Chemistry, 48*, 5331–5337.

Deguchi, Y., & Miyazak, K. (2010). Anti-hyperglycemic and anti-hyperlipidemic effects of guava leaf extract. *Nutrition and Metabolism, 7*, 9.

Del Rio, D., Rodriguez-Mateos, A., Spencer, J. P. E., Tognolini, M., Borges, G., & Crozier, A. (2013). Dietary (poly)phenolics in human health: Structures, bioavailability, and evidence of protective effects against chronic diseases. *Antioxidants & Redox Signaling, 8*, 1818–1892.

Deng, J., Xu, Z., Xiang, C., Liu, J., Zhou, L., Li, T., et al. (2017). Comparative evaluation of maceration and ultrasonic-assisted extraction of phenolic compounds from fresh olives. *Ultrasonics Sonochemistry, 37*, 328–334.

Devore, E. E., Kang, J. H., Breteler, M. M., & Grodstein, F. (2012). Dietary intakes of berries and flavonoids in relation to cognitive decline. *Annals of Neurology, 72*, 135–143.

Dewettinck, K., & Huyghebaert, A. (1999). Fluidized-bed-coating-in-food-technology. *Trends in Food Science and Technology, 10*, 163–168.

Dias, M. I., Ferreira, I. C. F. R., & Barreiro, M. F. (2015). Microencapsulation of bioactives for food applications. *Food & Function, 6*, 1035–1052.

Doerge, D. R., & Sheehan, D. M. (2002). Goitrogenic and estrogenic activity of soy isoflavones. *Environmental Health Perspectives, 110*, 349–353.

Duan, M. H., Fang, T., Ma, J. F., Shi, Q. L., Peng, Y., Ge, F. H., et al. (2018). Homogenate-assisted high-pressure disruption extraction for determination of phenolic acids in *Lonicerae japonicae* flos. *Journal of Chromatography B, 1097–1098*, 119–127.

Dunnick, J. K., & Halley, J. R. (1992). Toxicity and carcinogenicity studies of quercetin, a natural component of foods. *Toxicological Sciences, 19*, 423–431.

EFSA Panel on Food Additives and Nutrient Sources Added to Food. (2015). Risk assessment for peri- and post-menopausal women taking food supplements containing isolated isoflavones. *EFSA Journal, 13*, 4246.

EFSA Panel on Food Additives and Nutrient Sources Added to Food. (2018). Scientific opinion on the safety of green tea catechins. *EFSA Journal, 16*, 5239.

Eseberri, I., Lasa, A., Churruca, I., & Portillo, M. P. (2013). Resveratrol metabolites modify adipokine expression and secretion in 3T3-L1 pre-adipocytes and mature adipocytes. *PLoS One, 8*, e63918.

Espín, J. C., García-Conesa, M. T., & Tomás-Barberán, F. A. (2007). Nutraceuticals: Facts and fiction. *Phytochemistry, 68*, 2986–3008.

Etxeberria, U., Arias, N., Boqué, N., Macarulla, M. T., Portillo, M. P., Martínez, J. A., et al. (2015). Reshaping faecal gut microbiota composition by the intake of trans-resveratrol and quercetin in high-fat sucrose diet-fed rats. *The Journal of Nutritional Biochemistry, 26*, 651–660.

Faridi, A., Assadpour, E., & Mahdi, S. (2018). Improving the bioavailability of phenolic compounds by loading them within lipid-based nanocarriers. *Trends in Food Science and Technology*, *76*, 56–66.

Ferarsa, S., Zhang, W., Moulai-Mostefa, N., Ding, L., Jaffrin, M. Y., & Grimi, N. (2018). Recovery of anthocyanins and other phenolic compounds from purple eggplant peels and pulps using ultrasonic-assisted extraction. *Food and Bioproducts Processing*, *109*, 19–28.

Ferreira, I. C. F. R., Martins, N., & Barros, L. (2017). Phenolic compounds and its bioavailability: In vitro bioactive compounds or health promoters? In F. Toldrá (Ed.), *Advances in food and nutrition research* (pp. 1–44). Burlington: Academic Press.

Ferriola, P. C., Cody, V., & Middleton, E. (1989). Protein kinase C inhibition by plant flavonoids: Kinetic mechanisms and structure–activity relationships. *Biochemical Pharmacology*, *38*, 1617–1624.

Fiesel, A., Gessner, D. K., Most, E., & Eder, K. (2014). Effect of dietary polyphenol-rich plant products from grape or hop on pro-inflammatory gene expression in the intestine, nutrient digestibility and faecal microbiota of weaned pigs. *BMC Veterinary Research*, *10*, 196.

Fischer, U. A., Carle, R., & Kammerer, D. R. (2011). Identification and quantification of phenolic compounds from pomegranate (*Punica granatum* L.) peel, mesocarp, aril and differently produced juices by HPLC-DAD-ESI/MS(n). *Food Chemistry*, *127*, 807–821.

Fontana, A. R., Antoniolli, A., & Bottini, R. (2013). Grape pomace as a sustainable source of bioactive compounds: Extraction, characterization, and biotechnological applications of phenolics. *Journal of Agricultural and Food Chemistry*, *62*, 8987–9003.

Fowler, Z. L., & Koffas, M. A. G. (2009). Biosynthesis and biotechnological production of flavanones: Current state and perspectives. *Applied Microbiology and Biotechnology*, *83*, 799–808.

Freeman, B. L., Eggett, D. L., & Parker, T. L. (2010). Synergistic and antagonistic interactions of phenolic compounds found in navel oranges. *Journal of Food Science*, *75*, 570–576.

Galati, G., Lin, A., Sultan, A. M., & O'Brien, P. J. (2006). Cellular and in vivo hepatotoxicity caused by green tea phenolic acids and catechins. *Free Radical Biology and Medicine*, *40*, 570–580.

Galli, F. (2007). Interactions of polyphenolic compounds with drug disposition and metabolism. *Current Drug Metabolism*, *8*, 830–838.

García, P., Romero, C., & Brenes, M. (2018). Bioactive substances in black ripe olives produced in Spain and the USA. *Journal of Food Composition and Analysis*, *66*, 193–198.

García-Lomillo, J., González-San José, M. L., Pino-García, R., Rivero-Pérez, M. D., & Muñiz-Rodríguez, P. (2014). Antioxidant and antimicrobial properties of wine byproducts and their potential uses in the food industry. *Journal of Agricultural and Food Chemistry*, *62*, 12595–12602.

Ghafoor, K., Al-Juhaimi, F. Y., & Choi, Y. H. (2012). Supercritical fluid extraction of phenolic compounds and antioxidants from grape (*Vitis labrusca* B.) seeds. *Plant Foods for Human Nutrition*, *67*, 407–414.

Gil, M. I., Tomás-Barberán, F. A., Hess-Pierce, B., Holcroft, D. M., & Kader, A. A. (2000). Antioxidant activity of pomegranate juice and its relationship with phenolic composition and processing. *Journal of Agricultural and Food Chemistry*, *48*, 4581–4589.

Godos, J., Marventano, S., Mistretta, A., Galvano, F., & Grosso, G. (2017). Dietary sources of polyphenols in the mediterranean healthy eating, aging and lifestyle (MEAL) study cohort. *International Journal of Food Sciences and Nutrition*, *68*, 750–756.

Gonthier, M. P., Verny, M. A., Besson, C., Remesy, C., & Scalbert, A. (2003). Chlorogenic acid bioavailability largely depends on its metabolism by the gut microflora in rats. *Journal of Nutrition*, *133*, 1853–1859.

González-Díaz, C., Gil-González, D., & Ávarez-Dardet, C. (2018). Scientific evidence on functional food and its commercial communication: A review of legislation in Europe and the USA. *Journal of Food Science*, *83*, 2710–2717.

González-Paramás, A. M., Ayuda-Durán, B., Martínez, S., González-Manzano, S., & Santos-Buelga, C. (2018). The mechanisms behind the biological activity of flavonoids. *Current Medicinal Chemistry, 25*, 1–14.

González-Sarrías, A., García-Villalba, R., Romo-Vaquero, M., Alasalvar, C., Örem, A., Zafrilla, P., et al. (2017). Clustering according to urolithin metabotype explains the interindividual variability in the improvement of cardiovascular risk biomarkers in overweight-obese individuals consuming pomegranate: A randomized clinical trial. *Molecular Nutrition & Food Research, 61*, 1600830.

Granado-Serrano, A. B., Martín, M. A., Bravo, L., Goya, L., & Ramos, S. (2012). Quercetin modulates Nrf2 and glutathione-related defenses in HepG2 cells: Involvement of p38. *Chemico-Biological Interactions, 195*, 154–164.

Grand View Research. (2016). *Polyphenols market analysis by product (grape seed, green tea, apple), by application (functional food, functional beverages, dietary supplements) and segment forecasts to 2024*. Retrieved from http://www.grandviewresearch.com/industry-analysis/polyphenolsmarket-analysis (January, 2019).

Griffin, L. E., Witrick, K. A., Klotz, K., Dorenkott, M. R., Goodrich, K. M., Fundaro, G., et al. (2017). Alterations to metabolically active bacteria in the mucosa of the small intestine predict anti-obesity and anti-diabetic activities of grape seed extract in mice. *Food & Function, 8*, 3510–3522.

Grosso, G., Micek, A., Godos, J., Pajak, A., Sciacca, S., Galvano, F., et al. (2017). Dietary flavonoid and lignan intake and mortality in prospective cohort studies: Systematic review and dose-response meta-analysis. *American Journal of Epidemiology, 185*, 1304–1316.

Grosso, G., Stepaniak, U., Topor-Mądry, R., Szafraniec, K., & Pająk, A. (2014). Estimated dietary intake and major food sources of polyphenols in the Polish arm of the HAPIEE study. *Nutrition, 30*, 1398–1403.

Gry, J., Black, L., Eriksen, F. D., Pilegaard, K., Plumb, J., Rhodes, M., et al. (2007). EuroFIR-BASIS: A combined composition and biological activity database for bioactive compounds in plant-based foods. *Trends in Food Science and Technology, 18*, 434–444.

Gu, L., Kelm, M. A., Hammerstone, J. F., Beecher, G., Holden, J., Haytowitz, D., et al. (2004). Concentrations of proanthocyanidins in common foods and estimations of normal consumption. *The Journal of Nutrition, 134*, 613–617.

Guillamón, E., Pedrosa, M. M., Burbano, C., Cuadrado, C., De Sánchez, M. C., & Muzquiz, M. (2008). The trypsin inhibitors present in seed of different grain legume species and cultivar. *Food Chemistry, 107*, 68–74.

Guldiken, B., Linke, A., Capanoglu, E., Boyacioglu, D., Kohlus, R., Weiss, J., et al. (2019). Formation and characterization of spray dried coated and uncoated liposomes with encapsulated black carrot extract. *Journal of Food Engineering, 246*, 42–50.

Hagerman, A. E., & Butler, L. G. (1981). The specificity of proanthocyanidin-protein interactions. *Journal of Biological Chemistry, 256*, 4494–4497.

Hagerman, A. E., Rice, M. E., & Ritchard, N. T. (1998). Mechanisms of protein precipitation for two tannins, pentagalloyl glucose and epicatechin-(4,8)-catechin. *Journal of Agricultural and Food Chemistry, 46*, 2590–2595.

Hagiwara, A., Hirose, M., Takahashi, S., Ogawa, K., Shirai, T., & Ito, N. (1991). Forestomach and kidney carcinogenicity of caffeic acid in F344 rats and C57BL/6N × C3H/HeN F1 mice. *Cancer Research, 51*, 5655–5660.

Haining, Z., & Yongkun, M. (2017). Optimisation of high hydrostatic pressure assisted extraction of anthocyanins from rabbiteye blueberry pomace. *Czech Journal of Food Science, 35*, 180–187.

Han, R. M., Li, D. D., Chen, C. H., Liang, R., Tian, Y. X., Zhang, J. P., et al. (2011). Phenol acidity and ease of oxidation in isoflavonoid/β-carotene antioxidant synergism. *Journal of Agricultural and Food Chemistry, 59*, 10367–10372.

Haslam, E. (1998). *Practical polyphenolics: From structure to molecular recognition and physiological action*. Cambridge: Cambridge University Press.

Heleno, S. A., Diz, P., Prieto, M. A., Barros, L., Rodrigues, A., Barreiro, M. F., et al. (2016). Optimization of ultrasound-assisted extraction to obtain mycosterols from *Agaricus bisporus* L. by response surface methodology and comparison with conventional soxhlet extraction. *Food Chemistry*, *197*, 1054–1063.

Heleno, S. A., Prieto, M. A., Barros, L., Rodrigues, A., Barreiro, M. F., & Ferreira, I. C. F. R. (2016). Optimization of microwave-assisted extraction of ergosterol from *Agaricus bisporus* L. by-products using response surface methodology. *Food and Bioproducts Processing*, *100*, 25–35.

Heo, H. J., Kim, Y. J., Chung, D., & Kim, D. O. (2007). Antioxidant capacities of individual and combined phenolics in a model system. *Food Chemistry*, *104*, 87–92.

Hertog, M. G., Feskens, E. J., Hollman, P. C., Katan, M. B., & Kromhout, D. (1993). Dietary antioxidant flavonoids and risk of coronary heart disease: The Zutphen elderly study. *Lancet*, *342*, 1007–1011.

Hertog, M. G. L., Hollman, P. C. H., & Katan, M. B. (1992). Content of potentially anticarcinogenic flavonoids of 28 vegetables and 9 fruits commonly consumed in the Netherlands. *Journal of Agricultural and Food Chemistry*, *40*, 2379–2383.

Hertog, M. G., Kromhout, D., Aravanis, C., Blackburn, H., Buzina, R., Fidanza, F., et al. (1995). Flavonoid intake and long-term risk of coronary heart disease and cancer in the seven countries study. *Archives of Internal Medicine*, *155*, 381–386.

Hervert-Hernández, D., Pintado, C., Rotger, R., & Goñi, I. (2009). Stimulatory effect of grape pomace polyphenols on *Lactobacillus acidophilus* growth. *International Journal of Food Microbiology*, *136*, 119–122.

Hirose, M., Hoshiya, T., Mizoguchi, Y., Nakamura, A., Akagi, K., & Shirai, T. (2001). Green tea catechins enhance tumor development in the colon without effects in the lung or thyroid after pretreatment with 1,2-dimethylhydrazine or 2,2-dihydroxy-di-n-propylnitrosamine in male F344 rats. *Cancer Letters*, *168*, 23–29.

Hirvonen, T., Virtamo, J., Korhonen, P., Albanes, D., & Pietinen, P. (2001). Flavonol and flavone intake and the risk of cancer in male smokers (Finland). *Cancer Causes and Control*, *12*, 789–796.

Hoadley, J. E., & Rowlands, J. C. (2014). FDA perspectives on food label claims in the United States. In D. Bagchi (Ed.), *Nutraceutical and functional food regulations in the United States and around the world* (pp. 121–140). San Diego: Academic Press.

Hollman, P. C. H. (2014). Unravelling of the health effects of polyphenols is a complex puzzle complicated by metabolism. *Archives of Biochemistry and Biophysics*, *559*, 100–105.

Hollman, P. C. H., & Arts, I. C. W. (2000). Flavonols, flavones and flavanols—Nature, occurrence and dietary burden. *Journal of the Science of Food and Agriculture*, *80*, 1081–1093.

Howitz, K. T., Bitterman, K. J., Cohen, H. Y., Lamming, D. W., Lavu, S., Wood, J. G., et al. (2003). Small molecule activators of sirtuins extend *Saccharomyces cerevisiae* lifespan. *Nature*, *425*, 191–196.

Hu, J., Webster, D., Cao, J., & Shao, A. (2018). The safety of green tea and green tea extract consumption in adults. Results of a systematic review. *Regulatory Toxicology and Pharmacology*, *95*, 412–433.

Hugo, P. C., Gil-Chávez, J., Sotelo-Mundo, R. R., Namiesnik, J., Gorinstein, S., & González-Aguilar, G. A. (2012). Antioxidant interactions between major phenolic compounds found in "Ataulfo" mango pulp: Chlorogenic, gallic, protocatechuic and vanillic acids. *Molecules*, *17*, 12657–12664.

Hui, C., Qi, X., Qianyong, Z., Xiaoli, P., Jundong, Z., & Mantian, M. (2013). Flavonoids, flavonoid subclasses and breast cancer risk: A meta-analysis of epidemiologic studies. *PLoS One*, *8*, 515.

Hurrell, R., & Egli, I. (2010). Iron bioavailability and dietary reference values. *American Journal of Clinical Nutrition*, *91*, 1461S–1467S.

Hwang, E. I., Masafumi, K., Yasuo, O., & Sueharu, H. (2003). Production of plant-specific flavanones by *Escherichia coli* containing an artificial gene cluster. *Applied and Environmental Microbiology*, *69*, 2699–2706.

Jacques, P. F., Cassidy, A., Rogers, G., Peterson, J. J., Meigs, J. B., & Dwyer, J. T. (2013). Higher dietary flavonol intake is associated with lower incidence of type 2 diabetes. *Journal of Nutrition*, *143*, 1474–1480.

Jaganath, I. B., & Crozier, A. (2010). Dietary flavonoids and phenolic compounds. In C. G. Fraga (Ed.), *Plant phenolics and human health: Biochemistry, nutrition, and pharmacology* (pp. 1–49). Hoboken: John Wiley & Sons.

Jakobek, L., & Matić, P. (2019). Non-covalent dietary fiber—Polyphenol interactions and their influence on polyphenol bioaccessibility. *Trends in Food Science & Technology*, *83*, 235–247.

Janissen, B., & Huynh, T. (2018). Chemical composition and value-adding applications of coffee industry by-products: A review. *Resources, Conservation and Recycling*, *128*, 110–117.

Jiménez-López, C., Caleja, C., Prieto, M. A., Barreiro, M. F., Barros, L., & Ferreira, I. C. F. R. (2018). Optimization and comparison of heat and ultrasound assisted extraction techniques to obtain anthocyanin compounds from *Arbutus unedo* L. fruits. *Food Chemistry*, *264*, 81–91.

Jin, G., Asou, Y., Ishiyama, K., Okawa, A., Kanno, T., & Niwano, Y. (2018). Proanthocyanidin-rich grape seed extract modulates intestinal microbiota in ovariectomized mice. *Journal of Food Science*, *83*, 1149–1152.

Joven, J., Micol, V., Segura-Carretero, A., Alonso-Villaverde, C., & Menéndez, J. A. (2014). Polyphenols and the modulation of gene expression pathways: Can we eat our way out of the danger of chronic disease? *Critical Reviews in Food Science and Nutrition*, *54*, 985–1001.

Juan, M. E., Alfaras, I., & Planas, J. M. (2010). Determination of dihydroresveratrol in rat plasma by HPLC. *Journal of Agricultural and Food Chemistry*, *58*, 7472–7475.

Kafantaris, I., Kotsampasi, B., Christodoulou, V., Kokka, E., Kouka, P., Terzopoulou, Z., et al. (2017). Grape pomace improves antioxidant capacity and faecal microflora of lambs. *Journal of Animal Physiology and Animal Nutrition*, *101*, 108–121.

Kähkönen, M. P., Hopia, A. I., & Heinonen, M. (2001). Berry phenolics and their antioxidant activity. *Journal of Agricultural and Food Chemistry*, *49*, 4076–4082. Es este no?.

Kallscheuer, N., Vogt, M., & Marienhagen, J. (2017). A novel synthetic pathway enables microbial production of polyphenols independent from the endogenous aromatic amino acid metabolism. *ACS Synthetic Biology*, *6*, 410–415.

Kallscheuer, N., Vogt, M., Stenzel, A., Gätgens, J., Bott, M., & Marienhagen, J. (2016). Construction of a *Corynebacterium glutamicum* platform strain for the production of stilbenes and (2S)-flavanones. *Metabolic Engineering*, *38*, 47–55.

Kalra, E. K. (2003). Nutraceutical—Definition and introduction. *AAPS Pharm Science*, *5*, 27–28.

Karam, J., Bibiloni, M. D. M., & Tur, J. A. (2018). Polyphenol estimated intake and dietary sources among older adults from Mallorca Island. *PLoS One*, *13*, e0191573.

Katsuyama, Y., Matsuzawa, M., Funa, N., & Horinouchi, S. (2008). Production of curcuminoids by *Escherichia coli* carrying an artificial biosynthesis pathway. *Microbiology*, *154*, 2620–2628.

Kern, S. M., Bennett, R. N., Needs, P. W., Mellon, F. A., Kroon, P. A., & Garcia-Conesa, M. T. (2003). Characterization of metabolites of hydroxycinnamates in the in vitro model of human small intestinal epithelium caco-2 cells. *Journal of Agricultural and Food Chemistry*, *51*, 7884–7891.

Khedkar, S., Ciliberti, S., & Bröring, S. (2016). The EU health claims regulation: Implications for innovation in the EU food sector. *British Food Journal, 118*, 2647–2665.

Kim, J., Kim, S., Huh, K., Kim, Y., Joung, H., & Park, M. (2011). High serum isoflavone concentrations are associated with the risk of precocious puberty in Korean girls. *Clinical Endocrinology, 75*, 831–835.

King, T., Cole, M., Farber, J. M., Eisenbrand, G., Zabaras, D., Fox, E. M., et al. (2017). Food safety for food security: Relationship between global megatrends and developments in food safety. *Trends in Food Science and Technology, 68*, 160–175.

Knekt, P., Jarvinen, R., Reunanen, A., & Maatela, J. (1996). Flavonoid intake and coronary mortality in Finland: A cohort study. *British Medical Journal, 312*, 478–481.

Koopman, F., Beekwilder, J., Crimi, B., van Houwelingen, A., Hall, R. D., Bosch, D., et al. (2012). De novo production of the flavonoid naringenin in engineered *Saccharomyces cerevisiae. Microbial Cell Factories, 11*, 1–15.

Koponen, J. M., Happonen, A. M., Mattila, P. H., & Törrönen, A. R. (2007). Contents of anthocyanins and ellagitannins in selected foods consumed in Finland. *Journal of Agricultural and Food Chemistry, 55*, 1612–1619.

Kroon, P. A., Clifford, M. N., Crozier, A., Day, A. J., Donovan, J. L., Manach, C., et al. (2004). How should we assess the effects of exposure to dietary polyphenols in vitro? *The American Journal of Clinical Nutrition, 80*, 15–21.

Kroon, P. A., Faulds, C. B., Ryden, P., Robertson, J. A., & Williamson, G. (1997). Release of covalently bound ferulic acid from fiber in the human colon. *Journal of Agricultural and Food Chemistry, 45*, 661–667.

Kühnau, J. (1976). The flavonoids. A class of semi-essential food componentes: Their role in human nutrition. *World Review of Nutrition and Dietetics, 24*, 117–191.

Kurin, E., Atanasov, A. G., Donath, O., Heiss, E. H., Dirsch, V. M., & Nagy, M. (2012). Synergy study of the inhibitory potential of red wine polyphenols on vascular smooth muscle cell proliferation. *Planta Medica, 78*, 772–778.

Lafay, S., Gil-Izquierdo, A., Manach, C., Morand, C., Besson, C., & Scalbert, A. (2006). Chlorogenic acid is absorbed in its intact form in the stomach of rats. *Journal of Nutrition, 136*, 1192–1197.

Lakenbrink, C., Lapczynski, S., Maiwald, B., & Engelhardt, U. H. (2000). Flavonoids and other polyphenols in consumer brews of tea and other caffeinated beverages. *Journal of Agricultural and Food Chemistry, 48*, 2848–2852.

Lalor, F., & Wall, P. G. (2011). Health claims regulations. Comparison between USA, Japan and European Union. *British Food Journal, 113*, 298–313.

Lalor, F., & Wall, P. G. (2012). The EU's health claims legislation: How much evidence is enough? *CAB Reviews, 7*, 1–9.

Landete, J. M. (2011). Ellagitannins, ellagic acid and their derived metabolites: A review about source, metabolism, functions and health. *Food Research International, 44*, 1150–1160.

Laranjinha, J. (2010). Translation of chemical properties of polyphenols into biological activity with impact on human health. In C. Santos-Buelga, M. T. Escribano, & V. Lattanzio (Eds.), *Vol. 2. Recent advances in polyphenols research* (pp. 269–282). Chichester: Wiley-Blackwell.

Lattanzio, V., Kroon, P. A., Quideau, S., & Treutter, D. (2008). Plant phenolics—Secondary metabolites with diverse functions. In F. Daayf & V. Lattanzio (Eds.), *Recent advances in polyphenol research* (pp. 1–35). New Jersey: Blackwell Publishing Ltd.

Lee-Hilz, Y. Y., Boerboom, A. M. J. F., Westphal, A. H., van Berkel, W. J. H., Aarts, J. M. M. J. G., & Rietjens, I. M. C. M. (2006). Pro-oxidant activity of flavonoids induces EpRE-mediated gene expression. *Chemical Research in Toxicology, 19*, 1499–1505.

Leopoldini, M., Russo, N., & Toscano, M. (2011). The molecular basis of working mechanism of natural polyphenolic antioxidants. *Food Chemistry, 125*, 288–306.

Lesjak, M., Hoque, R., Balesaria, S., Skinner, V., Debnam, E. S., Srai, S. K. S., et al. (2014). Quercetin inhibits intestinal iron absorption and ferroportin transporter expression in vivo and in vitro. *PLoS One, 9*, e102900.

Letenneur, L., Proust-Lima, C., Le, G. A., Dartigues, J. F., & Barberger-Gateau, P. (2007). Flavonoid intake and cognitive decline over a 10-year period. *American Journal of Epidemiology, 165*, 1364–1371.

Li, J., Lin, S., Vanhoutte, P. M., Woo, C. W., & Xu, A. (2016). *Akkermansia muciniphila* protects against atherosclerosis by preventing metabolic endotoxemia-induced inflammation in Apoe−/− mice. *Circulation, 133*, 2434–2446.

Lila, M. A., Burton-Freeman, B., Grace, M., & Kalt, W. (2016). Unraveling anthocyanin bioavailability for human health. *Annual Review of Food Science and Technology, 7*, 375–393.

Lim, C. G., Wong, L., Bhan, N., Dvora, H., Xu, P., Venkiteswaran, S., et al. (2015). Development of a recombinant *Escherichia coli* strain for overproduction of the plant pigment anthocyanin. *Applied and Environmental Microbiology, 81*, 6276–6284.

Liu, Y. J., Zhan, J., Liu, X. L., Wang, Y., Ji, J., & He, Q. Q. (2014). Dietary flavonoids intake and risk of type 2 diabetes: A meta-analysis of prospective cohort studies. *Clinical Nutrition, 33*, 59–63.

López, N., Puértolas, E., Condón, S., Álvarez, I., & Raso, J. (2008). Effects of pulsed electric fields on the extraction of phenolic compounds during the fermentation of must of Tempranillo grapes. *Innovative Food Science and Emerging Technologies, 9*, 477–482.

López-Giral, N., González-Arenzana, L., González-Ferrero, C., López, R., Santamaría, P., López-Alfaro, I., et al. (2015). Pulsed electric field treatment to improve the phenolic compound extraction from Graciano, Tempranillo and Grenache grape varieties during two vintages. *Innovative Food Science and Emerging Technologies, 28*, 31–39.

Lores-Aguin, M., Garcia Jares, C., Alvarez Casas, M., & Llompart, M. (2014). Polyphenol extract from white-grape residue. WO2014013122A1.

Määttä-Riihinen, K. R., Kamal-Eldin, A., Mattila, P. H., González-Paramás, A. M., & Törrönen, A. R. (2004). Distribution and contents of phenolic compounds in eighteen Scandinavian berry species. *Journal of Agricultural and Food Chemistry, 52*, 4477–4486.

Maeda-Yamamoto, M. (2017). Development of functional agricultural products and use of a new health claim system in Japan. *Trends in Food Science and Technology, 69*, 324–332.

Maeda-Yamamoto, M., & Ohtani, T. (2018). Development of functional agricultural products utilizing the new health claim labeling system in Japan. *Bioscience, Biotechnology, and Biochemistry, 82*, 554–563.

Malongane, F., McGaw, L. J., & Mudau, F. N. (2017). The synergistic potential of various teas, herbs and therapeutic drugs in health improvement: A review. *Journal of the Science of Food and Agriculture, 97*, 4679–4689.

Manach, C., Scalbert, A., Morand, C., Rémésy, C., & Jiménez, L. (2004). Polyphenols: Food sources and bioavailability. *The American Journal of Clinical Nutrition, 79*, 727–747.

Maruyama, K., Kihara-Negishi, F., Ohkura, N., Nakamura, Y., Nasui, M., & Saito, M. (2017). Simultaneous determination of catechins and caffeine in green tea-based beverages and foods for specified health uses. *Food and Nutrition Sciences, 8*, 316–325.

Masella, R., Di Benedetto, R., Varì, R., Filesi, C., & Giovannini, C. (2005). Novel mechanisms of natural antioxidant compounds in biological systems: Involvement of glutathione and glutathione-related enzymes. *The Journal of Nutritional Biochemistry, 16*, 577–586.

Massounga-Bora, A. F., Ma, S., Li, X., & Liu, L. (2018). Application of microencapsulation for the safe delivery of green tea polyphenols in food systems: Review and recent advances. *Food Research International, 105*, 241–249.

Mazzanti, G., Di Sotto, A., & Vitalone, A. (2015). Hepatotoxicity of green tea: An update. *Archives of Toxicology, 89*, 1175–1191.

Mazzanti, G., Menniti-Ippolito, F., Moro, P. A., Cassetti, F., Raschetti, R., Santuccio, C., et al. (2009). Hepatotoxicity from green tea: A review of the literature and two unpublished cases. *European Journal of Clinical Pharmacology, 65*, 331–341.

McCullough, M. L., Peterson, J. J., Patel, R., Jacques, P. F., Shah, R., & Dwyer, J. T. (2012). Flavonoid intake and cardiovascular disease mortality in a prospective cohort of US adults. *The American Journal of Clinical Nutrition, 95*, 454–464.

McGhie, T. K., Ainge, G. D., Barnett, L. E., Cooney, J. M., & Jensen, D. J. (2003). Anthocyanin glycosides from berry fruit are absorbed and excreted unmetabolized by both humans and rats. *Journal of Agricultural and Food Chemistry, 51*, 4539–4548.

Medina-Torres, N., Ayora-Talavera, T., Espinosa-Andrews, H., Sánchez-Contreras, A., & Pacheco, N. (2017). Ultrasound assisted extraction for the recovery of phenolic compounds from vegetable sources. *Agronomy, 7*, 47.

Menéndez, C., Duenas, M., Galindo, P., González-Manzano, S., Jiménez, R., Moreno, L., et al. (2011). Vascular deconjugation of quercetin glucuronide: The flavonoid paradox revealed? *Molecular Nutrition & Food Research, 55*, 1780–1790.

Menéndez, J. A., Joven, J., Aragonés, G., Barrajón-Catalán, E., Beltrán-Debon, R., Borras-Linares, I., et al. (2013). Xenohormetic and anti-aging activity of secoiridoid polyphenols present in extra virgin olive oil: A new family of gerosuppressant agents. *Cell Cycle, 12*, 555–578.

Metodiewa, D., Jaiswal, A. K., Cenas, N., Dickancaité, E., & Segura-Aguilar, J. (1999). Quercetin may act as a cytotoxic prooxidant after its metabolic activation to semiquinone and quinoidal product. *Free Radical Biology & Medicine, 26*, 107–116.

Milke, L., Aschenbrenner, J., Marienhagen, J., & Kallscheuer, N. (2018). Production of plant-derived polyphenols in microorganisms: Current state and perspectives. *Applied Microbiology and Biotechnology, 102*, 1575–1585.

Miranda, A. M., Steluti, J., Fisberg, R. M., & Marchioni, D. M. (2016). Dietary intake and food contributors of polyphenols in adults and elderly adults of Sao Paulo: A population-based study. *British Journal of Nutrition, 115*, 1061–1070.

Mirón-Mérida, V. A., Yáñez-Fernández, J., Montáñez-Barragán, B., & Barragán, B. E. (2019). Valorization of coffee parchment waste (Coffee arabica) as a source of caffeine and phenolic compounds in antifungal gellan gum films. *Lebensmittel-Wissenschaft & Technologie, 101*, 167–174.

Mirosa, M., & Mangan-Walker, E. (2018). Young Chinese and functional foods for mobility health: Perceptions of importance, trust, and willingness to purchase and pay a premium. *Journal of Food Products Marketing, 24*, 216–234.

Monteiro, M., Farah, A., Perrone, D., Trugo, L. C., & Donangelo, C. (2007). Chlorogenic acid compounds from coffee are differentially absorbed and metabolized in humans. *Journal of Nutrition, 137*, 2196–2201.

Montgomery, A., Adeyeni, T., San, K. K., Heuertz, R. M., & Ezekiel, U. R. (2016). Curcumin sensitizes silymarin to exert synergistic anticancer activity in colon cancer cells. *Journal of Cancer, 7*, 1250–1257.

Moon, Y. J., Wang, X., & Morris, M. E. (2006). Dietary flavonoids: Effects on xenobiotic and carcinogen metabolism. *Toxicology In Vitro, 20*, 187–210.

Mortensen, A., Kulling, S. E., Schwartz, H., Rowland, I., Ruefer, C. E., Rimbach, G., et al. (2009). Analytical and compositional aspects of isoflavones in food and their biological effects. *Molecular Nutrition & Food Research, 53*, S266–S309.

Munir, M. T., Kheirkhah, H., Baroutian, S., Quek, S. Y., & Young, B. R. (2018). Subcritical water extraction of bioactive compounds from waste onion skin. *Journal of Cleaner Production, 183*, 487–494.

Murphy, K. J., Walker, K. M., Dyer, K. A., & Bryan, J. (2019). Estimation of daily intake of flavonoids and major food sources in middle-aged Australian men and women. *Nutrition Research*, *61*, 64–81.

Murray, N. J., Williamson, M. P., Lilley, T. H., & Haslam, E. (1994). Study of the interaction between proline-rich proteins and a polyphenol by 1H-NMR spectroscopy. *European Journal of Biochemistry*, *219*, 915–921.

Nabet, N., Gilbert-López, B., Madani, K., Herrero, M., Ibáñez, E., & Mendiola, J. A. (2019). Optimization of microwave-assisted extraction recovery of bioactive compounds from Origanum glandulosum and Thymus fontanesii. *Industrial Crops and Products*, *129*, 395–404.

Nagata, H., Takekoshi, S., Takagi, T., Honma, T., & Watanabe, K. (1999). Antioxidative action of flavonoids, quercetin and catechin, mediated by the activation of glutathione peroxidase. *The Tokai Journal of Experimental and Clinical Medicine*, *24*, 1–11.

Nascimento-Souza, M. A., de Paiva, P. G., Pérez-Jiménez, J., Castro Franceschini, S. C., & Ribeiro, A. Q. (2018). Estimated dietary intake and major food sources of polyphenols in elderly of Viçosa, Brazil: A population-based study. *European Journal of Nutrition*, *57*, 617–627.

Neunert, G., Górnaś, P., Dwiecki, K., Siger, A., & Polewski, K. (2015). Synergistic and antagonistic effects between alpha-tocopherol and phenolic acids in liposome system: Spectroscopic study. *European Food Research and Technology*, *241*, 749–757.

Neveu, V., Moussy, A., Rouaix, H., Wedekind, R., Pon, A., Knox, C., et al. (2017). Exposome-explorer: A manually-curated database on biomarkers of exposure to dietary and environmental factors. *Nucleic Acids Research*, *45*, 979–984.

Neveu, V., Pérez-Jiménez, J., Vos, F., Crespy, V., du Chaffaut, L., Mennen, L., et al. (2010). Phenol-explorer: An online comprehensive database on polyphenol contents in foods. *Database (Oxford)*, *2010*, bap024.

Nikmaram, N., Leong, S. Y., Koubaa, M., Zhu, Z., Barba, F. J., Greiner, R., et al. (2017). Effect of extrusion on the anti-nutritional factors of food products: An overview. *Food Control*, *79*, 62–73.

Nipornram, S., Tochampa, W., Rattanatraiwong, P., & Singanusong, R. (2018). Optimization of low power ultrasound-assisted extraction of phenolic compounds from mandarin *(Citrus reticulata Blanco cv. Sainampueng)* peel. *Food Chemistry*, *241*, 338–345.

O'Leary, K. A., Day, A. J., Needs, P. W., Sly, W. S., O'Brien, N. M., & Williamson, G. (2001). Flavonoid glucuronides are substrates for human liver β-glucuronidase. *FEBS Letters*, *503*, 103–106.

O'Sullivan, J. J., Norwood, E. A., O'Mahony, J. A., & Kelly, A. L. (2019). Atomisation technologies used in spray drying in the dairy industry: A review. *Journal of Food Engineering*, *243*, 57–69.

Oh, H. I., Hoff, J. E., Armstrong, G. S., & Haff, L. A. (1980). Hydrophobic interaction in tannin-protein complexes. *Journal of Agricultural and Food Chemistry*, *28*, 394–398.

Oidtmann, J., Schantz, M., Mäder, K., Baum, M., Berg, S., Betz, M., et al. (2012). Preparation and comparative release characteristics of three anthocyanin encapsulation systems. *Journal of Agricultural and Food Chemistry*, *60*, 844–851.

Oludemi, T., Barros, L., Prieto, M. A., Heleno, S. A., Barreiro, M. F., & Ferreira, I. C. F. R. (2018). Extraction of triterpenoids and phenolic compounds from Ganoderma lucidum: Optimization study using the response surface methodology. *Food & Function*, *9*, 209–226.

Ovaskainen, M. L., Törrönen, R., Koponen, J. M., Sinkko, H., Hellström, J., Reinivuo, H., et al. (2008). Dietary intake and major food sources of polyphenols in Finnish adults. *Journal of Nutrition*, *138*, 562–566.

Ozdal, T., Sela, D. A., Xiao, J., Boyacioglu, D., Chen, F., & Capanoglu, E. (2016). The reciprocal interactions between polyphenols and gut microbiota and effects on bioaccessibility. *Nutrients*, *8*, 78.

Pan, M. H., Lai, C. S., Wu, J. C., & Ho, C. T. (2013). Epigenetic and disease targets by polyphenols. *Current Pharmaceutical Design*, *19*, 6156–6185.

Panya, A., Kittipongpittaya, K., Laguerre, M., Bayrasy, C., Lecomte, J., Villeneuve, P., et al. (2012). Interactions between α-tocopherol and rosmarinic acid and its alkyl esters in emulsions: Synergistic, additive, or antagonistic effect? *Journal of Agricultural and Food Chemistry*, *60*, 10320–10330.

Pasquel-Reátegui, J. L., Machado, A. P. D. F., Barbero, G. F., Rezende, C. A., & Martínez, J. (2014). Extraction of antioxidant compounds from blackberry (*Rubus* sp.) bagasse using supercritical CO_2 assisted by ultrasound. *The Journal of Supercritical Fluids*, *94*, 223–233.

Pérez-Jiménez, J., Fezeu, L., Touvier, M., Arnault, N., Manach, C., Hercberg, S., et al. (2011). Dietary intake of 337 polyphenols in French adults. *The American Journal of Clinical Nutrition*, *93*, 1220–1228.

Pérez-Jiménez, J., & Saura-Calixto, F. (2015). Macromolecular antioxidants or non-extractable polyphenols in fruit and vegetables: Intake in four European countries. *Food Research International*, *74*, 315–323.

Petry, N., Egli, I., Zeder, C., Walczyk, T., & Hurrell, R. (2010). Polyphenols and phytic acid contribute to the low iron bioavailability from common beans in young women. *Journal of Nutrition*, *140*, 1977–1982.

Pettinato, M., Casazza, A. A., Ferrari, P. F., Palombo, D., & Perego, P. (2019). Eco-sustainable recovery of antioxidants from spent coffee grounds by microwave-assisted extraction: Process optimization, kinetic modeling and biological validation. *Food and Bioproducts Processing*, *114*, 31–42.

Phan, M. A. T., Paterson, J., Bucknall, M., & Arcot, J. (2018). Interactions between phytochemicals from fruits and vegetables: Effects on bioactivities and bioavailability. *Critical Reviews in Food Science and Nutrition*, *58*, 1310–1329.

Pimentel-Moral, S., Borrás-Linares, I., Lozano-Sánchez, J., Arráez-Román, D., Martínez-Férez, A., & Segura-Carretero, A. (2018). Microwave-assisted extraction for *Hibiscus sabdariffa* bioactive compounds. *Journal of Pharmaceutical and Biomedical Analysis*, *156*, 313–322.

Pimentel-Moral, S., Borrás-Linares, I., Lozano-Sánchez, J., Arráez-Román, D., Martínez-Férez, A., & Segura-Carretero, A. (2019). Supercritical CO_2 extractions of bioactive compounds from *Hibiscus sabdariffa*. *The Journal of Supercritical Fluids*. https://doi.org/10.1016/j.supflu.2018.11.005.

Pinela, J., Prieto, M. A., Barreiro, M. F., Carvalho, A. M., Oliveira, M. B. P. P., Curran, T. P., et al. (2017). Valorisation of tomato wastes for development of nutrient-rich antioxidant ingredients: A sustainable approach towards the needs of the today's society. *Innovative Food Science and Emerging Technologies*, *41*, 160–171.

Pinela, J., Prieto, M. A., Barros, L., Maria, A., Oliveira, M. B. P. P., Saraiva, J. A., et al. (2018). Cold extraction of phenolic compounds from watercress by high hydrostatic pressure: Process modelling and optimization. *Separation and Purification Technology*, *192*, 501–512.

Pinelo, M., Manzocco, L., Nuñez, M. J., & Nicoli, M. C. (2004). Interaction among phenols in food fortification: Negative synergism on antioxidant capacity. *Journal of Agricultural and Food Chemistry*, *52*, 1177–1180.

Pinto da Costa, J. (2017). A current look at nutraceuticals—Key concepts and future prospects. *Trends in Food Science & Technology*, *62*, 68–78.

Plazzotta, S., & Manzocco, L. (2018). Effect of ultrasounds and high pressure homogenization on the extraction of antioxidant polyphenols from lettuce waste. *Innovative Food Science and Emerging Technologies*, *50*, 11–19.

Plumb, G. W., Garcia-Conesa, M. T., Kroon, P. A., Rhodes, M., Ridley, S., & Williamson, G. (1999). Metabolism of chlorogenic acid by human plasma, liver, intestine and gut microflora. *Journal of the Science of Food and Agriculture*, *79*, 390–392.

Poquet, L., Clifford, M. N., & Williamson, G. (2009). Bioavailability of flavanols and phenolic acids. In C. G. Fraga (Ed.), *Plant phenolics and human health: Biochemistry, nutrition, and pharmacology* (pp. 51–89). Hoboken: John Wiley & Sons.

Poveda, J. M., Loarce, L., Alarcón, M., Díaz-maroto, M. C., & Alañón, M. E. (2018). Revalorization of winery by-products as source of natural preservatives obtained by means of green extraction techniques. *Industrial Crops and Products*, *112*, 617–625.

Pozuelo, M. J., Agis-Torres, A., Hervert-Hernández, D., López-Oliva, M. E., Muñoz-Martínez, E., Rotger, R., et al. (2012). Grape antioxidant dietary fiber stimulates lactobacillus growth in rat cecum. *Journal of Food Science*, *77*, H59–H62.

Procházková, D., Boušová, I., & Wilhelmová, N. (2011). Antioxidant and prooxidant properties of flavonoids. *Fitoterapia*, *82*, 513–523.

Qilong, R., Huabin, X., Zongb, B., Baogen, S., Qiwei, Y., Yiwen, Y., et al. (2013). Recent advances in separation of bioactive natural products. *Chinese Journal of Chemical Engineering*, *21*, 937–952.

Quideau, S., Deffieux, D., Douat-Casassus, C., & Pouysegu, L. (2011). Plant polyphenols: Chemical properties, biological activities and synthesis. *Angewandte Chemie International Edition*, *50*, 586–621.

Rashidinejad, A., Birch, E. J., Sun-Waterhouse, D., & Everett, D. W. (2014). Delivery of green tea catechin and epigallocatechin gallate in liposomes incorporated into low-fat hard cheese. *Food Chemistry*, *156*, 176–183.

Rauf, A., Imran, M., Suleria, H. A. R., Ahmad, B., Peters, D. G., & Mubarak, M. S. (2017). A comprehensive review of the health perspectives of resveratrol. *Food & Function*, *8*, 4284–4305.

Rechner, A. R., Smith, M. A., Kuhnle, G., Gibson, G. R., Debnam, E. S., Srai, S. K., et al. (2004). Colonic metabolism of dietary polyphenols: Influence of structure on microbial fermentation products. *Free Radical Biology & Medicine*, *36*, 212–225.

Requena, T., Monagas, M., Pozo-Bayón, M. A., Martín-Álvarez, P. J., Bartolomé, B., & Del Campo, R. (2010). Perspectives of the potential implications of wine polyphenols on human oral and gut microbiota. *Trends in Food Science and Technology*, *21*, 332–344.

Ribnicky, D. M., Roopchand, D. E., Oren, A., Grace, M., Poulev, A., Lila, M. A., et al. (2014). Effects of a high fat meal matrix and protein complexation on the bioaccessibility of blueberry anthocyanins using the TNO gastrointestinal model (TIM-1). *Food Chemistry*, *142*, 349–357.

Riciputi, Y., Diaz-de-cerio, E., Akyol, H., Capanoglu, E., Cerretani, L., Fiorenza, M., et al. (2018). Establishment of ultrasound-assisted extraction of phenolic compounds from industrial potato by-products using response surface methodology. *Food Chemistry*, *269*, 258–263.

Rienks, J., Barbaresko, J., & Nöthlings, U. (2017). Association of polyphenol biomarkers with cardiovascular disease and mortality risk: A systematic review and meta-analysis of observational studies. *Nutrients*, *9*, 415.

Rios, L. Y., Bennett, R. N., Lazarus, S. A., Remesy, C., Scalbert, A., & Williamson, G. (2002). Cocoa procyanidins are stable during gastric transit in humans. *The American Journal of Clinical Nutrition*, *76*, 1106–1110.

Rocchetti, B., Corrêa, M., Souza, D., Pací, M., Maria, R., Bezerra, N., et al. (2018). Combining pressurized liquids with ultrasound to improve the extraction of phenolic compounds from pomegranate peel (*Punica granatum* L.). *Ultrasonics Sonochemistry*, *48*, 151–162.

Rodriguez, A., Strucko, T., Stahlhut, S. G., Kristensen, M., Svenssen, D. K., Forster, J., et al. (2017). Metabolic engineering of yeast for fermentative production of flavonoids. *Bioresource Technology*, *245*, 1645–1654.

Romero, C., & Brenes, M. (2012). Analysis of total contents of hydroxytyrosol and tyrosol in olive oils. *Journal of Agricultural and Food Chemistry*, *60*, 9017–9022.

Romo-Vaquero, M., García-Villalba, R., González-Sarriás, A., Beltrán, D., Tomás-Barberán, F. A., & Espín, J. C. (2015). Interindividual variability in the human metabolism of ellagic acid: Contribution of gordonibacter to urolithin production. *Journal of Functional Foods*, *17*, 785–791.

Rothwell, J. A., Pérez-Jiménez, J., Neveu, V., Medina-Ramon, A., M'Hiri, N., Garcia Lobato, P., et al. (2013). Phenol-explorer 3.0: A major update of the phenol-explorer database to incorporate data on the effects of food processing on polyphenol content. *Database (Oxford)*, *2013*, bat070.

Rothwell, J. A., Urpi-Sarda, M., Boto-Ordoñez, M., Llorach, R., Farran-Codina, A., Barupal, D. K., et al. (2016). Systematic analysis of the polyphenol metabolome using the Phenol-Explorer database. *Molecular Nutrition & Food Research*, *60*, 203–211.

Roura, E., Andres-Lacueva, C., Estruch, R., & Lamuela-Raventos, R. M. (2006). Total polyphenol intake estimated by a modified Folin–Ciocalteu assay of urine. *Clinical Chemistry*, *52*, 749–752.

Russo, G. L., Vastolo, V., Ciccarelli, M., Albano, L., Macchia, P. E., & Ungaro, P. (2017). Dietary polyphenols and chromatin remodeling. *Critical Reviews in Food Science and Nutrition*, *57*, 2589–2599.

Rusznyak, S., & Szent-Gyorgyi, A. (1936). Vitamin P: Flavonols as vitamins. *Nature*, *138*, 27.

Samsami-Kor, M., Daryani, N. E., Asl, P. R., & Hekmatdoost, A. (2015). Anti-inflammatory effects of resveratrol in patients with ulcerative colitis: A randomized, double-blind, placebo-controlled pilot study. *Archives of Medical Research*, *46*, 280–285.

Santos, P., Aguiar, A. C., Barbero, G. F., Rezende, C. A., & Martínez, J. (2015). Supercritical carbon dioxide extraction of capsaicinoids from malagueta pepper (*Capsicum frutescens* L.) assisted by ultrasound. *Ultrasound Sonochemistry*, *22*, 78–88.

Santos, P. H., Baggio Ribeiro, D. H., Micke, G. A., Vitali, L., & Hense, H. (2019). Extraction of bioactive compounds from feijoa (*Acca sellowiana* (O. Berg) Burret) peel by low and high-pressure techniques. *The Journal of Supercritical Fluids*, *145*, 219–227.

Santos, S. A. O., Villaverde, J. J., Silva, C. M., Neto, C. P., & Silvestre, A. J. D. (2012). Supercritical fluid extraction of phenolic compounds from *Eucalyptus globulus* labill bark. *The Journal of Supercritical Fluids*, *71*, 71–79.

Santos-Buelga, C., & González-Paramás, A. M. (2014). Strategies in the analysis of flavonoids. In K. Hostettmann, H. Stuppner, A. Marston, & S. Chen (Eds.), *Handbook of chemical and biological plant analytical methods* (pp. 1–25). Chichester: John Wiley & Sons.

Santos-Buelga, C., & González-Paramás, A. M. (2019). Anthocyanins. In L. Melton, F. Shahidi, & P. Varelis (Eds.), *Encyclopedia of food chemistry* (pp. 10–21). Amsterdam: Elsevier.

Santos-Buelga, C., & Scalbert, A. (2000). Proanthocyanidins and tannin-like compounds—Nature, occurrence, dietary intake and effects on nutrition and health. *Journal of the Science of Food and Agriculture*, *80*, 1094–1117.

Saura-Calixto, F., Serrano, J., & Goñi, I. (2007). Intake and bioaccessibility of total polyphenols in whole diet. *Food Chemistry*, *101*, 492–501.

Seeram, N. P., Henning, S. M., Zhang, Y., Suchard, M., Li, Z., & Heber, D. (2006). Pomegranate juice ellagitannin metabolites are present in human plasma and some persist in urine for up to 48 hours. *Journal of Nutrition*, *136*, 2481–2485.

Selma, M. V., Espín, J. C., & Tomás-Barberán, F. A. (2009). Interaction between phenolics and gut microbiota: Role in human health. *Journal of Agricultural and Food Chemistry*, *57*, 6485–6501.

Serra, A., Macià, A., Romero, M. P., Reguant, J., Ortega, N., & Motilva, M. J. (2012). Metabolic pathways of the colonic metabolism of flavonoids (flavonols, flavones and flavanones) and phenolic acids. *Food Chemistry*, *130*, 383–393.

Setchell, K. D., Brown, N. M., & Lydeking-Olsen, E. (2002). The clinical importance of the metabolite equol-a clue to the effectiveness of soy and its isoflavones. *Journal of Nutrition, 132*, 3577–3584.

Setchell, K. D., & Clerici, C. (2010). Equol: History, chemistry, and formation. *Journal of Nutrition, 140*, 1355S–1362S.

Shimoi, K., Saka, N., Nozawa, R., Sato, M., Amano, I., Nakayama, T., et al. (2001). Deglucuronidation of a flavonoid, luteolin monoglucuronide, during inflammation. *Drug Metabolism and Disposition, 29*, 1521–1524.

Soler-Rivas, C., Espín, J. C., & Wichers, H. J. (2000). Oleuropein and related compounds. *Journal of the Science of Food and Agriculture, 80*, 1013–1023.

Song, M. R., & Im, M. (2018). Moderating effects of food type and consumers' attitude on the evaluation of food items labeled "additive-free" *Journal of Consumer Behaviour, 17*, e1–e12.

Souza, J. E. D., Casanova, L. M., & Costa, S. S. (2015). Bioavailability of phenolic compounds: A major challenge for drug development? *Revista Fitos, 9*, 55–67.

Stalmach, A., Mullen, W., Barron, D., Uchida, K., Yokota, T., Cavin, C., et al. (2009). Metabolite profiling of hydroxycinnamate derivatives in plasma and urine following the ingestion of coffee by humans: Identification of biomarkers of coffee consumption. *Drug Metabolism & Disposition, 37*, 1759–1768.

Stalmach, A., Steiling, H., Williamson, G., & Crozier, A. (2010). Bioavailability of chlorogenic acids following acute ingestion of coffee by humans with an ileostomy. *Archives of Biochemistry and Biophysics, 501*, 98–105.

Stevens, J. F., & Maie, C. S. (2016). The chemistry of gut microbial metabolism of polyphenols. *Phytochemistry Reviews, 15*, 425–444.

Suárez-Valles, B., Santamaria-Victorero, J., Mangas, J. J., & Blanco, D. (1999). High-performance liquid chromatography of the neutral phenolic compounds of low molecular weight in apple juice. *Journal of Agricultural and Food Chemistry, 42*, 2732–2736.

Sulaiman, S. F., Sajak, A. A. B., Ooi, K. L., Supriatno, & Seow, E. M. (2011). Effect of solvents in extracting polyphenols and antioxidants of selected raw vegetables. *Journal of Food Composition and Analysis, 24*, 506–515.

Szkudelska, K., Nogowski, L., & Szkudelski, T. (2009). The inhibitory effect of resveratrol on leptin secretion from rat adipocytes. *European Journal of Clinical Investigation, 39*, 899–905.

Tabasco, R., Sánchez-Patán, F., Monagas, M., Bartolomé, B., Moreno-Arribas, M. V., Peláez, C., et al. (2011). Effect of grape polyphenols on lactic acid bacteria and bifidobacteria growth: Resistance and metabolism. *Food Microbiology, 28*, 1345–1352.

Tamires, D., Pereira, V., Tarone, A. G., Baú, C., Cazarin, B., Fernández, G., et al. (2019). Pressurized liquid extraction of bioactive compounds from grape marc. *Journal of Food Engineering, 240*, 105–113.

Tang, S. Y., & Halliwell, B. (2010). Medicinal plants and antioxidants: What do we learn from cell culture and *Caenorhabditis elegans* studies? *Biochemical and Biophysical Research Communications, 394*, 1–5.

Teixeira, A., Baenas, N., Dominguez-Perles, R., Barros, A., Rosa, E., Moreno, D. A., et al. (2014). Natural bioactive compounds from winery by-products as health promoters: A review. *International Journal of Molecular Sciences, 15*, 15638–15678.

Terao, J. (2009). Flavonols: Metabolism, bioavailability, and health impacts. In C. G. Fraga (Ed.), *Plant phenolics and human health: Biochemistry, nutrition, and pharmacology* (pp. 185–196). Hoboken: John Wiley & Sons.

Terao, J., Murota, K., & Kawai, Y. (2011). Conjugated quercetin glucuronides as bioactive metabolites and precursors of aglycone in vivo. *Food & Function, 2*, 11–17.

Thilakarathna, S. H., & Rupasinghe, H. P. V. (2013). Flavonoid bioavailability and attempts for bioavailability enhancement. *Nutrients, 5*, 3367–3387.

Thompson, L. U., Robb, P., Serraino, M., & Cheung, F. (1991). Mammalian lignan production from various foods. *Nutrition and Cancer, 16*, 43–52.
Timilsena, Y. P., Akanbi, T. O., Khalid, N., Adhikari, B., & Barrow, C. J. (2019). Complex coacervation: Principles, mechanisms and applications in microencapsulation. *International Journal of Biological Macromolecules, 121*, 1276–1286.
Tomás-Barberán, F. A., & Clifford, M. N. (2000). Flavanones, chalcones and dihydrochalcones—Nature, occurrence and dietary burden. *Journal of the Science of Food and Agriculture, 80*, 1073–1080.
Tomás-Barberán, F. A., García-Villalba, R., González-Sarrías, A., Selma, M. V., & Espín, J. C. (2014). Ellagic acid metabolism by human gut microbiota: Consistent observation of three urolithin phenotypes in intervention trials, independent of food source, age, and health status. *Journal of Agricultural and Food Chemistry, 62*, 6535–6538.
Tomás-Barberán, F. A., Selma, M. V., & Espin, J. C. (2016). Interactions of gut microbiota with dietary polyphenols and consequences to human health. *Current Opinion in Clinical Nutrition and Metabolic Care, 19*, 471–476.
Tomás-Barberán, F. A., Selma, M. V., & Espin, J. C. (2018). Polyphenols' gut microbiota metabolites: Bioactives or biomarkers? *Journal of Agricultural and Food Chemistry, 66*, 3593–3594.
Tresserra-Rimbau, A., Guasch-Ferré, M., Salas-Salvadó, J., Toledo, E., Corella, D., Castañer, O., et al. (2016). Intake of total polyphenols and some classes of polyphenols is inversely associated with diabetes in elderly people at high cardiovascular disease risk. *Journal of Nutrition, 146*, 767–777.
Tresserra-Rimbau, A., Rimm, E. B., Medina-Remón, A., Martínez-González, M. A., de la Torre, R., Corella, D., et al. (2014). Inverse association between habitual polyphenol intake and incidence of cardiovascular events in the PREDIMED study. *Nutrition, Metabolism, and Cardiovascular Diseases, 24*, 639–647.
Tsai, P. Y., Ka, S. M., Chang, J. M., Chen, H. C., Shui, H. A., Li, C. Y., et al. (2011). Epigallocatechin-3-gallate prevents lupus nephritis development in mice via enhancing the Nrf2 antioxidant pathway and inhibiting NLRP3 inflammasome activation. *Free Radical Biology & Medicine, 51*, 744–754.
U.S. Department of Agriculture, Agricultural Research Service. (2008). *USDA database for the isoflavone content of selected foods*. Release 2.0. http://www.ars.usda.gov/Services/docs.htm?docid=6382.
U.S. Department of Agriculture, Agricultural Research Service. (2011). *USDA database for the flavonoid content of selected foods*. Release 3.0. http://www.ars.usda.gov/Services/docs.htm?docid=6231.
Ulaszewska, M., Weinert, C. H., Trimigno, A., Portmann, R., Andres-Lacueva, C., Badertscher, R., et al. (2019). Nutrimetabolomics: An integrative action for metabolomic analyses in human nutritional studies. *Molecular Nutrition & Food Research, 63*, 1800384.
Unwin, I., Jansen-van der Vliet, M., Westenbrink, S., Presser, K., Infanger, E., Porubska, J., et al. (2016). Implementing the EuroFIR document and data repositories as accessible resources of food composition information. *Food Chemistry, 193*, 90–96.
Vauzour, D., Rodriguez-Mateos, A., Corona, G., Oruna-Concha, M. J., & Spencer, J. P. (2010). Polyphenols and human health: Prevention of disease and mechanisms of action. *Nutrients, 2*, 1106–1131.
Viacava, G. E., Roura, S. I., & Agüero, M. V. (2015). Optimization of critical parameters during antioxidants extraction from butterhead lettuce to simultaneously enhance polyphenols and antioxidant activity. *Chemometrics and Intelligent Laboratory Systems, 146*, 47–54.
Vieira da Silva, B., Barreira, J. C. M., & Oliveira, M. B. P. P. (2016). Natural phytochemicals and probiotics as bioactive ingredients for functional foods: Extraction, biochemistry and protected-delivery technologies. *Trends in Food Science and Technology, 50*, 144–158.

Vieira, V., Prieto, M. A., Barros, L., Coutinho, J. A. P., Ferreira, O., & Ferreira, I. C. F. R. (2017). Optimization and comparison of maceration and microwave extraction systems for the production of phenolic compounds from *Juglans regia* L. for the valorization of walnut leaves. *Industrial Crops and Products, 107*, 341–352.

Vieira, V., Prieto, M. A., Barros, L., Coutinho, J. A. P., Ferreira, I. C. F. R., & Ferreira, O. (2018). Enhanced extraction of phenolic compounds using choline chloride based deep eutectic solvents from Juglans regia L. *Industrial Crops and Products, 115*, 261–271.

Vincekovi, M., Viskic, M., Juric, S., Giacometti, J., Bursac Kovacevic, D., Putnik, P., et al. (2017). Innovative technologies for encapsulation of Mediterranean plants extracts. *Trends in Food Science and Technology, 69*, 1–12.

Vitale, M., Masulli, M., Rivellese, A. A., Bonora, E., Cappellini, F., Nicolucci, A., et al. (2018). Dietary intake and major food sources of polyphenols in people with type 2 diabetes: The TOSCA.IT. Study. *European Journal of Nutrition, 57*, 679–688.

Viveros, A., Chamorro, S., Pizarro, M., Arija, I., Centeno, C., & Brenes, A. (2011). Effects of dietary polyphenol-rich grape products on intestinal microflora and gut morphology in broiler chicks. *Poultry Science, 90*, 566–578.

Walle, T. (2011). Bioavailability of resveratrol. *Annals of the New York Academy of Sciences, 1215*, 9–15.

Wang, S., Meckling, K. A., Marcone, M. F., Kakuda, Y., & Tsao, R. (2011). Synergistic, additive, and antagonistic effects of food mixtures on total antioxidant capacities. *Journal of Agricultural and Food Chemistry, 59*, 960–968.

Wang, X., Ouyang, Y. Y., Liu, J., & Zhao, G. (2014). Flavonoid intake and risk of CVD: A systematic review and meta-analysis of prospective cohort studies. *British Journal of Nutrition, 111*, 1–11.

Wang, Z. M., Zhao, D., Nie, Z. L., Zhao, H., Zhou, B., Gao, W., et al. (2014). Flavonol intake and stroke risk: A meta-analysis of cohort studies. *Nutrition, 30*, 518–523.

Williams, R. J., Spencer, J. P., & Rice-Evans, C. (2004). Flavonoids: Antioxidants or signalling molecules? *Free Radical Biology & Medicine, 36*, 838–849.

Williamson, G., Barron, D., Shimoi, K., & Terao, J. (2005). In vitro biological properties of flavonoid conjugates found in vivo. *Free Radical Research, 39*, 457–469.

Williamson, G., & Clifford, M. N. (2010). Colonic metabolites of berry polyphenols: The missing link to biological activity? *British Journal of Nutrition, 104*, S48–S66.

Williamson, G., & Clifford, M. N. (2017). Role of the small intestine, colon, and microbiota in determining the metabolic fate of polyphenols. *Biochemical Pharmacology, 139*, 24–39.

Williamson, G., Kay, C. D., & Crozier, A. (2018). The bioavailability, transport, and bioactivity of dietary flavonoids: A review from a historical perspective. *Comprehensive Reviews in Food Science and Food Safety, 17*, 1054–1112.

Wishart, D. S., Jewison, T., Guo, A. C., Wilson, M., Knox, C., Liu, Y., et al. (2013). HMDB 3.0—The human metabolome database in 2013. *Nucleic Acids Research, 41*, 801–807.

Wood, J. G., Rogina, B., Lavu, S., Howitz, K., Helfand, S. L., Tatar, M., et al. (2004). Sirtuin activators mimic caloric restriction and delay ageing in metazoans. *Nature, 430*, 686–689.

Wu, X., He, B., Liu, J., Feng, H., Ma, Y., Li, D., et al. (2016). Molecular insight into gut microbiota and rheumatoid arthritis. *International Journal of Molecular Sciences, 17*, 431.

Xiao, J. B., & Hogger, P. (2015). Dietary polyphenols and type 2 diabetes: Current insights and future perspectives. *Current Medicinal Chemistry, 22*, 23–38.

Xie, Y., Huang, S., & Su, Y. (2016). Dietary flavonols intake and risk of esophageal and gastric cancer: A meta-analysis of epidemiological studies. *Nutrients, 8*, 91.

Yamada, K., Sato-Mito, N., Nagata, J., & Umegaki, K. (2008). Health claim evidence requirements in Japan. *The Journal of Nutrition, 138*, 1192S–1198S.

Yan, Q. Y., & Bennick, A. (1995). Identification of histatins as tannin-binding proteins in human saliva. *Biochemical Journal, 311*, 341–347.

Yan, Y., Gao, Y. Y., Liu, B. Q., Niu, X. F., Zhuang, Y., & Wang, H. Q. (2010). Resveratrol-induced cytotoxicity in human Burkitt's lymphoma cells is coupled to the unfolded protein response. *BMC Cancer, 10*, 445.

Ye, Q., Georges, N., & Selomulya, C. (2018). Microencapsulation of active ingredients in functional foods: From research stage to commercial food products. *Trends in Food Science & Technology, 78*, 167–179.

Ying, W., Xiong, X., Chen, J., & Yang, J. (2013) Method for producing grape extract with high ORAC value and grape extract so produced. US 8,343,565 B2.

Zabaniotou, A., & Kamaterou, P. (2018). Food waste valorization advocating circular bioeconomy—A critical review of potentialities and perspectives of spent coffee grounds biorefinery. *Journal of Cleaner Production, 211*, 1553–1566.

Zamora-Ros, R., Agudo, A., Lujan-Barroso, L., Romieu, I., Ferrari, P., Knaze, V., et al. (2012). Dietary flavonoid and lignan intake and gastric adenocarcinoma risk in the European prospective investigation into cancer and nutrition (EPIC) study. *American Journal of Clinical Nutrition, 96*, 1398–1408.

Zamora-Ros, R., Forouhi, N. G., Sharp, S. J., González, C. A., Buijsse, B., Guevara, M., et al. (2014). Dietary intakes of individual flavanols and flavonols are inversely associated with incident type 2 diabetes in European populations. *Journal of Nutrition, 144*, 335–343.

Zamora-Ros, R., Knaze, V., Rothwell, J. A., Hémon, B., Moskal, A., Overvad, K., et al. (2016). Dietary polyphenol intake in Europe: The European prospective investigation into cancer and nutrition (EPIC) study. *European Journal of Nutrition, 55*, 1359–1375.

Zamora-Ros, R., Touillaud, M., Rothwell, J. A., Romieu, I., & Scalbert, A. (2014). Measuring exposure to the polyphenol metabolome in observational epidemiologic studies: Current tools and applications and their limits. *The American Journal of Clinical Nutrition, 100*, 11–26.

Zhang, X., Dong, D., Wang, H., Ma, Z., Wang, Y., & Wu, B. (2015). Stable knock-down of efflux transporters leads to reduced glucuronidation in UGT1A1-overexpressing HeLa cells: The evidence for glucuronidation-transport interplay. *Molecular Pharmaceutics, 12*, 1268–1278.

Zhang, N., & Mutilangi W. (2013). Zero calorie polyphenol aqueous dispersions. WO2013082065A1.

Zhang, L., Wang, Y., Li, D., Ho, C. T., Li, J., & Wan, X. (2016). The absorption, distribution, metabolism and excretion of procyanidins. *Food & Function, 7*, 1273–1281.

Zhang, H., Yu, D., Sun, J., Liu, X., Jiang, L., Guo, H., et al. (2014). Interaction of plant phenols with food macronutrients: Characterisation and nutritional-physiological consequences. *Nutrition Research Reviews, 27*, 1–15.

Zhao, Z., Egashira, Y., & Sanada, H. (2003). Digestion and absorption of ferulic acid sugar esters in rat gastrointestinal tract. *Journal of Agricultural and Food Chemistry, 51*, 5534–5539.

Zhao, Z., Egashira, Y., & Sanada, H. (2004). Ferulic acid is quickly absorbed from rat stomach as the free form and then conjugated mainly in liver. *Journal of Nutrition, 134*, 3083–3088.

Zhong, X. S., Ge, J., Chen, S. W., Xiong, Y. Q., Ma, S. J., & Chen, Q. (2016). Association between dietary isoflavones in soy and legumes and endometrial cancer: A systematic review and meta-analysis. *Journal of the Academy of Nutrition and Dietetics, 118*, 637–651.

Zhou, P., Wang, X., Liu, P., Huang, J., Wang, C., & Pan, M. (2018). Enhanced phenolic compounds extraction from *Morus alba* L. leaves by deep eutectic solvents combined with ultrasonic-assisted extraction. *Industrial Crops and Products, 120*, 147–154.

CHAPTER FIVE

Pigments and vitamins from plants as functional ingredients: Current trends and perspectives

Rúbia Carvalho Gomes Corrêa, Jéssica Amanda Andrade Garcia, Vanesa Gesser Correa, Tatiane Francielli Vieira, Adelar Bracht, Rosane Marina Peralta*

Postgraduate Program in Food Science, Department of Biochemistry, Laboratory of Biochemistry of Microorganisms and Food Science, State University of Maringa, Maringá, Paraná, Brazil
*Corresponding author: e-mail addresses: rmperalta@uem.br; rosanemperalta@gmail.com

Contents

Abstract

The food manufacturing industry has increasingly focused in the development of wholesome and safer products, including certified labeled "super foods," "healthy foods" and "functional foods," which are currently under great demand worldwide. Plant pigments and vitamins are amidst the most common additives incorporated to foodstuff, not only for improving their nutritional status but also for coloration, preservation, and even therapeutic purposes. The recovery of pigments from agro industrial wastes using green emerging approaches is a current trend and clearly the best alternative to ensure their sustainable obtainment and make these ingredients more popular, although still full of challenging aspects. Stability and bioavailability limitations of these active molecules in food matrices have been increasingly studied, and a number of methods

Advances in Food and Nutrition Research, Volume 90
ISSN 1043-4526
https://doi.org/10.1016/bs.afnr.2019.02.003

have been proposed to minimize these issues, among which the incorporation of a co-pigment, exclusion of O_2 during processing and storage, and above all, microencapsulation and nanoencapsulation techniques. The most recent advances and challenges in the application of natural pigments and vitamins in functional foods, considering only reports of the last 5 years, were the focus of this chapter.

1. Introduction

The food manufacturing industry has increasingly focused in the development of wholesome and safer products, including certified labeled "super foods," "healthy foods" and "functional foods," which are currently under great demand worldwide (Bigliardi & Galati, 2013). Phenolic compounds, vitamins and carotenoids, besides dietary fibers and minerals, are amidst the most commonly natural ingredients added to food products (Carocho, Barreiro, Morales, & Ferreira, 2014). These bioactive components are utilized to aggregate value, being incorporated to foodstuff not only for improving their nutritional status but also for coloration, preservation, and even therapeutic purposes, depending on the concentration employed (Martins & Ferreira, 2017).

Organoleptic characteristics greatly influence food acceptance, selection, and subsequent consumption. Color is one of the most impactful and delightful attributes of food products, which can instantly affect consumers' preference and eating desires, thus being crucial for its purchase (Martins, Roriz, Morales, Barros, & Ferreira, 2016, 2017). Though natural foods possess their own color intensities, storage conditions, manufacturing and processing practices commonly provoke marked alterations on their final coloration; thereby, the use of food additives constitutes an effective and promising strategy to mask unpleasant features (Carocho et al., 2014; Martins et al., 2016). Although artificial pigments display superior stability, more varied hue, besides vibrant color, their consumption has been related to negative outcomes on human health, such as attention deficit, hyperactivity, irritability, disturbed sleep, and aggressiveness in children, likewise several allergies and even carcinogenic responses on prolonged consumption (Carocho et al., 2014; Chhikara, Kushwaha, Sharma, Gat, & Panghal, 2019).

Owing to the health consciousness of modern consumers, colorants derived from natural sources are an appealing alternative that have gained increasing popularity (Carocho et al., 2014). To please consumers who request natural ingredients, leading-edge food and beverage companies, following the current "clean label" trend, have committed to diminish or eliminate artificial substances (including synthetic colorants) from their products (Cortez, Luna-Vital, Margulis, & Mejia, 2017). Thus, the replacement of artificial coloring ingredients by natural colorants is among the top concerns and challenges of food companies today (Pires et al., 2018). Extracts obtained from pigments such as anthocyanins, carotenoids, betalains, and chlorophylls, all very abundant in plant matrices, have been approved as coloring ingredients by FDA (Food and Drugs Administration), EFSA (European Food Safety Authority) and Codex (Ngamwonglumlert, Devahastin, & Chiewchan, 2017).

An appropriate and balanced nutrition is pivotal to maintain the ordinary body functions, prevent diseases, and age vigorously. However, there are a number of cases of malnutrition owed to deficient, excessive or imbalanced intake of a broad range of nutrients present in foods (Allen, Methven, & Gosney, 2013). These nutrition issues can have distinct origins such as medical and/or environmental causes, or particular necessities (Pereira, Barros, & Ferreira, 2017). Food fortification has been proved to be a powerful strategy for defeating vitamin deficiency, and micronutrient interactions strongly assert for multiple micronutrient fortification.

In the United States, for instance, vitamin B_9 (folic acid) is now added to supplemented grain products, thus being incorporated into the larger part of commercial breakfast cereals. Recent data show that the folic acid status in the U.S. population has improved remarkably, likely because of this fortification (Gironés-Vilaplana, Villaño, Marhuenda, Moreno, & García-Viguera, 2017). In the past few years, a new and promising market in the area of fortified processed foodstuff has been opened up, where the vitamin-fortified food products stand out (Talbot-Walsh, Kannar, & Selomulya, 2018).

Such impressive growth in the awareness of modern consumers to natural food additives and micronutrient needs is also a consequence of the extensive investigation efforts regarding these themes, which have greatly increased. The total number of scientific papers published in the past 5 years containing the term "pigments" in their titles is expressive, surpassing 7800 articles. Among these, >400 are located at the search domain *Food Science and Technology*, a total of 14 being review articles about plant pigments (obtained from *Web of Science*, July 2018). Likewise, the number of papers presenting the term "vitamins" in their titles raised almost twofold in the last decade.

Table 1 brings a compilation of the most relevant reviews on the topic of plant pigments and vitamins published during the past 5 years within the search domain *Food Science and Technology*. Considering this period and area, the

Table 1 Important reviews concerning pigments and vitamins from plants, their chemical features, bioactivities and applications, published in the last 5 years under the search domain of *Food Science and Technology*.

Source	Main contribution	Author's conclusion
Jäpelt and Jakobsen (2013)	Authors summarized the past year's evidence on sterol biosynthesis leading to provitamin D. They addressed the occurrence of vitamin D and its hydroxylated metabolites in higher plants, also discussing the limitations, advantages and trends with respect to the analytical methods employed in studies of vitamin D and related compounds	Not only animal foods and/or food products but also fruits and vegetables have the potential to serve as vitamin D sources. The Solanaceae family, to which belongs potato, tomato, and pepper, contains high amounts of vitamin D3. This information is of special interest considering the importance of this family in human nutrition
Watanabe, Yabuta, Tanioka, and Bito (2013)	Vit B12 is partially degraded and loses its bioactivity when foods are cooked and inadequately stored. The intrinsic factor-mediated gastrointestinal absorption system in humans has evolved to selectively assimilate active vit B12 from natural vit B12 compounds sources, including its degradation products and inactive corrinoids that are present in several foods	In order to prevent vit B12 deficiency in vegetarians and elderly subjects (besides other risk groups), it is essential to identify plant source foods that contain high levels of bioactive vit B12 and, in conjunction, to develop novel vit B12-fortified foods
Yang, Laillou, Smith, Schofield, and Moench-Pfanner (2013)	The vast majority of the reviewed papers showed that circulating vit D increased in a dose-dependent manner with increased intake of vit D-fortified foods. However, in some studies the extra intake was insufficient to augment vitamin D levels to 50 nmol/L	Fortification of largely consumed foods (such as edible oil) with vitamin D could be a good strategy to ameliorate vitamin D status in Southeast Asian countries. Intake modeling studies are required to establish the resulting additional intakes, and fortification of other food products should be considered

Card, Gorska, Cutler, and Harrington (2014)	In this review, authors discuss the prophylactic administration of vitamin K1 in term and preterm neonates, interactions between vitamins K and E, the industrial conversion of vitamin K to dihydrovitamin K in foods, tissue-specific conversion of vitamin K to menaquinone-4, the biological activity of the five and seven carbon metabolites of vitamin K and circadian variations	Vitamin K is an essential fat-soluble micronutrient that in humans is obtained mainly from plants such as phylloquinone. Research on vitamin K metabolism is crucial for understanding vitamin K biology in health and illness. Progress in this area, driven by knowledge of vitamin K and the availability of markers of vitamin K status, presented positive results in many areas of medicine and nutrition
Gengatharan, Dykes, and Choo (2015)	The authors reviewed the pharmacological attributes, such as antioxidant, anti-cancer, anti-lipidemic and antimicrobial capabilities of betalains obtained from several matrices such as red beetroot, amaranth, prickly pear and red pitaya, for potential application as functional foods	Betalains possess both esthetic values and positive health outcomes in food, also being water-soluble, what favors their incorporation into aqueous food systems. High yielding strain selection and application of biotechnological tools could facilitate the improvement of betalains production by known plant sources
Lo Piero (2015)	The most recent advances in red orange anthocyanins were reviewed, with special focus on their biosynthesis and regulation. Both the quantity and anthocyanins profile in red orange cultivars vary substantially depending on variety, maturity, region of cultivation, and manifold other environmental factors. Thus, the production of high anthocyanin content fruits remains limited to a few regions with characteristic climate conditions	Future research efforts should focus at identifying the genes involved in anthocyanin modification, elucidating the mechanism of vacuole compartmentation of pigments, and defining the role of either phyto-hormones or biotic and abiotic factors in inducing anthocyanin accumulation

Continued

Table 1 Important reviews concerning pigments and vitamins from plants, their chemical features, bioactivities and applications, published in the last 5 years under the search domain of *Food Science and Technology*.—cont'd

Source	Main contribution	Author's conclusion
Turturică, Oancea, Râpeanu, and Bahrim (2015)	This work summarizes anthocyanin content in fruits, their important role in human health, aspects of their biochemistry, bioavailability and distribution in some fruits, besides the biosynthetic pathway, manifold extraction, separation and purification mythologies, and identification methods	Advanced chromatography methods and environmental protection are essential for the development of non-conventional extraction methods. Maintenance of anthocyanin's bioactive properties during raw material processing represents a very important approach for establishing the features of anthocyanins under different physicochemical conditions, thus having a fundamental role in health-promoting food products
Gandía-Herrero, Escribano, and García-Carmona (2016)	Studies with multiple cancer cell lines have revealed the high chemopreventive potential of betalains, which finds in vitro support in a strong antiradical and antioxidant activity. Both in vivo and bioavailability experiments reinforce the potential chemoprotective action role played by betalains in the diet	All the bioactivities described are probably related to the high antiradical capacity of betalains' structural unit, betalamic acid. The use of extracts in the majority of the in vivo experiments limits the conclusions about the mechanisms involved and the therapeutic potential of the assays. Although studies with purified betalains remain scarce, they provide exciting conclusions
Martins et al. (2016)	This review presents an extensive approach on natural/synthetic food colorants currently allowed with established acceptable daily intake, describes the techniques that have been applied to optimize food attractiveness, shelf life and color stability, whereas displays the trends and future perspectives on this topic	Natural food colorants not only provide high quality, efficiency and organoleptic properties to food but also play a contributive role as health promoters. Anthocyanins, carotenoids, phenolic compounds, beet derivatives, annatto and some curcuminoids are amidst the most used for such purpose, however, strict regulatory practices have been applied looking for food quality assurance

Cortez et al. (2017)	In this review work authors evaluated, compared, and discussed the most recent information included in worldwide patents and in scientific papers concerning distinct methods for the stabilization of anthocyanins for their application as colorants in food systems	Decreasing the pH value to 2.8 in anthocyanin solutions can induce a structural shift to the flavylium cation, which confers greater stability to the anthocyanin molecules. Among the anthocyanin stabilizing strategies, addition of co-pigment compounds like polymers, phenolic compounds, and metals can be cited
Khoo, Azlan, Tang, and Lim (2017)	Several cell culture studies, animal models, and human clinical trials, evidence that anthocyanidins and anthocyanins possess antioxidant and antimicrobial properties, improve visual and neurological health, and protect against various non-communicable diseases, being these outcomes related to their potent antioxidant properties. Different mechanisms and pathways are involved in such protective outcomes	Summarily, free-radical scavenging, changes in blood biomarkers, COX and MAPKs pathways, as well as inflammatory cytokines signaling are the typical mechanisms of action of anthocyanidins and anthocyanins in the prevention of illnesses
Lachman, Martinek, Kotíková, Orsák, and Šulc (2017)	Flavonoids are mainly present in the outer layer of grains whereas carotenoids that are responsible for yellow color of grains are in the endosperm. Hence, accumulation of these pigments in the grain can represent an important target in breeding programs aimed at increasing the concentrations of these bioactives in grain and derivative products	Currently, wheat breeders are focused on the development of novel types of color-grained wheat with improved characteristics including quality, yield and higher pigment levels with potential beneficial outcomes on human health and nutrition. As most pigments are found in the outer layer of grains, they are present at higher contents in whole meal than in white breads

Continued

Table 1 Important reviews concerning pigments and vitamins from plants, their chemical features, bioactivities and applications, published in the last 5 years under the search domain of *Food Science and Technology*.—cont'd

Source	Main contribution	Author's conclusion
Low, Mwanga, Andrade, Carey, and Ball (2017)	Orange-fleshed sweet potato (OP) is a rich plant-based source of beta-carotene, which the body converts into vitamin A. Researchers have recognized the potential of OP varieties to address widespread vitamin A deficiency in sub-Saharan Africa, and since 1995, have been using an integrated agriculture-nutrition approach for this purpose	The ingestion of 100 g of orange-fleshed sweet potato can meet the daily vitamin A needs of a young child. Breeding in Africa was a requisite to obtain OP varieties competitive with local varieties. Integrating nutrition education was essential for impacting the young child vitamin A status
Martins and Ferreira (2017)	Bio-residues are valuable sources of carotenoids, mostly carotenes (from vegetal agro industrial wastes) and xanthophylls (from animal origin). Currently, the most common approach for carotenoids recovery is the extraction with organic solvents	Combined extraction methodologies, including emergent methods such as supercritical fluid extraction, microwave- and enzyme-assisted extractions, guarantee higher recovery yields of carotenoid pigments
Neri-Numa, Pessoa, Paulino, and Pastore (2017)	Genipin is a natural blue pigment obtained from *Gardenia* sp. and *Genipa americana* L. It presents potential to be used as an alternative food-grade ingredient. Likewise, its crosslinked biopolymers have potential applications in medical, pharmaceutical and industrial areas. Finally, the therapeutic properties of genipin against several illnesses were compiled	The greatest challenge concerning the industrial exploitation of genipin is its stability. However, considering the advantages of low toxicity, biocompatibility, permeability, and similarity with the extracellular matrix and intrinsic cellular interaction, it is worth to keep the efforts to overcome this drawback and to make its broad industrial application feasible

Ngamwonglumlert et al. (2017)	Authors wrote a comprehensive review of appropriate pre-treatment and extraction techniques for chlorophylls, carotenoids, betalains and anthocyanins, using the pigment stability and extraction yield as the assessment criteria	Further to the extraction yield and pigments stability, other factors such as investment and operating costs as well as applicability of the selected techniques must be considered for the selection of the appropriate approach. Combination of diverse extraction and stability enhancement procedures can be performed to raise both the pigment stability and extraction yield
Öztürk (2017)	Authors discussed the challenges regarding production methods and factors affecting the stability of lipophilic vitamins and nanoemulsion delivery systems. Recent investigations on bioavailability evaluation of vitamins A, D, E encapsulated in oil-in-water nanoemulsions were presented	Protective encapsulation techniques are essential during food fortification for preventing vitamins' degradation and ensuring their bioavailability in the human gastrointestinal system. For this purpose, oil-in-water nanoemulsions are promising delivery systems
Raddatz-Mota et al. (2017)	Authors reviewed the most recent literature on *Bixa orellana* L., a natural source of red pigment and vitamin E, with emphasis on bixin, norbixin, tocotrienols and tocopherols biosynthesis, industrial applications of annatto extracts, as well as its nutraceutical potential and its benefits for human health	Annatto extract is a natural colorant largely accepted and applied in the food industry. The usual industrial way for obtaining the annatto extract is using a KOH alkaline solution. However, this procedure presents a yield of only 6% of the processed tissue
Saghiri, Asatourian, Ershadifar, Moghadam, and Sheibani (2017)	Angiogenesis, the formation of new blood vessels, is a fundamental process in wound healing, tissue regeneration, and tumor growth. Depending on the physiological setting and the administered dose, vitamins (A, B, C, D, E and K) can display angiogenic action on blood vessels	Vitamin A can promote both anti-angiogenesis and angiogenesis outcomes, whereas vitamins B1, B3, and B12 mimic angiogenesis in body and, finally, some vitamins restrain angiogenesis

Continued

Table 1 Important reviews concerning pigments and vitamins from plants, their chemical features, bioactivities and applications, published in the last 5 years under the search domain of *Food Science and Technology*.—cont'd

Source	Main contribution	Author's conclusion
Samyor, Das, and Deka (2017)	Authors compiled the available information about pigmented rice (red rice, black, purple, brown, and brown red rice) regarding the bioactive compounds, their concentration, biological activities and potential benefits to human health. The antioxidant activity and scavenging capacity of the anthocyanins from pigmented rice could be explored as functional food ingredients	Further research and development work in pigmented rice and its constituents are needed to elucidate the mechanisms involved in its bioactivities, such as antioxidant and free-radical scavenging, anti-tumor, anti-atherosclerosis, anti-allergic, anti-influenza and anti-obesity properties
Soares, Carrascosa, and Raposo (2017)	Vitamin C possesses antioxidant properties that contribute to the heath beneficial outcomes of broccoli. This overview addressed the reduction of the secondary plant products contents in broccoli, such as glucosinolates and vitamin C, by the cooking process	Notwithstanding some controversy, the majority of studies shows that conventional cooking methods, such as boiling, steaming, and frying, and non-conventional ones, like microwaving, provoked significant degradation of vitamin C and glucosinolates. Steaming is the most suitable method for preserving these two compounds
Yeh, Barbano, and Drake (2017)	Several degradation products of vitamins A and D possess flavor/fragrance applications, however, only a few were explored regarding their possible flavor contributions to fluid milk. Vitamin concentrates can effect flavor and flavor stability to f this product. In this study, authors proposed mechanisms of off-flavor formation and addressed changes in flavor stability of fluid milk	Milk exposure to light may imply in vitamin destruction, being vitamin fortification another possible source of off-flavor in fluid milk. Unraveling the impact of vitamin addition and degradation in fluid milk will help the dairy industry to enhance fluid milk quality

Zhao et al. (2017)	This review summarizes, for the first time, the current information over the chemical implications as well as classic physiochemical effects of anthocyanin glycosyl acylation	Glycosyl acylation tends to increase, both in vitro and in vivo, the chemical stability of anthocyanins, being that the mechanisms essentially implicate physicochemical, stereochemical, photochemical, biochemical and environmental aspects under definite conditions. Not only the acylation sites but also the types and numbers of acyl groups affect the acylated anthocyanins' stability
Martins et al. (2017)	Betalain natural pigments represent a promising and safe alternative to synthetic dyes, but their chemical instability has limited their widespread use. Temperature, pH, water activity, oxygen, light, chelating agents, the presence of other compounds, pigment concentration, storage, and processing conditions are the most important factors affecting their stability	It is crucial to establish optimum processing conditions to maximize the stability of betalains and their extraction yields, focusing on their effective use as natural food colorants, functional ingredients and value-added food products
Polturak and Aharoni (2018)	This work discusses betalain metabolism in light of recent advances in the field, with a current survey of characterized genes and enzymes that take part in betalain biosynthesis, catabolism, and transcriptional regulation. Authors presented a broad view of currently used and potential new sources for betalains, including utilization of natural sources or metabolic engineering	The study of betalain biochemistry has seen major advances in recent years, eventually leading to the full elucidation of the core betalain biosynthetic pathway in plants. The increasing availability of transcriptomics and genomic data from Caryophyllales plants might enable the identification of additional genes and enzymes that take part in betalain biosynthesis, what will in turn permit progress in metabolic engineering possibilities, such as the production of betalains with varying hues or compounds with increased stability

principal research topics with respect to plant pigments were: recent advances on pre-treatment, extraction and stabilization methods (Cortez et al., 2017; Ngamwonglumlert et al., 2017; Zhao et al., 2017); production and selection of anthocyanin-rich food matrices (Lo Piero, 2015; Samyor et al., 2017); updates on bioactivities, among which the chemopreventive potential of betalains (Gandía-Herrero et al., 2016; Gengatharan et al., 2015; Khoo et al., 2017); recovery of pigments from agro industrial wastes and by-products, especially carotenoids (Martins & Ferreira, 2017); and, finally, the prospection of novel active colorant compounds (Neri-Numa et al., 2017; Raddatz-Mota et al., 2017). Regarding vitamins, important reviews on their biosynthesis in plants were published (Jäpelt & Jakobsen, 2013) as well as on the (increasingly common) biofortification of crops (Low et al., 2017). There are also papers dealing with their most recently explored biological activities (Saghiri et al., 2017), metabolism (Card et al., 2014), bioavailability (Watanabe et al., 2013) and stability (Soares et al., 2017). Lastly, revisions on the efficacy and controversies regarding the fortification of food products with vitamins have also been published (Yang et al., 2013; Yeh et al., 2017).

The importance of natural pigments and vitamins in functional foods justifies per se the inclusion of this chapter in this book. The main focus was the most recent advances and challenges in the area considering reports of the last 5 years.

2. General features of plant pigments and vitamins

2.1 Plant pigments

Natural pigments can be obtained from plants, microorganisms, or animal matrices. Structurally, they can be classified into distinct groups that encompass compounds with specific features, such as isoprenoid derivatives (e.g., carotenoids and iridoids), benzopyran derivatives (anthocyanins and others flavonoid pigments), quinones (benzoquinone, naphthoquinone, and anthraquinone), tetrapyrrole derivatives (such as chlorophylls and heme colors), *N*-heterocyclic compounds different from tetrapyrroles (purines, pterins, flavins, phenazines, phenoxazines, and betalains) and melanins (Neri-Numa et al., 2017). Amidst the pigments from vegetal sources, the most important are either water- or lipid-soluble compounds comprehending carotenoids, betalains, anthocyanins and chlorophylls, which are distinct in both structure and, consequently, in their biosynthetic pathways (Zhang, Butelli, & Martin, 2014). Fig. 1 shows the most investigated

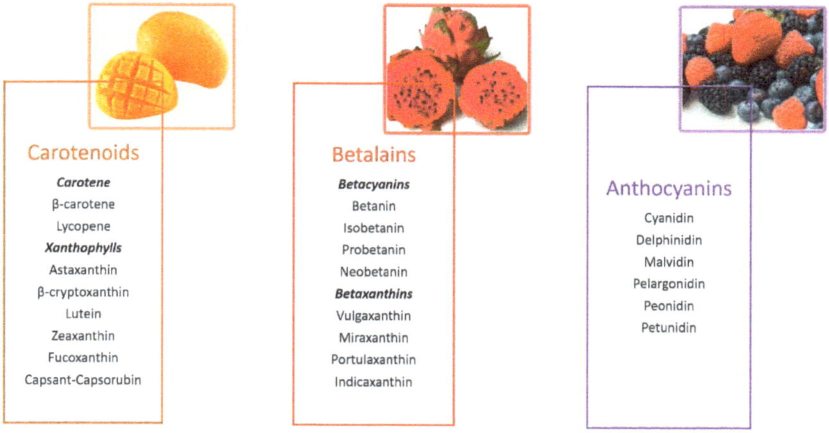

Fig. 1 The most investigated plant pigments in the past few years regarding bioactivities and potential application as food ingredients.

plant pigments in the past few years, regarding bioactivities and potential as food ingredients, which will be focused in this chapter.

Carotenoids include an extensive group of bioactive compounds, found above all in the vegetable kingdom but also in algae and some microorganisms. Being water-insoluble, middle soluble in organic solvents and fully fat-soluble, these pigments earn a typical coloration mainly conferred by the presence of xanthophylls (Martins & Ferreira, 2017). Belong to the tetraterpenes class, carotenoids are the most relevant group consisting of 40 carbon atoms, formed by the junction of 8 isoprene units. Holding >10 double conjugated linkages, these molecules possess the interesting ability of fixing monomolecular oxygen during the photochemical processes, a feature that justifies its yellow to yellow-orange color and its noted antioxidant potential (Rodriguez-Amaya, 2018). Thereby, and considering its ability in oxygen fixation, carotenoids may be divided into two different classes: carotenes, such as β-carotene and lycopene; and xanthophylls, like astaxanthin, β-cryptoxanthin, capsant-capsorubin, fucoxanthin, lutein and zeaxanthin (Fig. 1). Carotenes are reputable as non-oxygenated carotenoids, holding a characteristic hydrocarbon form, whereas xanthophylls are designated oxygenated carotenoids, being synthesized within the plastids of plants (Oroian & Escriche, 2015). Their main plant sources are guava (*Psidium guajava* L.), carrot (*Daucus carota* L.), tomato (*Solanum lycopersicum* L.), sweet potato [*Ipomoea batatas* (L.) Lam.], apricot (*Prunus* spp.), papaya (*Carica papaya* L.), squash (*Cucurbita* spp.), corn (*Zea mays* L.) and green plants (Corrêa, Haminiuk, Sora, Bergamasco, & Vieira, 2014; Martins & Ferreira, 2017).

Betalains are highly active, water-soluble nitrogenous pigments found exclusively in plants of the Caryophyllales order (Polturak & Aharoni, 2018). They occur in two structurally distinct forms, namely red-violet betacyanins and yellow-orange betaxanthins, being the color imputed to the structure's resonating double bonds (Chhikara et al., 2019). Betacyanins are derivatives of betanidin (an iminium adduct of betalamic acid and cyclo-dopa dihydroxyphenylalanine), and display absorbance spectra centered at wavelengths around 536 nm. Acyl-glycosylation of one or two hydroxyl groups are viable in betacyanins molecules, allowing the obtainment of complex pigment structures (Gandía-Herrero et al., 2016). In contrast, for betaxanthins, molecules formed by the condensation of α-amino acids or amines with betalamic acid, no glycosylation has ever been reported. Their absorbance spectra are centered at wavelengths around 480 nm (Gandía-Herrero et al., 2016; Gengatharan et al., 2015). Notwithstanding, both groups are promising food-grade colorants due to their non-toxic, non-carcinogenic and non-poisonous features (Chhikara et al., 2019). While betacyanins are found in great amounts in fruits, flowers, leaves and roots, betaxanthins are also present in tubers and do not commonly occur in leaves (Martins et al., 2017). Contrarily to the omnipresent anthocyanin and carotenoid classes of pigments, betalains are relatively rare in nature. Within the Caryophyllales order, betalains occur in a mutually exclusive fashion with anthocyanins, seeing that no plant was found to naturally synthetize both types of pigments. The vibrant colors of betalains make them to cherished ornamental plants, as exemplified by bougainvillea (*Bougainvillea* spp.), four o'clock (*Mirabilis jalapa* L.), cockscomb (*Celosia cristata* L.), moss rose (*Portulaca grandiflora* Hook.), and globe amaranth (*Gomphrena globosa* L.). Others are highly estimated food crops, such as beetroot and Swiss chard (*Beta vulgaris* L.), prickly pear (*Opuntia* spp.), and dragon fruit (*Hylocereus* spp.) (Polturak & Aharoni, 2018).

Anthocyanins are accountable for the red, blue, purple, and even black colors of fruits, vegetables, grains, flowers, and other plant tissues (Turturică et al., 2015). These bioactive compounds are water-soluble glycosides and acylglycosides of anthocyanidins in the form of polyhydroxylated and polymethoxylated heterosides derived from flavylium or 2-phenylbenzopyrilium ions (Dia et al., 2015). The most plentiful anthocyanidins are cyanidin, delphinidin, and pelargonidin followed by the profused malvidin, petunidin, and peonidin (Fig. 1). The coloration of anthocyanins relies upon their substitutions, and if they are acylated or non-acylated. Under acidic conditions, for instance, the color of non- and mono-acylated anthocyanins is set on greatly by substitutions in

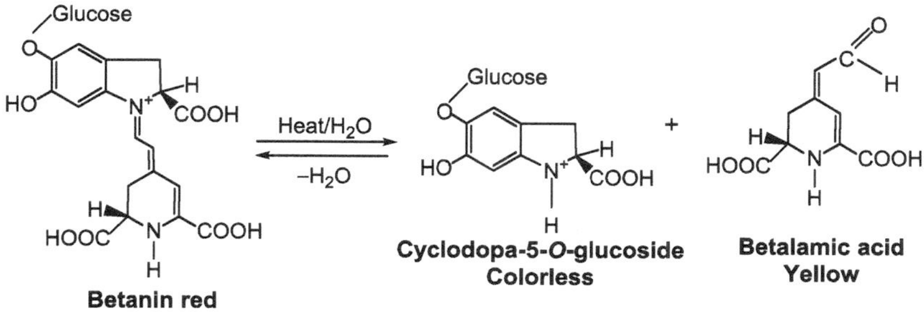

Fig. 2 Degradation reaction of the pigment betanin (red) when submitted to mild alkaline conditions, which produces the colorless compound cyclodopa-5-*O*-glucoside and betalamic acid.

the B-ring of the aglycon (Trouillas et al., 2016) (Fig. 2). Increased hydroxyl substitutions on the B-ring lead to a shift of the visible absorption maximum (λ_{max}) to longer wavelengths, generating a bathochromic shift to produce a bluer hue. Overall, it is established that acylated anthocyanins are more suitable for diverse applications, including food coloring, owing to their higher stability (Cortez et al., 2017). Among the wide variety of edible pigmented flowers there are anthocyanin red flowers such as hibiscus (*Hibiscus* spp.), rose (*Rosa* spp.), pineapple sage (*Salvia elegans* Vahl), red clover (*Trifolium pratense* L.), and pink blossom (*Prunus* spp.). Others are blue, such as cornflower (*Centaurea cyanus* L.), blue chicory (*Cichorium intybus* L.), and blue rosemary (*Rosmarinus officinalis* L.), and still others are purple, such as purple mint [*Perilla crispa* (Thunb.) Tanaka], purple passion flower (*Passiflora incarnata* L.), purple sage [*Salvia dorrii* (Kellogg) Abrams], common violet (*Viola* spp.), and lavender (*Lavandula* spp.). The anthocyanin-rich fruits include an endless list of berries, currants (*Ribes nigrum* L.), plums (*Prunus* spp.), grapes (*Vitis vinifera* spp.), pigmented sweet oranges [*Citrus sinensis* (L.) Osbeck.] and some tropical fruits. In addition, red to purplish blue-colored leafy vegetables, grains, roots, and tubers, like the anthocyanin-rich black carrot (*Daucus carota* ssp. *sativus* var. Atrorubens Alef.), eggplants (*Solanum melongena* L.), red cabbage (*Brassica oleracea* L.), and purple potato (*Solanum tuberosum* spp.), are all potential functional foods loaded with anthocyanins (Khoo et al., 2017; Turturică et al., 2015).

2.2 Plant vitamins

Vitamins are a broad class of organic compounds mandatory for the accurate maintenance of its normal functions. These micronutrients play numerous roles in intermediary metabolism and in the specialized metabolism of specific organs, being usually converted in the body into more complex molecules that function as co-enzymes; so much that inadequate vitamin intake gives rise to a variety of deficiency syndromes, such as the potentially fatal scurvy (vitamin C) and beriberi (vitamin D) (Campbell, 2017). Exclusively vitamin D is synthetized by the body; all other vitamins are acquired from food and/or supplementation strategies. Under normal circumstances, people are able to obtain the distinct vitamins through an adequate nutrition, but commonly minimum micronutrients requirements are not achieved, what means they have to be obtained through a supplementary source (Gironés-Vilaplana et al., 2017).

According to their solubility, vitamins can be classified into two categories: water-soluble vitamins and fat-soluble vitamins. The first category comprehends the B-group vitamins and vitamin C, whereas the second category comprises vitamin A, vitamin D, vitamin E, and vitamin K. The B-complex vitamins are a broad family that includes vitamin B1 (thiamin), vitamin B2 (riboflavin), vitamin B3 (nicotinamide), vitamin B5 (pantothenic acid), vitamin B6 (piridoxine), vitamin B8 (biotin), vitamin B9 (folacin), and vitamin B12 (cobalamins). Most vitamins can be synthetized by microorganisms and by some plants, and to a minor degree by animals. Complete information on the biosynthesis of vitamins can be found in the recent work of Gironés-Vilaplana et al. (2017); likewise, an overview of the latest discoveries regarding their health effects and bioavailability can be obtained in the review papers displayed in Table 1 (Card et al., 2014; Saghiri et al., 2017; Soares et al., 2017; Yang et al., 2013).

3. Applications in food industry

3.1 Plant pigments as food colorants

In regard to the application in foods, anthocyanins are the most widely explored pigments. Table 2, a compilation of the past 5-year reports on the development of food products and food packaging added with plant pigments, evidences this by the great number of works concerning these compounds. In addition to their well-established application as food colorants and functionalizing agents (Díaz-García et al., 2015; Zanoletti et al., 2017), anthocyanins are now being tested as indicator sensors in intelligent packing (Pereira Jr et al., 2015). In short, anthocyanin extracts that are added to food packaging formulations, shift color from red to green when pH increases, what commonly happens in food spoilage (Shukla et al., 2016). Furthermore, anthocyanin fractions of grape pomace, very rich in malvidin

Table 2 Experimental studies concerning the design, potential functionality and stability of food products and food packaging, enriched and/or added with plant pigments, published in the last 5 years.

Pigment plant source	Compound or compound group	Food application	Main findings	Reference
Grape (*V. vinifera* L.) marc extract	Malvidin-3-*O*-glucoside	Enrichment of durum wheat biscuits	The results evidenced the feasibility of producing anthocyanin-enriched durum wheat biscuits using extra-virgin olive oil, with both low fats and sugar levels. The developed functional product showed a satisfactory acceptability	Pasqualone et al. (2014)
Thyme (*Thymus moroderi* Pau ex Martínez)	Anthocyanins	Colorant for yogurts	A food colorant that presented good stability during storage, high color strength as well as high antioxidant capacity, can be obtained from the flowers and bracts of thyme, with potential to be used not only for coloring but also for fortifying foods	Díaz-García et al. (2015)
Red cabbage (*Brassica oleraceae* L.)	Anthocyanins	Ingredient in food packaging featuring time-temperature indicators (TTI)	The changes in the developed TTI's color supplies an inexpensive and practical way to indicate that a food has undergone transformations in its chemical profile, once the color shift is an outcome from pH changes that take place when a given food has been spoiled	Pereira Jr, de Arruda, and Stefani (2015)

Continued

Table 2 Experimental studies concerning the design, potential functionality and stability of food products and food packaging, enriched and/or added with plant pigments, published in the last 5 years.—cont'd

Pigment plant source	Compound or compound group	Food application	Main findings	Reference
Red cabbage (*Brassica oleraceae* L.) and rose flower (*Rosa* spp.)	Anthocyanins	Anthocyanin-based indicator sensor for intelligent food packaging	Anthocyanin extracts obtained from rose flowers and red cabbage were immobilized on filter paper as carrier to elaborate a colorimetric sensor for use in intelligent packaging. Indicator sensor was based on anthocyanin's color changing from red to green due to pH increase	Shukla, Kandeepan, Vishnuraj, and Soni (2016)
Grape (*Vitis vinifera* L.) pomace	Malvidin-3-O-glucoside	As an ingredient in active biodegradable films	Both gum arabic (GA) and maltodextrin (MD) showed effectiveness in encapsulating the grape pomace anthocyanins (up to 92%), with no difference between the treatments. However, a higher antioxidant activity was observed in GA powders, probably by virtue of their higher solubility in water	Stoll, Costa, Jablonski, Flôres, and de Oliveira Rios (2016)
Purple wheat (*Triticum aestivum* L.)	Anthocyanins	Production of fiber and anthocyanin-enriched pasta	A two-step debranning procedure was performed in purple wheat to obtain anthocyanin-rich fractions (F1 and F2), that were further used to produce enriched pasta. The enriched samples presented higher or comparable content in total and soluble fiber and higher ferric reducing-antioxidant power than the control sample, whereas the highest amount of anthocyanins was found in F1 (696 mg/g)	Zanoletti et al. (2017)

Goji berry (*Lycium barbarum* L.)	Carotenoids	Enrichment of extra-virgin olive oil	The enrichment of extra-virgin olive oil with goji carotenoid compounds showed to be a promising strategy to improve nutraceutical quality while promoting to some extent oil heat stability. Moreover, the direct extraction of goji health-promoting carotenoids using the vegetable oil as a solvent could be an alternative approach for industrial scale-up	Blasi et al. (2018)
Tomato (*Solanum lycopersicum* L.) peels industrial by-product	Lycopene-rich oleoresin	Supplementation of refined olive and sunflower oils	Tomato peels oleoresin showed high DPPH and ABTS antioxidant activities. It stabilized efficiently the refined olive and sunflower oils. Lycopene apparently promoted a pro-oxidation effect in refined olive oil at high concentrations. Primary oxidation indices of both supplemented oils were highly correlated to the lycopene content	Kehili, Choura, Zammel, Allouche, and Sayadi (2018)
Beetroot (*Beta vulgaris* L.) pomace extract	Betalains	Ingredient for the preparation of antioxidant-rich ginger candy	The optimum process conditions for the production of ginger candy enriched with antioxidants were established, namely the blanching time of 7.81 min and beetroot pomace extract concentration of 9.24% (with 0.905 desirability). This new information allows the improvement of the phytochemical potential of the product, a cost effective form of exploiting the beetroot pomace	Kumar, Kushwaha, Goyal, Tanwar, and Kaur (2018)

Continued

Table 2 Experimental studies concerning the design, potential functionality and stability of food products and food packaging, enriched and/or added with plant pigments, published in the last 5 years.—cont'd

Pigment plant source	Compound or compound group	Food application	Main findings	Reference
Globe flower (*Gomphrena globosa*)	Betacyanins	Coloring agent in ice-cream formulation	Overall, ice creams formulated with the *G. globosa* extract were similar, in nutritional, color, individual sugars and fatty acids profiles, to those prepared with beetroot extract. Moreover, the markers distribution in the linear discriminant analysis indicated that the positive outcomes produced by the *G. globosa* natural food-grade colorant were maintained throughout storage time	Roriz, Barreira, Morales, Barros, and Ferreira (2018)
Dry tomato (*Solanum lycopersicum* L.) waste	Carotenoids	Enrichment of various vegetable oils	The highest extraction yields were obtained when using ultrasound-assisted extraction by soaking for 7 days and microwave-assisted extraction for 50 min. In some oil samples, the enrichment with carotenoids improved their oxidative and thermal stability, whereas in others it caused an increase in the peroxide value and a decrease in the induction time	Nour, Corbu, Rotaru, Karageorgou, and Lalas (2018)
Sea buckhorn (*Hippophae rhamnoides* L.)	Carotenoid	Food-grade colorant and function ingredient for muffins	Carotenoids from sea buckthorn extract were successfully encapsulated by complex coacervation and freeze-drying, using whey proteins and gum acacia as carrier materials. Muffins formulated with the encapsulated powders had a satisfactory total carotenoids content and antioxidant activity, improved texture and better sensorial acceptance than the control sample	Ursache et al. (2018)

3-*O*-glucoside, have been tested for the functionalization of biscuits, and as antioxidant ingredients in active biodegradable films (Pasqualone et al., 2014; Stoll et al., 2016).

The carotenoid colorants curcumin (E 100), urucum (E 160b), and lutein (E 161b) have been increasingly adopted by the food industry as natural alternatives to tartrazine (E 102), a very popular synthetic food colorant, whose consumption was associated with irritability, restlessness and sleep disturbance in children (Carocho, Morales, & Ferreira, 2015). Curcumin, a hydrophobic yellow-orange polyphenol obtained from the rhizome of *Curcuma longa* L., is nowadays applied as a stabilizer (in jelly manufacture) or as a coloring additive in cheeses, pickles, mustards, cereals, soups, ice creams and yogurts. In addition to its coloring capacity, curcumin presents extensively proven antibacterial, anti-proliferative, anti-inflammatory, antioxidant and anti-carcinogenic capabilities (Mangolim et al., 2014).

Regarding betalain red-purple natural coloring ingredients (betacyanins), betanin (e.g., betanidin derivative) from beetroot (*Beta vulgaris* L.) have been the only ones approved (E 162) for being safely utilized as food additives and are presently used in the formulation of burgers, desserts, ice-cream, jams, jellies, soups, sauces, beverages, and several dairy products (Kumar et al., 2018; Martins et al., 2017).

3.2 Vitamins as fortifying and preservative agents

Not only government agencies but also food and beverages leading-edge companies have shown increasingly interest in foods fortified with essential micronutrients, above all, vitamins that promote good health in humans (Öztürk, 2017). As a result, the number of investigations on the fortification of vitamin-containing foods has grown significantly in recent years. The main experimental papers reporting the application of vitamins as food ingredients published in the past few years are summarized in Table 3. In all examples, synthetic forms of vitamins were used. No report on the addition of vitamins extracted from natural sources was found. Although differences in behavior and responses are expected only for those cases in which the synthetic and natural vitamin forms are chemically distinct, fruits and vegetables (and plant-derived foodstuff, especially those which have underwent milder processing) are composed of a number of nutrients and phytochemicals which seem to interact synergistically and have a positive effect on their bioavailability (Carr & Vissers, 2013).

Table 3 Experimental studies concerning the design, potential functionality and stability of food products fortified with vitamins, published in the last 5 years.

Vitamin	Common plant sources	Role in human health	Food fortification examples	Reference
Vitamin A (carotenoids, including beta-carotene) 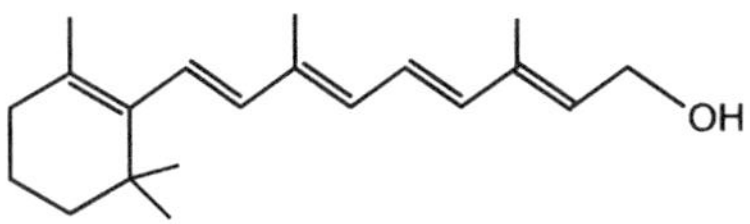 	Squash; sweet potato; kale and collard; carrot; spinach; apricot, mango, and cantaloupe melon	Vitamin A possesses three main components (retinol, retinoic acid, retinal), which play vital roles in human vision, normal embryonic development, growth, and resistance to infection	Fortification of soybean oil, fluid milk, milk for cheese making and pandesal bread	Hemery et al. (2015), Yeh et al. (2017), and Öztürk (2017)
Vitamin B5 (pantothenic acid) 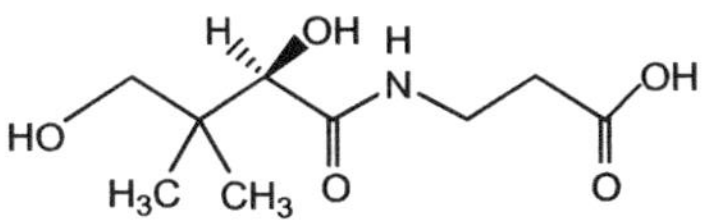 	Avocado; sunflower seed; broccoli, and whole grains (pantothenic acid is found in the outer layers of the grains)	Vitamin B5 displays its major biological function as part of coenzyme A, which itself plays a role in multiple steps of cellular metabolism, including the synthesis of several substances. Vitamin B5 is applied to treat sunburn, and conjunctivitis	Augmentation of vitamin B5 uptake capacity of milled rice via ultrasound application, a novel approach for rice fortification	Bonto, Camacho, and Camacho (2018)

Vitamin B6 (pyridoxine) 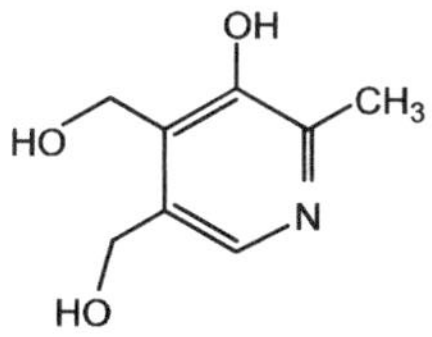	Banana; chickpea; potato; pistachio; and sesame seed	Vitamin B6 acts as a coenzyme in several reactions that are involved in amino acid, carbohydrates and lipid metabolism, and plays a role in neuronal signaling through the synthesis of neurotransmitters	Non-traditional wild rice flakes proved to be a representative source of vitamins, mainly pyridoxine, besides pantothenic and folic acids, niacin and thiamine	Sumczynski, Koubová, Šenkárová, and Orsavová (2018)
Vitamin B9 (folate or folic acid) 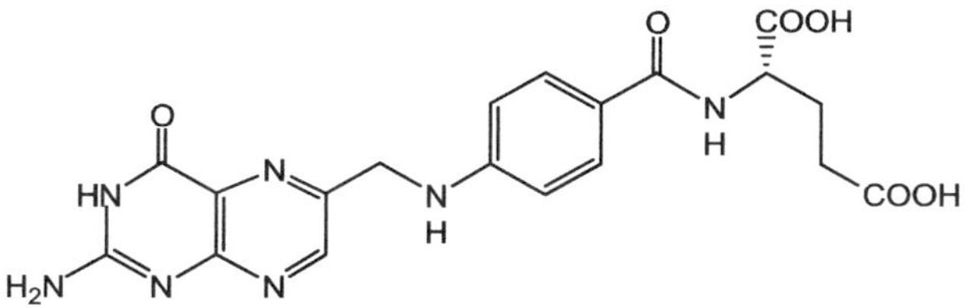	Dark leaf vegetables like broccoli and spinach; brussels sprouts; chickpea; avocado; lentil; and citrus fruits	Folate plays a key role in conjunction with B12 and B6 vitamins, primarily in nucleotide synthesis, methionine synthesis, methionine regeneration from homocysteine, and oxidation and reduction of one-carbon units required for normal cell division and growth	Folate fortification of white and whole-grain bread through the incorporation of Swiss chard and spinach	López-Nicolás et al. (2014) and Kim et al. (2018)

Continued

Table 3 Experimental studies concerning the design, potential functionality and stability of food products fortified with vitamins, published in the last 5 years.—cont'd

Vitamin	Common plant sources	Role in human health	Food fortification examples	Reference
Vitamin D2 (ergocalciferol) 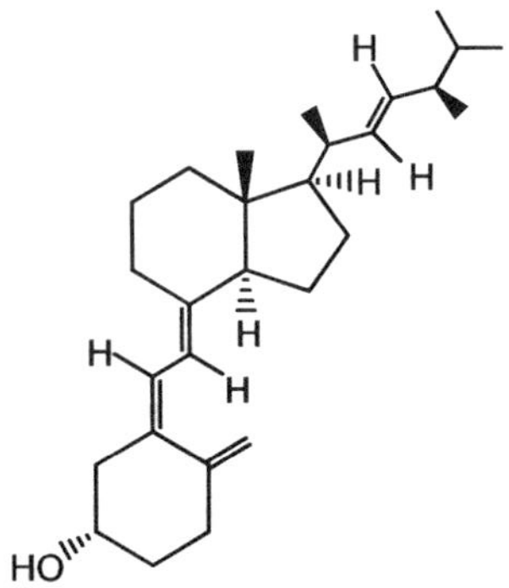	Tomato; potato; summer squash; waxyleaf nightshade; day-blooming cestrum; alfalfa; and pepper	Vitamin D comprises a group of fat-soluble vitamins, that is calciferol (D2) and cholecalciferol (D3), which exert bioactive roles in bone metabolism, boost the immune system and prevent rickets disease in children	Successful fortification of dahi and yogurt with the maintenance of products' characteristics. Fortification of yogurt as an accessible strategy to prevent diabetes	Jäpelt and Jakobsen (2013), Kaushik and Arora (2017), Kaushik, Sachdeva, and Arora (2017), and Mostafai et al. (2018)
Vitamin E (tocopherols) 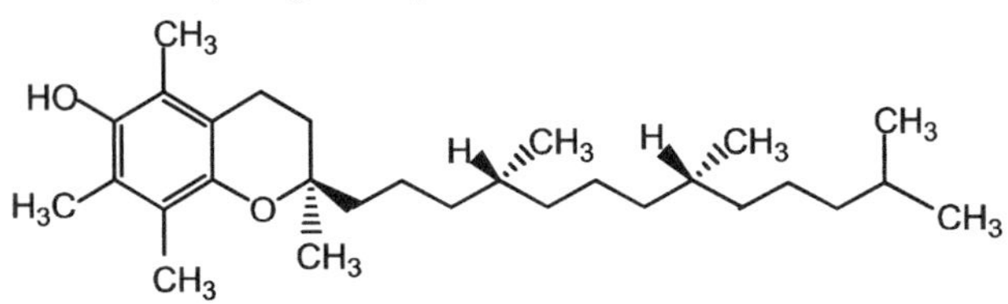	Wheat germ oil; sunflower seed; almond; hazelnut oil; mamey sapote; spinach; avocado; mango; kiwifruit; and olive oil	Vitamin E, also fat-soluble, has two main groups: tocopherols and tocotrienols. The bioactive components α-tocopherol and γ-tocotrienol possess high antioxidant activity displaying anti-carcinogenic effects	Fortification of chocolate milk using liposomes. Fortification of breakfast cereals with α-tocopherol acetate in oil in-water emulsion	Marsanasco, Calabró, Piotrkowski, Chiaramoni, and del Alonso (2016) and Öztürk (2017)

Moreover, currently vitamins are used as preservatives to increase the shelf life of foodstuff, especially because of their antioxidant potential (Gironés-Vilaplana et al., 2017). The use of ascorbic acid (E 300) and its salts sodium ascorbate (E 301) and calcium ascorbate (E 302), as well as of vitamin E in the forms of tocopherol-rich extract (E 306), alpha-tocopherol (E 307), gamma-tocopherol (E 308), delta-tocopherol (E 309), and of fatty esters of ascorbic acid (E 304) have been approved by the EU regulation 1129/2011 (EU Commission, 2011). Another application of vitamins as food additives is in the production of edible coatings, which are thin layers of edible material that cover food surface. Vitamin C, for instance, has been used as coating to maintain the nutritional quality of strawberries an increase product's microbial stability; moreover, ascorbic acid seems to create a favorable environment for probiotic bacteria (Sogvar, Saba, & Emamifar, 2016).

4. Challenges in the stabilization of bioactive molecules

Notwithstanding the intense search for plant and microbial colorant sources and the efforts to improve yield, relatively few natural coloring ingredients have reached the market and are currently being used by the food industry (Rodriguez-Amaya, 2018). Commercial anthocyanins, namely cyanidin-3-glucoside, pelargonidin 3-glucoside and peonidin-3-*O*-glucoside have been indeed applied, and their effectiveness has been increasingly evaluated (Martins et al., 2016).

It is widely known that external interferences such as pH, temperature, humidity, salinity, and storage conditions greatly influence anthocyanin coloration and stability, as well as the presence of enzymes, proteins, metallic ions, and other polyphenols, besides ascorbic acid, and sugar (Rodriguez-Amaya, 2018). Affected by all the above-cited factors, the color of the anthocyanin compounds may vary from red to purple and blue; at pH 1–2, for instance, the red flavylium cation predominates (Fig. 3). In consequence, the application of anthocyanin pigments as food colorants and functional ingredients has been limited by their low stability and interaction with other compounds in the food matrix (Ngamwonglumlert et al., 2017).

In the past few years, several strategies, procedures and techniques have been developed and increasingly applied to solve the stability issues involving plant pigments, concomitantly broadening their use as food additives. Cortez et al. (2017) comprehensively reviewed the main methods used for anthocyanins stabilization, namely the incorporation of co-pigment (e.g., polymers, phenolic compounds, and metals); exclusion of O_2 during processing and storage; besides microencapsulation and nanoencapsulation

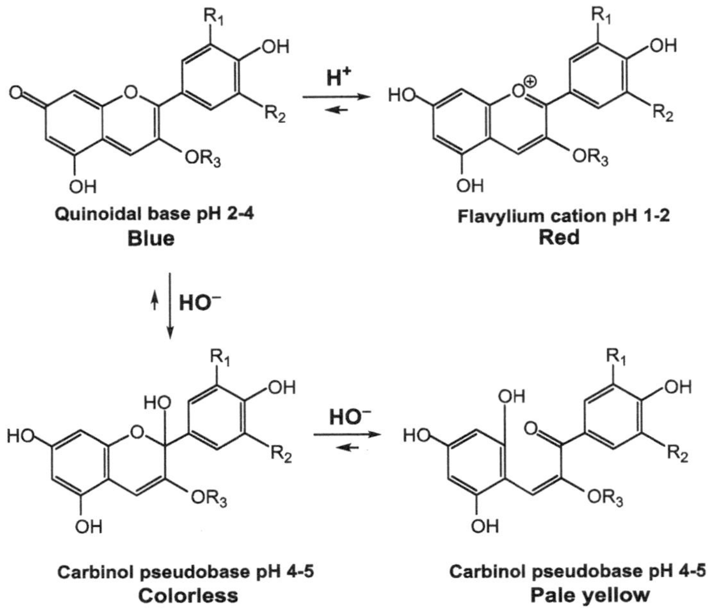

Fig. 3 The different structural forms, and consequently coloration, that anthocyanin compounds assume according to pH fluctuations.

methods. Encapsulation, and particularly spray drying, offers protection to anthocyanins, their controlled and targeted release in food products, thus enhancing stability during storage time. However, there is a lack of studies on other techniques for encapsulation. Although the employment of maltodextrin as a coating material for microencapsulation of anthocyanins has been reported by some authors, the search for other coating materials would be welcome (Yousuf, Gul, Wani, & Singh, 2016). García-Tejeda, Salinas-Moreno, and Martínez-Bustos (2015) studied the microencapsulation of anthocyanins from purple maize via spray drying and tested modified normal and waxy maize starches as wall materials. They reported that starch derivatization improved solubility as well as microencapsulation efficiency; in addition, esterified normal maize starch showed superior anthocyanin retention after storage. Zhao et al. (2017) have compiled the existing literature on the chemical implications of anthocyanin glycosyl acylation, the influence of acylation on the stability of acylated anthocyanins and the involved mechanisms (Table 1). According to these authors, glycosyl acylation improves both the in vitro and in vivo chemical stability of anthocyanins. The stability degree will depend on both the acylation site and on the type and the number of acyl groups.

Also very unstable compounds, betalains are stable in a pH ranging from 4 to 6; the chromophore, betalamic acid, is prone to complete dissociation at $pH \geq 5.32$. Overall, their stability is reduced by the increase of moisture content as well as by the presence of chelating agents and oxygen. Stability of betalains is enhanced, however, by their concentration in the plant matrices and/or food product and by the addition of antioxidant agents such as ascorbic acid. However, temperature (i.e., thermal treatment) is considered the most critical factor that directly affects their stability (Martins et al., 2017). Betanin, for instance, when submitted to mild alkaline conditions, either during heating of an acidic solution or during thermal processing, is decomposed into the colorless cyclodopa 5-*O*-glucoside and betalamic acid (Rodriguez-Amaya, 2018) (Fig. 2). Delia et al. (2019) reported the successful microencapsulation of Cactaceae *Escontria chiotilla* and *Stenocereus queretaroensis* fruits' betalains through a spray drying process employing only *Opuntia ficus-indica* mucilage as a wall agent.

Owing to their tendency to degrade upon exposure to high temperatures and light, and considering the latest achievements on betalain colorants stability, currently these compounds can be efficaciously employed as colorants of frozen foods, low temperature dairy products and even products with short-shelf life, such as yogurts, ice creams and sausages (Martins et al., 2017; Polturak & Aharoni, 2018).

The utilization of carotenoids as food and beverage functional ingredients can be difficult due to their insolubility in water, instability, and low bioavailability. Both solubility and instability issues can be minimized by the formulation of water-dispersable market products, such as colloidal suspensions, emulsions, or dispersions in suitable colloids (Rodriguez-Amaya, 2018). Latterly, research on this area has been focused on encapsulation and nanoencapsulation strategies. Almeida et al. (2018) assessed distinct formulations of curcumin, namely curcumin powder, water-dispersible curcumin and nanoencapsulated curcumin, as yogurt colorants. Both modified forms of the pigment achieved a higher color homogeneity; the nanoencapsulated curcumin displaying the strongest coloring capacity. The development of new water compatible formulations from hydrophobic colorants is definitely an important step in bringing these natural compounds to a wider industrial utilization, what indubitably benefits consumers.

Vitamin content of foods and foodstuff is susceptible to losses and can also suffer degradation through manufacture and storage processes. Most losses are due to their solubility in water that is dependent on the cooking method chosen. Nonetheless, some vitamins are subject to extra degradation processes. B-complex vitamins, for instance, are more labile to temperature and light, whereas fat-soluble vitamins (A, D, E and K) are more prone to degradation by oxygen. However, losses vary according to the type of food and its processing: thiamine, natural folates and vitamin C, for example, can be 100% degraded when submitted to certain concomitant cooking and processing conditions (de Lourdes Samaniego-Vaesken, Alonso-Aperte, & Varela-Moreiras, 2012). Thermal processing of food, such as freezing and cooking, decreases vitamin B5 (pantothenic acid) levels. So much that the refining of grains (one of the major sources of this vitamin) can incur in vitamin B5 losses of even 50% (Ota et al., 2018).

In the past years, vitamin-loaded nanocarriers have been studied as an alternative to minimize vitamins' stability issues in the production of fortified functional foods. Hasanvand, Fathi, Bassiri, Javanmard, and Abbaszadeh (2015), for instance, developed high amylose starch-based nanoparticles for entrapment of vitamin D3 in food systems, proving that the release from the encapsulate could be postponed in gastric media. Moreover, sensory analysis performed with fortified milk within the developed nanocarriers did not show any significant difference with respect to the blank milk sample and the nanoparticles strategy even masked the after taste of vitamin D3 and improved its solubility.

5. Promising functional ingredients

Natural pigments have drawn great attention in recent years, much more for their health-promoting biological functions than for their coloring properties (Benmeziane et al., 2018). Carotenoids have been the most studied in terms of health-promoting effects, comprising epidemiological, in vitro, animal, and human intervention studies. A wide range of bioactivities have been assigned to anthocyanins, based primarily on cell culture and animal studies (clinical trials are still lacking), whereas studies on the positive health effects of betacyanin and chlorophyll are in their initial stages (Rodriguez-Amaya, 2018).

Coloring compounds obtained from the seeds of achiote (*Bixa orellana* L.), a plant native to tropical America, probably to the Amazon basin in Brazil, have been used since pre-Hispanic times. However, in the past years, the active carotenoids from achiote, mainly bixin and norbixin (Fig. 4), in addition to tocotrienols and tocopherols, have gained much attention due to their

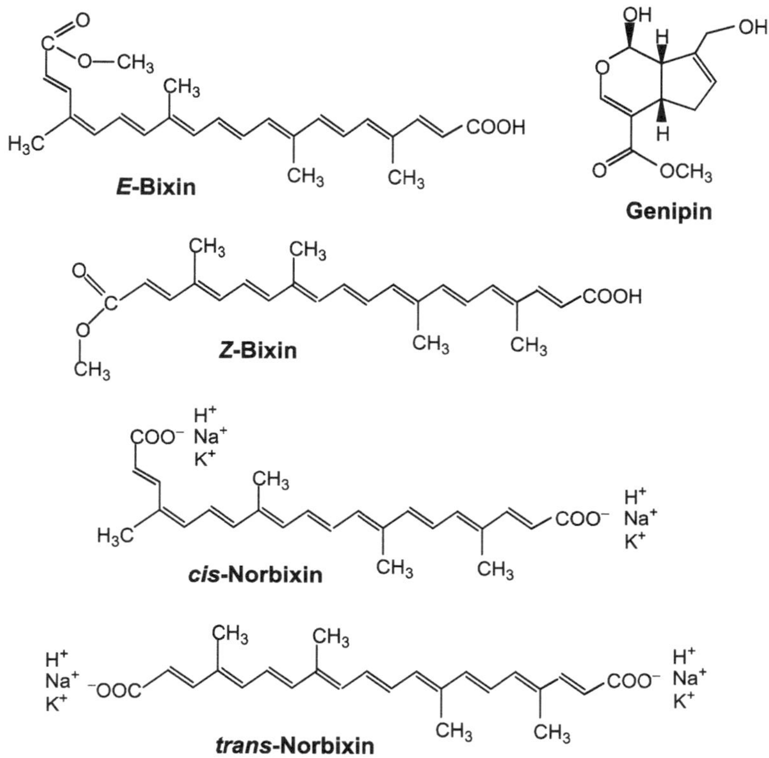

Fig. 4 Chemical structures of some emerging bioactive pigments: bixin and norbixin, dark-red and yellow colorants, respectively, extracted from achiote (*Bixa orellana* L., also known as anatto) (Raddatz-Mota et al., 2017); and genipin, a blue colorant isolated from *Gardenia* sp. and from genipap (*Genipa americana* L.) that has displayed potential health benefits for the food and pharmaceutical areas (Neri-Numa et al., 2017).

biological activities and benefits for human health, besides their presumed potential for industrial applications (Raddatz-Mota et al., 2017). Furthermore, carotenoids such as lycopene-rich oleoresin, have been widely added to vegetable oils to enhance both nutrition profile and stability due to their antioxidant effects (Blasi et al., 2018; Kehili et al., 2018; Nour et al., 2018) (Table 2).

By virtue of the relatively limited occurrence of betalain-producing species in nature, and of the even more restrict number of non-toxic edible sources, researchers have been focused in the improvement and development of new sources of betalains. Regarding the last approach, their efforts have been focused in discovering alternative plant species and cell cultures for betalain extraction, besides the development of novel betalain sources through metabolic engineering of plants and microbes (Martins et al., 2017; Polturak & Aharoni, 2018). Fortunately, not only beetroot is a source betalains but also several cactus fruits, mainly the ones belonging to the *Hylocereus* and *Opuntia* genera (Martins et al., 2016). The prickly pear (*Opuntia* spp.), widely distributed in Mexico, Latin America, South Africa and the Mediterranean, present diverse colors due to the combination of betacyanins and betaxanthins. The red pitaya, also called red dragon fruit (*Hylocereus polyrhizus* L.), on its turn, is famous for its esthetically amazing deep purple color pulp with numberless small soft seeds. The appealing exotic fruit that has conquered the worldwide market is rich in betacyanins, such as betanin, phyllocactin, hylocerenin, and their isomers (Gengatharan et al., 2015). Roriz et al. (2018) innovatively used betacyanin-rich extracts of globeflower (*Gomphrena globosa* L.) as ice-cream colorants, with very positive results (Table 2).

Although anthocyanins, betalains and carotenoids have been used in large scale for red, orange and yellow hues, natural green and blue colorants are scant. This motivates the worldwide search for novel sustainable colorant sources, such as still unexploited fruits, vegetables and edible flowers. Recently, genipin (Fig. 4), a natural blue pigment obtained from *Gardenia* sp. and *Genipa americana* L., have presented potential as a colorant for food and cosmetics. Furthermore, its appealing cross-linking and carrier agent features for clinical practices, make this pigment a suitable and eco-friendly option to artificial compounds (Table 1) (Neri-Numa et al., 2017).

6. Contribution in a biocircular economy

The high processing expenditure and the low accessibility of natural coloring matrices are the major hindrances for these products becoming more popular. Thus, the efforts to overcome these obstacles and reduce production costs have been mainly focused on the discovery of novel natural coloring substances, and in particular on the valorization of several colored agro industrial by-products and wastes (Baaka, Ksibi, & Mhenni, 2017). Although most bioactive pigments are broadly distributed in fruits, carotenoids are also found (and very commonly at higher amounts) in plant peels and seeds, which are in general discarded by regular consumers and industry. The tomato peel, for instance, contains almost fivefold more lycopene than the tomato pulp (Martins & Ferreira, 2017).

Generally, to manufacture a natural colorant, an extraction procedure is first needed to free the desired crude pigment from a matrix, which is most of the time a plant material. An appropriate extraction strategy not only increments the yield but also (and most importantly) prevents the degradation of the extracted pigments, leading to the production of natural colorant ingredients of superior quality (Ngamwonglumlert et al., 2017). Moreover, it is well established that combination of various extraction methods allows higher extraction efficiency and yield (Martins & Ferreira, 2017).

Now there is a trend to recover bioactive compounds, including pigments, from plant bio-residues by means of green extraction. This approach involves the utilization of environmental friendly solvents and renewable products and minimization of energy consumption. The overall purpose is to obtain suitable extracts in terms of safety and other quality parameters (Benmeziane et al., 2018). Table 4 describes recent studies on the recovery of plant pigments from plant bio-residues using different extraction approaches. In the past few years, emergent techniques such as supercritical fluid extraction (SFE) (Lima et al., 2018), pressurized liquid extraction (PLE) (Paes et al., 2014), microwave- and enzyme-assisted extractions (MAE and EAE) (Li et al., 2016; Strati et al., 2015), ohmic heating (Loypimai et al., 2015), ultrasound-assisted extraction (UAE) (Backes et al., 2018; Benmeziane et al., 2018; D'Alessandro et al., 2014), and pulsed electric fields (PEF) (Koubaa et al., 2016; Pataro et al., 2018; Zhou et al., 2015) have been increasingly applied to the recovery of pigments from plant sources and corresponding residues (Table 4). Recently, biosorption has been employed as a method for separation and concentration of anthocyanins from extracts of agro industrial wastes. This is a very attractive process due to its relative simplicity, easy handling and low cost, especially if the bio-sorbent is cheap and highly available (Kohno et al., 2014). Stafussa et al. (2016) have reported the application of microorganisms for the recovery of anthocyanins from grape pomace extracts via biosorption by waste yeast (*Saccharomyces cerevisiae*) from brewer activity (Table 4).

To the best of our best knowledge, there are no reports on the recovery and posterior application of vitamins from plant bio-residues. The main reason for this is the high instability of these compounds, which suffer degradation very easily during the stages of industrial processing and subsequent waste handling. In addition, the cost of the chemical synthesis of vitamins is quite attractive, making the hypothesis of recovering vitamins from natural wastes very unviable.

Table 4 Recovery of plant pigments from agro industrial by-product or wastes using diverse extraction technologies, reported in the past 5 years.

Bio-residue	Recovery strategy and adopted conditions	Results	Reference
Black bean (*Phaseolus vulgaris* L.) coats	Water-citric acid 2% extraction; stirring for 4 h at 40 °C. Furthermore, the stabilization of extracted anthocyanins by co-pigmentation with 2% β-cyclodextrin (β-CD) was investigated	The recovered anthocyanin-rich fractions (powders and aqueous extracts) were assessed as colorants in sport beverages. All anthocyanin fractions combined with β-CD presented increased half-life, higher *D*-values and fewer differences in colorimetric parameters under darkness and 4 °C conditions	Aguilera et al. (2016)
Black chokeberry (*Aronia melanocarpa* Michx. Elliott) wastes	Ultrasound-assisted extraction (UAE). Extraction kinetics study: temperature (20–70 °C), ethanol content in the solvent (0–50%) and ultrasound power (0–100 W)	An evident positive effect on the extraction of total polyphenols with the increase of temperature and ethanol content in solvent, being that the best outcomes were obtained in the beginning of the extraction processes and at low temperatures	D'Alessandro, Dimitrov, Vauchel, and Nikov (2014)
Black glutinous rice (*Oryza sativa* L.) bran	Ohmic heating (OHM) assisted solvent extraction; distinct levels of electric field strengths of 50–200 $V\,cm^{-1}$ were applied	The bran contain great levels of cyanidin-3-*O*-glucoside, delphinidin and pelargonidin. The utilization of OHM to assist solvent extraction of anthocyanins from black rice bran aiming the obtainment of a natural colorant powder was successful, inclusive in comparison with steam-assisted solvent extraction methods	Loypimai, Moongngarm, Chottanom, and Moontree (2015)
Blackberry (*Rubus* sp.) pulp residues	Maceration at room temperature (25 °C). A complete experimental design and response surface methodology were applied to estimate how the quantity of solvent (20–50 mL), number of extractions (1–5), and time (10–30 min) affected the recovery of anthocyanins	The optimal extraction conditions were achieved for 20 mL of acidified ethanol (0.1% HCI), 3 extractions, and 10 min, with a yield of 59%, thus resulting in the recovery of 25.9 mg of cyanidin-3-glucoside per 100 g of blackberry bagasse (dry basis)	de Vargas, Jablonski, Flôres, and Rios (2017)

Blackcurrant (*Ribes nigrum* L.) marc	Homogenate-microwave-assisted extraction (H-MAE)	The optimized conditions were ethanol volume fraction of 60%; homogenate time of 3 min; liquid-solid ratio of 28.3 mL/g; 0.3% of antioxidant *tert*-butylhydroquinone; pH of 2.5; microwave irradiation power of 551 W; microwave irradiation time of 16.4 min, with good yields of flavonols and anthocyanins (470 μg/g), with relatively short extraction time	Li et al. (2016)
Blueberry (*Vaccinium myrtillus* L.) residues	Pressurized liquid extraction (PLE) using water, ethanol and acetone at different proportions, with temperature, pressure and solvent flow rate kept constant at 40 °C, 20 MPa and 10 ml/min. Supercritical CO_2 extraction (SFE), with water, acidified water, and ethanol as modifiers Pulsed electric field (PEF) using response surface methodology (RSM) for optimizing the extraction procedure	The highest antioxidant activities and phenolic content values among PLE extracts were obtained when using pure ethanol and ethanol/water, with the best anthocyanins recovery yields obtained when using acidified water as solvent. In SFE, the best condition for all functional components assessed was at 90% CO_2, 5% water, and 5% ethanol The optimized parameters for extraction, with an anthocyanin recovery quantity of 220 mg/L, were ethanol solvent, 60% (acidified with 0.1% [*v*/v] hydrochloric acid); liquid to liquid ratio, 1:6 (mL/mL); pulse number, 10 ea.; and electric field strength, 20 kV/cm. In comparison with UAE, PEF increased the anthocyanin extraction yield demanding a milder extraction temperature and a shorter extraction time	Paes, Dotta, Barbero, and Martínez (2014) and Zhou, Zhao, and Huang (2015)
Caneberry (*Rubus* spp.) press residues	Extraction using 800 mL/L methanol aqueous solution with 0.5 mL/L acetic acid	HPLC-DAD-ESI/MS analysis allowed the identification of cyanidin glycosides in all press residues, being that cyanidin-3-glucoside was prevalent in blackberry samples (up to 1398 mg/kg) while cyanidin-3-sophoroside was prevalent in raspberry samples (up to 581.0 mg/kg). Antioxidant capacity (AC), assessed by ABTS, reducing power and α-glucosidase inhibitory potential assays, was superior in blackberry residues. AC was in good correlation with total anthocyanin content	Tumbas Šaponjac et al. (2014)

Continued

Table 4 Recovery of plant pigments from agro industrial by-product or wastes using diverse extraction technologies, reported in the past 5 years.—cont'd

Bio-residue	Recovery strategy and adopted conditions	Results	Reference
Cantaloupe melon (*Cucumis melo* L.) wastes	Ultrasound-assisted extraction was performed with the fixed frequency of 20 kHz and temperature of 21 °C. Several solvent mixtures, hexane contents in the solution, extraction times, amplitudes and solvent-powder ratios were tested. Response surface methodology was applied aiming the optimization of the extraction of carotenoids	The cantaloupe waste samples contained lutein and β-carotene as principal carotenoids. Scanning electron microscopy analysis showed noticeable microstructural changes after ultrasound extraction. K, Na, P, Mg, Ca, Fe, Cu, Mn and Zn were identified in sample wastes, K being the major one. The extract displayed antioxidant activity in the DPPH assay	Benmeziane et al. (2018)
Carrot (*Daucus carota* L.) peels	Supercritical CO_2 (S-CO_2) extraction employing ethanol as co-solvent. The optimal conditions for maximum mass yield were: 58.5 °C, 306 bar and 14.3% of ethanol, and at 59.0 °C, 349 bar and 15.5% ethanol for carotenoid recovery (86.1%)	The optimization of extraction conditions resulted in a 96.2% of carotenoid recovery. The kinetic studies revealed that supercritical CO_2 can extract carotenoid fractions from carrot peels speedily, while model fitting emphasized the rapid extraction trend and desorbing nature of carotenoids. Such findings can be applied for other vegetable co-product matrices with similar structure, aiming the recovery of carotenoids	Lima, Charalampopoulos, and Chatzifragkou (2018)
Chokeberry [*Aronia melanocarpa* (Michx.) Elliott]	Solid-state fermentation (SSF) with *Aspergillus niger* and *Rhizopus oligosporus*, with flasks incubated under static conditions at 30 °C during 12 days	The extractable phenolic compounds increased >1.7-fold during both fermentation processes, and a similar trend was found for total flavonoid contents. A longer fermentation time implied in greater loss of total anthocyanins. Furthermore, SSF not only improved the oil recovery rate but also resulted in the extraction of lipids with better nutritional quality characteristics	Dulf, Vodnar, Dulf, Diaconeasa, and Socaciu (2018)

Elderberry (*Sambucus nigra* L.) branches	For each 25 g of elderberry branches, 1 L of a water and ethanol (95%) solution was added and boiled during 15 min. After cooling, the extract was filtered and concentrated via rotary evaporation to 100 mL	Elderberry branches have significant quantities of anthocyanins, flavonols and cinnamate esters, with similar or even superior antioxidant activities in comparison to berries	Silva, Ferreira, and Nunes (2017)
Grape (*Vitis vinifera* L.) pomace extracts	Recovery of anthocyanins from grape pomace extracts by biosorption in brewer's yeast. The bio-sorbent consisted of residual biomass of *Saccharomyces cerevisiae*	The most expressive biosorption capacity of anthocyanins in *S. cerevisiae* biomass was observed for the Tannat grape pomace extract. Both Temkin and DeR models satisfactorily described the process and confirmed its chemisorption nature. The following functional groups were identified in yeast cells via FTIR method: carboxyl, amino/hydroxyl and amide groups, which were involved in anthocyanin's biosorption	Stafussa et al. (2016)
Fig fruit (*Ficus carica* L.) peel	Ultrasound-assisted extraction (UEA) was the most effective method, and the optimal extraction conditions were 21 min, 310 W, and 100% of ethanol	The authors performed a comparison of three different techniques (heat, microwave, and ultrasound) for anthocyanin extraction maximization. Furthermore, the joint effect of the identified relevant variables for each technique were described through the response surface methodology. UEA was the most potent method, yielding 3.82 mg of cyanidin-3-*O*-rutinoside per gram of extracted residue; an increased non-linear relationship was observed for concentrations in the range 5–200 g/L, being the optimal solution close to 150 g/L	Backes et al. (2018)

Continued

Table 4 Recovery of plant pigments from agro industrial by-product or wastes using diverse extraction technologies, reported in the past 5 years.—cont'd

Bio-residue	Recovery strategy and adopted conditions	Results	Reference
Juçara palm heart (*Euterpe edulis* Mart.) residues	PLE was assessed at 10 MPa and 40, 60 and 80 °C employing ethanol, water, acidified mixture of ethanol + water and acidified water as solvents. Afterward, the best PLE solvent was selected as co-solvent for SFE with CO_2	In the group of PLE extracts, the highest antioxidant activity and phenolics content were obtained with the acidified mixture of ethanol + water at 80 °C, while the highest anthocyanin content was observed for acidified water extract at 40 °C. The acidified mixture ethanol + water, the selected co-solvent for SFE, enhanced significantly the anthocyanin content of the extracts obtained by this method	del Pilar Garcia-Mendoza et al. (2017)
Pitaya [*Hylocereus undatus* (Haw.) Britton & Rose and *Hylocereus megalanthus* (K. Schum. ex Vaupel) Ralf Bauer) fruit by-products	Optimum extraction yields were achieved with $X_1 = 1/149.95$ g/mL, $X_2 = 72.27$ °C and $X_3 = 39.39$ min (white-fleshed red pitaya) and $X_1 = 1/148.96$ g/mL, $X_2 = 72.56$ °C and $X_3 = 5.02$ min (yellow pitaya), whereas a maximum betacyanin content were obtained with $X_1 = 1/150$ g/mL, $X_2 = 49.33$ °C and $X_3 = 5$ min	A green microwave-assisted extraction of phenolic compounds from pitaya peels was optimized via Box-Behnken design using three factors: solid/solvent ratio (X_1), temperature (X_2) and extraction time (X_3). Results evidence that the peels of yellow pitaya (*H. megalanthus*) and white-fleshed red pitaya (*H. undatus*) fruits are valuable sources of multiple bioactive phenolic compounds. This was the first identification of phenolic compounds in yellow pitaya using HPLC-DAD-ESI-MS^n. White-fleshed red pitaya peels are especially abundant in betacyanins, thus being suitable for the obtainment of food-grade colorants	Ferreres et al. (2017)

Pomegranate (*Punica granatum* L.) wastes	The optimum operating conditions for the green ultrasound extraction of carotenoids were established: temperature of 51.5 °C; peels/solvent ratio of 0.10; amplitude level of 58.8%; sunflower oil as solvent and extraction period of 30 min	Authors proposed a new approach for exploiting pomegranate peels in food industries. Carotenoids were ultrasound extracted from the fruit wastes using vegetable oils as solvents. As a result, an oil enriched with antioxidants was obtained. The optimum extraction yield was about 0.325 mg carotenoids/100 g of dry pomegranate peels	Goula, Ververi, Adamopoulou, and Kaderides (2017)
Prickly pear [*Opuntia ficus-indica* var. sanguigna (OS) and gialla (OG) and *Opuntia engelmannii* (OE)] peels	Hydroethanolic (ethanol:water, 80:20, v/v) extraction of the lyophilized peels. The sample (1 g) was extracted twice by stirring with 25 mL of hydroalcoholic solution (25 °C at 150 rpm) for 1 h	Twelve phenolic compounds were identified in the *Opuntia* spp. peels, being betanin the most abundant betacyanin in the analyzed materials. Among tested samples, *Opuntia engelmannii* was significantly richer in betacyanins. Besides presenting antioxidant potential, the hydroethanolic extracts of all species revealed to be more active than ampicillin when tested against 16 pathological strains	Melgar et al. (2017)
Purple eggplant (*Solanum melongena* L.) peels	Peels grinded in acidic water at 75 °C provided optimal extraction of total phenolics	Five antocyanins extracted from eggplant peels were identified by HPLC: delphinidin 3-*O*-rutinoside, delphinidin 3-*O*-rutinoside-5-*O*-glucoside, petunidin 3-*O*-rutinoside, where malvidin 3-*O*-rutinoside-5-*O*-glucoside and cyanidin 3-*O*-rutinoside. Phenolic compounds extraction was enhanced by the use of a ultrasonic probe, however, microscopic evaluation revealed cell denaturation after this procedure	Ferarsa et al. (2018)

Continued

Table 4 Recovery of plant pigments from agro industrial by-product or wastes using diverse extraction technologies, reported in the past 5 years.—cont'd

Bio-residue	Recovery strategy and adopted conditions	Results	Reference
Red prickly pear [(*Opuntia stricta* Haworth (Haw.)] peels	The PEF conditions for achieving the maximum betanin recovery yields were: electric field strength of 20 kV/cm, reaching a plateau (≈50 mg betanin/100 g fresh fruit) after 50 min diffusion at 20 °C	Pulsed electric fields (PEF) and ultrasounds (USN) were tested as pre-treatment methods. Betanin and isobetanin were identified in the extract via HPLC analysis. Both PEF and USN allowed higher recovery yields of red colorant compounds and less impurity. However, scanning electron microscopy pictures showed that PEF can induce cell wall permeabilization without destroying the cell tissue, thus making easier selective extraction of intracellular interest components	Koubaa et al. (2016)
Tomato (*Solanum lycopersicum* L.) wastes	Enzyme-assisted extraction with pectinase and cellulose, at 45 and 55 °C during 3 min. High pressure assisted solvent extraction performed at 700 MPa by using ($P < 0.05$) lower ratios of solvent: solid (6:1 and 4:1 mL:g) and reduced processing time (10 min) Pulsed electric fields (PEF) (0.25–0.75 kV/cm, 1 kJ/kg) combined with steam blanching (SB) (1 min at 50–70 °C)	Maximum total carotenoid and lycopene extraction yields were obtained for samples previously treated with enzymes and further extracted with ethyl lactate, corresponding to almost 6-fold and 10-fold increases, respectively, with respect to non-treated samples. High pressure assisted extraction led to higher extraction yields compared to conventional solvent extraction processes PEF was combined with steam blanching SB to intensify carotenoids extraction. The combined approach showed a synergistic effect on carotenoids recovery from tomato peels, being that all-*trans*-lycopene was the most abundant carotenoid in this waste material, and not producing any degradation/isomerization of lycopene	Strati, Gogou, and Oreopoulou (2015) and Pataro et al. (2018)

7. Conclusion and future prospective

The past 5 years studies on the potentialities of pigments and vitamins from plants as food additives reveal highly positive and promising prospects for future investigations, which undoubtedly benefit consumers. In order to make natural active pigments more popular in food industry, to reduce their processing costs is mandatory. In this sense, the recovery of pigments from agro industrial wastes using green emerging approaches is an irreversible tendency and clearly the best alternative to ensure their sustainable obtainment, although still full of challenging aspects. Future investigation efforts should aim to expand the information on the biochemical features of these molecules, not only for the development of strategies to solve their stability issues but also to optimize their exploitation as functional food ingredients.

Acknowledgments

R.C.G.C. thanks the Conselho Nacional de Desenvolvimento Científico e Tecnológico (CNPq) for the financial support provided for her postdoctoral research in State University of Maringá (Process number 167378/2017-1). R.M.P. (Project number 307944/2015-8) and A.B. (Project number 304090/2016-6) are CNPq research grant recipients.

References

Aguilera, Y., Mojica, L., Rebollo-Hernanz, M., Berhow, M., de Mejía, E. G., & Martín-Cabrejas, M. A. (2016). Black bean coats: New source of anthocyanins stabilized by β-cyclodextrin copigmentation in a sport beverage. *Food Chemistry*, *212*, 561–570.

Allen, V. J., Methven, L., & Gosney, M. A. (2013). Use of nutritional complete supplements in older adults with dementia: Systematic review and meta-analysis of clinical outcomes. *Clinical Nutrition*, *32*, 950–957.

Almeida, H. H., Barros, L., Barreira, J. C., Calhelha, R. C., Heleno, S. A., Sayer, C., et al. (2018). Bioactive evaluation and application of different formulations of the natural colorant curcumin (E100) in a hydrophilic matrix (yogurt). *Food Chemistry*, *261*, 224–232.

Baaka, N., El Ksibi, I., & Mhenni, M. F. (2017). Optimisation of the recovery of carotenoids from tomato processing wastes: application on textile dyeing and assessment of its antioxidant activity. *Natural Product Research*, *31*(2), 196–203.

Backes, E., Pereira, C., Barros, L., Prieto, M. A., Genena, A. K., Barreiro, M. F., et al. (2018). Recovery of bioactive anthocyanin pigments from *Ficus carica* L. peel by heat, microwave, and ultrasound based extraction techniques. *Food Research International*, *113*, 197–209.

Benmeziane, A., Boulekbache-Makhlouf, L., Mapelli-Brahm, P., Khodja, N. K., Remini, H., Madani, K., et al. (2018). Extraction of carotenoids from cantaloupe waste and determination of its mineral composition. *Food Research International*, *111*, 391–398.

Bigliardi, B., & Galati, F. (2013). Innovation trends in the food industry: The case of functional foods. *Trends in Food Science and Technology*, *31*, 118–129.

Blasi, F., Rocchetti, G., Montesano, D., Lucini, L., Chiodelli, G., Ghisoni, S., et al. (2018). Changes in extra-virgin olive oil added with *Lycium barbarum* L. carotenoids during frying: Chemical analyses and metabolomic approach. *Food Research International, 105*, 507–516.
Bonto, A. P., Camacho, K. S. I., & Camacho, D. H. (2018). Increased vitamin B5 uptake capacity of ultrasonic treated milled rice: A new method for rice fortification. *LWT-Food Science and Technology, 95*, 32–39.
Campbell, I. (2017). Macronutrients, minerals, vitamins and energy. *Anaesthesia & Intensive Care Medicine, 18*, 141–146.
Card, D. J., Gorska, R., Cutler, J., & Harrington, D. J. (2014). Vitamin K metabolism: Current knowledge and future research. *Molecular Nutrition & Food Research, 58*, 1590–1600.
Carocho, M., Barreiro, M. F., Morales, P., & Ferreira, I. C. F. R. (2014). Adding molecules to food, pros and cons: A review on synthetic and natural food additives. *Comprehensive Reviews in Food Science and Food Safety, 13*, 377–399.
Carocho, M., Morales, P., & Ferreira, I. C. F. R. (2015). Natural food additives: Quo vadis? *Trends in Food Science & Technology, 45*, 284–295.
Carr, A., & Vissers, M. (2013). Synthetic or food-derived vitamin C—Are they equally bioavailable? *Nutrients, 5*, 4284–4304.
Chhikara, N., Kushwaha, K., Sharma, P., Gat, Y., & Panghal, A. (2019). Bioactive compounds of beetroot and utilization in food processing industry: A critical review. *Food Chemistry, 272*, 192–200.
Corrêa, R. C., Haminiuk, C. W., Sora, G. T., Bergamasco, R., & Vieira, A. M. (2014). Antioxidant and rheological properties of guava jam with added concentrated grape juice. *Journal of the Science of Food and Agriculture, 94*, 146–152.
Cortez, R., Luna-Vital, D. A., Margulis, D., & Mejia, E. (2017). Natural pigments: Stabilization methods of anthocyanins for food applications. *Comprehensive Reviews in Food Science and Food Safety, 16*, 180–198.
D'Alessandro, L. G., Dimitrov, K., Vauchel, P., & Nikov, I. (2014). Kinetics of ultrasound assisted extraction of anthocyanins from *Aronia melanocarpa* (black chokeberry) wastes. *Chemical Engineering Research and Design, 92*, 1818–1826.
de Lourdes Samaniego-Vaesken, M., Alonso-Aperte, E., & Varela-Moreiras, G. (2012). Vitamin food fortification today. *Food & Nutrition Research, 56*, 5459.
de Vargas, E. F., Jablonski, A., Flôres, S. H., & Rios, A. D. O. (2017). Obtention of natural dyes from industrial blackberry pulp residues (*Rubus* sp). *Journal of Food Processing and Preservation, 41*, e12777.
del Pilar Garcia-Mendoza, M., Espinosa-Pardo, F. A., Baseggio, A. M., Barbero, G. F., Junior, M. R. M., Rostagno, M. A., et al. (2017). Extraction of phenolic compounds and anthocyanins from juçara (*Euterpe edulis* Mart.) residues using pressurized liquids and supercritical fluids. *The Journal of Supercritical Fluids, 119*, 9–16.
Delia, S. C., Chávez, G. M., Frank, M. L. M., Araceli, S. G. P., Irais, A. L., & Franco, A. A. (2019). Spray drying microencapsulation of betalain rich extracts from *Escontria chiotilla* and *Stenocereus queretaroensis* fruits using cactus mucilage. *Food Chemistry, 272*, 715–722.
Dia, V. P., Wang, Z., West, M., Singh, V., West, L., & De Mejia, E. G. (2015). Processing method and corn cultivar affected anthocyanin concentration from dried distillers grains with solubles. *Journal of Agricultural and Food Chemistry, 63*, 3205–3218.
Díaz-García, M. C., Castellar, M. R., Obón, J. M., Obón, C., Alcaraz, F., & Rivera, D. (2015). Production of an anthocyanin-rich food colourant from *Thymus moroderi* and its application in foods. *Journal of the Science of Food and Agriculture, 95*, 1283–1293.
Dulf, F. V., Vodnar, D. C., Dulf, E. H., Diaconeasa, Z., & Socaciu, C. (2018). Liberation and recovery of phenolic antioxidants and lipids in chokeberry (*Aronia melanocarpa*) pomace by solid-state bioprocessing using *Aspergillus niger* and *Rhizopus oligosporus* strains. *LWT-Food Science and Technology, 87*, 241–249.

EU Commission. (2011). Commission Regulation (EU) No 1129/2011 of 11 November 2011 amending Annex II to Regulation (EC) No 1333/2008 of the European Parliament and of the Council by establishing a Union list of food additives. *Official Journal of the European Union L, 295*, 12-11.

Ferarsa, S., Zhang, W., Moulai-Mostefa, N., Ding, L., Jaffrin, M. Y., & Grimi, N. (2018). Recovery of anthocyanins and other phenolic compounds from purple eggplant peels and pulps using ultrasonic-assisted extraction. *Food and Bioproducts Processing, 109*, 19–28.

Ferreres, F., Grosso, C., Gil-Izquierdo, A., Valentão, P., Mota, A. T., & Andrade, P. B. (2017). Optimization of the recovery of high-value compounds from pitaya fruit by-products using microwave-assisted extraction. *Food Chemistry, 230*, 463–474.

Gandía-Herrero, F., Escribano, J., & García-Carmona, F. (2016). Biological activities of plant pigments betalains. *Critical Reviews in Food Science and Nutrition, 56*, 937–945.

García-Tejeda, Y. V., Salinas-Moreno, Y., & Martínez-Bustos, F. (2015). Acetylation of normal and waxy maize starches as encapsulating agents for maize anthocyanins microencapsulation. *Food and Bioproducts Processing, 94*, 717–726.

Gengatharan, A., Dykes, G. A., & Choo, W. S. (2015). Betalains: Natural plant pigments with potential application in functional foods. *LWT-Food Science and Technology, 64*, 645–649.

Gironés-Vilaplana, A., Villaño, D., Marhuenda, J., Moreno, D. A., & García-Viguera, C. (2017). Vitamins. In C. M. Galanakis (Ed.), *Nutraceutical and functional food components: Effects of innovative processing techniques* (pp. 159–201): Academic Press.

Goula, A. M., Ververi, M., Adamopoulou, A., & Kaderides, K. (2017). Green ultrasound-assisted extraction of carotenoids from pomegranate wastes using vegetable oils. *Ultrasonics Sonochemistry, 34*, 821–830.

Hasanvand, E., Fathi, M., Bassiri, A., Javanmard, M., & Abbaszadeh, R. (2015). Novel starch based nanocarrier for vitamin D fortification of milk: Production and characterization. *Food and Bioproducts Processing, 96*, 264–277.

Hemery, Y. M., Fontan, L., Moench-Pfanner, R., Laillou, A., Berger, J., Renaud, C., et al. (2015). Influence of light exposure and oxidative status on the stability of vitamins A and D3 during the storage of fortified soybean oil. *Food Chemistry, 184*, 90–98.

Jäpelt, R. B., & Jakobsen, J. (2013). Vitamin D in plants: A review of occurrence, analysis, and biosynthesis. *Frontiers in Plant Science, 4*, 136.

Kaushik, R., & Arora, S. (2017). Effect of calcium and vitamin D 2 fortification on physical, microbial, rheological and sensory characteristics of yoghurt. *International Food Research Journal, 24*, 1744–1752.

Kaushik, R., Sachdeva, B., & Arora, S. (2017). Effect of calcium and vitamin D2 fortification on quality characteristics of dahi. *International Journal of Dairy Technology, 70*, 269–276.

Kehili, M., Choura, S., Zammel, A., Allouche, N., & Sayadi, S. (2018). Oxidative stability of refined olive and sunflower oils supplemented with lycopene-rich oleoresin from tomato peels industrial by-product, during accelerated shelf-life storage. *Food Chemistry, 246*, 295–304.

Khoo, H. E., Azlan, A., Tang, S. T., & Lim, S. M. (2017). Anthocyanidins and anthocyanins: Colored pigments as food, pharmaceutical ingredients, and the potential health benefits. *Food & Nutrition Research, 61*, 1361779.

Kim, G. B., Chen, Y., Kang, W., Guo, J., Payne, R., Li, H., et al. (2018). The critical chemical and mechanical regulation of folic acid on neural engineering. *Biomaterials, 178*, 504–516.

Kohno, Y., Haga, E., Yoda, K., Shibata, M., Fukuhara, C., Tomita, Y., et al. (2014). Adsorption behavior of natural anthocyanin dye on mesoporous silica. *Journal of Physics and Chemistry of Solids, 75*, 48–51.

Koubaa, M., Barba, F. J., Grimi, N., Mhemdi, H., Koubaa, W., Boussetta, N., et al. (2016). Recovery of colorants from red prickly pear peels and pulps enhanced by pulsed electric field and ultrasound. *Innovative Food Science & Emerging Technologies*, *37*, 336–344.

Kumar, V., Kushwaha, R., Goyal, A., Tanwar, B., & Kaur, J. (2018). Process optimization for the preparation of antioxidant rich ginger candy using beetroot pomace extract. *Food Chemistry*, *245*, 168–177.

Lachman, J., Martinek, P., Kotíková, Z., Orsák, M., & Šulc, M. (2017). Genetics and chemistry of pigments in wheat grain—A review. *Journal of Cereal Science*, *74*, 145–154.

Li, X., Chen, F., Li, S., Jia, J., Gu, H., & Yang, L. (2016). An efficient homogenate-microwave-assisted extraction of flavonols and anthocyanins from blackcurrant marc: Optimization using combination of Plackett-Burman design and Box-Behnken design. *Industrial Crops and Products*, *94*, 834–847.

Lima, M. A., Charalampopoulos, D., & Chatzifragkou, A. (2018). Optimisation and modelling of supercritical CO_2 extraction process of carotenoids from carrot peels. *The Journal of Supercritical Fluids*, *133*, 94–102.

Lo Piero, A. R. (2015). The state of the art in biosynthesis of anthocyanins and its regulation in pigmented sweet oranges [(*Citrus sinensis*) L. Osbeck]. *Journal of Agricultural and Food Chemistry*, *63*, 4031–4041.

López-Nicolás, R., Frontela-Saseta, C., González-Abellán, R., Barado-Piqueras, A., Perez-Conesa, D., & Ros-Berruezo, G. (2014). Folate fortification of white and whole-grain bread by adding Swiss chard and spinach. Acceptability by consumers. *LWT-Food Science and Technology*, *59*, 263–269.

Low, J. W., Mwanga, R. O., Andrade, M., Carey, E., & Ball, A. M. (2017). Tackling vitamin A deficiency with biofortified sweetpotato in sub-Saharan Africa. *Global Food Security*, *14*, 23–30.

Loypimai, P., Moongngarm, A., Chottanom, P., & Moontree, T. (2015). Ohmic heating-assisted extraction of anthocyanins from black rice bran to prepare a natural food colourant. *Innovative Food Science & Emerging Technologies*, *27*, 102–110.

Mangolim, C. S., Moriwaki, C., Nogueira, A. C., Sato, F., Baesso, M. L., Neto, A. M., et al. (2014). Curcumin–β-cyclodextrin inclusion complex: Stability, solubility, characterisation by FT-IR, FT-Raman, X-ray diffraction and photoacoustic spectroscopy, and food application. *Food Chemistry*, *153*, 361–370.

Marsanasco, M., Calabró, V., Piotrkowski, B., Chiaramoni, N. S., & del Alonso, S. V. (2016). Fortification of chocolate milk with omega-3, omega-6, and vitamins E and C by using liposomes. *European Journal of Lipid Science and Technology*, *118*, 1271–1281.

Martins, N., & Ferreira, I. C. F. R. (2017). Wastes and by-products: Upcoming sources of carotenoids for biotechnological purposes and health-related applications. *Trends in Food Science & Technology*, *62*, 33–48.

Martins, N., Roriz, C. L., Morales, P., Barros, L., & Ferreira, I. C. F. R. (2016). Food colorants: Challenges, opportunities and current desires of agro-industries to ensure consumer expectations and regulatory practices. *Trends in Food Science & Technology*, *52*, 1–15.

Martins, N., Roriz, C. L., Morales, P., Barros, L., & Ferreira, I. C. F. R. (2017). Coloring attributes of betalains: A key emphasis on stability and future applications. *Food & Function*, *8*, 1357–1372.

Melgar, B., Dias, M. I., Ciric, A., Sokovic, M., Garcia-Castello, E. M., Rodriguez-Lopez,-A. D., et al. (2017). By-product recovery of *Opuntia* spp. peels: Betalainic and phenolic profiles and bioactive properties. *Industrial Crops and Products*, *107*, 353–359.

Mostafai, R., Mohammadi, R., Nachvak, S. M., Rezaei, M., Pasdar, Y., Abdollahzad, H., et al. (2018). Fortified yogurt with vitamin D as a cost-effective food to prevent diabetes: A randomized double-blind clinical trial. *Journal of Functional Foods*, *42*, 137–145.

Neri-Numa, I. A., Pessoa, M. G., Paulino, B. N., & Pastore, G. M. (2017). Genipin: A natural blue pigment for food and health purposes. *Trends in Food Science & Technology, 67*, 271–279.

Ngamwonglumlert, L., Devahastin, S., & Chiewchan, N. (2017). Natural colorants: Pigment stability and extraction yield enhancement via utilization of appropriate pretreatment and extraction methods. *Critical Reviews in Food Science and Nutrition, 57*, 3243–3259.

Nour, V., Corbu, A. R., Rotaru, P., Karageorgou, I., & Lalas, S. (2018). Effect of carotenoids, extracted from dry tomato waste, on the stability and characteristics of various vegetable oils. *Grasas y Aceites, 69*, 238.

Oroian, M., & Escriche, I. (2015). Antioxidants: Characterization, natural sources, extraction and analysis. *Food Research International, 74*, 10–36.

Ota, A., Istenič, K., Skrt, M., Šegatin, N., Žnidaršič, N., Kogej, K., & Ulrih, N. P. (2018). Encapsulation of pantothenic acid into liposomes and into alginate or alginate–pectin microparticles loaded with liposomes. *Journal of Food Engineering, 229*, 21–31.

Öztürk, B. (2017). Nanoemulsions for food fortification with lipophilic vitamins: Production challenges, stability, and bioavailability. *European Journal of Lipid Science and Technology, 119*, 1500539.

Paes, J., Dotta, R., Barbero, G. F., & Martínez, J. (2014). Extraction of phenolic compounds and anthocyanins from blueberry (*Vaccinium myrtillus* L.) residues using supercritical CO_2 and pressurized liquids. *The Journal of Supercritical Fluids, 95*, 8–16.

Pasqualone, A., Bianco, A. M., Paradiso, V. M., Summo, C., Gambacorta, G., & Caponio, F. (2014). Physico-chemical, sensory and volatile profiles of biscuits enriched with grape marc extract. *Food Research International, 65*, 385–393.

Pataro, G., Carullo, D., Siddique, M. A. B., Falcone, M., Donsì, F., & Ferrari, G. (2018). Improved extractability of carotenoids from tomato peels as side benefits of PEF treatment of tomato fruit for more energy-efficient steam-assisted peeling. *Journal of Food Engineering, 233*, 65–73.

Pereira, C., Barros, L., & Ferreira, I. C. F. R. (2017). Dietary supplements: Foods, medicines, or both? A controversial designation with unspecific legislation. *Current Pharmaceutical Design, 23*, 2722–2730.

Pereira, V. A., Jr., de Arruda, I. N. Q., & Stefani, R. (2015). Active chitosan/PVA films with anthocyanins from *Brassica oleraceae* (Red Cabbage) as time–temperature indicators for application in intelligent food packaging. *Food Hydrocolloids, 43*, 180–188.

Pires, T. C., Dias, M. I., Barros, L., Barreira, J. C., Santos-Buelga, C., & Ferreira, I. C. F. R. (2018). Incorporation of natural colorants obtained from edible flowers in yogurts. *LWT-Food Science and Technology, 97*, 668–675.

Polturak, G., & Aharoni, A. (2018). "La Vie en Rose": Biosynthesis, sources, and applications of betalain pigments. *Molecular Plant, 11*, 7–22.

Raddatz-Mota, D., Pérez-Flores, L. J., Carrari, F., Mendoza-Espinoza, J. A., de León-Sánchez, F. D., Pinzón-López, L. L., et al. (2017). Achiote (*Bixa orellana* L.): A natural source of pigment and vitamin E. *Journal of Food Science and Technology, 54*, 1729–1741.

Rodriguez-Amaya, D. B. (2018). Natural food pigments and colorants. In *Bioactive molecules in food* (pp. 1–35). Cham: Springer.

Roriz, C. L., Barreira, J. C., Morales, P., Barros, L., & Ferreira, I. C. F. R. (2018). *Gomphrena globosa* L. as a novel source of food-grade betacyanins: Incorporation in ice-cream and comparison with beet-root extracts and commercial betalains. *LWT-Food Science and Technology, 92*, 101–107.

Saghiri, M. A., Asatourian, A., Ershadifar, S., Moghadam, M. M., & Sheibani, N. (2017). Vitamins and regulation of angiogenesis: [A, B1, B2, B3, B6, B9, B12, C, D, E, K]. *Journal of Functional Foods, 38*, 180–196.

Samyor, D., Das, A. B., & Deka, S. C. (2017). Pigmented rice a potential source of bioactive compounds: A review. *International Journal of Food Science & Technology, 52*, 1073–1081.

Shukla, V., Kandeepan, G., Vishnuraj, M. R., & Soni, A. (2016). Anthocyanins based indicator sensor for intelligent packaging application. *Agricultural Research*, *5*, 205–209.
Silva, P., Ferreira, S., & Nunes, F. M. (2017). Elderberry (*Sambucus nigra* L.) by-products a source of anthocyanins and antioxidant polyphenols. *Industrial Crops and Products*, *95*, 227–234.
Soares, A., Carrascosa, C., & Raposo, A. (2017). Influence of different cooking methods on the concentration of glucosinolates and vitamin C in broccoli. *Food and Bioprocess Technology*, *10*, 1387–1411.
Sogvar, O. B., Saba, M. K., & Emamifar, A. (2016). Aloe vera and ascorbic acid coatings maintain postharvest quality and reduce microbial load of strawberry fruit. *Postharvest Biology and Technology*, *114*, 29–35.
Stafussa, A. P., Maciel, G. M., da Silva Anthero, A. G., da Silva, M. V., Zielinski, A. A. F., & Haminiuk, C. W. I. (2016). Biosorption of anthocyanins from grape pomace extracts by waste yeast: Kinetic and isotherm studies. *Journal of Food Engineering*, *169*, 53–60.
Stoll, L., Costa, T. M. H., Jablonski, A., Flôres, S. H., & de Oliveira Rios, A. (2016). Microencapsulation of anthocyanins with different wall materials and its application in active biodegradable films. *Food and Bioprocess Technology*, *9*, 172–181.
Strati, I. F., Gogou, E., & Oreopoulou, V. (2015). Enzyme and high pressure assisted extraction of carotenoids from tomato waste. *Food and Bioproducts Processing*, *94*, 668–674.
Sumczynski, D., Koubová, E., Šenkárová, L., & Orsavová, J. (2018). Rice flakes produced from commercial wild rice: Chemical compositions, vitamin B compounds, mineral and trace element contents and their dietary intake evaluation. *Food Chemistry*, *264*, 386–392.
Talbot-Walsh, G., Kannar, D., & Selomulya, C. (2018). A review on technological parameters and recent advances in the fortification of processed cheese. *Trends in Food Science & Technology*, *81*, 193–202.
Trouillas, P., Sancho-García, J. C., De Freitas, V., Gierschner, J., Otyepka, M., & Dangles, O. (2016). Stabilizing and modulating color by copigmentation: Insights from theory and experiment. *Chemical Reviews*, *116*, 4937–4982.
Tumbas Šaponjac, V., Gironés-Vilaplana, A., Djilas, S., Mena, P., Ćetković, G., Moreno, D. A., et al. (2014). Anthocyanin profiles and biological properties of caneberry (*Rubus* spp.) press residues. *Journal of the Science of Food and Agriculture*, *94*, 2393–2400.
Turturică, M., Oancea, A., Râpeanu, G., & Bahrim, G. (2015). Anthocyanins: Naturally occurring fruit pigments with functional properties. *Annals of the University Dunarea de Jos of Galati Fascicle VI—Food Technology*, *39*, 9–24.
Ursache, F. M., Andronoiu, D. G., Ghinea, I. O., Barbu, V., Ioniţă, E., Cotârleţ, M., et al. (2018). Valorizations of carotenoids from sea buckthorn extract by microencapsulation and formulation of value-added food products. *Journal of Food Engineering*, *219*, 16–24.
Watanabe, F., Yabuta, Y., Tanioka, Y., & Bito, T. (2013). Biologically active vitamin B12 compounds in foods for preventing deficiency among vegetarians and elderly subjects. *Journal of Agricultural and Food Chemistry*, *61*, 6769–6775.
Yang, Z., Laillou, A., Smith, G., Schofield, D., & Moench-Pfanner, R. (2013). A review of vitamin D fortification: Implications for nutrition programming in Southeast Asia. *Food and Nutrition Bulletin*, *34*, S81–S89.
Yeh, E. B., Barbano, D. M., & Drake, M. (2017). Vitamin fortification of fluid milk. *Journal of Food Science*, *82*, 856–864.
Yousuf, B., Gul, K., Wani, A. A., & Singh, P. (2016). Health benefits of anthocyanins and their encapsulation for potential use in food systems: A review. *Critical Reviews in Food Science and Nutrition*, *56*, 2223–2230.
Zanoletti, M., Parizad, P. A., Lavelli, V., Cecchini, C., Menesatti, P., Marti, A., et al. (2017). Debranning of purple wheat: Recovery of anthocyanin-rich fractions and their use in pasta production. *LWT-Food Science and Technology*, *75*, 663–669.

Zhang, Y., Butelli, E., & Martin, C. (2014). Engineering anthocyanin biosynthesis in plants. *Current Opinion in Plant Biology*, *19*, 81–90.
Zhao, C. L., Yu, Y. Q., Chen, Z. J., Wen, G. S., Wei, F. G., Zheng, Q., et al. (2017). Stability-increasing effects of anthocyanin glycosyl acylation. *Food Chemistry*, *214*, 119–128.
Zhou, Y., Zhao, X., & Huang, H. (2015). Effects of pulsed electric fields on anthocyanin extraction yield of blueberry processing by-products. *Journal of Food Processing and Preservation*, *39*, 1898–1904.

Further reading

Altuner, E. M., & Tokuşoğlu, Ö. (2013). The effect of high hydrostatic pressure processing on the extraction, retention and stability of anthocyanins and flavonols contents of berry fruits and berry juices. *International Journal of Food Science & Technology*, *48*, 1991–1997.
Cavalcanti, R. N., Santos, D. T., & Meireles, M. A. A. (2011). Non-thermal stabilization mechanisms of anthocyanins in model and food systems—An overview. *Food Research International*, *44*, 499–509.
Fang, J. (2015). Classification of fruits based on anthocyanin types and relevance to their health effects. *Nutrition*, *31*, 1301–1306.
Flores, F. P., Singh, R. K., & Kong, F. (2016). Anthocyanin extraction, microencapsulation, and release properties during *in vitro* digestion. *Food Reviews International*, *32*, 46–67.
Li, X., Ma, H., Huang, H., Li, D., & Yao, S. (2013). Natural anthocyanins from phytoresources and their chemical researches. *Natural Product Research*, *27*, 456–469.
Li, D., Wang, P., Luo, Y., Zhao, M., & Chen, F. (2017). Health benefits of anthocyanins and molecular mechanisms: Update from recent decade. *Critical Reviews in Food Science and Nutrition*, *57*, 1729–1741.
Passeri, V., Koes, R., & Quattrocchio, F. M. (2016). New challenges for the design of high value plant products: Stabilization of anthocyanins in plant vacuoles. *Frontiers in Plant Science*, *7*, 153.
Robert, P., & Fredes, C. (2015). The encapsulation of anthocyanins from berry-type fruits. trends in foods. *Molecules*, *20*, 5875–5888.

CHAPTER SIX

Glucosinolates: Molecular structure, breakdown, genetic, bioavailability, properties and healthy and adverse effects

M.A. Prieto[a,b], Cecilia Jiménez López[a,b], Jesus Simal-Gandara[a,*]
[a]Nutrition and Bromatology Group, Department of Analytical and Food Chemistry, Faculty of Food Science and Technology, University of Vigo—Ourense Campus, Ourense, Spain
[b]Nutrition and Food Science Group, Department of Analytical and Food Chemistry, CITACA, CACTI, University of Vigo—Vigo Campus, Vigo, Spain
*Corresponding author: e-mail address: jsimal@uvigo.es

Contents

Advances in Food and Nutrition Research, Volume 90
ISSN 1043-4526
https://doi.org/10.1016/bs.afnr.2019.02.008

Abstract

Glucosinolates are a large group of plant secondary metabolites with nutritional effects and biologically active compounds. Glucosinolates are mainly found in cruciferous plants such as Brassicaceae family, including common edible plants such as broccoli (*Brassica oleracea* var. *italica*), cabbage (*B. oleracea* var. *capitata f. alba*), cauliflower (*B. oleracea* var. *botrytis*), rapeseed (*Brassica napus*), mustard (*Brassica nigra*), and horseradish (*Armoracia rusticana*). If cruciferous plants are consumed without processing, myrosinase enzyme will hydrolyze the glucosinolates to various metabolites, such as isothiocyanates, nitriles, oxazolidine-2-thiones, and indole-3-carbinols. On the other hand, when cruciferous are cooked before consumption, myrosinase is inactivated and glucosinolates could be partially absorbed in their intact form through the gastrointestinal mucosa. This review paper summarizes the glucosinolate molecular breakdown, their genetic aspects from biosynthesis to precursors, their bioavailability (assimilation, absorption, and elimination of these molecules), their sensory properties, identified healthy and adverse effects, as well as the impact of processing on their bioavailability.

1. Glucosinolate molecular breakdown

1.1 Glucosinolate molecular structure

Glucosinolates (GSLs) or mustard oil glucosides are secondary metabolites synthesized by numerous species in the order Capparales, which includes agriculturally important crop plants of the Brassicaceae family (also known as cruciferous, because of the shape arrangement of the four petals of the flower) (Barba et al., 2016; Bell & Wagstaff, 2014, 2017; Wittstock & Halkier, 2002). GSLs are anions formed in a generic chemical structure (Fig. 1) by thiohydroximate-*O*-sulfonate group linked to glucose, and an alkyl, aralkyl, or indolyl side chain (R) (Barba et al., 2016).

The first glucosinolate structures to be elucidated were the structure of sinigrin (2-propenyl) (SIN) and sinalbin in 1956 (Ettlinger & Lundeen, 1956),

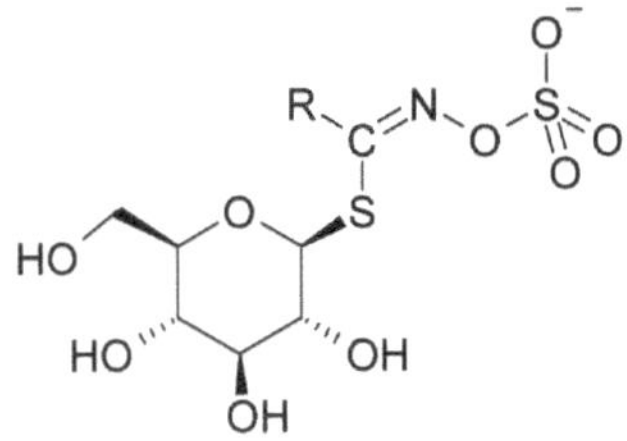

Fig. 1 Generic structure diagram of a GSL (the side group R varies). *Adapted from Redovniković, I. R., Glivetić, T., Delonga, K., & Vorkapić-Furač, J. (2008). Glucosinolates and their potential role in plant.* Periodicum Biologorum, 110*(4), 297–309.*

and the GSL term was used in 1968 (Ettlinger & Kjaer, 1968). Until now, >200 side-groups have been identified and cited in literature (Barba et al., 2016; Redovniković et al., 2008). The high number of glucosinolates is due to side chain modification elongation of the amino acid precursors prior to the formation of the glucosinolate core structure and from a wide range of secondary modifications, including oxidation, desaturation, hydroxylation, methoxylation, sulfation, and glucosylation (Agerbirk & Olsen, 2012), as well as substitutions with acyl conjugation on the sugar moieties. The R chain is derived from one of eight amino acids and can be aliphatic (alanine, leucine, isoleucine, methionine, or valine), aromatic (phenylalanine or tyrosine), or indole (tryptophan) (Redovniković et al., 2008; Wittstock & Halkier, 2002). Glucosinolates may be classified into subgroups according to many criteria. Fig. 2 shows a representative selection of well-known glucosinolate structures.

GLSs are prevalent throughout 15 botanical families of the order Capparales, such as the Brassicaceae, Capparaceae, and Resedaceae. The

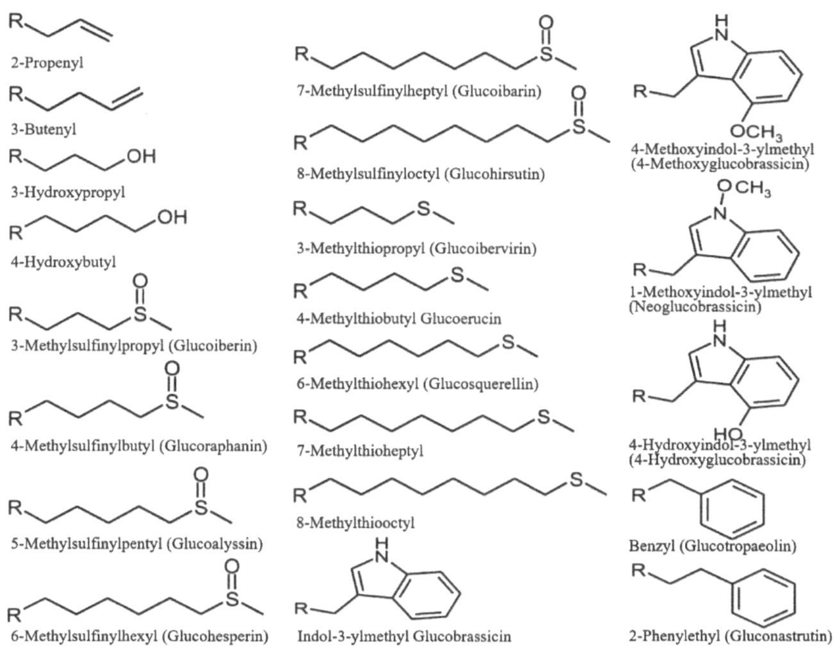

Fig. 2 Representative side chain structure of some GSLs known to date. R denotes the general structure of GSL. Common names, when available, are presented between brackets. *Adapted from Redovniković, I. R., Glivetić, T., Delonga, K., & Vorkapić-Furač, J. (2008). Glucosinolates and their potential role in plant.* Periodicum Biologorum, 110*(4), 297–309.*

majority of plants that contain GSLs belong to the family of Brassicaceae, that comprise >350 genera and 3000 species and are the most representative for the human diet. The "simplest" member of this family is the thale cress (*Arabidopsis thaliana*), the most extensively studied model organism in plant genetics, and the first plant to have its entire genome sequenced. The most commonly consumed edible plants from the Brassicaceae family include the vegetables (e.g., cabbage, broccoli, cauliflower, brussels sprouts), root vegetables (e.g., radish, turnip, swede), leaf vegetables (e.g., rocket salad), and relishes (e.g., wasabi, mustard) (Holst & Williamson, 2004). The content of GSLs can be low to moderate in foliage, ranging from 1000 ppm in some plants, up to 3000 ppm in Brussels sprouts. Concentrations of GSLs in roots and seeds can be higher, up to 30,000 ppm in horseradish root (*Armoracia rusticana* G. Gaertn., B. Mey. & Scherb.) and 60,000 ppm in mustard seed (*Brassica nigra* L.) (Agerbirk & Olsen, 2012).

1.2 Glucosinolate molecular breakdown

GSLs are stable molecular structures in plant cell and they are generally considered as non-toxic compounds. However, once the plant part comprising the glucosinolates fraction is broken (chewing, heating, or insect attack), a β-thioglucosidase (called myrosinase) is discharged (Wittstock & Halkier, 2002). Upon the tissue damage, myrosinases breakdown GSLs producing β-D-glucose and unstable aglucone (thiohydroximate-O-sulfonate). This last one can be reorganized in a variety of biologically active and/or toxic molecules (Fig. 3). GSLs occur throughout the tissues of all plant organs, whereas myrosinases are confined in scattered myrosin cells (expressed on the external surface of the plant cell wall), that appears to be GSL free. The enzyme is normally stored separately from GSLs in different cells, or in different intracellular compartments, depending on the plant species. The glucosinolate–myrosinase system provides plants with an effective defense system against herbivores and pathogens (Redovniković et al., 2008). They have different biological effects, ranging from antimicrobial and cancer-preventing to inflammatory and goitrogenic activities, and thus vegetables consumed by higher animals and humans have toxic as well as protective properties. The dual roles of glucosinolates and their degradation products as deterrents against generalist herbivores and as attractants to insects that are specialized feeders on glucosinolate-containing plants have been reviewed previously (Wittstock & Halkier, 2002).

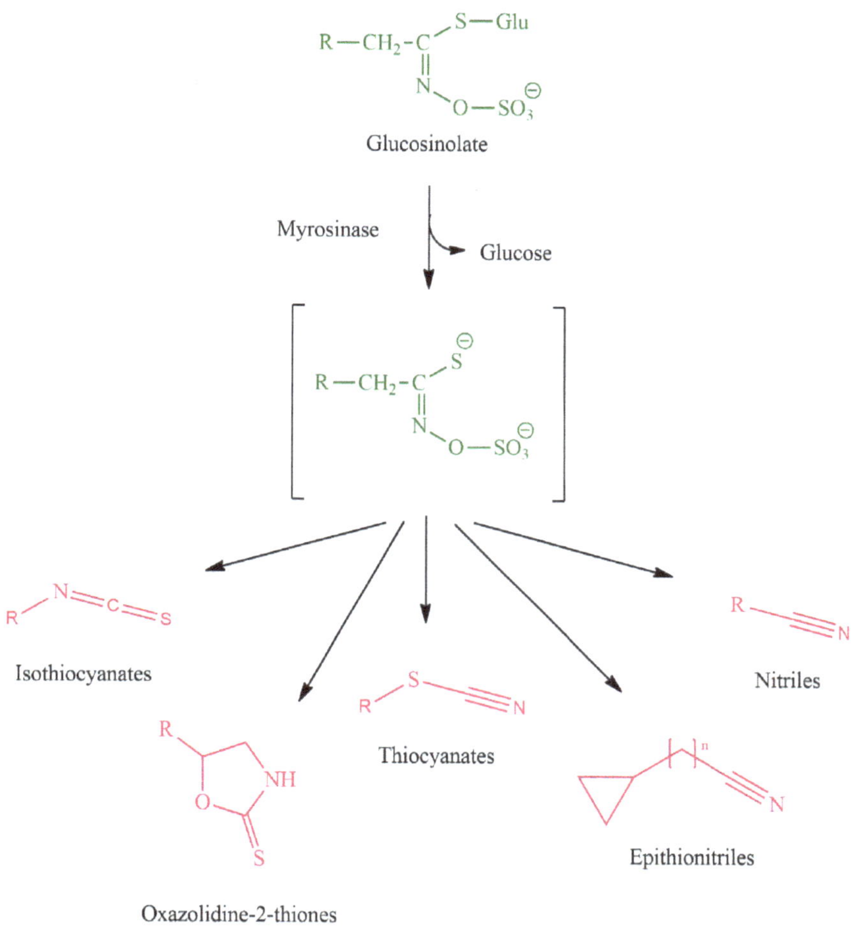

Fig. 3 Structure of possible GSL degradation products after enzymatic hydrolysis and their breakdown products. GSL structures are shown in green, rearrangement upon hydrolysis is shown in pink. Abbreviation: R, variable side chain. *Adapted from Redovniković, I. R., Glivetić, T., Delonga, K., & Vorkapić-Furač, J. (2008). Glucosinolates and their potential role in plant.* Periodicum Biologorum, 110*(4), 297–309 and Wittstock, U., & Halkier, B. A. (2002). Glucosinolate research in the Arabidopsis era.* Trends in Plant Science, *7(6), 263–270.*

The spontaneous reorganization of the unstable aglycone (chemical rearrangement of Lossen) (Fig. 3) results in the release of sulfate ion and in the formation of metabolites, the structures of which depend on the nature of the R chain of GSL, and the physicochemical conditions of the medium such as pH, the presence of ferrous ions (Fe^{2+}) and the presence or absence of protein factors such as epithiospecifier proteins (EPSs) (Fig. 3). The glucosinolate–myrosinase system (Barba et al., 2016; Sønderby, Geu-Flores, & Halkier, 2010; Wittstock & Halkier, 2002), once the damage of tissue is produced, can suffer different chemical structure processes depending on the factors described, as follows:

(1) At neutral pH favors the unstable aglycone rearranges to its isothiocyanates (ITCs) form. Most of the dietary ITCs absorbed by mammals from ingested plant material are formed by the action of myrosinase originating from the gastrointestinal bacteria. ITCs are highly reactive and present potent in vivo action as inducers of phase II enzymes (Barba et al., 2016). Numerous previous studies also reported their action as inhibitors of mitosis and stimulator of the apoptosis in human tumor cells. ITCs revealed also fungicidal, fungistatic, nematicidal, and bactericidal activities (Barba et al., 2016; Sønderby et al., 2010).

(2) If the GSL side chain is hydroxylated at carbon 3, spontaneous cyclization of the isothiocyanate results in the formation of an oxazolidine-2-thione.

(3) In the presence of an EPS nitriles are formed, normally favored at low pH ($pH < 3$). Nitriles might be directed against other pests or might attract natural herbivores opponents.

(4) If there is a terminal double bond in the side chain, the sulfur atom released during nitrile formation is captured by the double bond, resulting in the formation of epithionitriles.

(5) Some GSLs can be hydrolyzed to thiocyanates.

Modifications of the GSL R chain are of particular significant, because the physicochemical features and the biological relevance of the GSL degradation products are determined by the structure of the R chain. The biological properties related with GSLs and their derived products, particularly ITCs, is important to be comprehended, because the absorption routes of these molecules and their metabolism, if present, need to be taken into account in the processing parameters of food products (Rajan et al., 2016; Wu, Zhou, & Xu, 2009). For instance, the products formed are responsible for the characteristic flavor of Brassicaceous vegetables, but also their potential biological activity. This multiple set of parameters affecting the outcome of the hydrolysis gives rise to a complex profile of hydrolysis products (Holst & Williamson, 2004).

2. Genetic aspects of glucosinolates

Despite the great interest aroused by GSLs, due to their possible uses as enhancers of the defense mechanisms of crop plants (agricultural products) in different situations of stress, and despite the growing information that is being gathered about them, their extensive variety (>200 different structures of GSLs are known) makes it difficult to decrypt the biosynthesis mechanism of each of them completely (Frerigmann & Gigolashvili, 2014).

2.1 Glucosinolate biosynthesis

Initially, *Arabidopsis thaliana* (belonging to *Cruciferae* family) was chosen as a starting point for the study of the possible biosynthetic pathways of these compounds, due to its short genome and short life cycle (Redovniković, Textor, Lisnić, & Gershenzon, 2012), as well as the success obtained in previous trials about the place of synthesis and storage, and the cellular transport methods concerning the GSLs (Halkier, 2016). The availability of the *Arabidopsis* genome sequence has enabled functional genomics approaches and greatly facilitated quantitative trait locus (QTL) mapping to identify genes involved in GSL biosynthesis (Wittstock & Halkier, 2002). This plant is able to synthesize approximately 40 different types of GSLs, mainly derivatives of methionine and tryptophan, through the analysis of which the three basic steps that make up the general path of biosynthesis were elucidated: (a) elongation of the side chain, what means the production of the R group from amino acids, although this phase only occurs if the amino acids (aa) are methionine or phenylalanine, otherwise the aa does not need previous elongation; (b) production of the core GSL structure, by the addition of glucose and sulfur (Fig. 4); and (c) modification of the side chain, to give rise to the different derivatives that exist (Halkier & Du, 1997; Sønderby et al., 2010).

The synthesis occurs mainly in the cellular cytosol, with the participation of the chloroplasts in some reactions of steps (a) and (b). Thanks to the advance of biochemistry, as well as the analytical instruments and the techniques of using genetic markers, most of the enzymes involved in the different modifications that take place in this synthesis chain have been identified (Wang et al., 2011), as well as the genes that codify the information necessary for the synthesis of those catalyst proteins; however, because there are three large groups of GSLs (aliphatic, aromatic and indolic)

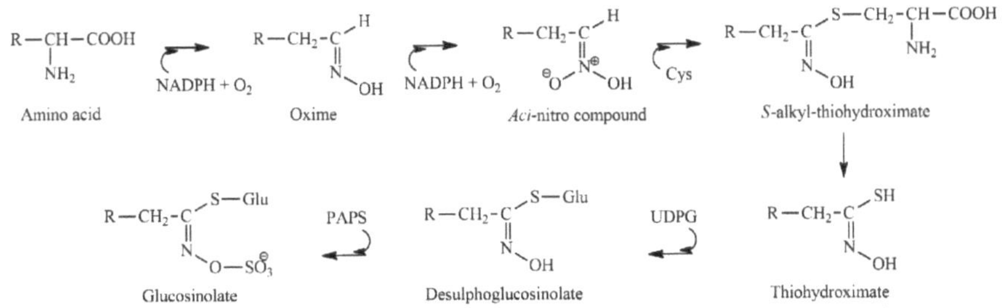

Fig. 4 Simplified general scheme of the formation of the core GSL structure. *Adapted from Halkier, B. A., & Du, L. (1997). The biosynthesis of glucosinolates.* Trends in Plant Science, *2(11), 425–431.*

(Frerigmann & Gigolashvili, 2014), the totality of genes, enzymes, and transcription factors involved in the synthesis of each type of them is variable, so some of them were assigned to the reactions by prediction, and others still remain unknown nowadays (Sønderby et al., 2010) (Fig. 5).

2.2 Genetic aspects

In the elongation phase (a), and in the case of methionine as precursor, methylthioalkylmalate synthase (MAM), bile acid-sodium symporter family protein 5 (BASS5), and branched-chain aminotransferases (BCATs) are involved (Sawada et al., 2009; Textor et al., 2007). Core GSL structure formation (b) takes place through oxidative decarboxylation mechanisms rolled by cytochromes P450 of CYP79 and CYP83, followed by C–S lyase, S-glucosyltransferase and sulfotransferase (Wittstock & Halkier, 2002). Coming up next, some loci such as GS-OX, GS-AOP, GS-OH, BZO1, and CYP81F2 are responsible of secondary modifications (c), which produce four derivatives in the case of indolic GSLs; and up to 12 in the case of aliphatic GSLs that come from methionine (Sønderby et al., 2010). In addition, some nuclear-localized regulators and R2R3-Myb transcription factors take part in glucosinolate biosynthesis and its regulation (Chun et al., 2018; Frerigmann & Gigolashvili, 2014; Gigolashvili, Berger, et al., 2007; Gigolashvili, Yatusevich, et al., 2007; Gigolashvili et al., 2008; Skirycz et al., 2006). There are several genes that also participate in co-substrate formation steps (Sønderby et al., 2010). Table 1 collects a summary of the genes and transcription factors known to date that encode the information necessary for the synthesis of GSLs.

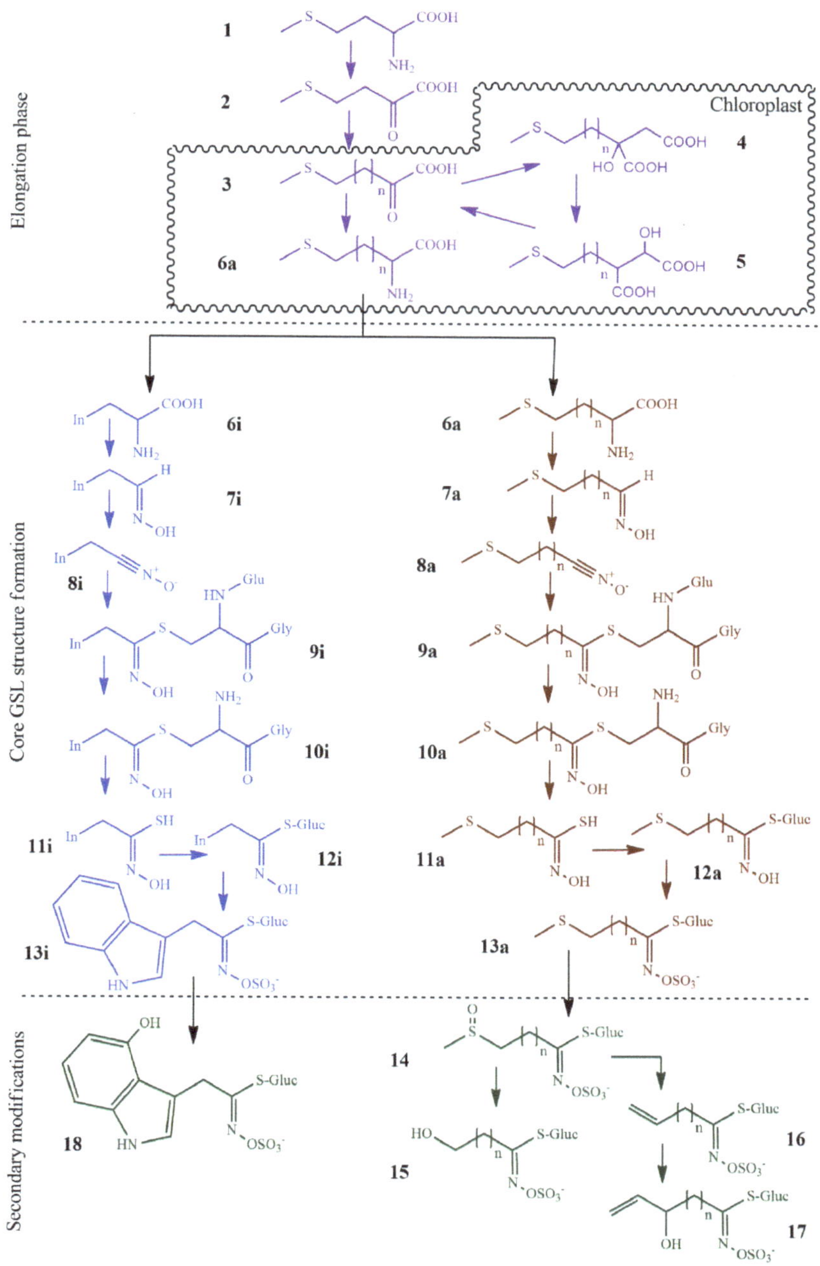

Fig. 5 General scheme of biosynthesis of aliphatic and indolic GSLs. *Adapted from Sønderby, I. E., Geu-Flores, F., & Halkier, B. A. (2010). Biosynthesis of glucosinolates—Gene discovery and beyond.* Trends in Plant Science, 15*(5), 283–290.*

Table 1 Inventory of transcription factors and genes involved in the GSLs biosynthesis.

Name	Other names	Reaction[a]	References
The aliphatic pathway			
BCAT4	MAAT-cytosol	1→2	Schuster et al. (2006)
BAT5[b]	BASS5	2→3	Sawada et al. (2009)
MAM1		3→4	Field et al. (2004); Textor et al. (2007)
MAM2		3→4	Benderoth et al. (2006); Kroymann, Donnerhacke, Schnabelrauch, and Mitchell-Olds (2003)
MAM3	MAM-L	3→4	Field et al. (2004); Textor et al. (2007)
IPMI LSU1	Aconitase, IPM-I/IPM-DHT, MAM-IL, IIL1, IPMI-L1, AtLeuC1	4→5	Knill et al. (2009); Wentzell et al. (2007)
IPMI SSU2[b]	Aconitase, AtLeuD1, IPMI2, MAM-IS, IPMI-S2	4→5	Knill et al. (2009); Wentzell et al. (2007)
IPMI SSU3[b]	Aconitase, AtLeuD2, IPMI1, MAM-IS, IPMI-S1	4→5	Knill et al. (2009); Wentzell et al. (2007)
BCAT3	MAAT-chloroplast	3→6	Knill et al. (2008)
CYP79F1	BUS1, SUPERSHOOT1, BUSHY1	6→7	Chen et al. (2003); Hansen et al. (2001)
CYP79F2		6→7	Chen et al. (2003); Hansen et al. (2001)
CYP83A1	REF2	7→8	Hemm, Ruegger, and Chapple (2003)
GSTF11[c]		8→9	Hirai et al. (2005); Wentzell et al. (2007)
GSTU20[c]		8→9	Hirai et al. (2005)
GGP1[b]		9→10	Geu-Flores et al. (2009)
SUR1	ALF1, HOOKLESS3, RTY1, C-S lyase	10→11	Mikkelsen, Hansen, Wittstock, and Halkier (2000)

Table 1 Inventory of transcription factors and genes involved in the GSLs biosynthesis.—cont'd

Name	Other names	Reaction[a]	References
UGT74C1[c]		11→12	Gachon, Langlois-Meurinne, Henry, and Saindrenan (2005)
SOT18	AtSTb	12→13	Hirai et al. (2005); Piotrowski et al. (2004)
SOT17	AtSTc	12→13	Piotrowski et al. (2004)
FMO-GSOX1		13→14	Hansen, Kliebenstein, and Halkier (2007)
FMO-GSOX2		13→14	Li et al. (2008); Wentzell et al. (2007)
FMO-GSOX3		13→14	Li et al. (2008); Wentzell et al. (2007)
FMO-GSOX4		13→14	Li et al. (2008); Wentzell et al. (2007)
FMO-GSOX5		13→14	Li et al. (2008)
AOP3		14→15	Kliebenstein (2001)
AOP2		14→16	Kliebenstein (2001)
GS-OH		16→17	Wentzell et al. (2007)
The indolic and benzenic pathways			
CYP79A2		6→7	Wittstock and Halkier (2000)
CYP79B2		6→7	Mikkelsen et al. (2000)
CYP79B3		6→7	Mikkelsen et al. (2000)
CYP83B1	SUR2	7→8	Naur et al. (2003)
GSTF9[c]		8→9	Wentzell et al. (2007)
GSTF10[c]		8→9	Wentzell et al. (2007)
GGP1[b]		9→10	Geu-Flores et al. (2009)
SUR1	C-S lyase	10→11	Mikkelsen, Naur, and Halkier (2004)
UGT74B1		11→12	Grubb et al. (2004)
SOT16	AtSTa	12→13	Piotrowski et al. (2004)
CYP81F2		13→18	Clay et al. (2009)

Continued

Table 1 Inventory of transcription factors and genes involved in the GSLs biosynthesis.—cont'd

Name	Other names	Reaction[a]	References
Transcription factors			
Dof1.1			Skirycz et al. (2006)
IQD1-1			Levy et al. (2005)
MYB28			Gigolashvili, Yatusevich, et al. (2007); Hirai et al. (2007)
MYB29			Gigolashvili et al. (2008); Hirai et al. (2007)
MYB34			Celenza et al. (2005)
MYB51			Gigolashvili, Berger, et al. (2007)
MYB76			Gigolashvili et al. (2008)
MYB122			Gigolashvili, Berger, et al. (2007)

[a]Numbers in this column refer to the numbered compounds in Fig. 5.
[b]Partially characterized enzyme.
[c]Predicted enzyme.
Adapted from Sønderby, I. E., Geu-Flores, F., & Halkier, B. A. (2010). Biosynthesis of glucosinolates—Gene discovery and beyond. *Trends in Plant Science, 15*(5), 283–290 and Wang, H., et al. (2011). Glucosinolate biosynthetic genes in Brassica rapa. *Gene, 487*(2), 135–142.

2.3 Complementary trials

Subsequently, other plants belonging to *Cruciferae* family have been studied with the intention of comparing and going deeper into the genome relative to the biosynthesis of GSLs. In the case of *Brassica rapa* (Chinese cabbage), 13 GSLs were identified and characterized, whose biosynthetic information was found in 102 genes, as orthologs of the 52 in *A. thaliana*; most of them present more than one copy, and they were present in 10 chromosomes. A high co-linearity was established between the synthetic pathways of both species, finding out that 93% of GSLs genes present similarity between *B. rapa* and *A. thaliana* (Wang et al., 2011). Another study carried out on *Eruca sativa* Mill. (arugula) shows that there is a high similarity (82–95%) in the sequence of homologous genes of this plant and other species of the Brassicaceae family. In fact, they determined that the genes present in *E. sativa* and in *Brassica* sp. are more phylogenetically similar to each other than they are to the corresponding *Arabidopsis* sequences (Katsarou et al., 2016).

In addition, as in the vast majority of organisms, there are factors that influence the expression of the genes responsible for GSL biosynthesis, such as the temperature at which plant growth occurs, resulting that moderate temperatures (15–21 °C) favor the production of GSLs, particularly of aliphatic GSLs (Kissen et al., 2016); the subspecies of the plants in question (Chun et al., 2018; Wang et al., 2012); the gamma radiation to which agricultural products are often subjected with conservation function, which, interestingly, favors the content of aliphatic GSLs (Banerjee, Rai, Penna, & Variyar, 2016); the plant organ where they are produced, as it has been demonstrated that smaller quantities are found in phloem, flowers and fruits (Redovniković et al., 2012), or the amount of N and S available to the plant, being normally proportional to the production of GSLs (Katsarou et al., 2016).

Future research focused on the regulation of GSLs biosynthesis in response to signaling molecules, turnover and translocation of GSLs in the plant, and the role of GSLs in plant–insect and plant–pathogen interactions is needed. With the increment of the information relative to genes and their regulation involved in the several different steps of GSLs biosynthesis, the improvement of nutritional quality and pest resistance of crop plants by genetic engineering of GSLs profiles is now a realistic possibility (Wittstock & Halkier, 2002).

3. Bioavailability of glucosinolates

To describe the concentration of a given compound or its metabolite at a target site, the term bioavailability was defined by the Food and Drug Administration (FDA) as "the rate and extent to which a therapeutic moiety is absorbed and becomes available to the site of drug action." When it comes to the bioavailability of a substance that does not need to be absorbed into the bloodstream, it is simply defined as "the rate and extent to which the active moiety becomes available at the site of action" (Chen et al., 2001).

Some biological properties have been associated with GSLs and their breakdown products, especially ITCs, for being the major hydrolysis product at physiological pH (Song, Morrison, Botting, & Thornalley, 2005). Due to that fact, understanding the absorption routes of these molecules and their metabolism is of great importance, but, compared to the existing knowledge on many dietary bioactive compounds, there are little data available on how, where and to what extent GLSs and their hydrolysis products are liberated, absorbed, distributed, metabolized and excreted in humans. In fact, most of the data have been derived from in vitro and animal studies (Barba et al., 2016).

Bioavailability of GSLs, or rather their bioactive hydrolysis products, is affected by numerous exogenous and endogenous parameters. It depends strongly on the:

- Nature of the plant material;
- Concentration of GSLs and their hydrolytic products in the plant material (Fernández-León et al., 2017);
- Concentration and stability of myrosinase in the plant material;
- Hydrolysis during storage and processing of the plant material;
- Particular solubility, stability and physicochemical characteristics of each GSL or derivative;
- The extent of cell disruption during mastication;
- Gastric digestion and small intestinal processes (Fernández-León et al., 2017);
- Colonic microbiota fermentation (Holst & Williamson, 2004).

3.1 Absorption in the human digestive tract

Once the GSLs are ingested, the absorption of a little portion of intact GSLs can occur directly in the stomach, although most go to the small intestine, where, according to some studies, a small fraction of intact GSLs can also be absorbed by the lining of the small intestine (Angelino & Jeffery, 2014; Clarke et al., 2011; Song et al., 2005). In vivo, this absorption results in the presence of native GSLs in urine up to 5% of the ingested dose (Barba et al., 2016; Sørensen et al., 2016). As it has been seen previously, the GSLs are hydrolyzed and suffer breakdown thanks to the action of the myrosinases, but, when cooking vegetables, most of the myrosinase activity is lost due to enzyme denaturalization by thermic treatments. However, the non-absorbed part of GSLs in the proximal gut can be hydrolyzed by the portion of no denatured myrosinase that remains in the plant consumed, and the breakdown products can be absorbed. Although mammalian tissues do not contain myrosinases, conversion of GSLs to ITCs, it still occurs in humans, mediated by the bacterial microflora of the colon (Dinkova-Kostova & Kostov, 2012). It has been proved that these metabolites are formed in the germ-free colon of rats, followed by colonization with human intestinal bacteria, and feeding with a pure GSL. Also,

Bifidobacterium strains from human intestinal microbiota are able to metabolize the GSLs to nitriles in vitro, as it can be seen in the case of *B. longum*, *B. pseudocatenulatum* and *B. adolescentis*, that were able to digest in vitro both GSLs, SIN, and Glucotropaeolin (benzylglucosinolate), causing a reduction in the medium pH. Consequently, the remaining non-hydrolyzed GSLs ingested will then transit to reach the colon where they are hydrolyzed by bacterial myrosinase activity, and the generated ITCs are absorbed or/and excreted (Fig. 6) (Barba et al., 2016; Cartea & Velasco, 2008).

Formation of other hydrolytic products such as nitriles and epithionitriles from GSLs by intestinal microbiota is really probable, but still poorly documented, so it requires more investigation. In addition, the individual diversity of the intestinal bacteria activities is associated with the generation of a wide range of metabolites (Barba et al., 2016).

The use of radiolabeled ITCs in rats indicates rapid absorption with a radioactive blood peak observed 3 h after ingestion (Conaway et al., 1999). However, a study conducted by Ye et al. (2002), in which four healthy non-smoking men were fed with ITC extracts obtained from broccoli sprouts, it is reported that the maximum peak of ITCs in blood is reached approximately 1 h after administration, although they begin to be detected in blood just 15 min after ingestion.

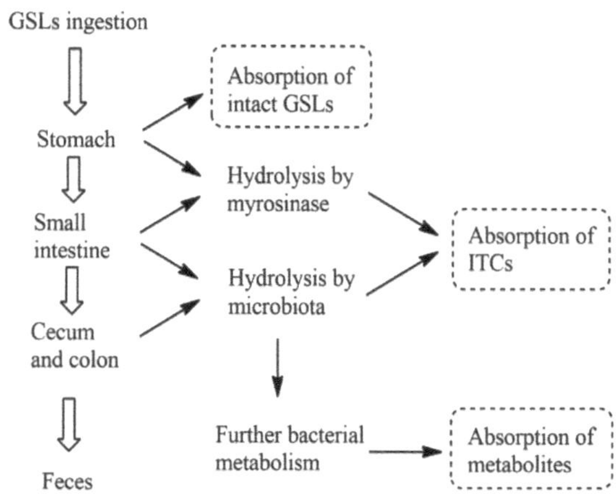

Fig. 6 Scheme of how the remaining non-hydrolyzed GSLs ingested will then transit to reach the colon where they are hydrolyzed by microbiota, which possess myrosinase activity, and the generated ITCs are absorbed or/and excreted. *Figure adapted from Barba, F. J., et al. (2016). Bioavailability of glucosinolates and their breakdown products: Impact of processing.* Frontiers in Nutrition, 3*(August), 1–12.*

3.2 Post-absorptive processes

Data on the distribution of GSL and their hydrolytic products are mostly derived from animal studies and are limited to few selective representatives, usually ITCs. Passage of ITCs across the enterocyte brush border is most likely by passive diffusion; but, once absorbed, ITCs can be secreted back into the gut through membrane-bound transporters, can pass into the plasma by diffusion or by other transporters, or can be metabolized in the enterocyte (Angelino & Jeffery, 2014). In rats fed with radiolabeled ITCs, radioactivity distribution is observed concentrated mainly in the intestinal mucosa, the liver, the kidneys, and bladder, followed by the lungs and spleen. However, the brain and the heart contain very low concentration of radioactivity (Bollard, Stribbling, Mitchell, & Caldwell, 1997; Conaway et al., 1999). More studies on the distribution in humans are required to determine, for example, binding behavior, intestinal membrane permeability, first pass metabolism, and GSL affinity. In addition, based on information on the distribution of glutathione (GSH), a prediction of the distribution of individual compounds might provide a good estimate of the in vivo distribution of GSLs derivatives (Shapiro, Stephenson, Fahey, & Wade, 1998).

To gain insight into the bioavailability of GSLs, specifically in that of ITCs, mercapturic acid has been used as a urinary biomarker, since it is the main elimination product of ITCs in humans generated after conjugation with glutathione, reaching urine values between 12% and 80% of the dose administered. The large variation between these values is mainly due to the amount of myrosinase present in the ingested plant (dependent on the plant in question and its processing), the gut microbiota of each individual and their ability to hydrolyze GSLs, and the structural properties of each GSL molecule (Rouzaud et al., 2004; Shapiro et al., 2001). When vegetables are ingested raw, greater excretion of mercapturic acid (17–88%) is always observed, since myrosinase is responsible for hydrolyzing the GSLs, whereas, if they are previously cooked and the hydrolysis is carried out by the intestinal microbiota, this amount does not exceed 20% (Barba et al., 2016). However, it seems that the frequent intake of *Brassica* sp. vegetables may favor the proliferation of bacteria that hydrolyze GSLs in the intestinal microbiota (Angelino & Jeffery, 2014). Another study conducted by Clarke et al. (2011) reveals that the bioavailability of ITCs in blood and the amount of them recovered in urine, quantified by UHPLC-MS/MS, is much higher when administered broccoli sprouts versus broccoli supplement, where myrosinase is inactivated. Furthermore, blood peak concentration is reached earlier (3 h versus 6 h, respectively), observing that the plasma clearance occurs following a kinetic order of 1, until the baseline level is regained at 24 h of intake (Angelino & Jeffery, 2014; Sørensen et al., 2016). Metabolites are excreted homogeneously suggesting no accumulation (Baenas, Suárez-martínez, García-viguera, & Moreno, 2017). From all this information it is deduced that the bioavailability of ITCs is proportional to the amount of myrosinase present, so it increases with the administration of raw cruciferous (Dinkova-Kostova & Kostov, 2012; Fowke, Fahey, Stephenson, & Hebert, 2001; Šamec, Pavlovi, Radoj, & Salopek-Sondi, 2018).

In other studies, quantifications of the dithiocarbamates present in different human samples, such as plasma or urine, were performed through the cyclocondensation with 1,2-benzenedithiol assay (Shapiro et al., 2001). This analysis enables the detection and quantification of ITCs and metabolites, not only in urine, but also in a variety of samples, including vegetable extracts, blood, cell lysates, and consequently enables pharmacokinetic studies in vivo (Barba et al., 2016; Dinkova-Kostova & Kostov, 2012). Basically, this method involves the addition of 1,2-benzenedithiol to the sample, so that it reacts with ITCs to form a cyclic condensation product, 1,3-benzodithiole-2-thione, which is easily quantifiable by spectroscopy, at 365 nm (Fig. 7). When the ITCs follow their main metabolic route of mercapturic acid, a series of ITCs-conjugates are produced, such as N-acetyl-L-cysteine-ITC, which are dithiocarbamates, so they are detectable by this method (Shapiro et al., 2001). The amount recovered during 8 h of urine collection was 58% of the amount of ITCs administered, so longer trials are needed to determine more accurately the quantity of ITCs metabolized (Ye et al., 2002).

Concerning the other glucosinolate breakdown products, their assimilation by the body is still poorly understood. Similar to ITCs, nitriles and epithionitriles could be metabolized and excreted in the urine as mercapturic acids (Barba et al., 2016). Another issue to consider is the variation of the responses of each individual to different xenobiotics. Therefore, much

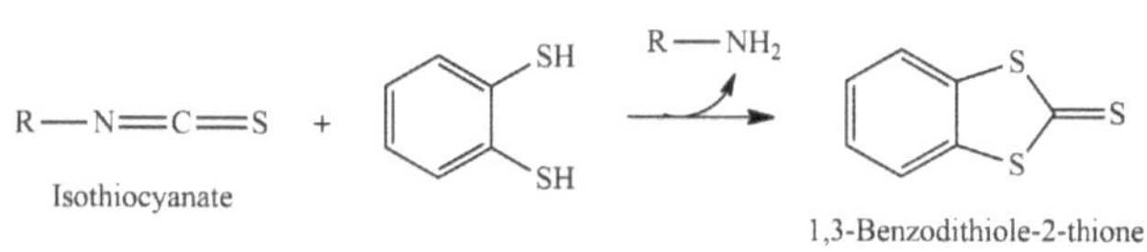

Fig. 7 Cyclocondensation of ITCs with 1,2-benzenedithiol gives rise to 1,3-benzodithiole-2-thione. *Adapted from Ye, L., et al. (2002). Quantitative determination of dithiocarbamates in human plasma, serum, erythrocytes and urine: Pharmacokinetics of broccoli sprout isothiocyanates in humans.* Clinica Chimica Acta, 316, *43–53.*

remains to be elucidated regarding the pharmacokinetics of these phytodrugs associated with such beneficial effects as chemoprevention (Cartea & Velasco, 2008).

4. Metabolism of glucosinolates

4.1 Metabolism in producing plants

As mentioned above, GSLs have a fundamental role in the defense of plants belonging to the order *Brassicales* (Agerbirk et al., 2018), that is why normally their elimination of the biological material is through the fulfillment of its defense function, that is, through the chain of reactions that make possible its breakdown in different active metabolites. Once the stress situation occurs, being it abiotic or induced by any living organism, GSLs and enzymes with β-thioglucosidase glucohydrolase activity (myrosinase) are released from the correspondent vacuoles in which they are stored separately. At this moment, when both substances come into contact, the catalysis of the GSLs begins, producing the hydrolysis of the β-D-glucose and giving rise to an unstable aglycone (thioamide), that derives in different metabolites such as ITCs, thiocyanate anions, nitriles, sulfates, and goitrins, depending on the aglycone structure, pH, ferrous ion concentration, and epithiospecifier proteins (Fig. 3) (Bischoff, 2016; Chen & Andreasson, 2001).

Classical myrosinases of the thioglucosidase group were for many years thought to be the unique enzymes catalyzing this reaction (Ahuja et al., 2016). However, other myrosinases responsible for turnover of glucosinolates in intact plants have been identified in recent years. Some unexpected, non-conventional degradation products have been reported, suggesting a varied and complex metabolism of glucosinolates in intact plants (Agerbirk et al., 2018). There may be other types of non-enzymatic catalysts, as seen in the breakdown of glucobarbarin and progoitrin (2-hydroxy-3-butenyl) (PRO) in the presence of high concentrations of ferrous salts or ferrous ions, where thioamides were detected in high or low amounts, respectively. It proposed ferrous ion as a possible catalyst of the turnover of β-hydroxyalkyl glucosinolates in intact plants, although little is still known about these mechanisms (Bellostas, Sørensen, Sørensen, & Sørensen, 2008). In addition, Agerbirk et al. (2018) suggest in their study the possibility of the formation of other degradation products of GSLs, as resedine, thanks to the initial action of myrosinase, but continuing with other novel catalysts, and without the formation of the thioamidic intermediate (Fig. 8).

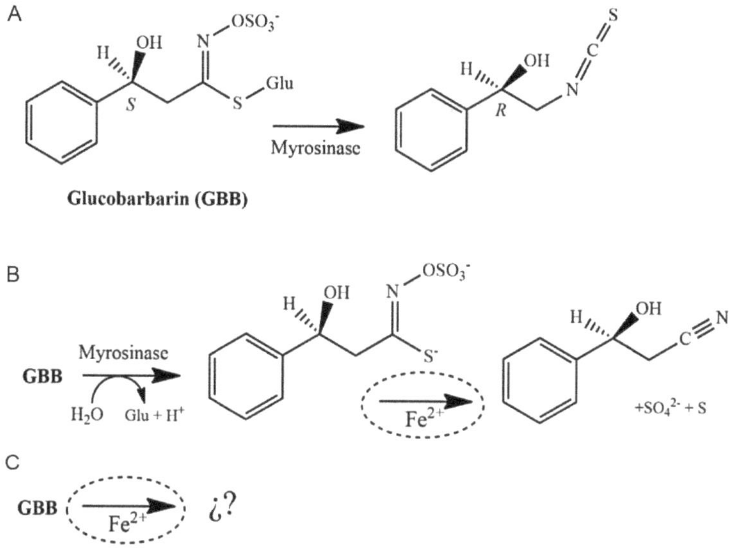

Fig. 8 Possible mechanisms of GSLs turnover. (A) Biochemical conversion of glucobarbarin (GBB) initiated by myrosinase. (B) An alternative of hydrolysis catalyzed by myrosinase, in the presence of ferrous ion. (C) Hypothetic, non-enzymatic conversion of GBB to a thioamide. This reaction has been suggested but has not been confirmed yet. *Adapted from Agerbirk, N., Matthes, A., Erthmann, P. Ø., Ugolini, L., Cinti, S., Lazaridi, E., et al. (2018). Glucosinolate turnover in Brassicales species to an oxazolidin-2-one, formed via the 2-thione and without formation of thioamide.* Phytochemistry, 153, *79–93.*

4.2 Metabolism in consumer organisms

All known natural ITCs are formed by rearrangement of the GSLs aglycone, and, at physiological pH, they are the major product of GSLs hydrolysis (Song et al., 2005). These compounds are regarded as the most toxic and common of the GSLs breakdown derivatives, and can cause delays in insect growth and development, and even death (Agrawal & Kurashige, 2003; Müller et al., 2010). For this reason, several studies have been carried out on the metabolism of said degradation products of GSLs in various herbivores. The reactive —N=C=S group of ITCs causes biological damage due to its high reactivity toward nucleophiles, functioning as an acceptor for thiol or amine side chains of proteins at physiological conditions. In this way, it is covalently bounded to proteins, modifying its tertiary and quaternary structures and triggering loss of functionality (Mi, di Pasqua, & Chung, 2011). However, it seems that they do not react directly with RNA or DNA (Xiao, Mi, Chung, & Veenstra, 2012). In the case of mammals, GSLs and their metabolites are not known to accumulate in muscle, fat, liver, or kidney, and are minimally detected in excreted in milk and eggs, but can cause off flavor.

The predominant metabolic pathway of ITCs is thanks to their electrophilic condition, which makes possible their detoxification by conjugation to the nucleophilic thiol (–SH) group of GSH promoted by glutathione transferases, a so-called phase-II detoxification reaction (Fig. 9) (Traka & Mithen, 2009). This detoxification takes place mainly in the liver and enterocytes, where the conjugated derivatives undergo a series of modifications, producing various intermediate dithiocarbamates, that can be excreted in the urine or complete the route to become a derivative of mercapturic acid (N-acetyl-S-(N-alkylthiocarbamoyl)-L-cysteine), also excreted via the urine and used as a biomarker (Ye et al., 2002). As an alternative route of the metabolism, a study with mice fed with radiolabeled ITCs showed that approximately 15% of the radioactivity was excreted in the respiratory process in the form of CO_2, or as unknown metabolites in the stool. Radioactivity was also detected in the bile, which means that there is circulation of metabolites between liver and intestine (Conaway et al., 1999). The information relative to other derivatives of the hydrolysis of

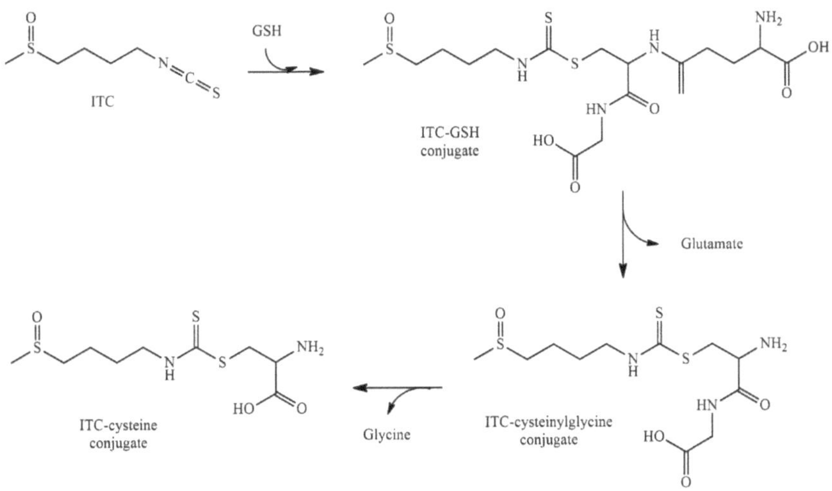

Fig. 9 Phase-II detoxification reaction in order to inactivate ITCs by conjugation to the nucleophilic thiol (–SH) group of glutathione (GSH). *Adapted from Jeschke, V., Gershenzon, J., & Vassão, D. G. (2016). A mode of action of glucosinolate-derived isothiocyanates: Detoxification depletes glutathione and cysteine levels with ramifications on protein metabolism in Spodoptera littoralis.* Insect Biochemistry and Molecular Biology, 71, *37–48.*

GSLs is rather scarce, so it is convenient to deepen the knowledge about them to get the most out of these potentially promising plant metabolites (Barba et al., 2016).

In the study conducted by (Schramm et al., 2012), they fed herbivorous insects of the Lepidoptera species, specifically *Spodoptera littoralis*, with GSLs tagged isotopically, concretely 4-methylsulfinylbutyl glucosinolate, because of being the plant's (*Arabidopsis thaliana*) major GSL. As they report, it was observed that 11% of GSLs are excreted in the form of ITC-GSH conjugate and its cysteinylglycine (CysGly) and cysteine (Cys) derivatives in fecal material. Approximately 66% of the GSLs ingested were excreted as unmodified ITC, but some of this pool was also conjugated to GSH and de-conjugated, re-releasing the free ITC upon passage through the insect. This suggestion arises from the increase of ITC-GSH conjugate excreted that is observed when adding Cys to the diet, a necessary aa for the biosynthesis of GSH (Jeschke et al., 2016). Further analysis of larval feces from several species of generalist lepidopterans (*Spodoptera exigua*, *S. littoralis*, *Mamestra brassicae*, *Trichoplusia ni* and *Helicoverpa armigera*) fed on different Brassicaceae revealed that GSH-, CysGly- and Cys-ITC-conjugates arise from a variety of aliphatic and aromatic ITCs derived from dietary GSLs (Schramm et al., 2012).

Precise understanding of glucosinolate enzymology and metabolons will be necessary for the successful alteration of glucosinolate profiles by metabolic engineering, in order to enhance plant defense and design functional foods, so that the possibility of nutritional cancer-prevention strategies can be contemplated (Grubb & Abel, 2006).

5. Sensory properties of glucosinolates

Once the main function of GSLs has been specified, clarified and explained as phytoprotector secondary metabolites against threats from a wide range of natures, this revision collects other additional properties that these compounds possess, and that contribute to the interest they awaken, since they make GSLs products unique and potentially useful for different uses and in different areas of investigation.

Obviously, the possibility of transforming them into future drugs or possible therapeutic agents is primordial and very striking, but, in addition to this value in medicine, they attract other sectors, such as the food area, since they are responsible for some sensorial characteristics of the vegetables of the Brassicaceae family, such as taste and smell (Redovniković et al., 2008).

Most of the hydrolysis products with volatile properties of GSLs, especially ITCs, produce pungent and bitter taste, as well as a sulfurous aroma when plant tissues are damaged (Rosa, Heaney, Fenwick, & Portas, 1997), that is, the breakdown is produced, unleashed by the GSLs coming into contact with the myrosinase enzymes, because the cells in which they were stored separately undergo a rupture (Fenwick, Heaney, Mullin, & VanEtten, 1983).

In the literature, it is described that the bitter effect is mainly produced by the ITCs obtained from SIN, gluconapin (3-butenyl) (GNA), and PRO, as well as glucobrassicin (3-indolylmethyl) (GBS) and neoglucobrassicin (1-methoxy-3-indolylmethyl) (NGBS), although each of them confers different flavor intensities. On the other hand, the alkyls GSLs, such as glucoerucin (4-methylthiobutyl) (GER), glucoiberverin (3-methylthiopropyl) (GIV), glucoiberin (3-methylsulfinylpropyl) (GIB), and glucoraphanin (4-methylsulfinylbutyl) (GRA), do not have the bitter taste ability attributed (Vig, Rampal, Thind, & Arora, 2009). In addition, other factors, such as the *Brassica* species in question, the cooking process (temperature, technique, time, etc.) or the part of the plant used, immensely influence the organoleptic behavior of the same compound. For example, in Brussels sprouts, ITCs formed from SIN and PRO have been related to bitter flavors (Fenwick et al., 1983; van Doorn et al., 1998), while in cauliflower, once cooked, GSLs NGBS and SIN are responsible for the bitter taste, so they have the ability to directly influence the taste and acceptance of consumers (Engel et al., 2002; Schonhof, Krumbein, & Bruckner, 2004). Another example is provided by cabbage, whose ITCs derived from PRO and gluconasturtiin (2-phenylethyl) (GST) are classified as pungent and intensely bitter (Fenwick et al., 1983; Rosa et al., 1997). Regarding the degree of maturity and the different varieties of the same species, such as turnip greens, it was concluded that, in sensory terms, they had a significant impact, as well as influencing the concentration of other compounds contained in vegetables, that is, in the nutritional composition (Jones & Sanders, 2002). Other factors, such as temperature, storage, the amount of nitrogen available to the plant during growth, affect the flavor, because they also directly influence the amount of GSLs that the plant synthesizes (Cools & Terry, 2018; Groenbaek et al., 2016; Helland et al., 2016; Johansen, Hagen, Bengtsson, & Mølmann, 2016; Mølmann et al., 2015).

Different studies focusing on the analysis of bitter taste respecting to the variety of concentrations of GSLs suggest that not only these compounds and their hydrolysis products are responsible for the organoleptic characteristics of *Brassica* sp. vegetables (Baik et al., 2003; Padilla et al., 2007), and propose that they are likely to be the product of a synergistic combination of several phytochemicals, such as flavonoids, phenolic compounds, and indole hydrolytic products (Cartea & Velasco, 2008).

However, the direct relationship between the content of GSLs and the sensory properties of a plant is very complex, because there are numerous factors and synergisms that influence these properties. Therefore, it is necessary that, in parallel to the medicinal use that could be given to them, it is deepened in the study of the organoleptic properties that these compounds give to food, since, with the passage of time and the increase of information available about the benefits and damages caused by the compounds contained in food, consumers become more demanding about what they eat (Cartea & Velasco, 2008). It is demonstrated that at low concentrations, GSLs have the property of appetite stimulation, as well as, of course, the ability to contribute to the taste of certain vegetables, such as horseradish, mustard, and black pepper (Bischoff, 2016). However, in other foods those flavors are not desirables, as in cauliflower or broccoli, so a study carried out by Müller-Maatsch, Gurtner, Carle, and Steingass (2019) contemplates the possibility of removing GSLs, the responsible elements of flavor, from certain foods, without affecting the content in other beneficial compounds, such as anthocyanins. Although they managed to eliminate volatile compounds and recover anthocyanins, the precursors of these volatile elements (intact GSLs) were more persistent and difficult to remove. These issues certainly concern the food industry, since they have the duty to satisfy consumers' demands.

6. Healthy and adverse effects of glucosinolates

6.1 Bioactivities of GSLs

While it is true that there is great information about the classification, structures, location, and even about some metabolic pathways of the GSLs, their possible beneficial effects on health have been left aside. However, these compounds possess certain properties that make them a possible and novel therapeutic tool. Therefore, in the last decade, more attention has been paid to GSLs in this aspect, demonstrating that a shift in their defensive role is possible, so that they can offer protection to human health (Johnson, Dinkova-Kostova, & Fahey, 2015; Vig et al., 2009).

Mainly, biological activities of GSLs can be attributed to their hydrolytic products, of which the ITCs are prominent examples. A summary is presented in Table 2. Although mammal tissues do not contain myrosinases,

Table 2 Summary of several healthy effects of some GSLs derivatives.

Compound	**Bioactivity**	**Reference**
Allyl-ITC	Fungicide Bactericide Antiproliferative	Chung, Huang, Huang, and Jen (2003); Nadarajah, Han, and Holley (2005); Xiao et al. (2003)
Allyl:benzyl:2-phenylethyl: phenyl ITCs (1:3.5:5.3:9.6)	Fungicide	Troncoso (2005)
Alkenyl aliphatic ITCs (methyl-ITC, propenyl-ITC, butenyl-ITC, pentenyl-ITC) (propenyl-ITC, ethyl-ITC)	Fungicide	Smolinska, Morra, Knudsen, and James (2003)
Benzyl-ITC	Fungicide Anti-proliferative	Smolinska et al. (2003); Kuroiwa et al. (2006)
Butenyl-ITC	Fungicide	Chung et al. (2003)
GER derived-ITC	Fungicide	Manici et al., 2000
GIB derived-ITC	Fungicide	(Manici et al. (2000)
GRA derived-ITC	Fungicide	Mari, Iori, Leoni, and Marchi (1996)
Glucotropaeolin derived-ITC	Fungicide	Manici, Lazzeri, and Palmieri (1997)
4-Hydroxybenzyl-ITC	Bactericide	Ekanayake et al. (2006)
Indole-3-carbinol (with tamoxifen)	Anti-proliferative	Cover et al. (1999)
3-Indolylacetonitrile	Fungicide	Smissman, Beck, and Boots (1961)
Indole ethyl-ITC	Anti-proliferative	Singh et al. (2007)
Methyl-ITC	Bactericide	Lin, Kim, Du, and Wei (2000)
4-(Methylsulfinyl)butyl isothiocyanate	Bactericide Anti-proliferative	Haristoy, Fahey, Scholtus, and Lozniewski (2005); Rose, Huang, Nam, and Whiteman (2005)
3-Methylsulfinylpropyl ITC	Fungicide	Manici et al. (1997)
7-Methylsulfinylheptyl-ITC	Anti-proliferative	Rose et al. (2005)

Table 2 Summary of several healthy effects of some GSLs derivatives.—cont'd

Compound	Bioactivity	Reference
Oxazolidinethiones	Bactericide	Schnug and Ceynowa (1990)
Phenyl-ITC	Bactericide Anti-proliferative	Bending and Lincoln (2000); Manesh and Kuttan (2005)
Propenyl-ITC	Fungicide Anti-proliferative	Jeong, Kim, Hu, and Kong (2004); Sexton, Kirkegaard, and Howlett (1999)
Phenylbenzyl-ITC	Anti-proliferative	Yu et al. (1998)
Phenylethyl-ITC	Fungicide	Angus, Gardner, Kirkegaard, and Desmarchelier (1994)
Phenylmethyl-ITC	Anti-proliferative	Yu et al. (1998)
Sinalbin (p-hydroxybenzylglucosinolate) derived-ITC	Fungicide	Fenwick et al. (1983)
SIN (2-propenylglucosinolate) derived-ITC	Fungicide	Sanchi, Odorizzi, Lazzeri, and Marciano (2005)
5-Vinyloxazolidine-2-thione	Fungicide	Smolinska, Knudsen, Morra, and Borek (1997)

Adapted from Vig, A. P., Rampal, G., Thind, T. S., & Arora, S. (2009). Bio-protective effects of glucosinolates—A review. *LWT—Food Science and Technology*, *42*(10), 1561–1572.

hydrolytic products of GSLs are achieved in humans thanks to the action of the intestinal flora, that is, our microbiota is capable of producing bioactive ITCs, among others, from GSLs (Dinkova-Kostova & Kostov, 2012). Then, ITCs are absorbed from the small bowel and colon, and their metabolites are detectable in human urine 2–3 h after consumption of *Brassica* sp. vegetables, once they have developed their biological function (Johnson, 2002).

The wide range of biological activities of products derived from the glucosinolate–myrosinase system is biologically and economically important. On one hand, hydrolytic products of GSLs have an important role in the plant defense against herbivores and other stress situations. On the other hand, these compounds have toxic (e.g., goitrogenic) as well as protective (e.g., cancer-preventing) effects in higher animals and humans. There is a strong interest in the ability to regulate and optimize the levels of individual GSLs tissue—specifically to improve the nutritional value and pest resistance of crops (Wittstock & Halkier, 2002). A study carried out by Wu et al. (2017) supports the need for a database in which the amount of GSLs contained in fresh vegetables is collected, as well as the amount that we actually ingest after processing the food, and shows that the use of techniques like microwaving and steaming, despite affecting the content in GSLs, gets a lower reduction of those compounds than other processing techniques such as blanching.

6.1.1 Biocidal effects

Thanks to a more in-depth investigation in the field of bioactivities associated with the GSLs that have been carried out in the last centenary, it has been discovered that they possess, among other capabilities, antifungal, antimicrobial, herbicidal, and even insecticidal or nematicidal activities. These biocidal effects are attributed to the hydrolysis products of GSLs, generated from the action of myrosinases due to the presence of some threat to the plant, in this case, of pathogens (Vig et al., 2009).

As a specific example, sulforaphane (SFP) (Fig. 10), extracted from broccoli, exhibits potential for treating *Helicobacter pylori*, bacteria responsible in gastritis, being associated with a marked increase in the risk of gastric cancer (Wu et al., 2017). Purified SFP showed inhibition of the growth and killed multiple strains of *H. pylori* in the test tube and in tissue culture, including antibiotic-resistant strains. However, in a small clinical trial, consumption of up to 56 g of GRA rich broccoli sprouts daily for a week was associated with *H. pylori* eradication in only three of nine gastritis patients tested, so further studies are needed before reaching a full conclusion (Herr & Büchler, 2010).

6.1.2 Chemopreventive effects

Recently, GSLs and their metabolic products have been identified as potent cancer-prevention agents in a wide range of animal models due to their ability to inhibit metabolic phase I by the suppression of cytochrome P450 enzymes, that metabolize (and activate) many carcinogenic agents

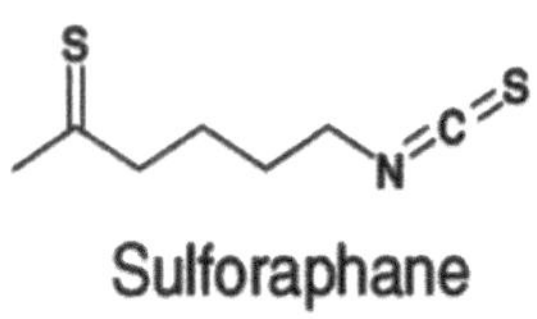

Fig. 10 Sulforaphane (SFP) structure.

(Gründemann & Huber, 2018; Herr & Büchler, 2010), and to induce phase II detoxification enzymes, such as quinone reductase, glutathione-S-transferase, and glucuronosyl transferases, as it has been also demonstrated through in vitro trials (Halkier & Gershenzon, 2006). One of the most extensively studied ITCs, SFP (Fig. 10), derivative of 4-methylsulfinylbutyl glucosinolate, was isolated from extracts of broccoli as a potent inducer of mammalian cytoprotective enzymes that block the cell cycle and promote apoptosis of cancerous cells (Dinkova-Kostova & Kostov, 2012; Zhang, Talalay, Chot, & Posnert, 1992). These effects raise the possibility that in addition to blocking DNA damage, ITCs may selectively inhibit the growth of tumor cells even after initiation by chemical carcinogens (Johnson, 2002).

Retrospective case–control studies have linked consumption of cruciferous vegetables to reduced risk of several cancers, including lung (Wu et al., 2015), gastric, breast (Bosetti et al., 2012), colorectal (Azeem et al., 2015), bladder (Al-Zalabani et al., 2016), and prostate cancer (Chan, Lok, & Woo, 2009; Wu et al., 2017).

These results are motivating efforts to increase the ITCs content of broccoli and to promote the health benefits of this family of vegetables (Halkier & Gershenzon, 2006). But, to define and exploit these potentially anti-carcinogenic effects it is important to understand and manipulate GSL chemistry and metabolism across the whole food-chain, from production and processing to consumption (Johnson, 2002).

6.1.3 Anti-inflammatory effects

ITCs can behave as modulators of inflammation, because they are able to reduce or even inhibit the activity of the nuclear factor "kappa-light-chain-enhancer" of activated B-cells (NF-kappaB) (Brunelli et al., 2010). It is known that NF-kappaB regulates the expression of cyclo-oxygenase 2 (COX-2), a pro-inflammatory enzyme responsible for elevated levels of prostaglandins and key inductor of inflammatory processes. It was shown that SFP suppresses both COX-2 mRNA and protein levels by inhibiting NF-kappaB-DNA-binding capacity via the PAP-kinase signaling pathway in human bladder and vascular endothelial cells (Shan et al., 2010). In another study, it was shown that TNF-α secretion was significantly inhibited at a concentration of 1 μM ($\geq$24% inhibition) in the presence of indole GSLs (Vo et al., 2014). That fact gains importance since inflammatory pathways play a crucial role in carcinogenesis, as well as other diseases of current importance (osteoarthritis) (Gründemann & Huber, 2018).

6.1.4 Other beneficial effects

It is known that mutations in the somatic cells are the key factors involved in the initiation and development of many diseases like cancer, atherosclerosis, degenerative heart diseases, cystic fibrosis, Huntington's disease, glaucoma, sickle cell anemia, phenylketonuria, and color-blindness (Poduri, Evrony, Cai, & Walsh, 2013). From the results obtained by Rampal et al. (2017) in a study carried out about the anti-mutagenic effects of 3 ITCs (allyl, benzyl, and 3-butenyl ITCs, individually and in binary combinations), it was observed that a combination of ITCs induced a stronger anti-mutagenic response even at relatively low concentrations, and without any signs of toxicity. Furthermore, it was discovered that ITCs showed more desmutagenic effect than bioantimutagenic, what means that they do not act on the repair and replication processes of the mutagen-damaged DNA, resulting in a decline in mutation frequency; but cause direct inactivation of the mutagens or their precursors (Rampal et al., 2017).

Fortunately, interest in health effects of the consumption of GSLs has recently been increasing, being found to interfere beneficially in the development of diseases of current interest, such as diabetes or cardiovascular diseases, that is, they have a protective function against the possibility of contracting these diseases. New studies are being carried out, so that in the not too distant future we can take advantage of the use of GSLs extracts (Dinkova-Kostova & Kostov, 2012; Fimognari et al., 2012). Table 2 collects some interesting derivatives of GSLs and their associated bioactivity.

6.2 Toxic effects

During years, exclusive or excessive feeding of vegetables and/or seeds from the Brassicaceae family has been associated with toxic effects. High levels of GSLs have been reported to cause some toxic effects including enlarged thyroid, reduced plasma thyroid hormone levels, some organ abnormalities (liver and kidney), decreased growth, decreased reproductive performance, and even mortality. Ruminants seem to be less sensitive to dietary GSLs, unlike pigs, which are the most severely affected by dietary GSLs compared to rabbit, poultry, and fish (Tripathi & Mishra, 2007).

It is demonstrated by clinical signs that allyl ITCs can cause irritant damage to the gastrointestinal tract when ingested in high levels, causing abdominal pain in ruminants and colic in horses. Treatment is symptomatic and includes a clean diet as well as pain control (Taljaard, 1993).

Thyroid enlargement has been associated with prolonged ingestion of vegetables from *Brassica* species containing goitrin and isothiocyanates, which block uptake of iodide by the thyroid, causing iodine depletion and, therefore, a T4 inhibition. Anti-thyroid effects of GSLs can result in subclinical signs, such as decreased reproductive performance and growth or, in more severe cases, clinically evident goiter (Taljaard, 1993). Thyroid enlargement and fetal deaths have been linked in experimental rodents. Thyroid hypertrophy has also been reported in poultry and decreased thyroid function has been reported in fish (Burel et al., 2000).

Anemia is also a common adverse effect of the overfeeding of livestock with the Brassicaceae family and also nitriles, another hydrolytic product of GSLs, have been associated with several hepatic effects, including bile duct hyperplasia, megalocytosis, zonal hepatocyte necrosis, and hepatic fibrosis. Renal megalocytosis has also been reported, while PRO has been associated with apoptosis and necrosis of pancreatic acinar cells (Collett, Stegelmeier, & Tapper, 2014).

Thus, as it can be seen, these data about toxicity refer to an excessive daily contribution of GSLs. To detect any toxic side effects of the sprout extracts supplied in therapeutic quantities (4.4 mg/kg per day in mice), indicators of liver (transaminases) and thyroid [thyroid-stimulating hormone, total triiodothyronine (T3), and free thyroxine (T4)] function were examined in detail. No significant or consistent subjective or objective abnormal events (toxicities) associated with any of the sprout extract ingestions were observed. Another study demonstrates improved cholesterol metabolism and reduction of multiple oxidative biomarkers by the broccoli sprout intake without obvious side effects (Herr & Büchler, 2010).

7. The fate of glucosinolates during processing of vegetables from *Brassica* species

Mainly, the health-beneficial effects of GSLs are attributed to their hydrolytic products, ITCs. Nevertheless, their formation depends on a wide variety of plant-intrinsic factors, such as the concentration of GSLs and the activities of myrosinases, and on numerous extrinsic postharvest factors, such as storage, industrial processing conditions, domestic preparation, mastication, and digestion (Barba et al., 2016; Oliviero, Verkerk, & Dekker, 2018).

Table 3 Principal GSLs identified in leaves of Brassica vegetable crops.

		White cabbage	Savoy cabbage	Red cabbage	Kale	Collard	Tronchuda cabbage	Broccoli	Brussels sprouts	Cauli flower	Kohlrabi	Turnip	Turnip greens	Turnip tops	Chinese cabbage	Swede	Leafrapej
Aliphatic glucosinolates	**GIB**	+	+	+	+	+	+	+	+	+	+	+	+	+	+	–	–
	PRO	+	+	+	+	+	+	+	+	+	+	+	+	+	+	+	+
	SIN	+	+	+	+	+	+	+	+	+	+	–	–	–	–	–	–
	GAL	+	–	–	–	–	+	+	–	–	+	–	+	–	–	–	+
	GRA	+	+	+	+	–	+	+	+	+	+	–	+	–	–	+	–
	GNA	+	+	+	+	–	+	+	+	–	+	+	+	+	+	–	+
	GBN	+	+	–	–	–	+	+	–	–	–	+	+	+	+	+	+
	GIV	+	+	+	+	+	–	–	+	+	+	–	+	+	–	–	+
Indole glucosinolates	**GER**	+	–	–	–	+	–	+	–	–	–	+	–	–	–	–	–
	GNL	–	–	–	–	–	+	+	–	–	–	+	+	–	+	+	+
	GBS	+	+	+	+	+	+	+	+	+	+	+	+	+	+	+	+
	NGBS	+	+	+	+	–	+	+	+	+	+	+	+	+	+	+	+
	4HGBS	+	–	–	+	–	+	+	–	–	+	+	+	+	–	+	–
	4MGBS	+	+	–	+	–	+	+	–	–	+	+	–	–	+	+	+
Aromatic	**GST**	+	+	–	+	–		+	–	–	–	+	+	+	+	+	+

Major glucosinolates found in each crop are shown in bold: ***GIB***: glucoiberin (3-methylsulfinylpropyl); ***PRO***: progoitrin (2-hydroxy-3-butenyl); ***SIN***: sinigrin (2-propenyl); ***GAL***: glucoalysiin (5-methylsulfinylpentyl); ***GRA***: glucoraphanin (4-methylsulfinylbutyl); ***GNA***: gluconapin (3-butenyl); ***GBN***: glucobrassicanapin (4-pentenyl); ***GIV***: glucoiberverin (3-methylthiopropyl); ***GER***: glucoerucin (4-methylthiobutyl); ***GNL***: gluconapoleiferin (2-hydroxy-4-pentenyl); ***GBS***: glucobrassicin (3-indolylmethyl); ***NGBS***: neoglucobrassicin (1-methoxy-3-indolylmethyl); ***4HGBS***: 4-hydroxyglucobrassicin (4-hydroxy-3-indolylmethyl); ***4MGBS***: 4-methoxyglucobrassicin (4-methoxy-3-indolylmethyl); ***GST***: gluconasturtiin (2-phenylethyl).

7.1 Glucosinolate composition of different vegetable *Brassica* species

Despite belonging to the same family, *Brassica* vegetables present a wide diversity in terms of the quantitative and qualitative composition of GSLs. Many authors have dedicated their work to the study of the distribution of different GSLs in the various species of *Cruciferous*. In Table 3 a summary of the main GSLs contained in different *Brassica* sp. vegetable crops is collected and organized according to the three large groups, in which the GSLs are classified taking into account their chemical structure (Cartea & Velasco, 2008).

The composition of GSLs in *B. rapa* crops turns out to be similar in all types, with few variations between Chinese cabbage and turnip, since both present GNA, glucobrassicanapin (4-pentenyl) (GBN), and their hydroxylated derivatives, PRO and gluconapoleiferin (2-hydroxy-4-pentenyl) (GNL), only that in turnip they are distributed between roots (PRO and GST), greens, and tops (GNA and GBN) (Padilla et al., 2007; Rosa, Heaney, Fenwick, & Portas, 2010).

In contrast, among the species of *B. oleracea*, considerable differences are observed regarding the composition of GSLs, although all of them contain GBS and GIB, and most also contain SIN, the amounts in which they accumulate are very variable. In the case of kales, SIN is the major GSL, while in cabbage leaves they are GIB or GB. Regarding broccoli, 50% of total GSLs are represented by GRA, although it also contains SIN, PRO, GNA, GBS, and NGBS. In other varieties, such as Brussels sprouts, collard, and cauliflower, the predominant GSLs are SIN, PRO and GBS (Baik et al., 2003; Padilla et al., 2007).

In vegetable crops of *B. napus*, leaf rape and swedes no major differences are observed. (Padilla et al., 2007) proved that the most abundant GSL in a variety of leaf rape called "nabicol" was GBN followed by PRO and GNA. In swedes, GBS, PRO and GST have been found as the major GSLs (Cartea & Velasco, 2008).

In order to acquire the full benefit of functional foods, it is necessary to know the natural variation in content of bioactive food components. Such variation might be regulated genetically or it might result from changes in the growing environment or from differences in post-harvest processing, storage or in food preparation conditions (Oliviero et al., 2018).

7.2 Influence of post-harvest treatments

Storage conditions strongly influence the content of GSLs in cruciferous vegetables (Banerjee, Variyar, Chatterjee, & Sharma, 2014). The main affecting parameters of post-harvest treatments on the quality and bioavailability of GSLs are the time, temperature, and packaging atmosphere (Jones, Faragher, & Winkler, 2006).

In the case of studies conducted using GRA as an object of analysis, it was found that its concentration in *B. oleracea* can vary significantly according to the conditions of post-harvest and packaging treatments, as it happens, for example, when storing the plant material at room temperature in open boxes for 3 days, losing 55% of the total content; or in plastic bags for 7 days, thus reducing 56% of the content (Rangkadilok et al., 2002). However, when the samples were stored at 4 °C for 10 days in modified atmosphere packaging, no significant differences were found, so they concluded that those were the best storage conditions for broccoli (Barba et al., 2016; Jones et al., 2006).

Another experimental work includes a curious study in which the conditions to which broccoli is subjected after harvesting are simulated, that is, it is transported and distributed at 1 °C for 7 days, and then it is exposed at 15 °C for 3 days. After this period of 10 days, the amount of GSLs had decreased between 70% and 80%, compared to the freshly harvested broccoli (Vallejo, Tomas-Barberan, & Garcia-Viguera, 2003). A similar study, but to which they added as a variant, the use of radiation (12 h/day), resulted in the fact that the period in which the samples remained between 0 and 4 °C did not alter the content of GSLs, but the biggest differences occurred during storage between 10 and 18 °C. The content of some molecules as 4-hydroxyglucobrassicin (4-hydroxy-3-indolylmethyl) (4HGBS) and aliphatic GSLs was increased after storage at 18 °C and applying a radiation treatment with visible light of 25 $\mu mol\, m^{-2}\, s^{-1}$, whereas for the vast majority of GSLs, the content was increased after storage at 10 °C, producing an increase in the content of indolyl 4HGBS and 4-methoxyglucobrassicin (4-methoxy-3-indolylmethyl) (4MGBS) when applying the same radiation conditions (Rybarczyk-Plonska et al., 2016).

Regarding the conservation atmosphere, different storage conditions of broccoli heads have been analyzed, concluding that an atmosphere of 5% CO_2 + 3% O_2 achieved an increase in the content of SFP and indole-3-carbinol after a period of 100 days at 0 °C (Badełek, Kosson, & Adamicki, 2012). Therefore, low storage temperatures, as well as the use of radiation and controlled atmosphere promote not only a good conservation of the GSLs present in the food, but also a possible increase in their concentration (Barba et al., 2016).

7.3 Influence of preparation and cooking conditions

Unfortunately, culinary processes can also modify the content, and, therefore, the bioavailability of GSLs and their derivatives. Some non-thermal processes can also affect them significantly, as demonstrated after analyzing *Brassica* species (broccoli, cauliflower, brussels sprouts, and green cabbage) finely shredded, showing a decrease of up to 75% in 6h. However, in another study the amount of GSLs was analyzed in chopped raw cabbage and broccoli after 48h of storage at room temperature, obtaining that most of the GSLs had reduced their content, with the exception of 4-hydroxy- and 4-methoxy-3-indolylmethyl GSLs, whose concentration had increased 15 times (Jones et al., 2006). The observed reductions in GSLs content are mainly due to the activity of myrosinase, which is altered in thermal processes, although the activity of that enzyme is not inhibited until subjected to temperatures higher than 80 °C, in the same way that it resists to pressures up to 30 MPa (Björkman & Lonnerdal, 1973; Ghawi, Methven, Rastall, & Niranjan, 2012). Therefore, submitting the plant samples to autoclave conditions would suppose the inactivation of the myrosinase. This would result in a higher content of GSLs in food, but also in a decrease in the amount of ITCs, which are the reported metabolites responsible for the beneficial activity associated with GSLs (Barba et al., 2016).

On the other hand, thermal treatments normally produce a significant modification not only of GSLs quantities, but also of other biomolecules, such as ascorbic acid (Oliviero et al., 2018). Table 4 shows some studies in which the effects on the content of GSLs of different processes under different conditions were analyzed. Among all the studied cooking processes, the one that most affects the content of GSLs is boiling. Boiling was more effective in reducing the levels of GSLs by thermal degradation as well as by the leaching of components into the boiling water (Nugrahedi, Verkerk, Widianarko, & Dekker, 2015; Verkerk et al., 2009). The thermal degradation of vegetables during boiling can cause GSLs losses of 5–75%, varying according to the structure of each GSL and the plant matrix in which it is found. In addition, inactivation of the myrosinase occurs by denaturing at such high temperatures, which inhibits the formation of ITCs (Jones et al., 2010). Authors concluded that avoiding boiling of vegetables could increase the bioavailability of both GSLs and ITCs (Oliviero et al., 2018).

Other cooking processes as steaming, microwaving, and stir-frying did not induce such drastic changes in the contents of GSLs. But the most harmless culinary thermal technique for GSLs is undoubtedly steaming. In addition to preserving content levels of GSLs, short times of treatment

Table 4 Studies related to the influence of culinary process applied in the amount of GSLs.

Treatment	Conditions	Results	Reference
Baking	200 °C, 5 min	↓ Total GSLs	Yuan, Sun, Yuan, and Wang (2009)
Blanching	10 min (cabbage)	↓ GSLs levels	Hwang and Thi (2015)
	66 or 76 °C, 145 s	↓ 92% Lipoxygenase, ↓ 18% myrosinase	Dosz and Jeffery (2013)
	86 or 96 °C, 145 s	Inactivated peroxidase, lipoxygenase, and myrosinase	Dosz and Jeffery (2013)
	30, 90 or 120 s (broccoli)	Just 120 s: ↓ 36% total GSLs	Park et al. (2013)
Boiling	10 min (cauliflower)	↓ 29.1% SIN	Girgin and El (2015)
	15 min	↓ 45–60% GSLs, ↓ 37–45% derivatives	Vieites-Outes, López-Hernández, and Lage-Yusty (2016)
	100 °C, 5, 15 or 30 min (Brussels sprouts)	Just 7 breakdown products found	Ciska, Drabińska, Honke, and Narwojsz (2015)
	5 min (red cabbage)	↓ Total GSLs	Xu et al. (2014)
	With cold start (25 °C)	↓ 50% Total GSLs	Bongoni et al. (2014)
	With hot start (100 °C)	↓ 41% Total GSLs	Bongoni et al. (2014)
	100 °C, 3.5 min (broccoli)	↓ 80% Total GSLs	Martínez-Hernández, Artés-Hernández, Gómez, and Artés (2013)
	12 min (Portuguese cabbage)	↓ 57% Total GSLs	Aires, Carvalho, and Rosa (2012)
	15 min (*Brassica rapa*)	↓ 81% Total GSLs	Aires et al. (2012)
	2 or 5 min	↓ Total GSLs	Jones et al. (2010)
	15 min	↓ 64% Total GSLs	Francisco et al. (2010)
Frying	180 °C, 5 min	↓ Total GSLs	Yuan et al. (2009)

Table 4 Studies related to the influence of culinary process applied in the amount of GSLs.—cont'd

Treatment	Conditions	Results	Reference
High pressure	7 min	↓ 20–33% GSLs, ↓ 18–23% derivatives	Vieites-Outes et al. (2016)
	15 min	↓ 64% Total GSLs	Francisco et al. (2010)
Microwaving	10 min (cabbage)	GSLs levels were preserved	Hwang and Thi (2015)
	450 W, 5 min (red cabbage)	↓ Total GSLs	Xu et al. (2014)
	900 W, 2.5 min (broccoli)	↓ 40% Total GSLs	Martínez-Hernández et al. (2013)
	800 W, 90 s (broccoli)	↓ 13–26% Total GSLs	Park et al. (2013)
	1100 W, 2 or 5 min	↓ Total GSLs	Jones et al. (2010)
	590 W, 5 min	↓ Total GSLs	Yuan et al. (2009)
Steaming	10 min (cauliflower)	↓ 9.6% SIN	Girgin and El (2015)
	10 min (cabbage)	GSLs levels were preserved	Hwang and Thi (2015)
	10 min	↓ 5–12% Aliphatic GSLs derivatives	Vieites-Outes et al. (2016)
	5 min (red cabbage)	↓ Total GSLs	Xu et al. (2014)
	100 °C, 8 min (broccoli)	↑ 127.9% Total GSLs	Fiore et al. (2013)
	100 °C, 0.02 MPa, 5 min (broccoli)	↓ 40% Total GSLs	Martínez-Hernández et al. (2013)
	12 min (Portuguese cabbage)	No significant modifications	Aires et al. (2012)
	2 or 5 min	No significant modifications	Jones et al. (2010)
	15 min	No significant modifications	Francisco et al. (2010)
	5 min	No significant modifications	Yuan et al. (2009)
Stir-frying	130 °C, 5 min	↓ Total GSLs	Xu et al. (2014)

SIN: sinigrin. GSLs: glucosinolates.

Adapted from Barba, F. J., et al. (2016). Bioavailability of glucosinolates and their breakdown products: Impact of processing. *Frontiers in Nutrition*, *3*(August), 1–12.

(<7 min) allow myrosinase activity, favoring the appearance of ITCs (Oliviero et al., 2018; Sarvan et al., 2017). Regarding microwaving, its effect depends strongly on the power and time used for the treatment. It seems like powers below 750 W alter relatively little the content of GSLs, obtaining, in addition, a partial inactivation of myrosinase, making possible the formation of ITCs. A good optimization of microwaving conditions would make this technique a promising instrument to cook food without losing in exchange important compounds such as GSLs (Rungapamestry, Duncan, Fuller, & Ratcliffe, 2006; Vallejo, Tomás-Barberán, & Garcia-Viguera, 2002; Verkerk & Dekker, 2004). Therefore, cooking vegetables from *Brassica* species and enjoying its beneficial effects for health is possible, as long as the characteristics of the particular plant and the GSLs contents are known, and an appropriate technique is chosen according to those characteristics, as well as the optimal conditions of its variables.

8. Main conclusions and future perspectives

It is well known that a regular consumption of vegetables from *Brassica* species is associated with several beneficial biological activities caused by the action of the breakdown products of GSLs. Anticarcinogenic effects might be the most aforementioned propriety, but recent studies have found other beneficial activities of GSLs, including regulatory functions in inflammation and stress response, antioxidant activities, and even direct antimicrobial properties. Future studies will undoubtedly find more benefits of these natural chemicals, as well as other studies have described their organoleptic characteristics, determining that strong flavors of vegetables tend to correlate with high GSLs concentrations.

Although there have been numerous studies on the assimilation and metabolism of GSLs and their hydrolysis products, the knowledge of their in vivo behavior should be deepened to really understand the interaction mechanisms between said molecules and the target tissues. Another important point to take into account is the investigation of the interaction between GSLs and their breakdown products, in the same way that it would be interesting to know how they interact with other food constituents of the whole diet.

On the other hand, the degradation rate of GSLs during food processing is insufficiently understood, as it is a complex process due to the simultaneous generation of breakdown products and the high influence of the food processing method applied. However, not all are advantages; some adverse effects found out in a variety of livestock species have been associated with the use of high quantities of *Brassica* vegetables to feed them. Examples of those undesired effects are gastrointestinal irritation, goiter, anemia, reduced growth, and hepatic lesions. In addition, high sulfur intake can be associated with trace mineral deficiencies and polioencephalomalacia, therefore, it is best to avoid overfeeding with species from this family.

References

Agerbirk, N., Matthes, A., Erthmann, P. Ø., Ugolini, L., Cinti, S., Lazaridi, E., et al. (2018). Glucosinolate turnover in Brassicales species to an oxazolidin-2-one, formed via the 2-thione and without formation of thioamide. *Phytochemistry*, *153*, 79–93.

Agerbirk, N., & Olsen, C. E. (2012). Glucosinolate structures in evolution. *Phytochemistry*, *77*, 16–45.

Agrawal, A. A., & Kurashige, N. S. (2003). A role for isothiocyanates in plant resistance against the specialist herbivore Pieris rapae. *Journal of Chemical Ecology*, *29*(6), 1403–1415.

Ahuja, I., et al. (2016). Arabidopsis myrosinases link the glucosinolate-myrosinase system and the cuticle. *Scientific Reports*, *6*, 1–14.

Aires, A., Carvalho, R., & Rosa, E. (2012). Glucosinolate composition of Brassica is affected by postharvest, food processing and myrosinase activity. *Journal of Food Processing and Preservation*, *36*(3), 214–224.

Al-Zalabani, A. H., et al. (2016). Modifiable risk factors for the prevention of bladder cancer: A systematic review of meta-analyses. *European Journal of Epidemiology*, *31*(9), 811–851.

Angelino, D., & Jeffery, E. (2014). Glucosinolate hydrolysis and bioavailability of resulting isothiocyanates: Focus on glucoraphanin. *Journal of Functional Foods*, 7(1), 67–76. https://doi.org/10.1016/j.jff.2013.09.029.

Angus, J. F., Gardner, P. A., Kirkegaard, J. A., & Desmarchelier, J. M. (1994). Biofumigation: Isothiocyanates released from Brassica roots inhibit growth of the take-all fungus. *Plant and Soil*, *162*(1), 107–112.

Azeem, S., et al. (2015). Diet and colorectal cancer risk in Asia—A systematic review. *Asian Pacific Journal of Cancer Prevention*, *16*, 5389–5396.

Badełek, E., Kosson, R., & Adamicki, F. (2012). The effect of storage in controlled atmosphere on the quality and health-promoting components of broccoli (Brassica oleracea bar. italica). *Vegetable Crops Research Bulletin*, 77, 89–100.

Baenas, N., Suárez-martínez, C., García-viguera, C., & Moreno, D. A. (2017). Bioavailability and new biomarkers of cruciferous sprouts consumption. *Food Research International*, *100*, 497–503.

Baik, H. Y., et al. (2003). Relating glucosinolate content and flavor of broccoli cultivars. *Journal of Food Science*, *68*(3), 1043–1050.

Banerjee, A., Rai, A. N., Penna, S., & Variyar, P. S. (2016). Aliphatic glucosinolate synthesis and gene expression changes in gamma-irradiated cabbage. *Food Chemistry*, *209*, 99–103.

Banerjee, A., Variyar, P. S., Chatterjee, S., & Sharma, A. (2014). Effect of post harvest radiation processing and storage on the volatile oil composition and glucosinolate profile of cabbage. *Food Chemistry*, *151*, 22–30.

Barba, F. J., et al. (2016). Bioavailability of glucosinolates and their breakdown products: Impact of processing. *Frontiers in Nutrition*, *3*, 1–12.

Bell, L., & Wagstaff, C. (2014). Glucosinolates, myrosinase hydrolysis products, and flavonols found in rocket (Eruca sativa and Diplotaxis tenuifolia). *Journal of Agricultural and Food Chemistry*, *62*(20), 4481–4492.

Bell, L., & Wagstaff, C. (2017). Enhancement of glucosinolate and isothiocyanate profiles in Brassicaceae crops: Addressing challenges in breeding for cultivation, storage, and consumer-related traits. *Journal of Agricultural and Food Chemistry, 65*(43), 9379–9403.

Bellostas, N., Sørensen, A. D., Sørensen, J. C., & Sørensen, H. (2008). Fe2+-catalyzed formation of nitriles and thionamides from intact glucosinolates. *Journal of Natural Products, 71*(1), 76–80.

Benderoth, M., Textor, S., Windsor, A. J., Mitchell-Olds, T., Gershenzon, J., & Kroymann, J. (2006). Positive selection driving diversification in plant secondary metabolism. *Proc Natl Acad Sci USA, 103*(24), 9118–9123.

Bending, G. D., & Lincoln, S. D. (2000). Inhibition of soil nitrifying bacteria communities and their activities by glucosinolate hydrolysis products. *Soil Biology and Biochemistry, 32*, 1261–1269.

Björkman, R., & Lonnerdal, B. O. (1973). Studies on myrosinases. 3. Enzymatic properties of myrosinases from *Sinapis alba* and *Brassica napus* seeds. *Biochimica et Biophysica Acta (BBA) - Enzymology, 327*, 121–131.

Bischoff, K. L. (2016). Glucosinolates. In R. C. Gupta (Ed.), *Nutraceuticals* (pp. 551–554). Elsevier Inc. https://doi.org/10.1016/B978-0-12-802147-7.00040-1.

Bollard, M., Stribbling, S., Mitchell, S., & Caldwell, J. (1997). The disposition of allyl isothiocyanate in the rat and mouse. *Food and Chemical Toxicology, 35*, 933–943.

Bongoni, R., et al. (2014). Evaluation of different cooking conditions on broccoli (Brassica oleracea var. italica) to improve the nutritional value and consumer acceptance. *Plant Foods for Human Nutrition, 69*(3), 228–234.

Bosetti, C., et al. (2012). Cruciferous vegetables and cancer risk in a network of case–control studies. *Annals of Oncology, 23*(8), 2198–2203.

Brunelli, D., et al. (2010). The isothiocyanate produced from glucomoringin inhibits NF-KB and reduces myeloma growth in nude mice in vivo. *Biochemical Pharmacology, 79*(8), 1141–1148.

Burel, C., et al. (2000). Dietary low-glucosinolate rapeseed meal affects thyroid status and nutrient utilization in rainbow trout (*Oncorhynchus mykiss*). *British Journal of Nutrition, 83*(6), 653–664.

Cartea, M. E., & Velasco, P. (2008). Glucosinolates in Brassica foods: Bioavailability in food and significance for human health. *Phytochemistry Reviews*, 7(2), 213–229.

Celenza, J. L., et al. (2005). The Arabidopsis ATR1 Myb transcription factor controls. *Plant Physiology, 137*, 253–262.

Chan, R., Lok, K., & Woo, J. (2009). Prostate cancer and vegetable consumption. *Molecular Nutrition and Food Research, 53*(2), 201–216.

Chen, S., & Andreasson, E. (2001). Update on glucosinolate metabolism and transport. *Plant Physiology and Biochemistry, 39*(9), 743–758.

Chen, M.-L., et al. (2001). Bioavailability and bioequivalence: An FDA regulatory overview. *Pharmaceutical Research, 18*(12), 1645–1650.

Chen, S., et al. (2003). CYP79F1 and CYP79F2 have distinct functions in the biosynthesis of aliphatic glucosinolates in Arabidopsis. *Plant Journal, 33*(5), 923–937.

Chun, J. H., et al. (2018). Molecular characterization of glucosinolates and carotenoid biosynthetic genes in Chinese cabbage (Brassica rapa L. ssp. pekinensis). *Saudi Journal of Biological Sciences, 25*(1), 71–82.

Chung, W. C., Huang, J. W., Huang, H. C., & Jen, J. F. (2003). Control, by Brassica seed pomace combined with Pseudomonas boreopolis, of damping-off of watermelon caused by Pythium sp. *Canadian Journal of Plant Pathology, 25*(3), 285–294.

Ciska, E., Drabińska, N., Honke, J., & Narwojsz, A. (2015). Boiled Brussels sprouts: A rich source of glucosinolates and the corresponding nitriles. *Journal of Functional Foods, 19*, 91–99.

Clarke, J. D., et al. (2011). Bioavailability and inter-conversion of sulforaphane and erucin in human subjects consuming broccoli sprouts or broccoli supplement in a cross-over study design. *Pharmacological Research*, *64*(5), 456–463.

Clay, N. K., et al. (2009). Glucosinolate metabolites required for an Arabidopsis innate immune response. *Science*, *323*(5910), 95–101.

Collett, M. G., Stegelmeier, B. L., & Tapper, B. A. (2014). Could nitrile derivatives of turnip (Brassica rapa) glucosinolates be hepato- or cholangiotoxic in cattle? *Journal of Agricultural and Food Chemistry*, *62*(30), 7370–7375.

Conaway, C. C., et al. (1999). Disposition and pharmacokinetics of phenethyl isothiocyanate and 6-phenylhexyl isothiocyanate in F344 rats. *Drug Metabolism and Disposition*, *27*(1), 13–20.

Cools, K., & Terry, L. A. (2018). The effect of processing on the glucosinolate profile in mustard seed. *Food Chemistry*, *252*, 343–348.

Cover, C. M., et al. (1999). Indole-3-carbinol and tamoxifen cooperate to arrest the cell cycle of MCF-7 human breast cancer cells. *Cancer Research*, *59*(6), 1244–1251.

Dinkova-Kostova, A. T., & Kostov, R. V. (2012). Glucosinolates and isothiocyanates in health and disease. *Trends in Molecular Medicine*, *18*(6), 337–347.

Dosz, E. B., & Jeffery, E. H. (2013). Modifying the processing and handling of frozen broccoli for increased sulforaphane formation. *Journal of Food Science*, *78*(9), H1459–H1463.

Ekanayake, A., et al. (2006). Isogard TM: A natural anti-microbial agent derived from white mustard seed. In *Acta horticulturae* (pp. 101–108). Leuven, Belgium: International Society for Horticultural Science (ISHS).

Engel, E., et al. (2002). Flavor-active compounds potentially implicated in cooked cauliflower acceptance. *Journal of Agricultural and Food Chemistry*, *50*(22), 6459–6467.

Ettlinger, M. G., & Kjaer, A. (1968). Sulfur compounds in plants. *Recent Advances in Phytochemistry*, *1*, 59–144.

Ettlinger, M. G., & Lundeen, A. J. (1956). The mustard oil of Limnanthes douglasii seed, m-methoxybenzyl isothiocyanate. *Journal of the American Chemical Society*, *78*(9), 1952–1954.

Fenwick, G. R., Heaney, R. K., Mullin, W. J., & VanEtten, C. H. (1983). Glucosinolates and their breakdown products in food and food plants. *Critical Reviews in Food Science and Nutrition*, *18*(2), 123–201.

Fernández-León, A. M., et al. (2017). Quantification and bioaccessibility of intact glucosinolates in broccoli 'Parthenon' and Savoy cabbage 'Dama'. *Journal of Food Composition and Analysis*, *61*, 40–46.

Field, B., et al. (2004). Glucosinolate and amino acid biosynthesis in Arabidopsis. *Plant Physiology*, *135*(2), 828–839.

Fimognari, C., et al. (2012). Natural isothiocyanates: Genotoxic potential versus chemoprevention. *Mutation Research*, *750*(2), 107–131.

Fiore, A., et al. (2013). Chemical profile and sensory properties of different foods cooked by a new radiofrequency oven. *Food Chemistry*, *139*(1–4), 515–520.

Fowke, J. H., Fahey, J. W., Stephenson, K. K., & Hebert, J. R. (2001). Using isothiocyanate excretion as a biological marker of Brassica vegetable consumption in epidemiological studies: Evaluating the sources of variability. *Public Health Nutrition*, *4*(3), 837–846.

Francisco, M., et al. (2010). Cooking methods of Brassica rapa affect the preservation of glucosinolates, phenolics and vitamin C. *Food Research International*, *43*(5), 1455–1463.

Frerigmann, H., & Gigolashvili, T. (2014). MYB34, MYB51, and MYB122 distinctly regulate indolic glucosinolate biosynthesis in Arabidopsis thaliana. *Molecular Plant*, 7(5), 814–828.

Gachon, C. M. M., Langlois-Meurinne, M., Henry, Y., & Saindrenan, P. (2005). Transcriptional co-regulation of secondary metabolism enzymes in Arabidopsis: Functional and evolutionary implications. *Plant Molecular Biology*, *58*(2), 229–245.

Geu-Flores, F., et al. (2009). Glucosinolate engineering identifies a γ-glutamyl peptidase. *Nature Chemical Biology*, *5*, 575.

Ghawi, S. K., Methven, L., Rastall, R. A., & Niranjan, K. (2012). Thermal and high hydrostatic pressure inactivation of myrosinase from green cabbage: A kinetic study. *Food Chemistry*, *131*(4), 1240–1247.

Gigolashvili, T., Berger, B., et al. (2007). The transcription factor HIG1/MYB51 regulates indolic glucosinolate biosynthesis in Arabidopsis thaliana. *Plant Journal*, *50*(5), 886–901.

Gigolashvili, T., Yatusevich, R., et al. (2007). The R2R3-MYB transcription factor HAG1/MYB28 is a regulator of methionine-derived glucosinolate biosynthesis in Arabidopsis thaliana. *Plant Journal*, *51*(2), 247–261.

Gigolashvili, T., et al. (2008). HAG2/MYB76 and HAG3/MYB29 exert a specific and coordinated control on the regulation of aliphatic glucosinolate biosynthesis in Arabidopsis thaliana. *New Phytologist*, *177*(3), 627–642.

Girgin, N., & El, S. N. (2015). Effects of cooking on in vitro sinigrin bioaccessibility, total phenols, antioxidant and antimutagenic activity of cauliflower (Brassica oleraceae L. var. Botrytis). *Journal of Food Composition and Analysis*, *37*, 119–127.

Groenbaek, M., et al. (2016). Nitrogen split dose fertilization, plant age and frost effects on phytochemical content and sensory properties of curly kale (Brassica oleracea L. var. sabellica). *Food Chemistry*, *197*, 530–538.

Grubb, C. D., & Abel, S. (2006). Glucosinolate metabolism and its control. *Trends in Plant Science*, *11*(2), 89–100.

Grubb, C. D., et al. (2004). Arabidopsis glucosyltransferase UGT74B1 functions in glucosinolate biosynthesis and auxin homeostasis. *Plant Journal*, *40*(6), 893–908.

Gründemann, C., & Huber, R. (2018). Chemoprevention with isothiocyanates—From bench to bedside. *Cancer Letters*, *414*, 26–33.

Halkier, B. A. (2016). General introduction to glucosinolates. *Advances in botanical research: Vol. 80*. Elsevier Ltd.

Halkier, B. A., & Du, L. (1997). The biosynthesis of glucosinolates. *Trends in Plant Science*, *2*(11), 425–431.

Halkier, B. A., & Gershenzon, J. (2006). Biology and biochemistry of glucosinolates. *Annual Review of Plant Biology*, *57*(1), 303–333.

Hansen, B. G., Kliebenstein, D. J., & Halkier, B. A. (2007). Identification of a flavin-monooxygenase as the S-oxygenating enzyme in aliphatic glucosinolate biosynthesis in Arabidopsis. *Plant Journal*, *50*(5), 902–910.

Hansen, C. H., et al. (2001). Cytochrome P450 CYP79F1 from Arabidopsis catalyzes the conversion of dihomomethionine and trihomomethionine to the corresponding aldoximes in the biosynthesis of aliphatic glucosinolates. *Journal of Biological Chemistry*, *276*(14), 11078–11085.

Haristoy, X., Fahey, J. W., Scholtus, I., & Lozniewski, A. (2005). Evaluation of the antimicrobial effects of several isothiocyanates on Helicobacter pylori. *Planta Medica*, *71*(4), 326–330.

Helland, H. S., et al. (2016). Storage of fresh-cut swede and turnip in modified atmosphere: Effects on vitamin C, sugars, glucosinolates and sensory attributes. *Postharvest Biology and Technology*, *111*, 150–160.

Hemm, M. R., Ruegger, M. O., & Chapple, C. (2003). The Arabidopsis ref2 mutant is defective in the gene encoding CYP83A1 and shows both phenylpropanoid and glucosinolate phenotypes. *The Plant Cell*, *15*(1), 179–194.

Herr, I., & Büchler, M. W. (2010). Dietary constituents of broccoli and other cruciferous vegetables: Implications for prevention and therapy of cancer. *Cancer Treatment Reviews*, *36*(5), 377–383.
Hirai, M. Y., et al. (2005). Elucidation of gene-to-gene and metabolite-to-gene networks in Arabidopsis by integration of metabolomics and transcriptomics. *Journal of Biological Chemistry*, *280*(27), 25590–25595.
Hirai, M. Y., et al. (2007). Omics-based identification of Arabidopsis Myb transcription factors regulating aliphatic glucosinolate biosynthesis. *Proceedings of the National Academy of Sciences of the United States of America*, *104*(15), 6478–6483.
Holst, B., & Williamson, G. (2004). A critical review of the bioavailability of glucosinolates and related compounds. *Natural Product Reports*, *21*(3), 425–447.
Hwang, E. S., & Thi, N. D. (2015). Impact of cooking method on bioactive compound content and antioxidant capacity of cabbage. *Korean Journal of Food Science and Technology*, *47*(2), 184–190.
Jeong, W. S., Kim, I. W., Hu, R., & Kong, A. N. T. (2004). Modulatory properties of various natural chemopreventive agents on the activation of NF-KB signaling pathway. *Pharmaceutical Research*, *21*(4), 661–670.
Jeschke, V., Gershenzon, J., & Vassão, D. G. (2016). A mode of action of glucosinolate-derived isothiocyanates: Detoxification depletes glutathione and cysteine levels with ramifications on protein metabolism in Spodoptera littoralis. *Insect Biochemistry and Molecular Biology*, *71*, 37–48.
Johansen, T. J., Hagen, S. F., Bengtsson, G. B., & Mølmann, J. A. B. (2016). Growth temperature affects sensory quality and contents of glucosinolates, vitamin C and sugars in swede roots (Brassica napus L. ssp. rapifera Metzg.). *Food Chemistry*, *196*, 228–235.
Johnson, I. T. (2002). Glucosinolates in the human diet. Bioavailability and implications for health. *Phytochemistry Reviews*, *1*(2), 183.
Johnson, T. L., Dinkova-Kostova, A. T., & Fahey, J. W. (2015). Glucosinolates from the Brassica vegetables and their health effects. In *Encyclopedia of food and health* (pp. 248–255). .
Jones, R. B., Faragher, J. D., & Winkler, S. (2006). A review of the influence of postharvest treatments on quality and glucosinolate content in broccoli (Brassica oleracea var. italica) heads. *Postharvest Biology and Technology*, *41*(1), 1–8.
Jones, G., & Sanders, O. G. (2002). A sensory profile of turnip greens as affected by variety and maturity. *Journal of Food Science*, *67*(8), 3126–3129.
Jones, R. B., et al. (2010). Cooking method significantly effects glucosinolate content and sulforaphane production in broccoli florets. *Food Chemistry*, *123*(2), 237–242.
Katsarou, D., et al. (2016). Glucosinolate biosynthesis in Eruca sativa. *Plant Physiology and Biochemistry*, *109*, 452–466.
Kissen, R., et al. (2016). Effect of growth temperature on glucosinolate profiles in Arabidopsis thaliana accessions. *Phytochemistry*, *130*, 106–118.
Kliebenstein, D. J. (2001). Gene duplication in the diversification of secondary metabolism: Tandem 2-oxoglutarate-dependent dioxygenases control glucosinolate biosynthesis in Arabidopsis. *The Plant Cell Online*, *13*(3), 681–693.
Knill, T., et al. (2008). Arabidopsis branched-chain aminotransferase 3 functions in both amino acid and glucosinolate biosynthesis. *Plant Physiology*, *146*(3), 1028–1039.
Knill, T., et al. (2009). Arabidopsis thaliana encodes a bacterial-type heterodimeric isopropylmalate isomerase involved in both Leu biosynthesis and the Met chain elongation pathway of glucosinolate formation. *Plant Molecular Biology*, *71*(3), 227–239.
Kroymann, J., Donnerhacke, S., Schnabelrauch, D., & Mitchell-Olds, T. (2003). Evolutionary dynamics of an Arabidopsis insect resistance quantitative trait locus. *Proceedings of the National Academy of Sciences of the United States of America*, *100*(Suppl. 2), 14587–14592.

Kuroiwa, Y., et al. (2006). Protective effects of benzyl isothiocyanate and sulforaphane but not resveratrol against initiation of pancreatic carcinogenesis in hamsters. *Cancer Letters*, *241*(2), 275–280.

Levy, M., et al. (2005). Arabidopsis IQD1, a novel calmodulin-binding nuclear protein, stimulates glucosinolate accumulation and plant defense. *Plant Journal*, *43*(1), 79–96.

Li, J., et al. (2008). Subclade of flavin-monooxygenases involved in aliphatic glucosinolate biosynthesis. *Plant Physiology*, *148*(3), 1721–1733.

Lin, C.-M., Kim, J., Du, W.-X., & Wei, C.-I. (2000). Bactericidal activity of isothiocyanate against pathogens on fresh produce. *Journal of Food Protection*, *63*(1), 25–30.

Manesh, C., & Kuttan, G. Ã. (2005). Effect of naturally occurring isothiocyanates in the inhibition of cyclophosphamide-induced urotoxicity. *Phytomedicine*, *12*(6–7), 487–493.

Manici, L. M., Lazzeri, L., & Palmieri, S. (1997). In vitro fungitoxic activity of some glucosinolates and their enzyme-derived products toward plant pathogenic fungi. *Journal of Agricultural and Food Chemistry*, *45*(7), 2768–2773.

Manici, L. M., et al. (2000). Suppressive activity of some glucosinolate enzyme degradation products on Pythium irregulare and Rhizoctonia solani in sterile soil. *Pest Management Science*, *56*(10), 921–926.

Mari, M., Iori, R., Leoni, O., & Marchi, A. (1996). Bioassays of glucosinolate-derived isothiocyanates against postharvest pear pathogens. *Plant Pathology*, *45*(4), 753–760.

Martínez-Hernández, G. B., Artés-Hernández, F., Gómez, P. A., & Artés, F. (2013). Induced changes in bioactive compounds of kailan-hybrid broccoli after innovative processing and storage. *Journal of Functional Foods*, *5*(1), 133–143.

Mi, L., di Pasqua, A. J., & Chung, F. L. (2011). Proteins as binding targets of isothiocyanates in cancer prevention. *Carcinogenesis*, *32*(10), 1405–1413.

Mikkelsen, M. D., Hansen, C. H., Wittstock, U., & Halkier, B. A. (2000). Cytochrome P450 CYP79B2 from Arabidopsis catalyzes the conversion of tryptophan to indole-3-acetaldoxime, a precursor of indole glucosinolates and indole-3-acetic acid. *Journal of Biological Chemistry*, *275*(43), 33712–33717.

Mikkelsen, M. D., Naur, P., & Halkier, B. A. (2004). Arabidopsis mutants in the C-S lyase of glucosinolate biosynthesis establish a critical role for indole-3-acetaldoxime in auxin homeostasis. *Plant Journal*, *37*(5), 770–777.

Mølmann, J. A. B., et al. (2015). Effects of temperature and photoperiod on sensory quality and contents of glucosinolates, flavonols and vitamin C in broccoli florets. *Food Chemistry*, *172*, 47–55.

Müller, R., et al. (2010). Differential effects of indole and aliphatic glucosinolates on lepidopteran herbivores. *Journal of Chemical Ecology*, *36*(8), 905–913.

Müller-Maatsch, J., Gurtner, K., Carle, R., & Steingass, C. B. (2019). Investigation into the removal of glucosinolates and volatiles from anthocyanin-rich extracts of red cabbage. *Food Chemistry*, *278*, 406–414.

Nadarajah, D., Han, J. H., & Holley, R. A. (2005). Use of mustard flour to inactivate *Escherichia coli* O157: H7 in ground beef under nitrogen flushed packaging. *International Journal of Food Microbiology*, *99*(3), 257–267.

Naur, P., et al. (2003). CYP83A1 and CYP83B1, two nonredundant cytochrome P450 enzymes metabolizing oximes in the biosynthesis of glucosinolates in Arabidopsis. *Plant Physiology*, *133*, 63–72.

Nugrahedi, P. Y., Verkerk, R., Widianarko, B., & Dekker, M. (2015). A mechanistic perspective on process-induced changes in glucosinolate content in Brassica vegetables: A review. *Critical Reviews in Food Science and Nutrition*, *55*(6), 823–838.

Oliviero, T., Verkerk, R., & Dekker, M. (2018). Isothiocyanates from Brassica vegetables—Effects of processing, cooking, mastication, and digestion. *Molecular Nutrition and Food Research*, *62*(18), e1701069.

Padilla, G., et al. (2007). Variation of glucosinolates in vegetable crops of Brassica rapa. *Phytochemistry*, *68*(1), 536–545.
Park, M.-H., et al. (2013). Variation of glucoraphanin and glucobrassicin: Anticancer components in Brassica during processing. *Food Science and Technology (Campinas)*, *33*(4), 624–631.
Piotrowski, M., et al. (2004). Desulfoglucosinolate sulfotransferases from Arabidopsis thaliana catalyze the final step in the biosynthesis of the glucosinolate core structure. *Journal of Biological Chemistry*, *279*(49), 50717–50725.
Poduri, A., Evrony, G. D., Cai, X., & Walsh, C. A. (2013). Somatic mutation, genomic variation, and neurological disease. *Science*, *341*(6141), 1237758.
Rajan, T. S., et al. (2016). Anticancer activity of glucomoringin isothiocyanate in human malignant astrocytoma cells. *Fitoterapia*, *110*, 1–7.
Rampal, G., et al. (2017). Synergistic antimutagenic effect of isothiocyanates against varied mutagens. *Food and Chemical Toxicology*, *109*, 879–887.
Rangkadilok, N., et al. (2002). The effect of post-harvest and packaging treatments on glucoraphanin concentration in broccoli (Brassica oleracea var. italica). *Journal of Agricultural and Food Chemistry*, *50*(25), 7386–7391.
Redovniković, I. R., Glivetić, T., Delonga, K., & Vorkapić-Furač, J. (2008). Glucosinolates and their potential role in plant. *Periodicum Biologorum*, *110*(4), 297–309.
Redovniković, I. R., Textor, S., Lisnić, B., & Gershenzon, J. (2012). Expression pattern of the glucosinolate side chain biosynthetic genes MAM1 and MAM3 of Arabidopsis thaliana in different organs and developmental stages. *Plant Physiology and Biochemistry*, *53*, 77–83.
Rosa, E. A. S., Heaney, R. K., Fenwick, G. R., & Portas, C. A. M. (1997). Glucosinolates in crop plants. *Horticultural Reviews*, *19*, 99–215. Wiley Online Books.
Rose, P., Huang, Q., Nam, C., & Whiteman, M. (2005). Broccoli and watercress suppress matrix metalloproteinase-9 activity and invasiveness of human MDA-MB-231 breast cancer cells. *Toxicology and Applied Pharmacology*, *209*(2), 105–113.
Rouzaud, G., et al. (2004). Hydrolysis of glucosinolates to isothiocyanates after ingestion of raw or microwaved cabbage by human volunteers. *Cancer Epidemiology Biomarkers & Prevention*, *13*(1), 125–131.
Rungapamestry, V., Duncan, A. J., Fuller, Z., & Ratcliffe, B. (2006). Changes in glucosinolate concentrations, myrosinase activity, and production of metabolites of glucosinolates in cabbage (Brassica oleracea var. capitata) cooked for different durations. *Journal of Agricultural and Food Chemistry*, *54*(20), 7628–7634.
Rybarczyk-Plonska, A., et al. (2016). Glucosinolates in broccoli (Brassica oleracea L. var. italica) as affected by postharvest temperature and radiation treatments. *Postharvest Biology and Technology*, *116*, 16–25.
Šamec, D., Pavlovi, I., Radoj, I., & Salopek-Sondi, B. (2018). Comparative analysis of phytochemicals and activity of endogenous enzymes associated with their stability, bioavailability and food quality in five Brassicaceae sprouts. *Food Chemistry*, *269*, 96–102.
Sanchi, S., Odorizzi, S., Lazzeri, L., & Marciano, P. (2005). Effect of Brassica carinata seed meal treatment on the Trichoderma harzianum T39-Sclerotinia species interaction. In *Acta horticulturae* (pp. 287–292). Leuven, Belgium: International Society for Horticultural Science (ISHS).
Sarvan, I., et al. (2017). Sulforaphane formation and bioaccessibility are more affected by steaming time than meal composition during in vitro digestion of broccoli. *Food Chemistry*, *214*, 580–586.
Sawada, Y., et al. (2009). Arabidopsis bile acid:Sodium symporter family protein 5 is involved in methionine-derived glucosinolate biosynthesis. *Plant & Cell Physiology*, *50*(9), 1579–1586.

Schnug, E., & Ceynowa, J. (1990). Phytopathological aspects of glucosinolates in oilseed rape. *Journal of Agronomy and Crop Science: Research on Crop Stress and Climate Change*, *165*, 319–328.

Schonhof, I., Krumbein, A., & Bruckner, B. (2004). Genotypic effects on glucosinolates and sensory properties of broccoli and cauliflower. *Die Nahrung*, *48*(1), 25–33.

Schramm, K., et al. (2012). Metabolism of glucosinolate-derived isothiocyanates to glutathione conjugates in generalist lepidopteran herbivores. *Insect Biochemistry and Molecular Biology*, *42*(3), 174–182.

Schuster, J., et al. (2006). Branched-chain aminotransferase is part of the chain elongation pathway in the biosynthesis of methionine-derived glucosinolates in Arabidopsis. *The Plant Cell*, *18*(10), 2664–2679.

Sexton, A. C., Kirkegaard, J. A., & Howlett, B. J. (1999). Glucosinolates in Brassica juncea and resistance to Australian isolates of Leptosphaeria maculans, the blackleg fungus. *Australasian Plant Pathology*, *28*(2), 95–102.

Shan, Y., et al. (2010). Protective effect of sulforaphane on human vascular endothelial cells against lipopolysaccharide-induced inflammatory damage. *Cardiovascular Toxicology*, *10*(2), 139–145.

Shapiro, A., Stephenson, K., Fahey, W., & Wade, L. (1998). Human metabolism and excretion of cancer chemoprotective glucosinolates and isothiocyanates of cruciferous vegetables. *Cancer Epidemiology, Biomarkers & Prevention*, 7(12), 1091–1100.

Shapiro, T. A., et al. (2001). Chemoprotective glucosinolates and isothiocyanates of broccoli sprouts: Metabolism and excretion in humans. *Cancer Epidemiology, Biomarkers & Prevention*, *10*(5), 501–508.

Singh, R. K., et al. (2007). Effect of indole ethyl isothiocyanates on proliferation, apoptosis, and MAPK signaling in neuroblastoma cell lines. *Bioorganic and Medicinal Chemistry Letters*, *17*(21), 5846–5852.

Skirycz, A., et al. (2006). DOF transcription factor AtDof1.1 (OBP2) is part of a regulatory network controlling glucosinolate biosynthesis in Arabidopsis. *Plant Journal*, *47*(1), 10–24.

Smissman, E. E., Beck, S. D., & Boots, M. R. (1961). Growth inhibition of insects and a fungus by indole-3-acetonitrile. *Science (New York, N.Y.)*, *133*(3451), 462.

Smolinska, U., Knudsen, G. R., Morra, M. J., & Borek, V. (1997). Inhibition of Aphanomyces euteiches f. sp. pisi by volatiles produced by hydrolysis of *Brassica napus* seed meal. *Plant Disease*, *81*(3), 288–292.

Smolinska, U., Morra, M. J., Knudsen, G. R., & James, R. L. (2003). Isothiocyanates produced by Brassicaceae species as inhibitors of Fusarium oxysporum. *Plant Disease*, *87*(4), 407–412.

Sønderby, I. E., Geu-Flores, F., & Halkier, B. A. (2010). Biosynthesis of glucosinolates—Gene discovery and beyond. *Trends in Plant Science*, *15*(5), 283–290.

Song, L., Morrison, J. J., Botting, N. P., & Thornalley, P. J. (2005). Analysis of glucosinolates, isothiocyanates, and amine degradation products in vegetable extracts and blood plasma by LC-MS/MS. *Analytical Biochemistry*, *347*(2), 234–243.

Sørensen, J. C., et al. (2016). Bioavailability and in vivo metabolism of intact glucosinolates. *Journal of Functional Foods*, *24*, 450–460.

Taljaard, T. L. (1993). Cabbage poisoning in ruminants. *Journal of the South African Veterinary Association*, *64*(2), 96–100.

Textor, S., et al. (2007). MAM3 catalyzes the formation of all aliphatic glucosinolate chain lengths in Arabidopsis. *Plant Physiology*, *144*(1), 60–71.

Traka, M., & Mithen, R. (2009). Glucosinolates, isothiocyanates and human health. *Phytochemistry Reviews*, *8*(1), 269–282.

Tripathi, M. K., & Mishra, A. S. (2007). Glucosinolates in animal nutrition: A review. *Animal Feed Science and Technology*, *132*(1–2), 1–27.

Troncoso, R. (2005). Analysis of the isothiocyanates present in cabbage leaves extract and their potential application to control Alternaria rot in bell peppers. *Food Research International, 38*(6), 701–708.

Vallejo, F., Tomás-Barberán, F. A., & Garcia-Viguera, C. (2002). Glucosinolates and vitamin C content in edible parts of broccoli florets after domestic cooking. *European Food Research and Technology, 215*(4), 310–316.

Vallejo, F., Tomas-Barberan, F., & Garcia-Viguera, C. (2003). Health-promoting compounds in broccoli as influenced by refrigerated transport and retail sale period. *Journal of Agricultural and Food Chemistry, 51*(10), 3029–3034.

van Doorn, H. E., et al. (1998). The glucosinolates sinigrin and progoitrin are important determinants for taste preference and bitterness of Brussels sprouts. *Journal of the Science of Food and Agriculture, 78*(1), 30–38.

Verkerk, R., & Dekker, M. (2004). Glucosinolates and myrosinase activity in red cabbage (Brassica oleracea L. var. Capitata f. rubra DC.) after various microwave treatments. *Journal of Agricultural and Food Chemistry, 52*(24), 7318–7323.

Verkerk, R., et al. (2009). Glucosinolates in Brassica vegetables: The influence of the food supply chain on intake, bioavailability and human health. *Molecular Nutrition and Food Research, 53*(Suppl. 2), 219–265.

Vieites-Outes, C., López-Hernández, J., & Lage-Yusty, M. A. (2016). Modificación de glucosinolatos en grelos (Brassica rapa subsp. rapa L.) sometidos a procesos térmicos culinarios. *CYTA—Journal of Food, 14*(4), 536–540.

Vig, A. P., Rampal, G., Thind, T. S., & Arora, S. (2009). Bio-protective effects of glucosinolates—A review. *LWT—Food Science and Technology, 42*(10), 1561–1572.

Vo, Q. V., et al. (2014). Bioorganic & medicinal chemistry synthesis and anti-inflammatory activity of indole glucosinolates. *Bioorganic & Medicinal Chemistry, 22*(2), 856–864.

Wang, H., et al. (2011). Glucosinolate biosynthetic genes in Brassica rapa. *Gene, 487*(2), 135–142.

Wang, J., et al. (2012). Genotypic variation of glucosinolates in broccoli (Brassica oleracea var. italica) florets from China. *Food Chemistry, 133*(3), 735–741.

Wentzell, A. M., et al. (2007). Linking metabolic QTLs with network and cis-EQTLs controlling biosynthetic pathways. *PLoS Genetics, 3*(9), 1687–1701.

Wittstock, U., & Halkier, B. A. (2000). Cytochrome P450 CYP79A2 from Arabidopsis thaliana L. catalyzes the conversion of L-phenylalanine to phenylacetaldoxime in the biosynthesis of benzylglucosinolate. *Journal of Biological Chemistry, 275*(19), 14659–14666.

Wittstock, U., & Halkier, B. A. (2002). Glucosinolate research in the Arabidopsis era. *Trends in Plant Science,* 7(6), 263–270.

Wu, X., Zhou, Q. H., & Xu, K. (2009). Are isothiocyanates potential anti-cancer drugs? *Acta Pharmacologica Sinica, 30*(5), 501–512.

Wu, Q. J., et al. (2015). Pre-diagnostic cruciferous vegetables intake and lung cancer survival among Chinese women. *Scientific Reports, 5,* 1–9.

Wu, X., et al. (2017). Journal of food composition and analysis challenges of developing a valid dietary glucosinolate database. *Journal of Food Composition and Analysis, 64,* 78–84.

Xiao, Z., Mi, L., Chung, F.-L., & Veenstra, T. D. (2012). Proteomic analysis of covalent modifications of tubulins by isothiocyanates. *Journal of Nutrition, 142*(7), 1377S–1381S.

Xiao, D., et al. (2003). Allyl isothiocyanate, a constituent of cruciferous vegetables, inhibits proliferation of human prostate cancer cells by causing G2/M arrest and inducing apoptosis. *Carcinogenesis, 24*(5), 891–897.

Xu, F., et al. (2014). Domestic cooking methods affect the nutritional quality of red cabbage. *Food Chemistry, 161,* 162–167.

Ye, L., et al. (2002). Quantitative determination of dithiocarbamates in human plasma, serum, erythrocytes and urine: Pharmacokinetics of broccoli sprout isothiocyanates in humans. *Clinica Chimica Acta, 316*(1–2), 43–53.

Yu, R., et al. (1998). Chemopreventive isothiocyanates induce apoptosis and caspase-3-like protease activity. *Cancer Research*, *388*(312), 402–408.
Yuan, G.-f., Sun, B., Yuan, J., & Wang, Q.-m. (2009). Effects of different cooking methods on health-promoting compounds of broccoli. *Journal of Zhejiang University Science B*, *10*(8), 580–588.
Zhang, Y., Talalay, P., Chot, C.-G., & Posnert, G. H. (1992). A major inducer of anticarcinogenic protective enzymes from broccoli: Isolation and elucidation of structure. *Proceedings of the National Academy of Sciences of the United States of America*, *89*(6), 2399–2403.

Further reading

Cheng, D.-L., Hashimoto, K., & Uda, Y. (2004). In vitro digestion of sinigrin and glucotropaeolin by single strains of Bifidobacterium and identification of the digestive products. *Food and Chemical Toxicology*, *42*(3), 351–357.
Ibdah, M., Chen, Y.-T., Wilkerson, C. G., & Pichersky, E. (2009). An aldehyde oxidase in developing seeds of Arabidopsis converts benzaldehyde to benzoic acid. *Plant Physiology*, *150*(1), 416–423.
Ibdah, M., & Pichersky, E. (2009). Arabidopsis Chy1 null mutants are deficient in benzoic acid-containing glucosinolates in the seeds. *Plant Biology*, *11*(4), 574–581.
Kliebenstein, D. J., et al. (2007). Characterization of seed-specific benzoyloxyglucosinolate mutations in Arabidopsis thaliana. *Plant Journal*, *51*(6), 1062–1076.
Mugford, S. G., et al. (2009). Disruption of adenosine-5'-phosphosulfate kinase in Arabidopsis reduces levels of sulfated secondary metabolites. *The Plant Cell*, *21*(3), 910–927.
Schlaeppi, K., et al. (2008). The glutathione-deficient mutant Pad2-1 accumulates lower amounts of glucosinolates and is more susceptible to the insect herbivore Spodoptera littoralis. *Plant Journal*, *55*(5), 774–786.
Sokol, J., Stegman, Z., Liebmann, J. M., & Ritch, R. (1996). Location of the iris insertion in pigment dispersion syndrome. *Ophthalmology*, *103*(2), 289–293.

CHAPTER SEVEN

Phytoestrogens, phytosteroids and saponins in vegetables: Biosynthesis, functions, health effects and practical applications

Francesco Di Gioia[a], Spyridon A. Petropoulos[b,*]
[a]Department of Plant Science, Pennsylvania State University, University Park, PA, United States
[b]Department of Crop Production and Rural Environment, University of Thessaly, Volos, Greece
*Corresponding author: e-mail address: fangio57gr@gmail.com

Contents

Abstract

Phytoestrogens are non-steroidal secondary metabolites with similarities in structure and biological activities with human estrogens divided into various classes of compounds, including lignans, isoflavones, ellagitannins, coumestans and stilbenes. Similarly, phytosteroids are steroidal compounds of plant origin which have estrogenic effects and can act as agonists, antagonists, or have a mixed agonistic/antagonistic activity to animal steroid receptors. On the other hand, saponins are widely distributed plant glucosides divided into triterpenoid and steroidal saponins that contribute to plant defense mechanism against herbivores. They present a great variation from a structural point of view, including compounds from different classes. In this chapter, the main vegetable sources of these compounds will be presented, while details regarding their biosynthesis and plant functions will be also discussed. Moreover, considering the significant bioactive properties that these compounds exhibit, special focus will be given on their health effects, either beneficial or adverse. The practical applications of these compounds in agriculture and phytomedicine will be also demonstrated, as well as the future prospects for related research.

Advances in Food and Nutrition Research, Volume 90
ISSN 1043-4526
https://doi.org/10.1016/bs.afnr.2019.02.004

1. Introduction

Vegetables are considered rich sources of phytochemicals, while certain species contain significant amounts of bioactive compounds including phytoestrogens, phytosteroids or saponins. Phytoestrogens are considered as non-steroidal compounds that have similar chemical structure with endogenous estrogens. Due to this similarity they can mimic or antagonize mammalian steroids and they can be either of plant origin (xenoestrogens) or derived after metabolism of precursors of plant origin (Bedell, Nachtigall, & Naftolin, 2014; Fang et al., 2001). Phytoestrogens can be divided into different classes of compounds, namely lignans, flavonoids, stilbenes, ellagitannins, and coumarin derivatives (e.g., coumestrols), which are present in several food products of plant origin (Landete et al., 2016; Nikolić, Savić-Gajić, Tačić, & Savić, 2017) (Fig. 1). Especially for flavonoids, they are wide spread throughout the plant kingdom and are considered as the most common compounds with estrogenic effects, including flavones, flavanones, isoflavones and isoflavanes (Lecomte, Demay, Ferrière, & Pakdel, 2017).

Phytosteroids are plant-derived steroids, which may present agonistic, antagonistic, or have a mixed activity to human and animal steroid receptors, depending on their structure and concentration, the receptor, and the cell type (Dean, Murphy, & Burdette, 2017). Plant steroids include several compound classes such as androstanes, cardiac glycosides, ecdysteroids, estranes, pregnanes, steroid alkaloids, steroid saponins and withanolides (Kreis & Müller-Uri, 2010) (Fig. 1). Moreover, phytosteroids may interact with

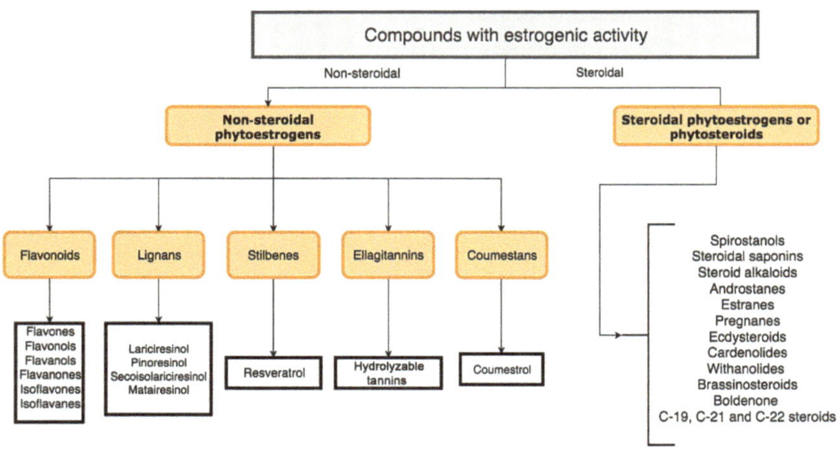

Fig. 1 Classification of compounds of plant origin with estrogenic activity.

several steroid receptors and steroid metabolizing enzymes, therefore they can affect the animal endocrine and reproductive system in multiple ways (Dean et al., 2017). This class includes various compounds such as digoxin, digitoxin, diosgenin, guggulsterone, phytosterols like β-sitosterol, and phytoestrogens like isoflavones (Savage, 2003). Brassinosteroids are also an important family of steroidal compounds with hormonal activity similar to animal steroids, including many polyhydroxysteroids (Shahzad et al., 2018). Other widespread phytocompounds such as apigenin have been identified as phytoprogestins which bind with progesterone receptors and may therefore affect progesterone-sensitive systems, e.g., breast and uterus cancer, while phytoestrogens such as genistein may also act as androgens through binding with androgen receptors (Dean et al., 2017).

Saponins are widely distributed plant glucosides (sapogenols or sapogenins), classified as anticipins, with strong foam-forming properties, which consist of an aglycone unit (triterpenoid and steroid or steroidal) bound with one or more carbohydrate moieties (González-Lamothe et al., 2009; Savage, 2003; Singh & Chaudhuri, 2018) (Fig. 2). These compounds have been detected in >100 families and include >11 classes of compounds with significant bioactive properties, including cycloartanes, cucurbitanes, dammaranes, tirucallanes, hopanes, lanostanes, lupanes, oleananes, taraxasteranes, and ursanes, as well as steroidal saponins (spirostanols, furostanols, open-chain saponins, and alkaloids) (Man, Gao, Zhang, Huang, & Liu, 2010).

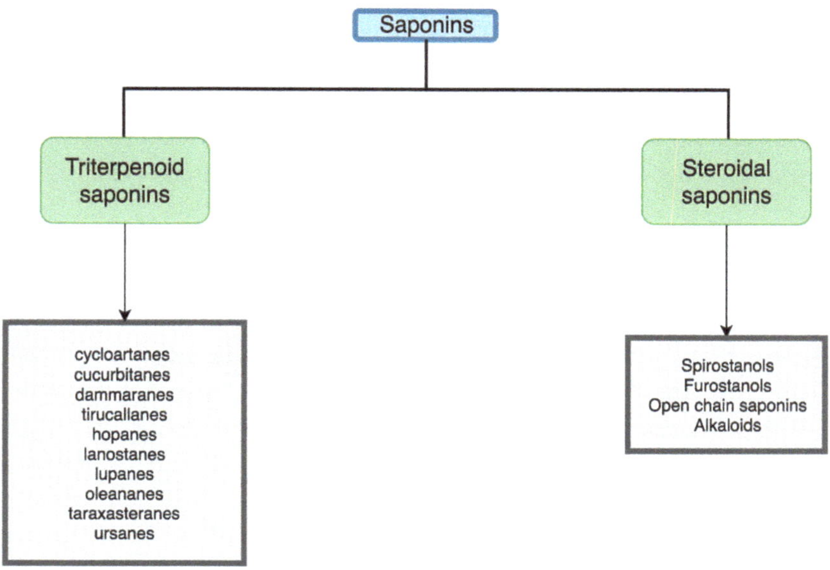

Fig. 2 Classification of saponins.

2. Vegetable sources of phytoestrogens, phytosteroids and saponins

Phytoestrogens include a broad spectrum of chemical compounds with flavonoids being the most commonly detected ones in various edible plants (Champ, 2002). Legumes are considered a good source of flavonoids, including flavones, flavonols, flavanols (Table 1), isoflavones (Table 2) and proanthocyanidins, as well as a good source of lignans and neolignans (Champ, 2002; Wilkinson, Wähälä, & Williamson, 2002). Isoflavones, such as genistein, and daidzein in particular, have a strong estrogenic activity and can be found in various legumes species consumed either as pulses or as green vegetables (Key, 2011; Wang et al., 2018). Not only pulses but green pods of legumes are also good sources of isoflavones, since according to Wang et al. (2018) nine isoflavones, including daidzein, daidzin, genistein, formononetin, ononin, isoerythrinin A, and two newly detected compounds, 2′-hydroxyerythrin A and daidzein-7-*O*-β-D-{6″-[(*E*)-but-2-enoyl]}glycoside, were isolated from green soy beans. Flavones, flavonols and isoflavones can be found also in the edible portions of several vegetable species belonging to some of the most important vegetable families including Amaryllidaceae, Apiaceae, Asparagaceae, Asteraceae, Brassicaceae, Cucurbitaceae, and Solanaceae (Tables 1 and 2).

Other sources of phytoestrogens include berry fruits (*Fragaria* × *ananassa* Duchesne, *Vaccinium macrocarpon*, *Vaccinium myrtillus*, *Rubus idaeus fruticosus*), onion (*Allium cepa* L.), garlic (*Allium sativum* L.), cabbage (*Brassica oleracea* L. var. *capitata*), broccoli (*Brassica oleracea* L. var. *italica* Plenck), pumpkin (*Cucurbita maxima* Duchesne), carrot (*Daucus carota* L.) and beetroot (*Beta vulgaris* L.) which mostly contain lignans and secoisolariciresinol in particular (Bacciottini et al., 2007), but also matairesinol, lariciresinol, pinoresinol, syringaresinol, and medioresinol (Table 3). A study conducted in United Kingdom regarding the consumption of foods that contain high amounts of phytoestrogens, reported that asparagus (*Asparagus officinalis* L.), okra (*Abelmoschus esculentus* (L.) Moench), parsnip (*Pastinaca sativa* L.), cabbage (*Brassica oleracea* L. var. *capitata*), carrots (*Daucus carota* L.), pumpkins (*Cucurbita maxima* Duchesne), sweet potato (*Ipomoea batatas* (L.) Lam.), watercress (*Nasturtium officinale* W.T. Aiton), broccoli (*Brassica oleracea* L. var. *italica* Plenck), and Brussels sprouts (*Brassica oleracea* L. var. *gemmifera* DC.) also contain high amounts of secoisolariciresinol and lesser amounts of matairesinol, while parsley (*Petroselinum crispum* (Mill.) Fuss.) contained

Table 1 Flavones and flavonols content (mg/kg of dry weight) of various vegetable species.

Species	Scientific name	Plant part	Flavones		Flavonols			References
			Apigenin derivatives	Luteolin derivatives	Kempferol derivatives	Quercetin derivatives	Myricetin	
Artichoke	*Cynara cardunculus* var. *scòlymus* (L.) Fiori	Head—outer bracts	270–985	0–74	—	—	—	Lombardo et al. (2010)
		Head—inner bracts	525–1723	0–288	—	—	—	
		Head—receptacle	712–6298	60–1683	—	—	—	
Bishop's weed	*Ammi majus* L.	Fruit	3889	—	n.d.	268.5	—	Shawky, Abou, and Kheir (2018)
		Shoot/leaves	2174	689	n.d.	221.7	—	
Dill	*Anethum graveolens* L.	Fruit	1037	827	n.d.	205.4	—	
		Shoot/leaves	736	—	n.d.	72.5	—	
Celery	*Apium graveolens* L.	Fruit	3026	2047	91.2	303.6	—	
		Shoot/leaves	825	—	n.d.	93.2	—	
Caraway	*Carum carvi* L.	Fruit	1884	—	n.d.	184.1	—	
Cilantro	*Coriandrum sativum* L.	Fruit	1523	—	45.2	104.2	—	
		Shoot/leaves	958	—	n.d.	47.2	—	
Cumin	*Cuminum cyminum* L.	Fruit	2964	1021	289.3	351.7	—	
Fennel	*Foeniculum vulgare* Mill.	Fruit	1452	1789	n.d.	204.5	—	
Parsley	*Petroselinum crispum* (Mill.) Fuss.	Fruit	982	3864	76.3	207.2	—	

Continued

Table 1 Flavones and flavonols content (mg/kg of dry weight) of various vegetable species.—cont'd

Species	Scientific name	Plant part	Flavones		Flavonols			References
			Apigenin derivatives	Luteolin derivatives	Kempferol derivatives	Quercetin derivatives	Myricetin	
		Shoot/leaves	582	341	n.d.	85.8	—	
Anise	*Pimpinella anisum* L.	Fruit	1982	—	n.d.	213.3	—	
Bell pepper	*Capsicum annuum* L.	Fruit	272	—	—	448.5	171.5	Miean and Mohamed (2001)
Broccoli	*Brassica oleracea* L. var. *italica* Plenck	Head	—	74.5	—	60	62.5	
Cabbage	*Brassica oleracea* L. var. *capitata*	Leaves	—	—	—	—	147.5	
Carrot	*Daucus carota* L.	Root	—	37.5	140	55	—	
Cauliflower	*Brassica oleracea* L. var. *botrytis*	Head	—	—	—	219	—	
Celery	*Apium graveolens* L.	Leaves	338.5	80.5	—	—	—	
Chili pepper	*Capsicum frutescens* L.	Fruit	—	1035	—	392	236	
Chinese cabbage	*Brassica oleracea* L. var. *chinensis* (L.) Prain	Leaves	187	—	—	—	31	
Garlic	*Allium sativum* L.	Clove	217	—	—	—	693	
Welsh Onion	*Allium fistulosum* L.	Leaves	—	391	832	1497.5	—	
Pumpkin	*Cucurbita maxima* Duchesne	Fruit	—	—	371	—	—	

Table 2 Isoflavones content (μg/100 g of fresh weight) of various vegetable species.

Species	Scientific name	Daidzein	Genistein	Glycitein	Biochanin A	Formononetin	Daizin	Genistin	Ononin	Sissotrin	References
Red clover—sprouts[a]	*Trifolium pratense* L.	2–109	2–268		6–268	2–1072	0.5–15	6–212	29–720	2–72	Budryn et al. (2018)
Onion	*Allium cepa* L.	1	4	2	—	—	—	—	—	—	Hu et al. (2014)
Garlic	*Allium sativum* L.	2	—	—	—	—	—	—	—	—	
Celery	*Apium graveolens* L.	3	13	3	—	—	—	—	—	—	
Cauliflower	*Brassica oleracea* L. var. *botrytis*	2	4	1	—	—	—	—	—	—	
Cabbage	*Brassica oleracea* L. var. *capitata*	3	1	—	—	—	—	—	—	—	
Cayenne pepper	*Capsicum annuum* L.	2	3	—	—	—	—	—	—	—	
Green pepper	*Capsicum annuum* L.	2	15	—	—	—	—	—	—	—	
Cucumber	*Cucumis sativus* L.	4	2	—	—	—	—	—	—	—	
Carrot	*Daucus carota* L.	24	64	3	—	—	—	—	—	—	
Kidney bean	*Phaseolus vulgaris* L.	8	11	5	—	—	—	—	—	—	
Long bean	*Phaseolus vulgaris* L.	11	22	7	—	—	—	—	—	—	
Tomato	*Solanum lycopersicum* L.	2	51	22	—	—	—	—	—	—	
Eggplant	*Solanum melongena* L.	4	1	2	—	—	—	—	—	—	
Potato	*Solanum tuberosum* L.	8	35	—	—	—	—	—	—	—	
Spinach	*Spinacia oleracea* L.	—	2	99	—	—	—	—	—	—	

Continued

Table 2 Isoflavones content (µg/100 g of fresh weight) of various vegetable species.—cont'd

Species	Scientific name	Daidzein	Genistein	Glycitein	Biochanin A	Formononetin	Daizin	Genistin	Ononin	Sissotrin	References
Okra	*Abelmoschus esculentus* (L.) Moench	—	<1	<1	1	<1	—	—	—	—	Kuhnle et al. (2009)
Leek	*Allium ampeloprasum* L.	<1	<1	<1	1	<1	—	—	—	—	
Onion	*Allium cepa* L.	—	<1	<1	<1	<1	—	—	—	—	
Garlic	*Allium sativum* L.	—	—	<1	2	<1	—	—	—	—	
Celery	*Apium graveolens* L.	<1	<1	<1	—	—	—	—	—	—	
Celery—root	*Apium graveolens* L.	<1	<1	<1	<1	<1	—	—	—	—	
Asparagus	*Asparagus officinalis* L.	—	<1	<1	2	<1	—	—	—	—	
Beetroot—raw	*Beta vulgaris* L.	—	<1	<1	1	—	—	—	—	—	
Cauliflower	*Brassica oleracea* L. var. *botrytis*	<1	<1	<1	<1	<1	—	—	—	—	
Cabbage	*Brassica oleracea* L. var. *capitata*	—	<1	<1	<1	<1	—	—	—	—	
Brussels sprouts	*Brassica oleracea* L. var. *gemmifera* DC.	<1	—	<1	<1	<1	—	—	—	—	
Broccoli	*Brassica oleracea* L. var. *italica* Plenck	<1	<1	<1	<1	—	—	—	—	—	
Turnip	*Brassica rapa* L. var. *rapa*	—	<1	<1	—	—	—	—	—	—	
Pepper—red	*Capsicum annuum* L.	—	<1	<1	4	<1	—	—	—	—	
Chicory	*Cichorium intybus* L.	<1	<1	<1	—	—	—	—	-	-	
Chick peas—dried	*Cicer arietinum* L.	16	79	9	441	62					

Watermelon	*Citrullus lanatus* (Thunb.) Matsum. & Nakai	<1	<1	<1	<1	<1	—	—	—	—
Melon, Cantaloupe	*Cucumis melo* L. var. *cantalupo* Ser.	<1	<1	<1	<1	<1	—	—	—	—
Melon, Galia	*Cucumis melo* L.	<1	<1	<1	<1	<1	—	—	—	—
Melon, Honeydew	*Cucumis melo* L. *Inodorus Group*	1	1	<1	<1	<1	—	—	—	—
Cucumber	*Cucumis sativus* L.	<1	<1	<1	—	<1	—	—	—	—
Pumpkin	*Cucurbita maxima* Duchesne	<1	<1	<1	—	<1	—	—	—	—
Courgette	*Cucurbita pepo* L.	<1	<1	<1	<1	—	—	—	—	—
Carrot	*Daucus carota* L.	2	1	<1	1	<1	—	—	—	—
Fennel	*Foeniculum vulgare* Mill.	<1	<1	<1	<1	—	—	—	—	—
Sweet potato	*Ipomoea batatas* (L.) Lam.	<1	<1	<1	<1	—	—	—	—	—
Lettuce, Iceberg	*Lactuca sativa* L.	<1	<1	<1	<1	—	—	—	—	—
Cress	*Lepidium sativum* L.	1	1	<1	—	<1	—	—	—	—
Watercress	*Nasturtium officinale* W.T. Aiton	<1	<1	<1	<1	<1	—	—	—	—
Parsnip	*Pastinaca sativa* L.	<1	<1	<1	5	—	—	—	—	—
Parsley	*Petroselinum crispum* (Mill.) Fuss.	—	57	<1	<1	<1	—	—	—	—

Continued

Table 2 Isoflavones content (μg/100 g of fresh weight) of various vegetable species.—cont'd

Species	Scientific name	Daidzein	Genistein	Glycitein	Biochanin A	Formononetin	Daizin	Genistin	Ononin	Sissotrin	References
Butter bean—dried	*Phaseolus limensis* L.	24	21	6	—	—	—	—	—	—	
French bean	*Phaseolus vulgaris* L.	12	35	2	1	<1	—	—	—	—	
Haricot bean—dried	*Phaseolus vulgaris* L.	6	14	<1	—	—	—	—	—	—	
Kidney bean—dried	*Phaseolus vulgaris* L.	15	26	32	—	—	—	—	—	—	
Green beans—frozen	*Phaseolus vulgaris* L.	3	16	<1	<1	<1	—	—	—	—	
Tomato	*Solanum lycopersicum* L.	<1	<1	<1	<1	—	—	—	—	—	
Eggplant	*Solanum melongena* L.	<1	<1	<1	<1	<1	—	—	—	—	
Broad bean—fresh	*Vicia faba* L.	<1	<1	<1	<1	<1	—	—	—	—	
Bean sprouts	*Vigna radiata* (L.) R. Wilczek	110	225	16	—	<1	—	—	—	—	
Mung bean—dried	*Vigna radiata* (L.) R. Wilczek	6	15	2	8	<1	—	—	—	—	

[a]Values for red clover sprouts are expressed in mg 100/g of dry weight.

Table 3 Lignans content (μg/100g of fresh weight) of various vegetable species.

Species	Scientific name	SECO[a]	MAT	LARI	PINO	SYR	MED	References
Asparagus	*Asparagus officinalis* L.	183	2	47	49	58	5	Yashin et al. (2018)
Pumpkin	*Cucurbita maxima* Duchesne	971	3	3913	235	0	0	
Artichoke	*Cynara cardunculus* var. *scòlymus* (L.) Fiori	171	0	154	3480	22	57	
Bean	*Phaseolus vulgaris* L.	1267	108	1771	295	12,967	336	
Cauliflower	*Brassica oleracea* L. var. *botrytis*	104	—	—	—	—	—	Konar, Poyrazoĝlu, Demir, and Artik (2012)
White cabbage	*Brassica oleracea* L. var. *capitata*	346	—	—	—	—	—	
Artichoke	*Cynara cardunculus* var. *scòlymus* (L.) Fiori	245	—	—	—	—	—	
Carrot	*Daucus carota* L.	235	—	—	—	—	—	
Iceberg lettuce	*Lactuca sativa* L. var. *capitata* L.	248	—	—	—	—	—	
Green bean	*Phaseolus vulgaris* L.	914	—	—	—	—	—	
Leek	*Allium ampeloprasum* L.	38	0	37	3	—	—	Milder, Arts, van de Putte, Venema, and Hollman (2005)
Onion	*Allium cepa* L.	18	0	19	0	—	—	
Garlic	*Allium sativum* L.	29	0	220	24	—	—	
Beetroot—boiled	Beta vulgaris L. var. conditiva Alef.	1	0	3	0	—	—	
Curly kale	*Brassica oleracea* L. var. *acephala* DC	19	12	599	1691	—	—	

Continued

Table 3 Lignans content (μg/100 g of fresh weight) of various vegetable species.—cont'd

Species	Scientific name	SECO[a]	MAT	LARI	PINO	SYR	MED	References
Cauliflower	*Brassica oleracea* L. var. *botrytis*	4	0	124	58	—	—	
Red cabbage	*Brassica oleracea* L. var. *capitata*	9	0	178	90	—	—	
Sauerkraut	*Brassica oleracea* L. var. *capitata*	67	0	116	133	—	—	
White cabbage	*Brassica oleracea* L. var. *capitata*	8	0	212	568	—	—	
Brussels sprouts	*Brassica oleracea* L. var. *gemmifera* DC	34	0	493	220	—	—	
Broccoli	*Brassica oleracea* L. var. *italica* Plenck	38	0	972	315	—	—	
Sweet pepper—green	*Capsicum annuum* L.	7	0	164	1	—	—	
Sweet pepper—red	*Capsicum annuum* L.	7	0	106	1	—	—	
Chicory	*Cichorium intybus* L.	17	0	6	25	—	—	
Endive	*Cichorium endivia* L.	14	0	15	9	—	—	
Cucumber	*Cucumis sativus* L.	8	0	59	1	—	—	
Courgette	*Cucurbita pepo* L.	18	0	64	37	—	—	
Carrot	*Daucus carota* L.	93	0	60	19	—	—	
Lettuce	*Lactuca sativa* L.	8	0	5	4	—	—	
Iceberg lettuce	*Lactuca sativa* L. var. *capitata* L.	9	0	2	0	—	—	

French bean	*Phaseolus vulgaris* L.	29	0	220	24	—	—	
Pea—in jars	*Pisum sativum* L.	0	0	14	20	—	—	
Tomato	*Solanum lycopersicum* L.	2	0	42	14	—	—	
Potato—nicola, boiled	*Solanum tuberosum* L.	2	0	17	0	—	—	
Potato—redstar, boiled	*Solanum tuberosum* L.	1	0	8	0	—	—	
Spinach—frozen	*Spinacia oleracea* L.	2	0	68	12	—	—	
Sweet corn—in jars	*Zea mays* L.	5	0	2	0	—	—	
Onion	*Allium cepa* L.	43	—	—	—	—	—	Hu et al. (2014)
Garlic	*Allium sativum* L.	58	—	—	—	—	—	
Celery	*Apium graveolens* L.	15	—	—	—	—	—	
Cauliflower	*Brassica oleracea* L. var. *botrytis*	19	—	—	—	—	—	
Cabbage	*Brassica oleracea* L. var. *capitata*	16	—	—	—	—	—	
Cayenne pepper	*Capsicum annuum* L.	24	—	—	—	—	—	
Green pepper	*Capsicum annuum* L.	16	—	—	—	—	—	
Cucumber	*Cucumis sativus* L.	56	—	—	—	—	—	
Pumpkin	*Cucurbita maxima* Duchesne	11	—	—	—	—	—	

Continued

Table 3 Lignans content (μg/100 g of fresh weight) of various vegetable species.—cont'd

Species	Scientific name	SECO[a]	MAT	LARI	PINO	SYR	MED	References
Carrot	*Daucus carota* L.	624	—	—	—	—	—	
Sweet potato	*Ipomoea batatas* (L.) Lam.	12	—	—	—	—	—	
Kidney beans	*Phaseolus vulgaris* L.	25	—	—	—	—	—	
Long beans	*Phaseolus vulgaris* L.	17	—	—	—	—	—	
Tomato	*Solanum lycopersicum* L.	15	—	—	—	—	—	
Eggplant	*Solanum melongena* L.	11	—	—	—	—	—	
Potato	*Solanum tuberosum* L.	18	—	—	—	—	—	
Spinach	*Spinacia oleracea* L.	16	—	—	—	—	—	
Okra	*Abelmoschus esculentus* (L.) Moench	84	<1	—	—	—	—	Kuhnle et al. (2009)
Leek	*Allium ampeloprasum* L.	65	<1	—	—	—	—	
Onion	*Allium cepa* L.	30	<1	—	—	—	—	
Garlic	*Allium sativum* L.	93	4	—	—	—	—	
Celery	*Apium graveolens* L.	6	<1	—	—	—	—	
Celery—root	*Apium graveolens* L.	10	<1	—	—	—	—	
Asparagus	*Asparagus officinalis* L.	149	3	—	—	—	—	
Beetroot—raw	*Beta vulgaris* L.	6	1	—	—	—	—	
Cauliflower	*Brassica oleracea* L. var. *botrytis*	14	<1	—	—	—	—	

Cabbage	*Brassica oleracea* L. var. *capitata*	10	<1	—	—	—	—
Brussels sprouts	*Brassica oleracea* L. var. *gemmifera* DC.	74	<1	—	—	—	—
Broccoli	*Brassica oleracea* L. var. *italica* Plenck	71	<1	—	—	—	—
Turnip	*Brassica rapa* L. var. *rapa*	12	<1	—	—	—	—
Pepper	*Capsicum annuum* L.	11	<1	—	—	—	—
Chicory	*Chicorum intybus* L.	18	<1	—	—	—	—
Watermelon	*Citrullus lanatus* (Thunb.) Matsum. & Nakai	34	<1	—	—	—	—
Melon, Cantaloupe	*Cucumis melo* L. var. *cantalupo* Ser.	16	<1	—	—	—	—
Melon, Galia	*Cucumis melo* L.	11	<1	—	—	—	—
Melon, Honeydew	*Cucumis melo* L. *Inodorus Group*	21	<1	—	—	—	—
Cucumber	*Cucumis sativus* L.	12	<1	—	—	—	—
Pumpkin	*Cucurbita maxima* Duchesne	153	<1	—	—	—	—
Courgette	*Cucurbita pepo* L.	35	<1	—	—	—	—
Carrot	*Daucus carota* L.	114	7	—	—	—	—
Fennel	*Foeniculum vulgare* Mill.	59	12	—	—	—	—
Sweet potato	*Ipomoea batatas* (L.) Lam.	136	<1	—	—	—	—
Lettuce, Iceberg	*Lactuca sativa* L.	7	1	—	—	—	—

Continued

Table 3 Lignans content (μg/100 g of fresh weight) of various vegetable species.—cont'd

Species	Scientific name	SECO[a]	MAT	LARI	PINO	SYR	MED	References
Cress	*Lepidium sativum* L.	15	<1	—	—	—	—	
Watercress	*Nasturtium officinale* W.T. Aiton	45	<1	—	—	—	—	
Parsnip	*Pastinaca sativa* L.	36	<1	—	—	—	—	
Parsley	*Petroselinum crispum* (Mill.) Fuss.	137	<1	—	—	—	—	
Butter bean—dried	*Phaseolus limensis* L.	141	2	—	—	—	—	
French bean	*Phaseolus vulgaris* L.	94	<1	—	—	—	—	
Haricot bean	*Phaseolus vulgaris* L.	106	—	—	—	—	—	
Kidney bean—dried	*Phaseolus vulgaris* L.	88	<1	—	—	—	—	
Green beans—frozen	*Phaseolus vulgaris* L.	38	—	—	—	—	—	
Tomato	*Solanum lycopersicum* L.	4	<1	—	—	—	—	
Eggplant	*Solanum melongena* L.	8	<1	—	—	—	—	
Broad bean—fresh	*Vicia faba* L.	20	<1	—	—	—	—	
Bean sprouts	*Vigna radiata* (L.) R. Wilczek	86	<1	—	—	—	—	
Mung bean—dried	*Vigna radiata* (L.) R. Wilczek	289	—	—	—	—	—	

[a] *SECO*, secoisolariciresinol; *MAT*, matairesinol; *LARI*, lariciresinol; *PINO*, pinoresinol; *SYR*, syringaresinol; *MED*, medioresinol.

both isoflavones and lignans (Kuhnle et al., 2009). According to the same study, with the exception of legumes the most abundant phytoestrogens in vegetable species are lignans and secoisolariciresinol in particular.

In a similar study carried out in Denmark for food products of plant origin, flaxseed was the richest source of lignans, while regarding the vegetable sources *Brassica* and *Allium* vegetables contained mostly lariciresinol and pinoresinol (Milder et al., 2005). Similar results regarding the lignans composition of carrot, cauliflower, cabbage, artichoke and iceberg lettuce were reported by Konar et al. (2012) and Hu et al. (2014) who also suggested secoisolariciresinol as the main lignan regardless of the extraction method. Yashin et al. (2018) reported that pumpkin, artichoke, and beans are also good sources of lariciresinol, pinoresinol, and syrigaresinol, respectively. Agradi et al. (2006) have also reported that several food products and flavoring agents typical in the Mediterranean diet present significant *in vitro* estrogenic activity which contributes to the overall health beneficial effects. Licorice root extracts in particular, contain several compounds demonstrating estrogenic effects, with isoliquiritigenin and liquiritigenin being the most important ones and having anti-proliferative activities against breast cancer (Maggiolini et al., 2002).

In addition, leguminous species are considered as good sources of coumestans and coumestrol in particular, which are mainly detected in seed sprouts (Table 4).

Among the various compounds classified as phytosteroids, phytosterols are one of the most widespread classes (Table 5). Another important class of steroidal compounds is brassinosteroids (BRs), which have been detected in various plant species from different families, although their generic name comes from *Brassica napus*, which was the first species where these compounds were isolated from Grove et al. (1979). BRs include various compounds with castasterone, brassinolide, typhasterol, 6-deoxocastasterone, teasterone and 28-norcastasterone being the most common ones (Fedina, Yarin, Mukhitova, Blufard, & Chechetkin, 2017; Tang, Han, & Chai, 2016) (Fig. 3).

From the above-mentioned compounds, castasterone and brassinolide are the most highly distributed and most biologically active BRs among the plant species (Pavlović et al., 2018; Schmidt, Yokota, Adam, & Takahashi, 1991), while 24-epibrassinolide and 28-homobrassinolide are the most studied compounds due to their commercial availability (Fedina et al., 2017). According to Duran et al. (2017), biological activity of BRs is depended on its structure, with 2α,3α-dihydroxy-7-oxa-6-ketone moiety

Table 4 Coumestrol content of leguminous sprouts.

Species	Scientific name	Coumestrol content[a]	References
Alfalfa sprout	*Medicago sativa* L.	4680 μg/100 g of FW	Franke, Custer, Cerna, and Narala (1994)
Alfalfa sprouts—freeze dried		72,010 μg/100 g of DW	
Red clover sprouts	*Trifolium pratense* L.	28,060 μg/100 g of FW	
Red clover sprouts—freeze dried		561,140 μg/100 g of DW	
Red clover	*Trifolium pratense* L.	105–1570 μg/100 g of DW	Mazur (1998)
Mung bean sprouts	*Vigna radiata* (L.) R. Wilczek	1000 μg/100 g of DW	
Soybean sprouts	*Glycine* max (L.) Merr.	7100 μg/100 g of FW	Ibarreta, Daxenberger, and Meyer (2001)
Soybean sprouts	*Glycine max* (L.) Merr.	13,400 μg/100 g of FW	Nakamura et al. (2001)
Soybean sprouts	*Glycine max* (L.) Merr.	2030 μg/100 g of FW	Surh, Kim, Koh, Kim, and Kwon (2006)
Mung bean sprouts	*Vigna radiata* (L.) R. Wilczek	340 μg/100 g of FW	
Bean sprouts	*Vigna radiata* (L.) R. Wilczek	361 μg/100 g of FW	Kuhnle et al. (2009)
Soybean sprouts	*Glycine max* (L.) Merr.	910 μg/100 g of DW	Oshima, Mine, Nakada, and Yanase (2016)
Red clover sprouts	*Trifolium pratense* L.	n.d.—7520 μg/100 g of DW	Budryn et al. (2018)
Soybean sprouts	*Glycine max* (L.) Merr.	16 μg/100 g of FW	Hu et al. (2014)
Mung bean sprouts	*Vigna radiata* (L.) R. Wilczek	283 μg/100 g of FW	

[a]*FW*, fresh weight; *DW*, dry weight.

Table 5 Sterols content (mg/100 g of fresh weight) of various vegetable species.

Species	Scientific name	β-Sitosterol	Campesterol	Stigmasterol	β-Sitostanol	Campestanol	Brassicasterol	Δ^5-Δ^7-Avenasterols	Stanols	Other sterols	Total sterols	References
		mg 100/g of FW										
Leek	*Allium ampeloprasum* L.	7.3	0.61	0.06	—	0.09	—	—	—	—	8.1	Normén, Johnsson, Andersson, Van Gameren, and Dutta (1999)
Onion	*Allium cepa* L.	7.0	0.82	0.57	—	—	—	—	—	—	8.4	
Celery	*Apium graveolens* L.	7.3	2.7	7.0	0.13	—	—	—	—	—	17	
Celeriac	*Apium graveolens* L. var. *rapaceum*	8.9	2.7	8.6	—	—	—	—	—	—	20	
Kale—boiled	*Brassica oleracea* L. var. *acephala*	7.4	0.91	0.38	0.11	0.07	—	—	—	—	8.8	
Cauliflower	*Brassica oleracea* L. var. *botrytis*	26	9.5	3.7	0.06	—	—	—	—	—	40	
White cabbage	*Brassica oleracea* L. var. *capitata*	9.4	2.8	0.2	0.35	—	—	—	—	—	13	
Chinese cabbage	*Brassica oleracea* L. var. *chinensis* (L.) Prain	6.8	1.6	0.03	0.06	0.03	—	—	—	—	8.5	

Continued

Table 5 Sterols content (mg/100 g of fresh weight) of various vegetable species.—cont'd

Species	Scientific name	β-Sitosterol	Campesterol	Stigmasterol	β-Sitostanol	Campestanol	Brassicasterol	Δ^5-Δ^7-Avenasterols	Stanols	Other sterols	Total sterols	References
		mg 100/g of FW										
Brussels sprouts	*Brassica oleracea* L. var. *gemmifera* DC.	34	8.0	0.38	—	—	—	—	—	—	43	
Broccoli	*Brassica oleracea* L. var. *italica* Plenck	31	6.9	1.1	0.08	0.10	—	—	—	—	39	
Swedish turnip	*Brassica rapa* L. var. *rapa*	14	3.3	0.26	0	0	—	—	—	—	17	
Pepper green	*Capsicum annuum* L.	4.9	2.0	0.33	—	—	—	—	—	—	7.2	
Carrot	*Daucus carota* L.	11	2.2	2.8	0.08	—	—	—	—	—	16	
Fennel	*Foeniculum vulgare* Mill.	5.1	0.35	4.3	0.02	—	—	—	—	—	9.8	
Parsnip	*Pastinaca sativa* L.	18	2.8	7.3	0.19	0	—	—	—	—	27	
Radish	*Raphanus sativus* L.	4.4	—	—	—	—	—	—	—	—	9.0	
Tomato	*Solanum lycopersicum* L.	2.4	0.28	1.7	0.23	0.05	—	—	—	—	4.7	

Potato—boiled	*Solanum tuberosum* L.	2.7	0.23	0.38	0.56	—	—	—	—	—	3.8	
Onion bulb	*Allium cepa* L.	6.2	0.4	0.7	0.1	—	—	—	—	—	7.4	Han, Yang, and Feng (2008)
Scallion	*Allium fistulosum* L.	16.2	5.1	—	0.7	—	—	—	—	—	22.0	
Garlic stalk	*Allium sativum* L.	10.6	1.3	0.6	—	—	—	—	—	—	12.5	
Garlic	*Allium sativum* L.	8.7	2.0	0.5	—	—	—	—	—	—	11.2	
Chinese chive	*Allium tuberosum* Rottler ex Spreng.	11.9	0.6	—	0.5	—	—	—	—	—	13.0	
Celery	*Apium graveolens* L.	8.9	0.5	4.2	0.5	0.1	—	—	—	—	14.2	
Wax gourd	*Benincasa hispida* (Thunb.) Cogn.	0.8	0.1	0.2	—	—	—	—	—	—	1.1	
Cauliflower	*Brassica oleracea* L. var. *botrytis*	40.8	0.9	1.1	—	—	—	—	—	—	42.8	
Cabbage,	*Brassica oleracea* L. var. *capitata*	10.4	2.0	1.2	—	—	—	—	—	—	13.6	

Continued

Table 5 Sterols content (mg/100 g of fresh weight) of various vegetable species.—cont'd

Species	Scientific name	β-Sitosterol	Campesterol	Stigmasterol	β-Sitostanol	Campestanol	Brassicasterol	Δ^5-Δ^7-Avenasterols	Stanols	Other sterols	Total sterols	References
		mg 100/g of FW										
Mini Chinese cabbage	*Brassica oleracea* L. var. *chinensis* (L.) Prain	12.2	0.3	2.5	0.2	—	—	—	—	—	15.2	
Chinese cabbage	*Brassica oleracea* L. var. *chinensis* (L.) Prain	9.6	0.5	2.1	0.5	—	—	—	—	—	12.7	
Broccoli	*Brassica oleracea* L. var. *italica* Plenck	34.5	5.3	0.7	0.4	—	—	—	—	—	40.9	
Rape	*Brassica rapa* L. var. *rapa*	8.5	1.4	—	0.3	—	—	—	—	—	10.2	
Hot pepper green	*Capsicum annuum* L.	8.1	2.8	0.8	1.1	0.4	—	—	—	—	13.2	
Sweet pepper	*Capsicum annuum* L.	2.1	1.1	—	—	—	—	—	—	—	3.2	
Endive	*Cichorium endivia* L. var. *latifolium* Lam.	10.4	1.3	4.3	1.3	—	—	—	—	—	17.3	
Coriander leaf	*Coriandrum sativum* L.	9.3	1.1	7.9	0.4	—	—	—	—	—	18.7	

Cucumber	*Cucumis sativus* L.	3.8	0.2	2.9	0.3	0.1	—	—	—	—	7.3
Zucchini	*Cucurbita pepo* L.	7.3	0.4	8.4	0.4	0.2	—	—	—	—	16.7
Carrot	*Daucus carota* L.	14.0	2.0	3.0	0.2	0.2	—	—	—	—	19.4
Soybean sprout	*Glycine max* (L.) Merr.	7.5	1.8	5.4	0.5	—	—	—	—	—	15.2
Sweet potato	*Ipomoea batatas* (L.) Lam.	6.6	2.3	0.6	0.4	—	—	—	—	—	9.9
Hyacinth bean green	*Lablab purpureus* (L.) Sweet	8.6	1.4	3.9	0.7	—	—		—	—	14.6
Romaine lettuce	*Lactuca sativa* L.	16.7	3.3	7.9	3.1	0.2	—	—	—	—	30.9
Pea	*Pisum sativum* L.	41.4	5.6	4.0	2.7	—	—	—	—	—	53.7
Snow pea	*Pisum sativum* L. var. *saccharatum*	14.0	1.4	2.0	0.3	—	—	—	—	—	17.7
White radish	*Raphanus sativus* L.	3.8	1.2	0.1	–	—	—	—	—	—	5.1
Tomato	*Solanum lycopersicum* L.	2.9	0.6	1.9	0.7	—	—	—	—	—	6.1

Continued

Table 5 Sterols content (mg/100 g of fresh weight) of various vegetable species.—cont'd

Species	Scientific name	β-Sitosterol	Campesterol	Stigmasterol	β-Sitostanol	Campestanol	Brassicasterol	Δ^5-Δ^7-Avenasterols	Stanols	Other sterols	Total sterols	References
		mg 100/g of FW										
Eggplant,	*Solanum melongena* L.	2.0	0.2	0.6	0.1	—	—	—	—	—	2.9	
Potato	*Solanum tuberosum* L.	1.8	0.2	0.9	0.7	—	—	—	—	—	3.6	
Spinach	*Spinacia oleracea* L.	5.4	0.8	2.9	0.8	0.7	—	—	—	—	10.6	
Cowpea	*Vigna unguiculata* (L.) Walp.	19.4	3.8	6.0	0.6	—	—	—	—	—	29.8	
Ginger	*Zingiber officinale* Roscoe	9.9	1.4	1.5	2.0	0.2	—	—	—	—	15.0	
Leek	*Allium ampeloprasum* L.	16.6	1.3	—	—	—	—	tr.	—	1.4	19.4	Piironen, Toivo, Puupponen-Pimiä, and Lampi (2003)
Onion—yellow	*Allium cepa* L.	7.0	0.6	1.2	—	—	tr.	—	—	0.4	9.3	
Dill	*Anethum graveolens* L.	15.5	1.6	13.3	—	—	0.2	0.2	0.6	1.2	32.5	
Red beet	*Beta vulgaris* L.	9.1	0.6	5.7	—	—	tr.	0.2	tr.	1.2	17.1	
Sweden	*Brassica napus* L.	10.3	2.7	—	—	—	—	—	—	tr.	13.2	

Cauliflower	*Brassica oleracea* L. var. *botrytis*	21.6	7.2	1.6	—	—	tr.	tr.	tr.	0.3	31.0
White cabbage	*Brassica oleracea* L. var. *capitata*	11.4	3.1	—	—	—	—	—	0.2	tr.	14.8
Chinese cabbage	*Brassica oleracea* L. var. *chinensis* (L.) Prain	10.3	2.2	2.0	—	—	—	—	tr.	tr.	13.0
Brussels	*Brassica oleracea* L. var. *gemmifera* DC.	27.7	7.1	0.3	—	—	0.2	—	0.8	0.9	37.0
Broccoli	*Brassica oleracea* L. var. *italica* Plenck	28.5	6.7	0.8	—	—	0.3	0.2	—	0.2	36.7
Sweet pepper—red	*Capsicum annuum* L.	16.4	4.2	0.2	—	—	tr.	—	0.5	0.6	22.0
Cucumber	*Cucumis sativus* L.	tr.	—	—	—	—	—	0.2	—	7.6	78.1
Carrot	*Daucus carota* L.	10.4	2.0	2.7	—	—	—	—	tr.	0.2	15.3
Lettuce	*Lactuca sativa* L.	3.7	0.6	2.4	—	—	—	0.2	0.5	1.1	8.5

Continued

Table 5 Sterols content (mg/100 g of fresh weight) of various vegetable species.—cont'd

Species	Scientific name	β-Sitosterol	Campesterol	Stigmasterol	β-Sitostanol	Campestanol	Brassicasterol	Δ^5-Δ^7-Avenasterols	Stanols	Other sterols	Total sterols	References
		mg 100/g of FW										
Parsley	*Petroselinum crispum* (Mill.) Fuss.	13.6	1.2	11.5	—	—	0.2	—	1.8	0.5	28.8	
Bean—frozen	*Phaseolus vulgaris* L.	5.4	0.8	3.6	—	—	—	—	1.2	1.4	12.4	
Pea	*Pisum sativum* L.	21.2	3.6	1.2	—	—	—	1.0	0.5	2.2	29.7	
Tomato	*Solanum lycopersicum* L.	3.3	0.5	1.6	—	—	—	0.4	0.8	0.9	7.4	
Potato	*Solanum tuberosum* L.	3.2	tr.	0.3	—	—	—	0.2	—	1.4	5.1	
Spinach—frozen	*Spinacia oleracea* L.	tr.	0.2	tr.	—	—	0.3	0.4	tr.	9.2	10.2	
Sweet corn—frozen	*Zea mays* L. convar. *saccharata* var. *rugosa*	15.2	5.4	1.5	—	—	—	1.5	3.2	1.7	28.5	

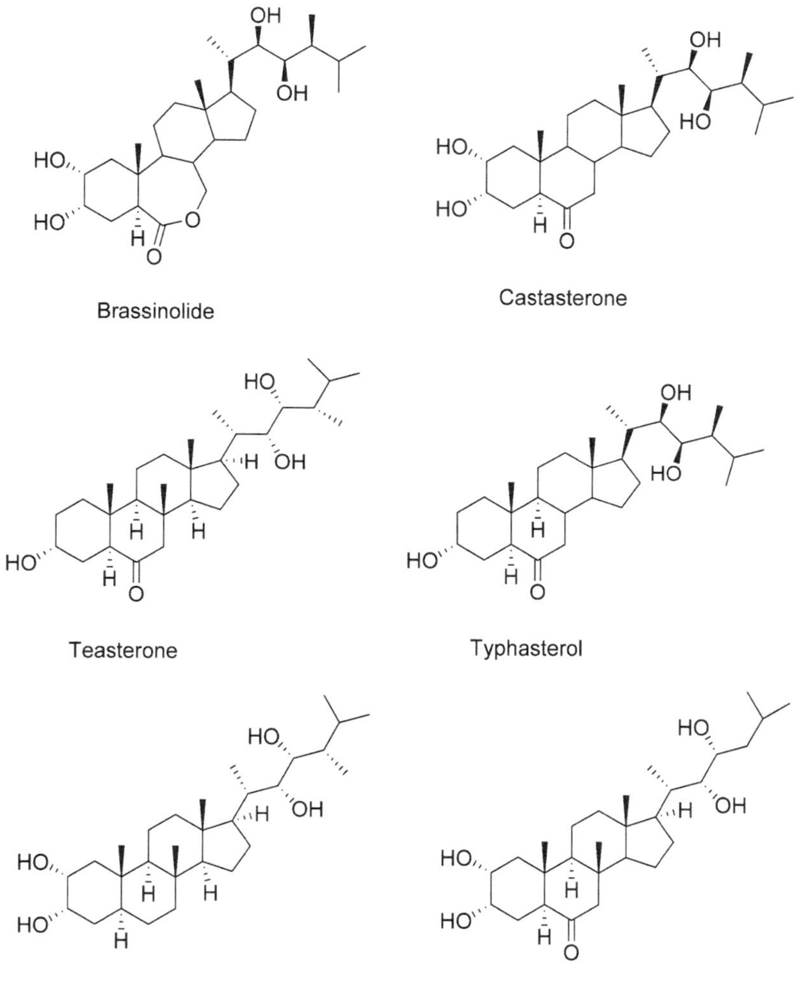

Fig. 3 Chemical structures of the main brassinosteroids.

being essential for such effects. BRs have been isolated from different plant parts and organs and have been attributed with hormonal and plant growth regulatory functions (Bajguz & Tretyn, 2003; Khripach, Zhabinskii, & de Groot, 1999). Withanolides are steroidal lactones which are usually found in species of the Solanaceae family such as tomatillo (*Physalis philadelphica*) which contains ixocarpalactone A and B, physalin B, ixocarpanolide, withaphysacarpin, and philadelphicalactone A, B, and D among others (Maldonado, Pérez-Castorena, Garcés, & Martínez, 2011; Xu et al., 2018).

Solanum species contain significant amounts of steroidal glykoalkaloids such as α-solanine, α-chaconine, solamargine, and solasonine which are associated with significant toxic effects (Al Sinani & Eltayeb, 2017; Wang, Panter, Gaffield, Evans, & Bunch, 2005). Cucurbitacins and cucurbitanes are typical triterpenes commonly found in *Cucurbita* species which exhibit significant bioactive properties (Aeri, Kaushik, & Mir, 2015; Tzortzakis, Chrysargyris, & Petropoulos, 2018). Various kinds of cucurbitacins (>18) have been isolated so far, all of which contain the same cucurbitane skeleton (Dhiman, Gupta, Sharma, Gill, & Goyal, 2012), while distinguishment from each other is based on oxygen's position (Saboo, Thorat, Tapadiya, & Khadabadi, 2013). Cucurbitacin E and cucurbitacin L 2-*O*-β-glucoside have been isolated from seeds of *Cucurbita* species (Elhadi et al., 2014), while cucurbitacin E was also detected in zucchini fruit (Hutt & Herrington, 1985). Two more cucurbitane glucosides, namely cucurbitacin L 2-*O*-β-D-glucopyranoside and cucurbitacin K 2-*O*-β-D-glucopyranoside, and two hexanorcucurbitane glycosides (2,16-dihydroxy-22,23,24,25,26,27-hexanorcucurbit-5-en-11,20-dione 2-*O*-β-D-glucopyranoside and 16-hydroxy-22,23,24,25,26,27-hexanorcucurbit-5-en-11,20-dione 3-*O*-α-L-rhamnopyranosyl-(1 → 2)-β-D-glucopyranoside) have been identified in fruit of *C. pepo* cv. *dayangua* (Wang, Pan, Deng, Xiang, & Gao, 2007).

Isoflavones are considered as analogs to mammalian estrogens compounds and they are usually present in soy bean, as well as in other species of the Fabaceae family. According to Boué et al. (2003), the extracts of seven legumes species showed a preferential estrogenic activity toward estrogen receptor β (ERβ), with kudzu root (*Pueraria lobata* L.) and red clover (*Trifolium pratense* L.) flowers and roots demonstrating the highest antagonistic activity due to isoflavones such as puerarin, daidzin, genistin, daidzein, and genistein. On the other hand, stilbenes are usually found in small berry fruits, grapes and peanuts, as well as in lotus roots (*Nelumbo nucifera*) and taro (*Colocasia esculenta*) plants (El Khawand, Courtois, Valls, Richard, & Krisa, 2018; Li et al., 2013). However, many other food products of vegetable origin may also contain significant amounts of stilbenes (Table 6), including celery, cucurbitaceous and solanaceous species, chives, and radishes among others (Peng et al., 2015).

Moreover, a novel alliospiroside A (Cepa2) has been isolated from shallot dry roots (Abdelrahman, Mahmoud, El-Sayed, Tanaka, & Tran, 2017), while species of *Allium* genus are considered a good source of steroidal saponins with important biological activities, including spirostanols, furostanols, and open-chain or cholestane saponins (Challinor & De Voss, 2013;

Table 6 Stilbenes content of various vegetable species.

Species	Scientific name	*trans*-Piceid	*trans*-Resveratrol	Piceid	Resveratrol	5,4′-Dihydroxy-3-methoxystilbene	3,5-Dihydroxy-4′-methoxystilbene	References
		(μg/100g fresh weight)		(μg/g dry weight)				
Red dock—root	*Rumex bucephalophorus* L.	—	—	—	165	204	239	Kerem, Regev-Shoshani, Flaishman, and Sivan (2003)
Tomato	*Solanum lycopersicum* L.	—	—	—	0.2	–	—	Sebastià, Montoro, León, and Soriano (2017)
Asparagus lettuce	*Lactuca sativa* L. var. *augustana*	n.d.—0.80	n.d.	—	—	—	—	Peng, Xu, Sun, Ying, and Hao (2015)
Cauliflower	*Brassica oleracea* L. var. *botrytis*	n.d.	n.d.—1.14	—	—	—	—	
Celery	*Apium graveolens* L.	43.04–783.29	n.d.—23.12	—	—	—	—	
Chinese cabbage	*Brassica oleracea* L. var. *pekinensis* L.	n.d.—4.80	n.d.—0.79	—	—	—	—	
Amaranth	*Amaranthus*	n.d.—1.30	n.d.—3.87	—	—	—	—	
Leaf lettuce	*Lactuca sativa L. crispa* L.	n.d.—9.22	n.d.—4.28	—	—	—	—	
Leaf mustard	*Brassica juncea* (L.) Czern.	0.10–22.80	n.d.—0.23	—	—	—	—	
Lettuce	*Lactuca sativa* L. var. *capitata* L.	n.d.—0.70	n.d.—0.33	—	—	—	—	

Continued

Table 6 Stilbenes content of various vegetable species.—cont'd

Species	Scientific name	*trans*-Piceid	*trans*-Resveratrol	Piceid	Resveratrol	5,4′-Dihydroxy-3-methoxystilbene	3,5-Dihydroxy-4′-methoxystilbene	References
		(μg/100 g fresh weight)		(μg/g dry weight)				
Pakchoi	*Brassica oleracea* L. var. *chinensis* (L.) Prain	2.42–8.70	n.d.—8.10	—	—	—	—	
Bell pepper	*Capsicum annuum* L.	n.d.—36.22	n.d.—2.60	—	—	—	—	
Bitter gourd	*Momordica charantia* L.	n.d.—5.92	n.d.—19.13	—	—	—	—	
Chili pepper	*Capsicum annuum* L.	n.d.—14.47	n.d.—24.06	—	—	—	—	
Cucumber	*Cucumis sativus* L.	n.d.—6.20	n.d.—2.00	—	—	—	—	
Eggplant	*Solanum melongena* L.	1.40–10.51	n.d.—4.58	—	—	—	—	
Green pepper	*Capsicum annuum* L.	n.d.—89.2	n.d.—42.30	—	—	—	—	
Sponge gourd	*Luffa aegyptiaca* Mill.	n.d.—16.90	n.d.—6.98	—	—	—	—	
Tomato	*Solanum lycopersicum* L.	1.10–12.00	n.d.—7.33	—	—	—	—	
Wax gourd	*Benincasa hispida* (Thunb.) Cogn.	0.53–13.68	n.d.—4.47	—	—	—	—	
Chives	*Allium schoenoprasum* L.	n.d.—58.74	n.d.—0.27	—	—	—	—	

Garlic bulb	*Allium sativum* L.	n.d.—2.00	n.d.—11.27	—	—	—	—	
Garlic-flowering stalk	*Allium sativum* L.	n.d.—0.55	n.d.	—	—	—	—	
Scallion	*Allium fistulosum* L.	n.d.	n.d.—1.20	—	—	—	—	
Carrot	*Daucus carota* L.	n.d.—1.88	n.d.—9.99	—	—	—	—	
Chufa	*Cyperus esculentus* L.	n.d.—3.20	n.d.—0.30	—	—	—	—	
Red radish	*Raphanus sativus* L.	18.16–194.40	n.d.—0.62	—	—	—	—	
White radish	*Raphanus sativus* L.	10.95–35.10	n.d.	—	—	—	—	
Japanese knotweed—sprouts	*Reynoutria japonica* Sieb. & Zucc	—	—	683	64	—	—	Vrchotová, Śerá, and Tříska (2007)
Bohemian knotweed—sprouts	*Reynoutria × bohemica* Chrtek & Chrtková	—	—	215	23	—	—	
Giant knotweed—sprouts	*Reynoutria sachalinensis* (Honda) Nakai	—	—	502	29	—	—	

Sobolewska, Michalska, Podolak, & Grabowska, 2016). Regarding spirostanols, >130 different compounds have been identified in *Allium* species so far, while the most common detected compounds are diosgenin, tigogenin, gitogenin, agigenin, alliogenin, and β-chlorogenin, with the latter being suggested as a chemical marker for identification of garlic presence in food products, since the organosulfuric compounds that are responsible for the characteristic aroma of garlic are very labile and cannot be used for chemical fingerprinting of food products (Ikeda, Tsumagari, & Nohara, 2000; Itakura et al., 2001). For furostanols, >140 compounds have been isolated from various *Allium* species, such as ascalonicoside B, ceparoside C, and chinenoside II (Carotenuto, Fattorusso, Lanzotti, & Magno, 1999; Fattorusso, Iorizzi, Lanzotti, & Taglialatela-Scafati, 2002; Sobolewska et al., 2016), while cholestane saponins are less widespread and only 18 individual compounds have been reported for 10 *Allium* species (Sobolewska et al., 2016). Moreover, Sadeghi, Zolfaghari, Senatore, and Lanzotti (2013) identified all types of saponins (spirostanols, furostanols, and cholestanol) in Persian leek (*A. ampeloprasum* subsp. *persicum* nom. nod), namely two new spirostanols (persicoside A and B), four new furostanols (a mixture of persicoside C1/c2 and D1/D2), and one cholestane (persicoside E).

Saponins have been detected in various edible legumes and vegetable species (Table 7) such as asparagus, lupins, lentils, soy bean, chickpeas, peas, potatoes and various types of beans (Dini, Schettino, Simioli, & Dini, 2001; Murphy, Marques-Lopes, & Sánchez-Tainta, 2018; Sparg, Light, & Van Staden, 2004). Moreover, Fujiwara, Yahara, Ikeda, Ono, and Nohara (2003) isolated a new steroidal saponin from tomato fruit, namely esculeoside, which presented significant cytotoxicity against MCF-7 cancer cells. In another study, Iorizzi, Lanzotti, Ranalli, De Marino, and Zollo (2002) isolated three novel furostanol saponins (capsicoside E, capsicoside F, and capsicoside G), as well as six already known oligoglycosides (capsicoside A and one derivative, a capsicoside G derivative, and capsicoside D and two derivatives) from *Capsicum annuum* L. var. *acuminatum* which presented high antimicrobial activity against yeasts. Moreover, in the study of Huang and Kong (2006) sarsasapogenin M and sarsasapogenin N were isolated for the first time from *Asparagus officinalis* roots, while seven already known steroidal saponins were also detected, namely (25*S*)-5β-spirostan-3β-ol-3-*O*-β-D-glucopyranosyl-(1,2)-[β-D-xylopyranosyl-(1,4)]-β-D-glucopyranoside, (25*S*)-5β-spirostan-3β-ol-3-*O*-β-D-glucopyranosyl-(1,2)-β-D-glucopyranoside, (25*S*)-5β-spirostan-3β-ol-3-*O*-α-L-rhamnopyranosyl-(1,2)-[α-L-rhamnopyranosyl-(1,4)]-β-D-glucopyranoside, (25*S*)26-*O*-β-D-glucopyranosyl-5β-furost-20

Table 7 Saponins content of various vegetable species.

Species	Scientific name	Plant portion	Saponine content	Unit[a]	References
Prickly nightshade	*Solanum torvum* Sw.	Mature fruit	8.60 ± 2.6	mg/g of DW	Koomson, Kwakye, Darkwah, Odum, and Asante (2018)
Prickly nightshade	*Solanum torvum* Sw.	Immature fruit	16.90 ± 9.4	mg/g of DW	
Asparagus	*Asparagus officinalis* L.	Roots	9.2–17.1	mg/g of DW	Zhang et al. (2018)
Garlic	*Allium sativum* L.	Garlic clove	2.9	mg/g of DW	Fenwick and Oakenfull (1983)
Asparagus	*Asparagus officinalis* L.	Stem	15	mg/g of DW	
Swiss chard	*Beta vulgaris* L.	Leaves	58	mg/g of DW	
Chick peas	*Cicer arietinum* L.	Beans	56	mg/g of DW	
Soya beans	*Glycine max* (L.) Merr.	Beans	43	mg/g of DW	
Alfalfa	*Medicago sativa* L.	Raw beans	56	mg/g of DW	
Alfalfa	*Medicago sativa* L.	Sprouts	87	mg/g of DW	
Green beans	*Phaseolus vulgaris* L.	Beans	13	mg/g of DW	
Haricot beans	*Phaseolus vulgaris* L.	Beans	19	mg/g of DW	
Kidney beans	*Phaseolus vulgaris* L.	Beans	16	mg/g of DW	
Green peas	*Pisum sativum* L.	Beans	11	mg/g of DW	
Spinach	*Spinacia oleracea* L.	Leaves and stalks	47	mg/g of DW	
Broad beans	*Vicia faba* L.	Beans	3.5	mg/g of DW	

Continued

Table 7 Saponins content of various vegetable species.—cont'd

Species	Scientific name	Plant portion	Saponine content	Unit[a]	References
Mung beans	*Vigna radiata* (L.) R. Wilczek	Beans	5.7	mg/g of DW	
Mung beans	*Vigna radiata* (L.) R. Wilczek	Shoots	27	mg/g of DW	
Alligator weed	*Alternanthera philoxeroides* Griseb	Leaves	176.0	mg/100 g FW	Singh et al. (2015)
Amaranth	*Amaranthus tricolor* L.	Leaves	169.0	mg/100 g FW	
Amaranth	*Amaranthus viridis* L.	Leaves	164.0	mg/100 g FW	
Malabar spinach	*Basella alba* L.	Leaves	261.7	mg/100 g FW	
Savoy beet	*Beta vulgaris* L. var. *bengalensis* Hort.	Leaves	286.5	mg/100 g FW	
Chick peas	*Cicer arietinum* L.	Leaves	123.0	mg/100 g FW	
Taro	*Colocasia esculent* (L.) Schott	Leaves	368.5	mg/100 g FW	
Pumpkin	*Cucurbita moschata* Duchesne	Flower	159.0	mg/100 g FW	
Pumpkin	*Cucurbita moschata* Duchesne	Leaves	143.5	mg/100 g FW	
Mexican coriander	*Eryngium foetidum* L.	Leaves	256.5	mg/100 g FW	

Water spinach	*Ipomea aquatica* Forsk	Leaves	229.5–436.5	mg/100 g FW	
Wild mint	*Mentha arvensis* L.	Leaves	128.5	mg/100 g FW	
Common purslane	*Portulaca oleracea* L.	Leaves	166.5	mg/100 g FW	
Arrowleaf elephant ear	*Xanthosoma mafafa*	Leaves	481	mg/100 g FW	Wallace, Marfo, and Plahar (1998)
Frog's gourd	*Ipomoea involucrata*	Leaves	424	mg/100 g FW	
African lettuce	*Launaea taxaracifolia*	Leaves	386	mg/100 g FW	
Asthma plant	*Euphorbia hirta*	Leaves	6.7	mg/100 g FW	
Wild asparagus	*Asparagus brachiphyllus* Turcz.	Spears	242.2	mg/100 g FW	Jaramillo-Carmona et al. (2017)
Wild asparagus	*Asparagus maritimus* (L.) Mill.	Spears	16.8–142.7	mg/100 g FW	
Wild asparagus	*Asparagus officinalis* L.	Spears	2.2	mg/100 g FW	
Wild asparagus	*Asparagus prostrates* Dumort	Spears	51.4	mg/100 g FW	
Wild asparagus	*Asparagus pseudoscaber* Grecescu	Spears	88.3	mg/100 g FW	

[a]*FW*, fresh weight; *DW*, dry weight.

(22)-ene-3β,26-diol-3-*O*-β-D-glucopyranosyl-(1,2)-β-D-glucopyranoside, yamogenin, β-sitosterol, and sitosterol-β-D-glucoside. *Asparagus* species have been also reported to contain various furostanol and spirostanol steroidal saponins such as aspaoligonins, protodioscin, pseudo-protodioscin and dioscin (Jaiswal, Liang, Ho, Chen, & Zhao, 2014; Kim et al., 2005). For example, *Allium filicinus* contains asparagusin A, aspafilioside A and B, and filiasparosides A, B, C, E, F, and G (Wu et al., 2010; Zhou et al., 2007), while dried roots of *Allium cochinchinensis* contain asparacoside asparacosins A and B, asparenydiol, and nyasol among other spirostanol saponins (Zhang et al., 2004). Spirostanol saponins have been also detected in *Allium porrum* bulbs by Carotenuto et al. (1999), while Abdelrahman et al. (2017) isolated Cepa2 from shallot (*Allium cepa* L. Aggregatum group). Other saponins detected in *Allium* species, include minutosides (A–C), and alliogenin and neoagigenin which were isolated from *A. minutiflorum* bulbs by Barile et al. (2007) and demonstrated significant antifungal properties against various soil-borne and air-borne plant pathogens. *Solanum* species contain various steroidal alkaloids and steroidal alkaloid saponins, such as solasodine and tomatidine (Lu, Luo, & Kong, 2011; Sahu, Banerjee, Mondal, & Mandal, 2008).

Phytoecdysteroids are another class of steroidal compounds with >400 identified compounds detected in various plant species, while the most commonly found compounds are 20-hydroxyecdysone and polypodine B (Cao, Riu, & Boo, 2018; Dinan, Harmatha, Volodin, & Lafont, 2009). Phytoecdysteroids are used from plants as deterrent agents against herbivorous insects and have been isolated in various amounts in 16 species of the genus *Asparagus*, including *A. officinalis* (Dinan, Savchenko, & Whiting, 2001). In addition, other vegetable species also contain phytoecdysteroids including spinach and other species of the Chenopodiaceae family (Dinan, Whiting, & Scott, 1998; Rothová et al., 2014), while quinoa seeds are also considered a rich source of such compounds (Kumpun et al., 2011; Zhu et al., 2001).

2.1 Biosynthesis and functions

Phytoestrogens may act directly after ingestion of food products or after biotransformation through gut microbes to more bioavailable and active compounds such as equols, urolithins, and enterolignans in the case of isoflavones, ellagitanins, and lignans, respectively (Landete et al., 2016). Phenolic compounds content in plants and by extension phytoestrogens content depends upon several pre- and postharvest factors, including genetic material, and agronomic and postharvest factors (Tomás-Barberán & Espín, 2001). Flavonoids are the most common phenolic compounds being responsible for plant defense against various biotic and abiotic stress factors (Calani, Dall'Asta, Bruni, & Del Rio, 2014). They are synthesized *via* the phenylpropanoid pathway (Petrussa et al., 2013), while lignans are also produced after the oxidative dimerization of two phenylpropanoid units (Imai, Nomura, & Fukushima, 2006). The main lignans detected in edible plants include lariciresinol, pinoresinol, secoisolariciresinol, and matairesinol, which are usually found as glycosidic conjugates (Sok, Cui, & Kim, 2009; Thompson, Boucher, Liu, Cotterchio, & Kreiger, 2006) (Fig. 4).

Ellagitannins are produced through the shikimate biosynthetic pathway and play a significant role in plant × insect interaction presenting diverse mechanisms of action depending on ellagitannins structure (Salminen, 2014). Stilbenes derived from the phenylpropanoid pathway and occur in a limited number of plant families, since the main enzyme responsible for stilbenes biosynthesis (stilbenes synthase) is not expressed in many plant species (Chong, Poutaraud, & Hugueney, 2009; Rivière, Pawlus, & Mérillon, 2012).

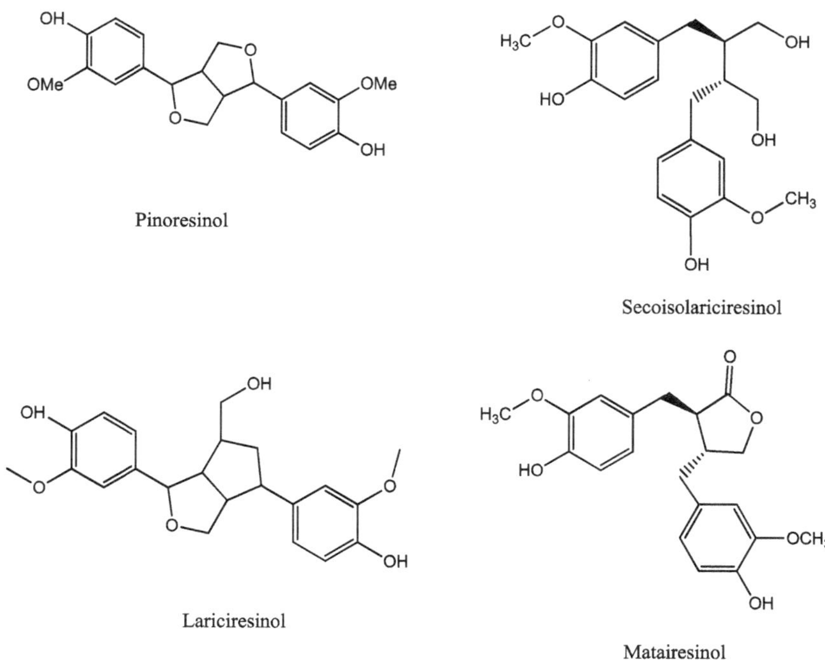

Fig. 4 The chemical structures of the main lignans.

The main function of stilbenes is related with defense mechanisms of plants against pathogens and herbivores, and abiotic stress factors, while they have been also attributed with health effects (Chong et al., 2009). Stilbenes naturally occur in the form of *E* and *Z* stereoisomers, with *Z* form showing higher biological activities than *E* isomers (Roupe, Remsberg, Yáñez, & Davies, 2006).

Phytosteroids are metabolic products of phytosterols and sapogenins and based on the primary substrate several steroidal compounds may be produced (Fig. 5), including spirostanols, saponins and cardenolides from cholesterol, withanolides from 24(28)-methylene cholesterol, brassinosteroids and boldenone from campesterol, C-19 and C-22 steroids from β-sitosterol, and C-21 steroids from sapogenins (Nes, 2011; Wang, Yao, & Wei, 2011).

Catabolism of phytosterols takes place through microbial transformation initiated by cholesterol oxidases (Aparicio & Martín, 2008). All steroids

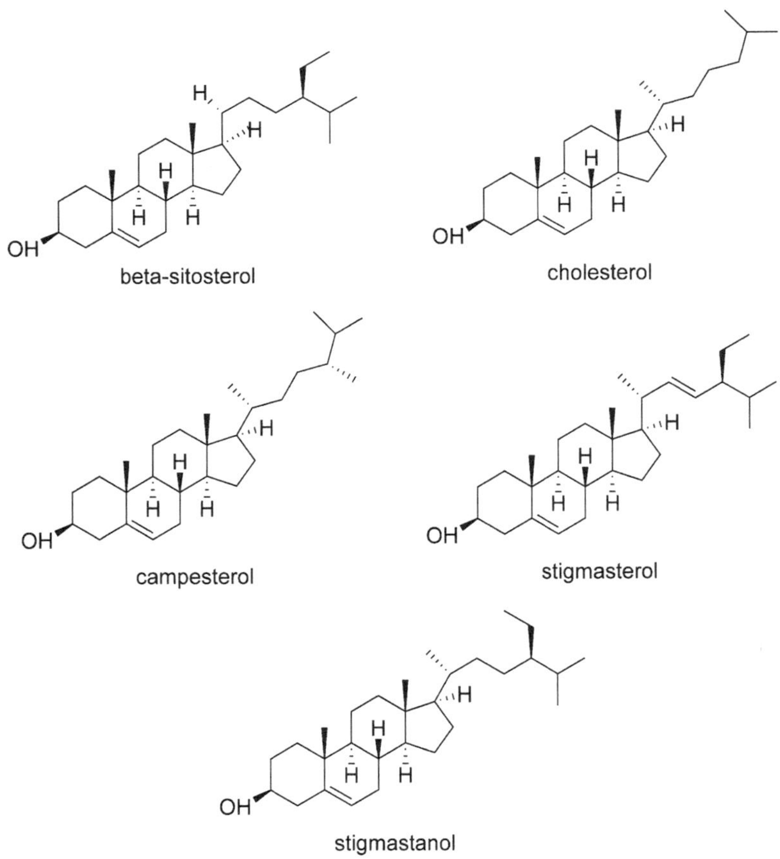

Fig. 5 Chemical structures of the main phytosterols.

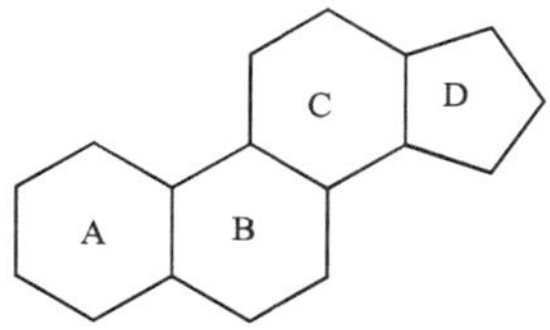

Fig. 6 Typical ring skeleton (1,2-cyclo-pentanoperhydrophenanthrene) of steroidal compounds.

contain the parental sterol frame (1,2-cyclopentanoperhydrophenanthre ne ring skeleton, Fig. 6) and differences in chemical structure refer to the side chain and the degrees of unsaturation and substitution (Nes, 2011). Particularly in spinach, the phytoecdysteroids (20-hydroxyecdysone, 20*E*) pools' turnover was studied in order to investigate its stability and its involvement in plant defense mechanisms (Schmelz, Grebenok, Ohnmeiss, & Bowers, 2000). The results of that study demonstrated that 20*E* levels increased after root damage through the octadecanoic acid pathway and the regulation of endogenous jasmonate biosynthesis, in a manner similar to synthesis of other compounds that contribute to plant defense.

According to Dinan, Savchenko, Whiting, and Sarker (1999), these compounds may act as agonists or antagonists to insect ecdysteroid receptor complex, therefore they may be used as control agents of invertebrate pests by disrupting their hormones (Dinan et al., 2001). In the study of Cao et al. (2018), it was pointed out that there are differences in 20*E* biosynthesis between male and female spinach plants, especially during the reproductive stage where ecdysteroids have an active role in protection of reproductive organs against insect attacks. Moreover, apart from protection against insect pests ecdysteroids may have an effect on CO_2 binding proteins and water cleavage (Kamlar et al., 2015), as well as in leaf senescence through the regulation of lipid composition of leaves cells (Fedina et al., 2017). Other compounds such as cucurbitacins and cucurbitanes, are commonly found in species of Cucurbitaceae family, are associated with bitter taste of fruit and are responsible for plant defense against pests and pathogens (Izawa, Amino, Kohmura, Ueda, & Kuroda, 2010; Ujváry, 2010).

Brassinosteroids are highly active steroidal compounds with significant hormonal functions which regulate several developmental and metabolic processes, as well as plant response to environmental (biotic and abiotic) stress conditions either independently or in conjunction with other plant hormones (Choudhary, Yu, Yamaguchi-Shinozaki, Shinozaki, & Tran, 2012; Hasan, Hayat, & Ahmad, 2011; Janeczko, Koscielniak, Pilipowicz,

Szarek-Lukaszewska, & Skoczowski, 2005; Shahzad et al., 2018). They are classified according to the length and the substitutions of their side chain, with >70 compounds being isolated and identified in various plant species (Bajguz & Tretyn, 2003). So far several studies reported the role of BRs in plant resistance against various stress factors, including salinity, drought, extreme temperatures, heavy metals toxicity and combined stresses (Ali, 2017; Anuradha & Rao, 2007; Pociecha, Dziurka, Oklestkova, & Janeczko, 2016; Sharma, Kaur, & Pati, 2017). According to Pavlović et al. (2018), drought tolerance is correlated with increased levels of typhasterole, and brassinolide and castasterone, in kale (*Brassica oleracea* var. *acephala* cv. IJK9) and Chinese cabbage (*B. rapa* ssp. *pekinensis*), respectively. Moreover, castasterone has been reported to increase plant tolerance against heavy metals toxicity through the restoration of photosynthetic activity (Kaur et al., 2017) and regulation of antioxidant enzymes responses (Kaur et al., 2018; Yadav et al., 2018).

Saponins are highly involved in plant defense mechanisms against plant diseases (González-Lamothe et al., 2009). Steroidal saponins have been also reported to present antifungal properties with efficacy depended on the saponin type and the chemical structure and number sugar moiety (Teshima et al., 2013; Yang et al., 2006), while the possible mechanisms of action involve the reduction in catalase activity and the content of proteins in pathogens, as well as the decreased rate of glucose usage and the destruction of fungi cell membranes (Zhang et al., 2006). A typical example of the defensive role of saponins in vegetable crops is α-tomatine which provides protection to tomato plants against various pathogenic fungi and bacteria (González-Lamothe et al., 2009), while virulence of pathogens is depended on the production of tomatine-detoxifying enzymes which help overcome plant defense barriers (Bouarab, Melton, Peart, Baulcombe, & Osbourn, 2002; Osbourn, 1996). Saponins are abundant in *Allium* species and as mentioned in the study of Itakura et al. (2001), although the most characteristic compounds of garlic are the organosulfuric ones, sapogenins and β-chlorogenin in particular may be a useful means for distinguishing garlic from other *Allium* species.

2.2 Health effects

Phytoestrogens are very common in human diet since several food products of plant origin contain various amounts of polyphenols with estrogenic activity (Di Carlo, Mascolo, Angelo, & Capasso, 1999; Orlikova,

Tasdemir, Golais, Dicato, & Diederich, 2011). When phytoestrogens are digested and pass through the digestive track, they are metabolized in the gut through various enzymes into bioactive by-products that present weak estrogenic activity due to their similarity with estrogens (Bedell et al., 2014). They can also present anti-estrogenic effects since they usually antagonize estrogens for binding with receptors (Landete, 2012).

There are several clinical and epidemiological studies which suggest that intake of phytoestrogens may have protective effects against several chronic diseases such as cancer, cardiovascular disease, menopause syndrome, and osteoporosis among others (Eden, 2012; Kuo, 1996; Landete et al., 2016) (Table 8).

Therefore, plant extracts can be used in hormone replacement therapies providing the positive effects against climacteric complaints and without having adverse side effects (Seidlova-Wuttke, Jarry, & Wuttke, 2013). For example, broccoli extracts have shown anti-estrogenic activity resulting in *in vivo* and *in vitro* apoptosis of MCF-7 breast cancer cells (Eman, Abeer, Samah, Amr, & Walaa, 2017). A high polyphenol intake on a daily basis through consumption of various products has been associated with low cardiovascular disease risk factors through improvement of blood lipid profile in several studies from various countries (Li et al., 2013; Ock, Sang, & Song, 2007; Ovaskainen et al., 2008; Pérez-Jiménez et al., 2011). Flavonoids are the most commonly polyphenols found in vegetables species, with isoflavones having high estrogenic activity due to its chemical resemblance with estrogens (Seleem et al., 2017). Isoflavones such as genistein have been reported to exhibit protective effects against prostate cancer through its high affinity to androgens signaling and/or modulation of estradiol metabolism (Nelles et al., 2011), as well as against BG-1 ovarian cancer through the suppression of signaling pathways of estrogen and insulin-like growth factor-1 receptors Hwang et al., 2013). Moreover, phytoestrogen supplementation has been associated with heart protective effects (Campesi, Romani, Marino, & Franconi, 2014). However food supplements rich in estrogens should be consumed with caution especially from individuals with heart disease problems (Haines et al., 2012). Other isoflavones (biochanin A and formonetin) have been suggested to have a protective effect on lumbar spine of post-menopausal women (Atkinson, Compston, Day, Dowsett, & Bingham, 2004). Moreover, the epidemiological study of Touillaud et al. (2009) demonstrated that a high phytoestrogens and isoflavones in particular diet through a high consumption of fruit and vegetables may affect estrogen receptors of breast cancer in premenopausal women. Other compounds

Table 8 The main health effects of phytoestrogens.

Compound	Health effect	Mechanism of action	References
Flavones			
Apigenin	Inhibition of human mammary carcinoma	Strong cytotoxic agent against breast cancer cells with a mechanism independent of estrogen receptors	Clarke and Wiseman (2004)
	Suppress 12-*O*-tetradecanoylphorbol-3-acetate (TPA)-mediated tumor promotion of mouse skin	Possibly through suppression of protein kinase C activity and nuclear oncogene expression	Gupta and Prakash (2014)
	Antibacterial, anti-inflammatory, diuretic, and hypotensive, promotes also smooth muscle relaxation		Gupta and Prakash (2014)
	Anti-*Candida* activity	Through inhibition of efflux pumps, which results in induction of cell death or apoptosis	Seleem, Pardi, and Murata (2017)
Luteolin	Anti-breast cancer	Potent inhibitors of aromatase activity. Aromatase is the main enzyme that participates in the transformation of testosterone into estradiol and is involved in breast cancer pathology. Luteolin downregulated also aromatase gene expression	Lecomte et al. (2017)
	Anti-inflammatory, anti-mutagenic, and antibacterial activities		Gupta and Prakash (2014)
	Anti-*Candida* activity	Through inhibition of efflux pumps, which results in induction of cell death or apoptosis	Seleem et al. (2017)

	Antioxidant: Inhibition of oxidative DNA damage induced by UV light irradiation in HL-60 cells with $IC_{50}\backslash 1\ \mu M$	The *in vitro* antioxidant activity of flavonoids is related to the number and position of hydroxyl groups	Arroo et al. (2014)
	Anti-mutagenic, anti-inflammatory and antibacterial activities		Dillard and German (2000)
Flavonols			
Kempferol	Inhibits *Candida* species development	Through inhibition of efflux pumps, which results in induction of cell death or apoptosis	Seleem et al. (2017)
	Anti-inflammatory	Cause inhibition of inducible nitric oxide synthase, cyclooxygenase-2 and reactive C-protein, and down-regulation of the nuclear factor kappaB pathway in Chang liver cells	García-Mediavilla et al. (2007)
	Anticancer, antioxidant and anti-inflammatory	Acts on a range of intracellular and extracellular targets involved in the cell signaling pathways that in turn are known to regulate the hallmarks of cancer growth progressions like apoptosis, cell cycle, invasion or metastasis, angiogenesis and inflammation	Kashyap et al. (2018)
	Anticancer	Capable of inhibiting binding of agonist and agonist-induced formation of the AHR/ARNT DNA-binding complex and upregulation of the AHR target gene, CYP1A1	Puppala, Gairola, and Swanson (2007)

Continued

Table 8 The main health effects of phytoestrogens.—cont'd

Compound	Health effect	Mechanism of action	References
Quercetin	Antioxidant, anti-tumor, and anti-inflammatory activity	Cancer cell-specific inhibition of proliferation	Jeong, An, Kwon, Rhee, and Lee (2009)
	Cancer prevention	Quercetin aglycone has been shown to interact with some receptors, particularly an aryl hydrocarbon receptor, which is involved in the development of cancers induced by certain chemicals	Murakami, Ashida, and Terao (2008)
Lignans			
Secoisolariciresinol	Antioxidant	Decreases the production of both primary and secondary oxidation products	Hano et al. (2017)
	Estrogenic activity, anti-inflammatory and apoptotic effects	Plant lignans can be converted by a consortium of intestinal bacteria to enterolignans which have tissue-specific estrogen receptor activation, anti-inflammatory and apoptotic effects	Yoder, Lancaster, Hullar, and Lampe (2015)
	Weak estrogenic or anti-estrogenic effects, antioxidant activity, anti-carcinogenic	Ability to induce phase 2 proteins and/or inhibit the activity of certain enzymes	Adolphe, Whiting, Juurlink, Thorpe, and Alcorn (2010)
Matairesinol	Antioxidant activity; protective effects in the estrogen-dependent conditions and estrogen-dependent diseases, positive effect on insomnia and cognitive functions, i.e., neural disorders, such as Alzheimer's disease	Scavenger of free radicals or by the formation of chelating complexes with metal ions	Nikolić et al. (2017)

Lariciresinol, Syringaresinol, Medioresinol	Anti-carcinogenic effects; anticancer		Adlercreutz (2007)
Medioresinol	Antifungal effects, leishmanicidal activity, cardiovascular disease risk reduction	Medioresinol has effects on mitochondria and induce the accumulation of ROS in *Candida albicans* cells	J. H. Hwang, Hwang, Liu, Woo, and Lee (2012)
Isoflavones			
Daidzein, Genistein, Glycitein	Estrogenic or anti-estrogenic, antioxidative, anti-proliferative, antiviral, antibacterial, insecticidal or fungistatic, cardioprotective, antiatherogenic, hypocholesterolemic, bone-maintaining, cancer protective, anti-carcinogenic Beneficial effects on breast and prostate cancers, menopausal symptoms, osteoporosis, atherosclerosis and stroke, and neurodegeneration	Bind both estrogen receptors (ER) a and b, and to mimic estrogenic actions	Branca and Lorenzetti (2005) and Cederroth, Zimmermann, and Nef (2012)
Daidzein, Genistein, Biochanin A, Formononetin	Antioxidant activity, protective effects in the estrogen-dependent conditions and estrogen-dependent diseases, positive effect on insomnia and cognitive functions, i.e., neural disorders, such as Alzheimer's disease	Scavenger of free radicals or formation of chelating complexes with metal ions	Nikolić et al. (2017)
Daidzein, Daidzin, Genistein, Formononetin, Ononin, Isoerythrinin A	Antimicrobial and antioxidant activity		Wang et al. (2018)

Continued

Table 8 The main health effects of phytoestrogens.—cont'd

Compound	Health effect	Mechanism of action	References
Daidzein, Genistein, Glycitein, Biochanin A, Formononetin, Daizin, Genistin, Ononin, Sissotrin	Potential adverse health effects	May act as endocrine disruptor and have (anti) estrogenic effects by binding to the estrogen receptors	Rietjens, Louisse, and Beekmann (2017)
Genistein	Protective effects against prostate cancer	High affinity to androgens signaling and/or modulation of estradiol metabolism	Nelles, Hu, and Prins (2011)
	BG-1 ovarian cancer	Suppression of signaling pathways of estrogen and insulin-like growth factor-1 receptors	Hwang et al. (2013)
Stilbenes			
Resveratrol	Anti-aging, anti-carcinogenic, anti-inflammatory, and antioxidant effect which may be relevant to chronic diseases and/or longevity in humans		Smoliga, Baur, and Hausenblas (2011)
	Antioxidant, anti-aging, anti-inflammatory effect, cancer prevention, cholesterol-lowering effect, enhanced insulin sensitivity, and increased lifespan		Rimando and Suh (2008)
Coumestans			
Coumestrol	Reduce the risk of breast and prostate cancer	Estrogen-like activity, by bonding to the same receptors)	Markaverich, Webb, Densmore, Rebecca, and Gregory (1995)

such as chalcones have also demonstrated significant chemopreventive and chemotherapeutic activities against carcinogenesis (Orlikova et al., 2011), while Calliste et al. (2001) highlighted the importance of conformation and position of hydroxy group for their estrogenic activity and antioxidant properties. *In vitro* studies with murine models have suggested that phytoestorgens administration may improve bone formation and suppress bone resorption, while they may also increase mineral density of bones contents of α1(I) collagen, alkaline phosphatase, osteocalcin, and osteopontin (Chiang & Pan, 2013).

Epidemiological studies confirmed chemopreventive activity of stilbenes against colon cancer, especially of resveratrol and pterostilbene which demonstrated high biological activity (Moretón-Lamas, Lago-Crespo, Lage-Yusty, & López-Hernández, 2017; Rimando & Suh, 2008), while Roupe et al. (2006) have also reported anti-inflammatory and antioxidant activities. However, the great diversity in chemical structure of stilbenes with various substituents and degrees of polymerization may have a detrimental role on bioavailability and *in vivo* efficiency of stilbene compounds (El Khawand et al., 2018). Other health effects of phytoestrogens include antimicrobial activity against *Candida albicans* and various bacterial strains (Seleem et al., 2017; Wang et al., 2018).

However, discrepancies in scientific literature exist with reports from many *in vivo* and *in vitro* studies reporting negative effects depending on human or animal model status (Humfrey, 1998; Patisaul & Jefferson, 2010; Whitten, Lewis, Russell, & Naftolin, 1995), while there are great concerns regarding the long-term exposure to these compounds and the interpretation of the meta-analysis of epidemiological studies results (Jargin, 2014). Moreover, there are singular reports of modifications in gender-related behavior through feminization and adverse effects on the reproductive system after soy consumption, while addition of soy products in animal rations has been reported to affect fertility and result in derangement of sexual development and behavior (Bateman & Patisaul, 2008; Chandrareddy, Muneyyirci-Delale, Mcfarlane, & Murad, 2008; Jefferson et al., 2009; Santell, Chang, Nair, & Helferich, 1997).

Phytosteroids have been also associated with important health effects since they interact with steroids receptors (Table 9). According to Gorelick-Feldman et al. (2008), phytoecdysteroids increased *in vivo* protein synthesis in mouse skeletal muscle cells, while they also increased grip strength *in vivo*. Moreover, *in vitro* studies with steroids from *Asparagus officinalis* roots demonstrated significant cytotoxicity against human A2780, HO-8910,

Table 9 The main health effects of phytosteroids.

Class of compounds	Health effect	Mechanism of action	References
Steroidal saponins	Natural agent for cancer therapy	Cytotoxicity to cancer cells by induction of apoptosis	Tong et al. (2012)
	Antifungal activity	Associated with their aglycone moieties and the number and structure of monosaccharide units in their sugar chains	Yang et al. (2006)
Spirostanols	Anti-inflammatory	Inhibition of superoxide anion generation and elastase release	Lee et al. (2014)
	Cytotoxic activity		Quan, Koyanagi, Komada, and Saito (2005)
Furostanols	Anti-proliferative	Cytotoxic activities against FaDu (human hypopharyngeal carcinoma) and Detroit 562 (human metastatic pharyngeal squamous cell carcinoma)	Xiang, Wang, Yi, and He (2016)
	Anti-inflammatory	Inhibitory effects on nitric oxide production	
Steroid alkaloids with an intact furan ring	Teratogenic	A basic nitrogen positioned α with respect to the steroidal plane and at appropriate distance beyond the D ring confers the teratogenicity on the molecule	Keeler (1978)
Steroid alkaloids	Toxic	Act as neurotoxins	Koleva, van Beek, Soffers, Dusemund, and Rietjens (2012)

Ecdysteroids	Potent chemical agents to prevent or delay both collagenase-related skin damages and oxidative stress	Potency to inhibit this matrix metalloproteinase and to chelate the iron ion acting as an electron donor	Nsimba, Kikuzaki, and Konishi (2008)
	Beneficial effects on fat and muscle tissues, may be able to prevent the metabolic syndrome and sarcopenia	By a non-estrogenic mechanism	Seidlova-Wuttke, Ehrhardt, and Wuttke (2010)
Cardenolides	Decrease rate and increased intensity of heartbeat, over dosage leads to cardiotoxicity	It is associated with special structure of aglycone is modified in regard to solubility and the nature of conjugated residue	Mueen Ahmed, Khan, and Shivananda (2009)
Withanolides	Chemotherapeutic agents	Induction of apoptosis at nanomolar concentrations of withanolides	Subramanian et al. (2014)
Brassinosteroids	Anti-proliferative activity, pro-apoptotic activity, and cell growth inhibitory responses in several human cancer cell lines with no effects on normal non-tumor cell growth	Causing cell cycle blockade and apoptosis of both hormone-sensitive and -insensitive human breast cancer cells	Steigerová et al. (2010, 2012)
	Anti-proliferative, anticancer, antiangiogenic, antiviral and antibacterial properties in animal cell systems, inhibit replication of viruses in confluent human cell cultures, sometimes with high selectivity indices, inducing cytotoxic effects in various types of cancer cells but not normal human cells		Oklestkova, Rárová, Kvasnica, and Strnad (2015)

Continued

Table 9 The main health effects of phytosteroids.—cont'd

Class of compounds	Health effect	Mechanism of action	References
	Antiviral effect against RNA and DNA viruses	Limiting virus protein synthesis and mature viral particle formation	Wachsman and Castilla (2012)
	Immunomodulatory and neuroprotecting activity	Blocked HSV-1-induced activation of NF-κB by inhibiting its translocation to the nucleus of infected corneal and conjunctival cells *in vitro*, as well as significantly reduced the secretion of TNF-α in infected NHC cells	Michelini, Berra, and Alché (2008)
	Antioxidant and neuroprotective activities in a mammalian neuronal cell line	Reduces the levels of intracellular reactive oxygen species and modulates superoxide dismutase, catalase, and glutathione peroxidase activities	Carange, Longpré, Daoust, and Martinoli (2011)
	In vitro anti-proliferative effects in animal cells	Cytotoxicity in PC-3 cells; activating polyamine catabolic machinery in prostate cancer cells	Obakan, Arisan, Coker-Gurkan, and Palavan-Unsal (2014) and Wu and Lou (2007)
Boldenone	Cause infertility	Reduction in proliferating cell nuclear antigen-positive spermatogonia	Tousson, El-Moghazy, Massoud, and Akel (2012)

Eca-109, MGC-803, CNE, LTEP-a-2, KB, and mouse L1210 cancer cell lines (Huang, Lin, & Kong, 2008). Due to their structural similarity with mammalian steroids, phytosteroids may also interfere with endocrine system (Dean et al., 2017), while compound A isolated from species of the *Salsola* genus exhibited anti-inflammatory and anticancer activities (Lesovaya et al., 2015). Steroids from *Allium chinense* bulbs exhibited protective effects against oxidative injury of rat cardiac H9C2 cells indicating cardioprotective effects (Ren, Qiao, Yang, & Zhou, 2010), while epibrassinolide was effective in *in vitro* conditions against small-cell lung carcinoma cell lines (Sadava & Kane, 2017). Moreover, steroidal compounds isolated from *Solanum xanthocarpum* and *Asparagus racemosus* were characterized from significant cytotoxic activity against colon carcinoma (HCT116) cell line (Bhutani, Paul, Fayad, & Linder, 2010). Steroid glycosides of *Hibiscus sabdariffa* such as β-sitosteroid glycoside showed protective effects against oxidative damage in rat primary hepatocytes (Tseng et al., 1997). In addition, *Solanum* species contain steroid glykoalkaloids such as solamargine and solasonine, which are considered anti-nutritional factors and can be toxic to human when consumed in high amounts (Al Sinani & Eltayeb, 2017; Friedman, 2006). Regarding cucurbitacins, Muruganantham, Solomon, and Senthamilselvi (2016) have reported significant anticancer activities of cucumber flowers against HePG2 cell lines attributed to its high content in cucurbitacins, lignans and flavonoids, while Aeri et al. (2015) associated with the high intake of cucurbitacins with several beneficial health effects, including anti-tumor, anti-diabetic, and anti-inflammatory properties among others. However, Ho, Ho, Ho, and Ho (2014) pointed out that despite the beneficial properties of these compounds, extremely bitter gourd fruit may result in adverse side effects such as diarrhea, gastrointestinal bleeding, hypotension and vomiting.

Saponins are important phytochemicals which have been associated with significant antioxidant properties (Table 10), as in the case of *Asparagus laricinus* roots where aqueous extracts showed promising antioxidant properties due to the presence of saponins among other compounds (Fuku, Al-Azzawi, Madamombe, & Mashele, 2013). Similar antioxidant properties have been reported for *Asparagus racemosus* Willd. root extracts, which also exhibited antimicrobial activity against *Staphylococcus aureus* (Tripathi, Tiwari, Anjum, Tewari, & Division, 2015), while *A. filicinus* root extracts demonstrated significant cytotoxic effects against human breast adenocarcinoma MDA-MB-231 cell line (Wu et al., 2010). Similarly, nyasol a saponin detected in *Asparagus gobicus* N.A. Ivanova ex Grubov and *A. cochinchinensis* roots was very effective against proliferation of human ovarian carcinoma

Table 10 The main health effects of saponins.

Health effect	Mechanism of action	References
Reduce plasma cholesterol levels, toxic in high doses	Saponins interact with dietary cholesterol, producing an insoluble complex that prevents cholesterol from being absorbed. Some saponins seem to affect cholesterol metabolism indirectly by interacting with bile acids	Savage (2003)
Antioxidant properties	Binding cholesterol and preventing cholesterol oxidation in the colon	Shi et al. (2004)
Hypocholesterolemic activity	Reduce plasma cholesterol and cholesterol concentration in the liver	
Cancer inhibition	Through antioxidant effects, direct and select cytotoxic effects against cancer cells, immunomodulation, acid and neutral sterol metabolism, and regulation of cell proliferation	
Antidiabetic action	Multiple mechanisms including restoration of the insulin response, increase of plasma insulin levels, and induction of the release of insulin from the pancreas	Marrelli et al. (2016)
Anti-obesity	Lipase inhibition	
Toxic if given intravenously	Hemolytic activity on human erythrocytes, which depends on the type of aglycone and sugar chains. This property is due to the interaction with sterols present in the erythrocyte membrane, which lead to an increase of membrane permeability and the consequent loss of hemoglobin	

Table 10 The main health effects of saponins.—cont'd

Health effect	Mechanism of action	References
Cytotoxic activity	For all saponins both aglycone and sugar structural features play an important role in determining cytotoxic activity	Podolak, Galanty, and Sobolewska (2010)
Cardioprotective effect	Limiting oxidative stress	Lai et al. (2010) and Ren et al. (2010)
Anticancer effects against several human and animal cancer cell lines	Cytotoxic activity	Mskhiladze et al. (2008) and Sobolewska et al. (2016)

(HO-8910) and human hepatoma (Bel-7402) cells (Yang, Huang, Yang, & Jia, 2004). Moreover, several saponin classes have been reported to present significant anticancer properties through various pathways due to diversity in its chemical structure (Man et al., 2010). According to Abdelrahman et al. (2017), Cepa2 a steroidal saponin which was isolated from dry roots of shallot showed significant *in vitro* efficiency against P3U1 myeloma cancer cell line. Other health effects of saponins include blood and tissue cholesterol reduction directly through binding with dietary cholesterol which inhibits cholesterol absorption or indirectly by interacting with bile acids which assist lipid absorption (Savage, 2003). Moreover, saponin-rich species such as *Allium* species have shown significant cardioprotective properties (Lai et al., 2010; Ren et al., 2010), as well as *in vitro* anticancer effects against several human and animal cancer cell lines, including human melanoma cell line (IGR-1), promyelotic leukemia cells (HL-60), human colorectal cancer cell lines (HCT-116, HT-29, and SW480), human colon adenocarcinoma (DLD-1), lung cancer cell line (HA549), human large-cell lung carcinoma (NCI-H460), and carcinoma (A549), human glioblastoma (SF-268), human breast adenocarcinoma (MCF-7), human hepatocellular liver carcinoma cell line (HepG2), murine fibrosarcoma cell line (WEHI 164), murine monocyte/macrophage cell line (J-774), murine leukemia cell lines (P-388 and L-1210), and normal skin fibroblast (WS1) (Mskhiladze et al., 2008; Sobolewska et al., 2016). The most potent compounds were dioscin (Podolak et al., 2010), eruboside B, leucospiroside A, yayoisaponin C, and aginoside (Mskhiladze et al., 2008). Moreover, CAY-1, a saponin detected in cayenne pepper showed significant synergistic action with amphotericin B and itraconazole against three *Aspergillus* species and *Candida albicans* (De Lucca et al., 2006).

Other effects of saponins include significant antifungal properties and several studies have investigated such effects. For example, Yang et al. (2006) tested 22C-27 saponins and 6 steroidal saponins from plant sources against pathogens such as *Candida albicans*, *C. glabrata*, *C. krusei*, *Cryptococcus neoformans*, and *Aspergillus fumigatus* and reported significant antifungal potency, especially for tigogenin saponins.

Apart from beneficial effects, saponins have been attributed with anti-nutritional and toxic effects due to its hemolytic properties; however these effects have been described only after intravenous application and not through oral administration (Savage, 2003). This is due to its low absorption rate after ingestion and their degradation by gut microflora (Savage, 2003). Other negative effects, include reports from *in vivo* studies where feeding monogastric animals with *Medicago sativa* saponins inhibited feeding efficiency and reduced animal growth rates (Hill, 2003). Genotoxic effects have been reported for phytoestrogens and coumestrol and genistein in particular, although the concentrations that caused such effects exceeded by much the amounts commonly ingested through diet (Andersen, Lauridsen, & Skibsted, 2003). Prolonged consumption of flaxseed resulted in an increase of 17β-estradiol mostly due to its high content in lignans which is 100 times higher comparing to other food products (Cardozo et al., 2012). Therefore, concerning the interference of phytoestrogens with endocrine system their role in endocrine disruption remains unclear and further studies are required, especially when considering the significant amounts of synthetic estrogen-like compounds in food products and their cumulative effects (Behr, Oehlmann, & Wagner, 2011).

3. Practical applications

Nowadays, plant-derived estrogens mainly from soy beans are used as dietary supplements to compensate for hormonal deficiencies during menopausal period (Jargin, 2014). However, several compounds isolated from other species have shown promising properties and have found practical applications in the farming and pharmaceutical industry.

Several saponins are already being used in clinical practice for the treatment of various types of cancer since they are associated with the elimination of cancer cells through apoptotic activities (Man et al., 2010). However, further studies are needed to elucidate the interaction of saponins with conventional drugs, especially with the intention to overcome incidences of drug resistance. So far, several practical applications of exogenous applied brassinosteroids (BRs) in horticultural crops production have been reported (Liu et al., 2017). In particular, yield, fruit ripening and pre- and postharvest quality of tomato fruit may be beneficially affected by 24-epibrassinolide (EBL) and 28-homobrassinolide (HBL) application (Aghdam, Asghari, Farmani, Mohayeji, & Moradbeygi, 2012; Liu et al., 2014; Montoya et al., 2005). Foliar application of 24-epibrassinolide has been suggested to improve nitrogen metabolism and antioxidant status of chickpea plants grown under Cd and/or saline conditions by affecting several physiological processes and enzymes activities (Wani, Tahir, Ahmad, Dar, & Nisar, 2017). According to Samira, Mansour-Gueddes, Dridi-mouhandes, and Denden (2012), foliar application of EBL on pepper plants improved significantly flower and fruit number and total yield, while immersion of tomato fruit in brassinolide (0–6 μM) enhanced tolerance to chilling injury under storage (Aghdam et al., 2012). Similarly, exogenous application of BRs on cucumber plants increased tolerance against photo-oxidative, and cold stress, as well as increased resistance against Cucumber Mosaic Virus (Xia et al., 2009). Moreover, immersion of potato tubers in EBL increased dormancy period (Korableva, Platonova, Dogonadze, & Evsunina, 2002), while spraying of plants increased tuber yield by 20% (Khripach, Zhabinskii, & De Groot, 2000). BRs application may also induce female flower production and early fruit development in cucumber plants (Fu et al., 2008; Papadopoulou & Grumet, 2005), while beneficial effects have been reported for beetroot yield under stress (Schilling, Schiller, & Otto, 1991), watermelon yield (Wang, Luo, Xu, & Zhao, 1994), growth, yield and nutritional value of faba and phaseolus bean (Rady, 2011; Talaat & Abdallah, 2010), and growth and yield of pepper plants (Serna, Hernández, Coll, Coll, & Amorós, 2012).

Steroidal saponins have been reported to present antifungal properties against various plant pathogens, therefore they could be used as alternative means for pathogens control in agriculture (Barile et al., 2007; Sadeghi et al., 2013). Generally, spirostanols were more efficient than furostanols and various compounds such as ceposides, tropeosides, and pericosides showed significant inhibitory effects against soil and air-borne pathogens (Lanzotti, Romano, Lanzuise, Bonanomi, & Scala, 2012; Sadeghi et al., 2013). Other compounds with antifungal activity, include alliospirosides A and

B (Teshima et al., 2013), minutosides A–C (Barile et al., 2007), and agigenin-3-*O*-trisaccharide (Lanzotti, Barile, Antignani, Bonanomi, & Scala, 2012).

In addition, several phytoestrogens are ingredients of many over the counter dietary supplement which are widely sold without significant research regarding their long-term health effects. For example, many of these supplements contain extracts or powder of licorice roots which although in high concentration show a great potential as *in vitro* chemopreventive agents against MCF7 breast cancer cells, long-term exposure to low concentrations may induce the proliferation of estrogen-dependent breast cancer (Maggiolini et al., 2002).

Ecdysteroids show also a great potential as natural control agents for insect pests, since they are highly involved in hormonal disruption of insects through the interaction with ecdysteroid receptors (Savchenko, Whiting, Germade, & Dinan, 2000). Several compounds may exhibit ecdysteroid antagonist activity, including withanolides, cucurbitacins, brassinosteroids, and oligostilbenes which could be further exploited in pest control management (Dinan, Nakagawa, & Hormann, 2012).

4. Conclusions

Phytoestrogens, phytosteroids and saponins are widely distributed compounds within the plant kingdom, while several vegetable species containing significant amounts of these compounds are contributing to a high dietary intake on a daily basis. Although several studies point out the beneficial health effects of these compounds, the fact that several dietary supplements with such ingredients are available in the market as over the counter products, needs further attention in order to postulate a positive risk-benefit recommendation for these products. Therefore, future research is needed to: (a) elucidate and characterize the mechanisms of actions for these compounds, (b) identify toxicity thresholds and side effects related to long-term exposure to such compounds, and (c) formulate regulations for safe consumption and marketing of these products.

References

Abdelrahman, M., Mahmoud, H. Y. A. H., El-Sayed, M., Tanaka, S., & Tran, L. S. (2017). Isolation and characterization of Cepa2, a natural alliospiroside A, from shallot (*Allium cepa* L. Aggregatum group) with anticancer activity. *Plant Physiology and Biochemistry*, *116*, 167–173. https://doi.org/10.1016/j.plaphy.2017.05.006.

Adlercreutz, H. (2007). Lignans and human health. *Critical Reviews in Clinical Laboratory Sciences*, *44*(5–6), 483–525. https://doi.org/10.1080/10408360701612942.

Adolphe, J. L., Whiting, S. J., Juurlink, B. H. J., Thorpe, L. U., & Alcorn, J. (2010). Health effects with consumption of the flax lignan secoisolariciresinol diglucoside. *British Journal of Nutrition, 103*(7), 929–938. https://doi.org/10.1017/S0007114509992753.

Aeri, V., Kaushik, U., & Mir, S. (2015). Cucurbitacins—An insight into medicinal leads from nature. *Pharmacognosy Reviews, 9*(17), 12. https://doi.org/10.4103/0973-7847.156314.

Aghdam, M. S., Asghari, M., Farmani, B., Mohayeji, M., & Moradbeygi, H. (2012). Impact of postharvest brassinosteroids treatment on PAL activity in tomato fruit in response to chilling stress. *Scientia Horticulturae, 144,* 116–120. https://doi.org/10.1016/j.scienta.2012.07.008.

Agradi, E., Vegeto, E., Sozzi, A., Fico, G., Regondi, S., & Tomè, F. (2006). Traditional healthy Mediterranean diet: Estrogenic activity of plants used as food and flavoring agents. *Phytotherapy Research, 20,* 670–675. https://doi.org/10.1002/ptr.

Al Sinani, S., & Eltayeb, E. (2017). The steroidal glycoalkaloids solamargine and solasonine in solanum plants. *South African Journal of Botany, 112,* 253–269. https://doi.org/10.1016/j.sajb.2017.06.002.

Ali, B. (2017). Practical applications of brassinosteroids in horticulture—Some field perspectives. *Scientia Horticulturae, 225,* 15–21. https://doi.org/10.1016/j.scienta.2017.06.051.

Andersen, M. L., Lauridsen, R. K., & Skibsted, L. H. (2003). Optimising the use of phenolic compounds in foods. In I. Johnson & G. Williamson (Eds.), *Phytochemical functional foods* (pp. 315–346). Cambridge: CRC Press. https://doi.org/10.1016/B978-1-85573-672-6.50020-4.

Anuradha, A., & Rao, S. S. R. (2007). The effect of brassinosteroids on radish (*Raphanus sativus* L.) seedlings growing under cadmium stress. *Plant, Soil and Environment, 53*(11), 465–472. https://doi.org/10.17221/2307-PSE.

Aparicio, J. F., & Martín, J. F. (2008). Microbial cholesterol oxidases: Bioconversion enzymes or signal proteins? *Molecular BioSystems, 4,* 804–809. https://doi.org/10.1039/b717500k.

Arroo, R. R. J., Beresford, K., Bhambra, A. S., Boarder, M., Budriesi, R., Cheng, Z., et al. (2014). Phytoestrogens as natural prodrugs in cancer prevention: Towards a mechanistic model. *Phytochemistry Reviews, 13*(4), 853–866. https://doi.org/10.1007/s11101-014-9355-3.

Atkinson, C., Compston, J. E., Day, N. E., Dowsett, M., & Bingham, S. A. (2004). The effects of phytoestrogen isoflavones on bone density in women: A double blind, randomized, placebo-controlled trial. *American Journal of Clinical Nutrition, 79,* 326–333. https://doi.org/10.1093/ajcn/79.2.326.

Bacciottini, L., Falchetti, A., Pampaloni, B., Bartolini, E., Carossino, A., & Brandi, M. L. (2007). Phytoestrogens: Food or drug? *Clinical Cases in Mineral and Bone Metabolism, 4*(2), 123–130.

Bajguz, A., & Tretyn, A. (2003). The chemical structures and occurence of brassinosteroids in plants. In S. Hayat & A. Ahmad (Eds.), *Brassinosteroids: Bioactivity and crop productivity* (pp. 1–44). Dordrecht, Netherlands: Springer.

Barile, E., Bonanomi, G., Antignani, V., Zolfaghari, B., Sajjadi, S. E., Scala, F., et al. (2007). Saponins from *Allium minutiflorum* with antifungal activity. *Phytochemistry, 68*(5), 596–603. https://doi.org/10.1016/j.phytochem.2006.10.009.

Bateman, H. L., & Patisaul, H. B. (2008). Disrupted female reproductive physiology following neonatal exposure to phytoestrogens or estrogen specific ligands is associated with decreased GnRH activation and kisspeptin fiber density in the hypothalamus. *Neurotoxicology, 29*(6), 988–997. https://doi.org/10.1016/j.neuro.2008.06.008.

Bedell, S., Nachtigall, M., & Naftolin, F. (2014). The pros and cons of plant estrogens for menopause. *Journal of Steroid Biochemistry and Molecular Biology, 139,* 225–236. https://doi.org/10.1016/j.jsbmb.2012.12.004.

Behr, M., Oehlmann, J., & Wagner, M. (2011). Estrogens in the daily diet: In vitro analysis indicates that estrogenic activity is omnipresent in foodstuff and infant formula. *Food and Chemical Toxicology, 49*(10), 2681–2688. https://doi.org/10.1016/j.fct.2011.07.039.

Bhutani, K. K., Paul, A. T., Fayad, W., & Linder, S. (2010). Apoptosis inducing activity of steroidal constituents from *Solanum xanthocarpum* and *Asparagus racemosus*. *Phytomedicine*, *17*(10), 789–793. https://doi.org/10.1016/j.phymed.2010.01.017.

Bouarab, K., Melton, R., Peart, J., Baulcombe, D., & Osbourn, A. (2002). A saponin-detoxifying enzyme mediates suppression of plant defences. *Nature*, *418*, 889. Retrieved from https://doi.org/10.1038/nature00950.

Boué, S. M., Wiese, T. E., Nehls, S., Burow, M. E., Elliott, S., Carter-Wientjes, C. H., et al. (2003). Evaluation of the estrogenic effects of legume extracts containing phytoestrogens. *Journal of Agricultural and Food Chemistry*, *51*(8), 2193–2199. https://doi.org/10.1021/jf021114s.

Branca, F., & Lorenzetti, S. (2005). Health effects of phytoestrogens. *Forum of Nutrition*, *57*, 100–111. https://doi.org/10.1159/000083773.

Budryn, G., Gałązka-Czarnecka, I., Brzozowska, E., Grzelczyk, J., Mostowski, R., Żyżelewicz, D., et al. (2018). Evaluation of estrogenic activity of red clover (*Trifolium pratense* L.) sprouts cultivated under different conditions by content of isoflavones, calorimetric study and molecular modelling. *Food Chemistry*, *245*, 324–336. https://doi.org/10.1016/J.FOODCHEM.2017.10.100.

Calani, L., Dall'Asta, M., Bruni, R., & Del Rio, D. (2014). Flavonoid occurrence, bioavailability, metabolism, and protective effects in humans: Focus on flavan-3-ols and flavonols. In A. Romani, V. Lattanzio, & S. Quideau (Eds.), *Vol. 4. Recent advances in polyphenol research*. (pp. 239–279). New Jersey: John Wiley & Sons. https://doi.org/10.1002/9781118329634.ch8.

Calliste, C., Le Bail, J., Trouillas, P., Pouget, C., Habrioux, G., Chulia, A., et al. (2001). Chalcones: Structural requirements for antioxidant, estrogenic and antiproliferative activities. *Anticancer Research*, *21*(6A), 3949–3956.

Campesi, I., Romani, A., Marino, M., & Franconi, F. (2014). Phenolic compounds from a sex-gender perspective. In A. Romani, V. Lattanzio, & S. Quideau (Eds.), *Vol. 4. Recent advances in polyphenol research*. (pp. 327–339). New Jersey: John Wiley & Sons. https://doi.org/10.1002/9781118329634.ch10.

Cao, V. D., Riu, K. Z., & Boo, K. H. (2018). Biosynthesis and accumulation of 20-hydroxyecdysone in individual male and female spinach plants during the reproductive stage. *Plant Physiology and Biochemistry*, *129*, 394–399. https://doi.org/10.1016/j.plaphy.2018.06.027.

Carange, J., Longpré, F., Daoust, B., & Martinoli, M.-G. (2011). 24-Epibrassinolide, a phytosterol from the brassinosteroid family, protects dopaminergic cells against MPP+-induced oxidative stress and apoptosis. *Journal of Toxicology*, *2011*(392859), 1–13. https://doi.org/10.1155/2011/392859.

Cardozo, L. F., Boaventura, G. T., Brant, L. H. C., Pereira, V. A., Velarde, L. G. C., & Chagas, M. A. (2012). Prolonged consumption of flaxseed flour increases the 17β-estradiol hormone without causing adverse effects on the histomorphology of Wistar rats' penis. *Food and Chemical Toxicology*, *50*(11), 4092–4096. https://doi.org/10.1016/j.fct.2012.08.039.

Carotenuto, A., Fattorusso, E., Lanzotti, V., & Magno, S. (1999). Spirostanol saponins of *Allium porrum* L. *Phytochemistry*, *51*, 1077–1082. https://doi.org/10.1016/S0031-9422(98)00712-2.

Cederroth, C. R., Zimmermann, C., & Nef, S. (2012). Soy, phytoestrogens and their impact on reproductive health. *Molecular and Cellular Endocrinology*, *355*(2), 192–200. https://doi.org/10.1016/j.mce.2011.05.049.

Challinor, V. L., & De Voss, J. J. (2013). Open-chain steroidal glycosides, a diverse class of plant saponins. *Natural Product Reports*, *30*(3), 429–454. https://doi.org/10.1039/c3np20105h.

Champ, M. M.-J. (2002). Non-nutrient bioactive substances of pulses. *British Journal of Nutrition, 88*(S3), S307–S319. https://doi.org/10.1079/BJN2002721.

Chandrareddy, A., Muneyyirci-Delale, O., Mcfarlane, S. I., & Murad, O. M. (2008). Adverse effects of phytoestrogens on reproductive health: A report of three cases. *Complementary Therapies in Clinical Practice, 14*(2), 132–135. https://doi.org/10.1016/j.ctcp.2008.01.002.

Chiang, S. S., & Pan, T. M. (2013). Beneficial effects of phytoestrogens and their metabolites produced by intestinal microflora on bone health. *Applied Microbiology and Biotechnology, 97*(4), 1489–1500. https://doi.org/10.1007/s00253-012-4675-y.

Chong, J., Poutaraud, A., & Hugueney, P. (2009). Metabolism and roles of stilbenes in plants. *Plant Science, 177*(3), 143–155. https://doi.org/10.1016/j.plantsci.2009.05.012.

Choudhary, S. P., Yu, J. Q., Yamaguchi-Shinozaki, K., Shinozaki, K., & Tran, L. S. P. (2012). Benefits of brassinosteroid crosstalk. *Trends in Plant Science, 17*(10), 594–605. https://doi.org/10.1016/j.tplants.2012.05.012.

Clarke, D., & Wiseman, H. (2004). Phytoestrogens. *Annual Review of Plant Biology, 55*, 225–261. https://doi.org/10.1002/9781444302288.ch7.

De Lucca, A. J., Bland, J. M., Boue, S., Vigo, C. B., Cleveland, T. E., & Walsh, T. J. (2006). Synergism of CAY-1 with amphotericin B and itraconazole. *Chemotherapy, 52*(6), 285–287. https://doi.org/10.1159/000095959.

Dean, M., Murphy, B. T., & Burdette, J. E. (2017). Phytosteroids beyond estrogens: Regulators of reproductive and endocrine function in natural products. *Mollecular and Cellular Endocrinology, 442*, 98–105. https://doi.org/10.1017/S0950268814002131.Tuberculosis.

Dhiman, K., Gupta, A., Sharma, D. K., Gill, N. S., & Goyal, A. (2012). A review on the medicinally important plants of the family Cucurbitaceae. *Asian Journal of Clinical Nutrition, 4*(1), 16–26.

Di Carlo, G., Mascolo, N., Angelo, A., & Capasso, F. (1999). Flavonoids: Old and new aspects of a class of natural therapeutic drugs. *Life Sciences, 65*(4), 337–353. https://doi.org/10.1016/S0024-3205(99)00120-4.

Dillard, C., & German, J. (2000). Phytochemicals: Nutraceuticals and human health. *Journal of the Science of Food and Agriculture, 80*, 1744–1756. Retrieved from http://ucanr.edu/datastoreFiles/608-54.pdf.

Dinan, L., Harmatha, J., Volodin, V., & Lafont, R. (2009). Phytoecdysteroids: Diversity, biosynthesis and distribution. In G. Smagghe (Ed.), *Ecdysone: Structures and functions* (pp. 3–45). Dordrecht: Springer. https://doi.org/10.1007/978-1-4020-9112-4_1.

Dinan, L., Nakagawa, Y., & Hormann, R. E. (2012). Structure-activity relationships of ecdysteroids and non-steroidal ecdysone agonists. In T. S. Dhadialla (Ed.), *Vol. 43. Advances in insect physiology* (1st ed., pp. 251–298). Elsevier Ltd. https://doi.org/10.1016/B978-0-12-391500-9.00003-6.

Dinan, L., Savchenko, T., & Whiting, P. (2001). Phytoecdysteroids in the genus *Asparagus* (Asparagaceae). *Phytochemistry, 56*(6), 569–576. https://doi.org/10.1016/S0031-9422(00)00438-6.

Dinan, L., Savchenko, T., Whiting, P., & Sarker, S. D. (1999). Plant natural products as insect steroid receptor agonists and antagonists. *Pesticide Science, 55*(3), 331–335. https://doi.org/10.1002/(SICI)1096-9063(199903)55:3<331::AID-PS889>3.0.CO;2-1.

Dinan, L., Whiting, P., & Scott, A. J. (1998). Taxonomic distribution of phytoecdysteroids in seeds of members of the chenopodiaceae. *Biochemical Systematics and Ecology, 26*(5), 553–576. https://doi.org/10.1016/S0305-1978(98)00005-2.

Dini, I., Schettino, O., Simioli, T., & Dini, A. (2001). Studies on the constituents of *Chenopodium quinoa* seeds: Isolation and characterization of new triterpene saponins. *Journal of Agricultural and Food Chemistry, 49*(2), 741–746. https://doi.org/10.1021/jf000971y.

Duran, M. I., González, C., Acosta, A., Olea, A. F., Díaz, K., & Espinoza, L. (2017). Synthesis of five known brassinosteroid analogs from hyodeoxycholic acid and their activities as plant-growth regulators. *International Journal of Molecular Sciences*, *18*(3), 1–13. https://doi.org/10.3390/ijms18030516.

Eden, J. A. (2012). Phytoestrogens for menopausal symptoms: A review. *Maturitas*, *72*(2), 157–159. https://doi.org/10.1016/j.maturitas.2012.03.006.

El Khawand, T., Courtois, A., Valls, J., Richard, T., & Krisa, S. (2018). A review of dietary stilbenes: Sources and bioavailability. *Phytochemistry Reviews*, *17*, 1007–1029. https://doi.org/10.1007/s11101-018-9578-9.

Elhadi, I. M., Koko, W. S., Dahab, M. M., Mohamed, Y., Imam, E., Elmonem, M. A., et al. (2014). Antigiardial activity of some *Cucurbita* species. *Journal of Forest Products and Industries*, *2*(4), 43–47.

Eman, R. Y., Abeer, Y. I., Samah, A. E.-N., Amr, M. M. I., & Walaa, A. E. K. (2017). Modulatory effect of *Brassica oleracea* L. var. *italica* extract in chemically induced mammary carcinomas in rats. *Bioscience Research*, *14*(2), 331–346.

Fang, H., Tong, W., Shi, L. M., Blair, R., Perkins, R., Branham, W., et al. (2001). Structure-activity relationships for a large diverse set of natural, synthetic, and environmental estrogens. *Chemical Research in Toxicology*, *14*(3), 280–294. https://doi.org/10.1021/tx000208y.

Fattorusso, E., Iorizzi, M., Lanzotti, V., & Taglialatela-Scafati, O. (2002). Chemical composition of shallot (*Allium ascalonicum* Hort.). *Journal of Agricultural and Food Chemistry*, *50*(20), 5686–5690. https://doi.org/10.1021/jf020396t.

Fedina, E., Yarin, A., Mukhitova, F., Blufard, A., & Chechetkin, I. (2017). Brassinosteroid-induced changes of lipid composition in leaves of *Pisum sativum* L. during senescence. *Steroids*, *117*, 25–28. https://doi.org/10.1016/j.steroids.2016.10.009.

Fenwick, D. E., & Oakenfull, D. (1983). Saponin content of food plants and some prepared foods. *Journal of the Science of Food and Agriculture*, *34*, 186–191.

Franke, A. A., Custer, L. J., Cerna, C. M., & Narala, K. K. (1994). Quantitation of phytoestrogens in legumes by HPLCt. *Journal of Agricultural and Food Chemistry*, *42*, 1905–1913. https://doi.org/10.1021/jf00045a015.

Friedman, M. (2006). Potato glycoalkaloids and metabolites: Roles in the plant and in the diet. *Journal of Agricultural and Food Chemistry*, *54*(23), 8655–8681. https://doi.org/10.1021/jf061471t.

Fu, F. Q., Mao, W. H., Shi, K., Zhou, Y. H., Asami, T., & Yu, J. Q. (2008). A role of brassinosteroids in early fruit development in cucumber. *Journal of Experimental Botany*, *59*(9), 2299–2308. https://doi.org/10.1093/jxb/ern093.

Fujiwara, Y., Yahara, S., Ikeda, T., Ono, M., & Nohara, T. (2003). Cytotoxic major saponin from tomato fruits. *Chemical and Pharmaceutical Bulletin*, *51*(2), 234–235. https://doi.org/10.1248/cpb.51.234.

Fuku, S., Al-Azzawi, A., Madamombe, I., & Mashele, S. (2013). Phytochemistry and free radical scavenging activity of *Asparagus laricinus*. *International Journal of Pharmacology*, *9*(5), 312–317.

García-Mediavilla, V., Crespo, I., Collado, P. S., Esteller, A., Sánchez-Campos, S., Tuñón, M. J., et al. (2007). The anti-inflammatory flavones quercetin and kaempferol cause inhibition of inducible nitric oxide synthase, cyclooxygenase-2 and reactive C-protein, and down-regulation of the nuclear factor kappaB pathway in Chang liver cells. *European Journal of Pharmacology*, *557*(2–3), 221–229. https://doi.org/10.1016/J.EJPHAR.2006.11.014.

González-Lamothe, R., Mitchell, G., Gattuso, M., Diarra, M. S., Malouin, F., & Bouarab, K. (2009). Plant antimicrobial agents and their effects on plant and human pathogens. *International Journal of Molecular Sciences*, *10*(8), 3400–3419. https://doi.org/10.3390/ijms10083400.

Gorelick-Feldman, J., MacLean, D., Ilic, N., Poulev, A., Lila, M. A., Cheng, D., et al. (2008). Phytoecdysteroids increase protein synthesis in skeletal muscle cells. *Journal of Agricultural and Food Chemistry, 56*(10), 3532–3537. https://doi.org/10.1021/jf073059z.

Grove, M. D., Spencer, G. F., Rohwedder, W. K., Mandava, N., Worley, J. F., Warthen, J. D., Jr., et al. (1979). Brassinolide, a plant growth-promoting steroid isolated from *Brassica napus* pollen. *Nature, 281*, 216. Retrieved from https://doi.org/10.1038/281216a0.

Gupta, C., & Prakash, D. (2014). Phytonutrients as therapeutic agents. *Journal of Complementary and Integrative Medicine, 11*(3), 151–169. https://doi.org/10.1515/jcim-2013-0021.

Haines, C. D., Harvey, P. A., Luczak, E. D., Barthel, K. K. B., Konhilas, J. P., Watson, P. A., et al. (2012). Estrogenic compounds are not always cardioprotective and can be lethal in males with genetic heart disease. *Endocrinology, 153*(9), 4470–4479. https://doi.org/10.1210/en.2012-1391.

Han, J.-H., Yang, Y.-X., & Feng, M.-Y. (2008). Contents of phytosterols in vegetables and fruits commonly consumed in China. *Biomedical and Environmental Sciences, 21*, 449–453. https://doi.org/10.1016/S0895-3988(09)60001-5.

Hano, C., Corbin, C., Drouet, S., Quéro, A., Rombaut, N., Savoire, R., et al. (2017). The lignan (+)-secoisolariciresinol extracted from flax hulls is an effective protectant of linseed oil and its emulsion against oxidative damage. *European Journal of Lipid Science and Technology, 119*(1600219), 1–9. https://doi.org/10.1002/ejlt.201600219.

Hasan, S. A., Hayat, S., & Ahmad, A. (2011). Brassinosteroids protect photosynthetic machinery against the cadmium induced oxidative stress in two tomato cultivars. *Chemosphere, 84*(10), 1446–1451. https://doi.org/10.1016/j.chemosphere.2011.04.047.

Hill, G. (2003). Plant antinutritional factors. Characteristics. In B. Caballero, P. Finglas, & F. Toldra (Eds.), *Encyclopedia of food sciences and nutrition* (pp. 4578–4587). Cambridge: Academic Press. https://doi.org/10.1016/B0-12-227055-X/01318-3.

Ho, C. H., Ho, M. G., Ho, S. P., & Ho, H. H. (2014). Bitter bottle gourd (*Lagenaria siceraria*) toxicity. *Journal of Emergency Medicine, 46*(6), 772–775. https://doi.org/10.1016/j.jemermed.2013.08.106.

Hu, X. J., Song, W. R., Gao, L. Y., Nie, S. P., Eisenbrand, G., & Xie, M. Y. (2014). Assessment of dietary phytoestrogen intake via plant-derived foods in China. *Food Additives and Contaminants, Part A: Chemistry, Analysis, Control, Exposure and Risk Assessment, 31*(8), 1325–1335. https://doi.org/10.1080/19440049.2014.930562.

Huang, X., & Kong, L. (2006). Steroidal saponins from roots of *Asparagus officinalis*. *Steroids, 71*(2), 171–176. https://doi.org/10.1016/j.steroids.2005.09.005.

Huang, X. F., Lin, Y. Y., & Kong, L. Y. (2008). Steroids from the roots of *Asparagus officinalis* and their cytotoxic activity. *Journal of Integrative Plant Biology, 50*(6), 717–722. https://doi.org/10.1111/j.1744-7909.2008.00651.x.

Humfrey, C. D. N. (1998). Phytoestrogens and human health effects: Weighing up the current evidence. *Natural Toxins, 6*(2), 51–59. https://doi.org/10.1002/(SICI)1522-7189(199804)6:2<51::AID-NT11>3.0.CO;2-9.

Hutt, T. F., & Herrington, M. E. (1985). The determination of bitter principles in zucchinis. *Journal of the Science of Food and Agriculture, 36*, 1107–1112.

Hwang, J. H. H., Hwang, I., Liu, Q.-H., Woo, E.-R., & Lee, D. G. G. G. (2012). (+)-Medioresinol leads to intracellular ROS accumulation and mitochondria-mediated apoptotic cell death in *Candida albicans*. *Biochimie, 94*(8), 1784–1793. https://doi.org/10.1016/j.biochi.2012.04.010.

Hwang, K. A., Park, M. A., Kang, N. H., Yi, B. R., Hyun, S. H., Jeung, E. B., et al. (2013). Anticancer effect of genistein on BG-1 ovarian cancer growth induced by 17 β-estradiol or bisphenol A via the suppression of the crosstalk between estrogen receptor alpha and insulin-like growth factor-1 receptor signaling pathways. *Toxicology and Applied Pharmacology, 272*(3), 637–646. https://doi.org/10.1016/j.taap.2013.07.027.

Ibarreta, D., Daxenberger, A., & Meyer, H. H. D. (2001). Possible health impact of phytoestrogens and xenoestrogens in food. *APMIS, 109*(3), 161–184. https://doi.org/10.1034/j.1600-0463.2001.090301.x.
Ikeda, T., Tsumagari, H., & Nohara, T. (2000). Steroidal oligoglycosides from the seeds of *Allium tuberosum*. *Chemical and Pharmaceutical Bulletin, 48*(3), 362–365.
Imai, T., Nomura, M., & Fukushima, K. (2006). Evidence for involvement of the phenylpropanoid pathway in the biosynthesis of the norlignan agatharesinol. *Journal of Plant Physiology, 163*(5), 483–487. https://doi.org/10.1016/j.jplph.2005.08.009.
Iorizzi, M., Lanzotti, V., Ranalli, G., De Marino, S., & Zollo, F. (2002). Antimicrobial furostanol saponins from the seeds of *Capsicum annuum* L. var. *acuminatum*. *Journal of Agricultural and Food Chemistry, 50*(15), 4310–4316. https://doi.org/10.1021/jf0116911.
Itakura, Y., Ichikawa, M., Mori, Y., Okino, R., Udayama, M., & Morita, T. (2001). How to distinguish garlic from the other *Allium* vegetables. *The Journal of Nutrition, 131*(3), 963S–967S. https://doi.org/10.2307/1940703.
Izawa, K., Amino, Y., Kohmura, M., Ueda, Y., & Kuroda, M. (2010). Human–environment interactions—Taste. In L. Mander & H.-W. Liu (Eds.), *Comprehensive natural products II: Chemistry and biology* (pp. 631–671). Oxford: Elsevier. https://doi.org/10.1016/B978-008045382-8.00108-8.
Jaiswal, Y., Liang, Z., Ho, A., Chen, H., & Zhao, Z. (2014). A comparative tissue-specific metabolite analysis and determination of protodioscin content in *Asparagus* species used in traditional Chinese medicine and Ayurveda by use of laser microdissection, UHPLC-QTOF/MS and LC-MS/MS. *Phytochemical Analysis, 25*(6), 514–528. https://doi.org/10.1002/pca.2522.
Janeczko, A., Koscielniak, J., Pilipowicz, M., Szarek-Lukaszewska, G., & Skoczowski, A. (2005). Protection of winter rape photosystem 2 by 24-epibrassinolide under cadmium stress. *Photosynthetica, 43*(2), 293–298. https://doi.org/10.1007/s11099-005-0048-4.
Jaramillo-Carmona, S., Rodriguez-Arcos, R., Jiménez-Araujo, A., López, S., Gil, J., Moreno, R., et al. (2017). Saponin profile of wild *Asparagus* species. *Journal of Food Science, 82*(3), 638–646. https://doi.org/10.1111/1750-3841.13628.
Jargin, S. v. (2014). Soy and phytoestrogens: Possible side effects. *German Medical Science: GMS e-Journal, 12*, 1–5. https://doi.org/10.3205/000203.
Jefferson, W. N., Padilla-Banks, E., Goulding, E. H., Lao, S.-P. C., Newbold, R. R., & Williams, C. J. (2009). Neonatal exposure to genistein disrupts ability of female mouse reproductive tract to support preimplantation embryo development and implantation. *Biology of Reproduction, 80*(3), 425–431. https://doi.org/10.1095/biolreprod.108.073171.
Jeong, J.-H., An, J. Y., Kwon, Y. T., Rhee, J. G., & Lee, Y. J. (2009). Effects of low dose quercetin: Cancer cell-specific inhibition of cell cycle progression. *Journal of Cellular Biochemistry, 106*(1), 73–82. https://doi.org/10.1002/jcb.21977.
Kamlar, M., Rothova, O., Salajkova, S., Tarkowska, D., Drasar, P., Kocova, M., et al. (2015). The effect of exogenous 24-epibrassinolide on the ecdysteroid content in the leaves of *Spinacia oleracea* L. *Steroids, 97*, 107–112. https://doi.org/10.1016/j.steroids.2014.12.024.
Kashyap, D., Sharma, A., Tuli, H. S., Sak, K., Garg, V. K., Buttar, H. S., et al. (2018). Apigenin: A natural bioactive flavone-type molecule with promising therapeutic function. *Journal of Functional Foods, 48*, 457–471. https://doi.org/10.1016/j.jff.2018.07.037.
Kaur, R., Yadav, P., Sharma, A., Kumar Thukral, A., Kumar, V., Kaur Kohli, S., et al. (2017). Castasterone and citric acid treatment restores photosynthetic attributes in *Brassica juncea* L. under Cd(II) toxicity. *Ecotoxicology and Environmental Safety, 145*, 466–475. https://doi.org/10.1016/j.ecoenv.2017.07.067.
Kaur, R., Yadav, P., Thukral, A. K., Sharma, A., Bhardwaj, R., Alyemeni, M. N., et al. (2018). Castasterone and citric acid supplementation alleviates cadmium toxicity by

modifying antioxidants and organic acids in *Brassica juncea*. *Journal of Plant Growth Regulation, 37*(1), 286–299. https://doi.org/10.1007/s00344-017-9727-1.

Keeler, R. F. R. (1978). Cyclopamine and related steroidal alkaloid teratogens: Their occurrence, structural relationship, and biologic effects. *Lipids, 13*(10), 708–715. https://doi.org/10.1007/BF02533750.

Kerem, Z., Regev-Shoshani, G., Flaishman, M. A., & Sivan, L. (2003). Resveratrol and two monomethylated stilbenes from israeli *Rumex bucephalophorus* and their antioxidant potential. *Journal of Natural Products, 66*(9), 1270–1272. https://doi.org/10.1021/np030087c.

Key, T. J. (2011). Fruit and vegetables and cancer risk. *British Journal of Cancer, 104*(1), 6–11. https://doi.org/10.1038/sj.bjc.6606032.

Khripach, V. A., Zhabinskii, V. N., & de Groot, A. E. (1999). Introduction. In V. A. Khripach, V. N. Zhabinskii, & A. E. de Groot (Eds.), *Brassinosteroids—A new class of plant hormones* (pp. 1–5). San Diego: Academic Press. https://doi.org/10.1016/B978-012406360-0/50002-2.

Khripach, V., Zhabinskii, V., & De Groot, A. (2000). Twenty years of brassinosteroids: Steroidal plant hormones warrant better crops for the XXI century. *Annals of Botany, 86*(3), 441–447. https://doi.org/10.1006/anbo.2000.1227.

Kim, G.-S., Kim, H.-T., Seong, J.-D., Oh, S.-R., Lee, C.-O., Bang, J.-K., et al. (2005). Cytotoxic steroidal saponins from the rhizomes of *Asparagus oligoclonos*. *Journal of Natural Products, 68*(5), 766–768. https://doi.org/10.1021/np040128k.

Koleva, I. I., van Beek, T. A., Soffers, A. E. M. F., Dusemund, B., & Rietjens, I. M. C. M. (2012). Alkaloids in the human food chain—Natural occurrence and possible adverse effects. *Molecular Nutrition & Food Research, 56*(1), 30–52. https://doi.org/10.1002/mnfr.201100165.

Konar, N., Poyrazoğlu, E. S., Demir, K., & Artik, N. (2012). Effect of different sample preparation methods on isoflavone, lignan, coumestan and flavonoid contents of various vegetables determined by triple quadrupole LC-MS/MS. *Journal of Food Composition and Analysis, 26*(1–2), 26–35. https://doi.org/10.1016/j.jfca.2012.01.002.

Koomson, D. A., Kwakye, B. D., Darkwah, W. K., Odum, B., & Asante, M. (2018). Phytochemical constituents, total saponins, alkaloids, flavonoids and vitamin C contents of ethanol extracts of five *Solanum torvum* fruits. *Pharmacognosy Journal, 10*(5), 946–950. https://doi.org/10.5530/pj.2018.5.160.

Korableva, N., Platonova, T., Dogonadze, M., & Evsunina, A. (2002). Brassinolide effect on growth of apical meristems, ethylene production, and abscisic acid content in potato tubers. *Biologia Plantarum, 45*(1), 39–43.

Kreis, W., & Müller-Uri, F. (2010). Biochemistry of sterols, cardiac glycosides, brassinosteroids, phytoecdysteroids and steroid saponins. In M. Wink (Ed.), *Biochemistry of plant secondary metabolism: Vol. 40. Annual plant reviews* (2nd ed., pp. 304–363). New Jersey: Blackwell Publishing Ltd. https://doi.org/10.1002/9781444320503.ch6.

Kuhnle, G. G. C., Dell, C., Aspinall, S. M., Runswick, S. A., Joosen, A. M. C. P., Mulligan, A. A., et al. (2009). Phytoestrogen content of fruits and vegetables commonly consumed in the UK based on LC–MS and 13C-labelled standards. *Food Chemistry, 116*(2), 542–554. https://doi.org/10.1016/j.foodchem.2009.03.002.

Kumpun, S., Maria, A., Crouzet, S., Evrard-Todeschi, N., Girault, J. P., & Lafont, R. (2011). Ecdysteroids from *Chenopodium quinoa* Willd., an ancient andean crop of high nutritional value. *Food Chemistry, 125*(4), 1226–1234. https://doi.org/10.1016/j.foodchem.2010.10.039.

Kuo, S. M. (1996). Antiproliferative potency of structurally distinct dietary flavonoids on human colon cancer cells. *Cancer Letters, 110*(1–2), 41–48. https://doi.org/10.1016/S0304-3835(96)04458-8.

Lai, W., Wu, Z., Lin, H., Li, T., Sun, L., Chai, Y., et al. (2010). Anti-ischemia steroidal saponins from the seeds of *Allium fistulosum*. *Journal of Natural Products*, *73*(6), 1053–1057. https://doi.org/10.1021/np900815p.

Landete, J. M. (2012). Plant and mammalian lignans: A review of source, intake, metabolism, intestinal bacteria and health. *Food Research International*, *46*(1), 410–424. https://doi.org/10.1016/j.foodres.2011.12.023.

Landete, J. M., Arqués, J., Medina, M., Gaya, P., de Las Rivas, B. D., & Muñoz, R. (2016). Bioactivation of phytoestrogens: Intestinal bacteria and health. *Critical Reviews in Food Science and Nutrition*, *56*(11), 1826–1843. https://doi.org/10.1080/10408398.2013.789823.

Lanzotti, V., Barile, E., Antignani, V., Bonanomi, G., & Scala, F. (2012). Antifungal saponins from bulbs of garlic, *Allium sativum* L. var. Voghiera. *Phytochemistry*, *78*, 126–134. https://doi.org/10.1016/j.phytochem.2012.03.009.

Lanzotti, V., Romano, A., Lanzuise, S., Bonanomi, G., & Scala, F. (2012). Antifungal saponins from bulbs of white onion, *Allium cepa* L. *Phytochemistry*, *74*, 133–139. https://doi.org/10.1016/j.phytochem.2011.11.008.

Lecomte, S., Demay, F., Ferrière, F., & Pakdel, F. (2017). Phytochemicals targeting estrogen receptors: Beneficial rather than adverse effects? *International Journal of Molecular Sciences*, *18*(7), 1–19. https://doi.org/10.3390/ijms18071381.

Lee, C.-L., Hwang, T.-L., Yang, J.-C., Cheng, H.-T., He, W.-J., Yen, C.-T., et al. (2014). Anti-inflammatory spirostanol and furostanol saponins from *Solanum macaonense*. *Journal of Natural Products*, 77(8), 1770–1783. https://doi.org/10.1021/np500057b.

Lesovaya, E., Yemelyanov, A., Swart, A. C., Swart, P., Haegeman, G., & Budunova, I. (2015). Discovery of compound A—A selective activator of the glucocorticoid receptor with anti-inflammatory and anti-cancer activity. *Oncotarget*, *6*(31), 30730–30744. https://doi.org/10.18632/oncotarget.5078.

Li, G., Zhu, Y., Zhang, Y., Lang, J., Chen, Y., & Ling, W. (2013). Estimated daily flavonoid and stilbene intake from fruits, vegetables, and nuts and associations with lipid profiles in Chinese adults. *Journal of the Academy of Nutrition and Dietetics*, *113*(6), 786–794. https://doi.org/10.1016/j.jand.2013.01.018.

Liu, L., Jia, C., Zhang, M., Chen, D., Chen, S., Guo, R., et al. (2014). Ectopic expression of a BZR1-1D transcription factor in brassinosteroid signalling enhances carotenoid accumulation and fruit quality attributes in tomato. *Plant Biotechnology Journal*, *12*(1), 105–115. https://doi.org/10.1111/pbi.12121.

Liu, J., Zhang, D., Sun, X., Ding, T., Lei, B., & Zhang, C. (2017). Structure-activity relationship of brassinosteroids and their agricultural practical usages. *Steroids*, *124*, 1–17. https://doi.org/10.1016/j.steroids.2017.05.005.

Lombardo, S., Pandino, G., Mauromicale, G., Knödler, M., Carle, R., & Schieber, A. (2010). Influence of genotype, harvest time and plant part on polyphenolic composition of globe artichoke [*Cynara cardunculus* L. var. *scolymus* (L.) Fiori]. *Food Chemistry*, *119*(3), 1175–1181. https://doi.org/10.1016/J.FOODCHEM.2009.08.033.

Lu, Y., Luo, J., & Kong, L. (2011). Steroidal alkaloid saponins and steroidal saponins from *Solanum surattense*. *Phytochemistry*, *72*(7), 668–673. https://doi.org/10.1016/j.phytochem.2011.01.028.

Maggiolini, M., Statti, G., Vivacqua, A., Gabriele, S., Rago, V., Loizzo, M., et al. (2002). Estrogenic and antiproliferative activities of isoliquiritigenin in MCF7 breast cancer cells. *Journal of Steroid Biochemistry and Molecular Biology*, *82*(4–5), 315–322. https://doi.org/10.1016/S0960-0760(02)00230-3.

Maldonado, E., Pérez-Castorena, A. L., Garcés, C., & Martínez, M. (2011). Philadelphicalactones C and D and other cytotoxic compounds from *Physalis philadelphica*. *Steroids*, *76*(7), 724–728. https://doi.org/10.1016/j.steroids.2011.03.018.

Man, S., Gao, W., Zhang, Y., Huang, L., & Liu, C. (2010). Chemical study and medical application of saponins as anti-cancer agents. *Fitoterapia*, *81*(7), 703–714. https://doi.org/10.1016/j.fitote.2010.06.004.

Markaverich, B. M., Webb, B., Densmore, C. L., Rebecca, R., & Gregory, R. R. (1995). Effects of coumestrol on estrogen receptor function and uterine growth in ovariectomized rats. *Environmental Health Perspectives*, *103*(6), 574–581. https://doi.org/10.1289/ehp.95103574.

Marrelli, M., Conforti, F., Araniti, F., Statti, G., Marrelli, M., Conforti, F., et al. (2016). Effects of saponins on lipid metabolism: A review of potential health benefits in the treatment of obesity. *Molecules*, *21*(1404), 1–20. https://doi.org/10.3390/molecules21101404.

Mazur, W. (1998). Phytoestrogen content in foods. *Baillière's Clinical Endocrinology and Metabolism*, *12*(4), 729–742. https://doi.org/10.1016/S0950-351X(98)80013-X.

Michelini, F. M., Berra, A., & Alché, L. E. (2008). The *in vitro* immunomodulatory activity of a synthetic brassinosteroid analogue would account for the improvement of herpetic stromal keratitis in mice. *The Journal of Steroid Biochemistry and Molecular Biology*, *108*(1–2), 164–170. https://doi.org/10.1016/J.JSBMB.2007.10.002.

Miean, K. H., & Mohamed, S. (2001). Flavonoid (myricetin, quercetin, kaempferol, luteolin, and apigenin) content of edible tropical plants. *Journal of Agricultural and Food Chemistry*, *49*(6), 3106–3112. https://doi.org/10.1021/jf000892m.

Milder, I. E. J., Arts, I. C. W., van de Putte, B., Venema, D. P., & Hollman, P. C. H. (2005). Lignan contents of Dutch plant foods: A database including lariciresinol, pinoresinol, secoisolariciresinol and matairesinol. *British Journal of Nutrition*, *93*, 393. https://doi.org/10.1079/BJN20051371.

Montoya, T., Nomura, T., Yokota, T., Farrar, K., Harrison, K., Jones, J. G. D., et al. (2005). Patterns of Dwarf expression and brassinosteroid accumulation in tomato reveal the importance of brassinosteroid synthesis during fruit development. *The Plant Journal*, *42*(2), 262–269. https://doi.org/10.1111/j.1365-313X.2005.02376.x.

Moretón-Lamas, E., Lago-Crespo, M., Lage-Yusty, M. A., & López-Hernández, J. (2017). Comparison of methods for analysis of resveratrol in dietary vegetable supplements. *Food Chemistry*, *224*, 219–223. https://doi.org/10.1016/j.foodchem.2016.12.060.

Mskhiladze, L., Legault, J., Lavoie, S., Mshvildadze, V., Kuchukhidze, J., Elias, R., et al. (2008). Cytotoxic steroidal saponins from the flowers of *Allium leucanthum*. *Molecules*, *13*(12), 2925–2934. https://doi.org/10.3390/molecules13122925.

Mueen Ahmed, K. K., Khan, M. Y., & Shivananda, B. G. (2009). Cardiovascular diseases and role of medicinal plants as a re-emerging health aid. *Pharmacognosy Reviews*, *3*(5), 8–14.

Murakami, A., Ashida, H., & Terao, J. (2008). Multitargeted cancer prevention by quercetin. *Cancer Letters*, *269*(2), 315–325. https://doi.org/10.1016/J.CANLET.2008.03.046.

Murphy, K. J., Marques-Lopes, I., & Sánchez-Tainta, A. (2018). Cereals and legumes. In A. Sánchez-Villegas & A. Sánchez-Tainta (Eds.), *The prevention of cardiovascular disease through the Mediterranean diet* (pp. 111–132). Cambridge: Academic Press. https://doi.org/10.1016/B978-0-12-811259-5.00007-X.

Muruganantham, N., Solomon, S., & Senthamilselvi, M. M. (2016). Anti-cancer activity of *Cucumis sativus* (cucumber) flowers against human liver cancer. *International Journal of Pharmaceutical and Clinical Research*, *8*(1), 39–41. Retrieved from https://www.scopus.com/inward/record.uri?eid=2-s2.0-84965137172&partnerID=40&md5=cdd4adf6ad17e7d20d5d823a3efde2e5.

Nakamura, Y., Kaihara, A., Yoshii, K., Tsumura, Y., Ishimitsu, S., & Tonogai, Y. (2001). Content and composition of isoflavonoids in mature or immature beans and bean sprouts consumed in Japan. *Journal of Health Science*, *47*(4), 394–406.

Nelles, J. L., Hu, W. Y., & Prins, G. S. (2011). Estrogen action and prostate cancer. *Expert Review of Endocrinology and Metabolism*, *6*(3), 437–451. https://doi.org/10.1586/eem.11.20.

Nes, W. D. (2011). Biosynthesis of cholesterol and other sterols. *Chemical Reviews*, *111*(10), 6423–6451. https://doi.org/10.1021/cr200021m.
Nikolić, I. L., Savić-Gajić, I. M., Tačić, A. D., & Savić, I. M. (2017). Classification and biological activity of phytoestrogens: A review. *Advanced Technologies*, *6*(2), 96–106. https://doi.org/10.5937/savteh1702096n.
Normén, L., Johnsson, M., Andersson, H., Van Gameren, Y., & Dutta, P. (1999). Plant sterols in vegetables and fruits commonly consumed in Sweden. *European Journal of Nutrition*, *38*, 84–89.
Nsimba, R. Y., Kikuzaki, H., & Konishi, Y. (2008). Ecdysteroids act as inhibitors of calf skin collagenase and oxidative stress. *Journal of Biochemical and Molecular Toxicology*, *22*(4), 240–250. https://doi.org/10.1002/jbt.20234.
Obakan, P., Arisan, E. D., Coker-Gurkan, A., & Palavan-Unsal, N. (2014). Epibrassinolide-induced apoptosis regardless of p53 expression via activating polyamine catabolic machinery, a common target for androgen sensitive and insensitive prostate cancer cells. *The Prostate*, *74*(16), 1622–1633. https://doi.org/10.1002/pros.22879.
Ock, K. C., Sang, J. C., & Song, W. O. (2007). Estimated dietary flavonoid intake and major food sources of U.S. adults. *Journal of Nutrition*, *137*(5), 1244–1252. Retrieved from https://doi.org/10.1093/jn/137.5.1244.
Oklestkova, J., Rárová, L., Kvasnica, M., & Strnad, M. (2015). Brassinosteroids: Synthesis and biological activities. *Phytochemistry Reviews*, *14*(6), 1053–1072. https://doi.org/10.1007/s11101-015-9446-9.
Orlikova, B., Tasdemir, D., Golais, F., Dicato, M., & Diederich, M. (2011). Dietary chalcones with chemopreventive and chemotherapeutic potential. *Genes & Nutrition*, *6*(2), 125–147. https://doi.org/10.1007/s12263-011-0210-5.
Osbourn, A. (1996). Performed antimicrobial compounds and plant defense against funga attack. *The Plant Cell*, *8*, 1821–1831. https://doi.org/10.1186/s12889-017-4886-3.
Oshima, A., Mine, W., Nakada, M., & Yanase, E. (2016). Analysis of isoflavones and coumestrol in soybean sprouts. *Bioscience, Biotechnology, and Biochemistry*, *80*(11), 2077–2079. https://doi.org/10.1080/09168451.2016.1196577.
Ovaskainen, M. L., Törrönen, R., Koponen, J. M., Sinkko, H., Hellström, J., Reinivuo, H., et al. (2008). Dietary intake and major food sources of polyphenols in Finnish adults. *The Journal of Nutrition*, *138*(3), 562–566. https://doi.org/10.1093/jn/138.3.562.
Papadopoulou, E., & Grumet, R. (2005). Brassinosteriod-induced femaleness in cucumber and relationship to ethylene production. *Hortscience*, *40*(6), 1763–1767.
Patisaul, H. B., & Jefferson, W. (2010). The pros and cons of phytoestrogens. *Frontiers in Neuroendocrinology*, *31*(4), 400–419. https://doi.org/10.1016/j.yfrne.2010.03.003.The.
Pavlović, I., Petřik, I., Tarkowská, D., Lepeduš, H., Bok, V. V., Brkanac, S. R., et al. (2018). Correlations between phytohormones and drought tolerance in selected *Brassica* crops: Chinese cabbage, white cabbage and kale. *International Journal of Molecular Sciences*, *19*(2866), 1–23. https://doi.org/10.3390/ijms19102866.
Peng, X. L., Xu, J., Sun, X. F., Ying, C. J., & Hao, L. P. (2015). Analysis of trans-resveratrol and trans-piceid in vegetable foods using high-performance liquid chromatography. *International Journal of Food Sciences and Nutrition*, *66*(7), 729–735. https://doi.org/10.3109/09637486.2015.1088934.
Pérez-Jiménez, J., Fezeu, L., Touvier, M., Arnault, N., Manach, C., Hercberg, S., et al. (2011). Dietary intake of 337 polyphenols in French adults. *American Journal of Clinical Nutrition*, *93*(6), 1220–1228. https://doi.org/10.3945/ajcn.110.007096.
Petrussa, E., Braidot, E., Zancani, M., Peresson, C., Bertolini, A., Patui, S., et al. (2013). Plant flavonoids-biosynthesis, transport and involvement in stress responses. *International Journal of Molecular Sciences*, *14*(7), 14950–14973. https://doi.org/10.3390/ijms140714950.
Piironen, V., Toivo, J., Puupponen-Pimiä, R., & Lampi, A.-M. (2003). Plant sterols in vegetables, fruits and berries. *Journal of the Science of Food and Agriculture*, *83*(4), 330–337. https://doi.org/10.1002/jsfa.1316.

Pociecha, E., Dziurka, M., Oklestkova, J., & Janeczko, A. (2016). Brassinosteroids increase winter survival of winter rye (*Secale cereale* L.) by affecting photosynthetic capacity and carbohydrate metabolism during the cold acclimation process. *Plant Growth Regulation*, *80*(2), 127–135. https://doi.org/10.1007/s10725-016-0149-z.

Podolak, I., Galanty, A., & Sobolewska, D. (2010). Saponins as cytotoxic agents: A review. *Phytochemistry Reviews*, *9*(3), 425–474. https://doi.org/10.1007/s11101-010-9183-z.

Puppala, D., Gairola, C. G., & Swanson, H. I. (2007). Identification of kaempferol as an inhibitor of cigarette smoke-induced activation of the aryl hydrocarbon receptor and cell transformation. *Carcinogenesis*, *28*(3), 639–647. https://doi.org/10.1093/carcin/bgl169.

Quan, H.-J., Koyanagi, J., Komada, F., & Saito, S. (2005). Preparations of vitamin D analogs, spirostanols and furostanols from diosgenin and their cytotoxic activities. *European Journal of Medicinal Chemistry*, *40*(7), 662–673. https://doi.org/10.1016/J.EJMECH.2005.02.003.

Rady, M. M. (2011). Effect of 24-epibrassinolide on growth, yield, antioxidant system and cadmium content of bean (*Phaseolus vulgaris* L.) plants under salinity and cadmium stress. *Scientia Horticulturae*, *129*(2), 232–237. https://doi.org/10.1016/j.scienta.2011.03.035.

Ren, G., Qiao, H. X., Yang, J., & Zhou, C. X. (2010). Protective effects of steroids from *Allium chinense* against H2O2-induced oxidative stress in rat cardiac H9C2 cells. *Phytotherapy Research*, *24*, 404–409. https://doi.org/10.1002/ptr.

Rietjens, I. M. C. M., Louisse, J., & Beekmann, K. (2017). The potential health effects of dietary phytoestrogens. *British Journal of Pharmacology*, *174*(11), 1263–1280. https://doi.org/10.1111/bph.13622.

Rimando, A. M., & Suh, N. (2008). Biological/chemopreventive activity of stilbenes and their effect on colon cancer. *Planta Medica*, *74*(13), 1635–1643.

Rivière, C., Pawlus, A. D., & Mérillon, J. M. (2012). Natural stilbenoids: Distribution in the plant kingdom and chemotaxonomic interest in vitaceae. *Natural Product Reports*, *29*(11), 1317–1333. https://doi.org/10.1039/c2np20049j.

Rothová, O., Holá, D., Kočová, M., Tůmová, L., Hnilička, F., Hniličková, H., et al. (2014). 24-Epibrassinolide and 20-hydroxyecdysone affect photosynthesis differently in maize and spinach. *Steroids*, *85*, 44–57. https://doi.org/10.1016/j.steroids.2014.04.006.

Roupe, K. A., Remsberg, C. M., Yáñez, J. A., & Davies, N. M. (2006). Pharmacometrics of stilbenes: Seguing towards the clinic. *Current Clinical Pharmacology*, *1*(1), 81–101. https://doi.org/10.2174/157488406775268246.

Saboo, S. S., Thorat, P. K., Tapadiya, G. G., & Khadabadi, S. S. (2013). Ancient and recent medicinal uses of cucurbitaceae family. *International Journal of Therapeutic Applications*, (9), 11–19.

Sadava, D., & Kane, S. E. (2017). The effect of brassinolide, a plant steroid hormone, on drug resistant small-cell lung carcinoma cells. *Biochemical and Biophysical Research Communications*, *493*(1), 783–787. https://doi.org/10.1016/j.bbrc.2017.08.094.

Sadeghi, M., Zolfaghari, B., Senatore, M., & Lanzotti, V. (2013). Spirostane, furostane and cholestane saponins from Persian leek with antifungal activity. *Food Chemistry*, *141*(2), 1512–1521. https://doi.org/10.1016/j.foodchem.2013.04.009.

Sahu, N., Banerjee, S., Mondal, N., & Mandal, D. (2008). Steroidal saponins. In A. Kinghorn, H. Falk, & J. Kobayasi (Eds.), *Progress in the chemistry of organic natural products* (pp. 45–141). Wien: Springer-Verlag.

Salminen, J. P. (2014). The chemistry and chemical ecology of ellagitannins in plant-insect interactions: From underestimated molecules to bioactive plant constituents. In A. Romani, V. Lattanzio, & S. Quideau (Eds.), *Vol. 4. Recent advances in polyphenol research*. (pp. 83–113). New Jersey: Wiley & Sons. https://doi.org/10.1002/9781118329634.ch4.

Samira, I. M., Mansour-Gueddes, S. B., Dridi-mouhandes, B., & Denden, M. (2012). 24-Epibrassinolide enhances flower and fruit production of pepper (*Capsicum annuum* L.) under salt stress. *Journal of Stress Physiology & Biochemistry*, *8*(3), 224–233.

Santell, R. C., Chang, Y. C., Nair, M. G., & Helferich, W. G. (1997). Dietary genistein exerts estrogenic effects upon the uterus, mammary gland and the hypothalamic/pituitary axis in rats. *The Journal of Nutrition*, *127*(2), 263–269. https://doi.org/10.1093/jn/127.2.263.

Savage, G. P. (2003). Saponins. In B. Caballero, P. Finglas, & F. Toldra (Eds.), *Encyclopedia of food sciences and nutrition* (2nd ed., pp. 5095–5098). Cambridge: Academic Press. https://doi.org/10.1016/B0-12-227055-X/01050-6.

Savchenko, T., Whiting, P., Germade, A., & Dinan, L. (2000). Ecdysteroid agonist and antagonist activities in species of the solanaceae. *Biochemical Systematics and Ecology*, *28*(5), 403–419. https://doi.org/10.1016/S0305-1978(99)00082-4.

Schilling, G., Schiller, C., & Otto, S. (1991). Influence of brassinosteroids on organ relations and enzyme activities of sugar-beet plants. In H. G. Cutler, Y. Takao, & A. Günter (Eds.), *Brassinosteroids* (pp. 208–219). Washington, DC: ACS Symposium Series. https://doi.org/10.1021/bk-1991-0474.ch018.

Schmelz, E. A., Grebenok, R. J., Ohnmeiss, T. E., & Bowers, W. S. (2000). Phytoecdysteroid turnover in spinach: Long-term stability supports a plant defense hypothesis. *Journal of Chemical Ecology*, *26*(12), 2883–2896. https://doi.org/10.1023/A:1026454213510.

Schmidt, J., Yokota, T., Adam, G., & Takahashi, N. (1991). Castasterone and brassinolidde in *Raphanus sativus* seeds. *Phytochemistry*, *30*(1), 364–365. https://doi.org/10.1016/0031-9422(91)84159-P.

Sebastià, N., Montoro, A., León, Z., & Soriano, J. M. (2017). Searching trans-resveratrol in fruits and vegetables: A preliminary screening. *Journal of Food Science and Technology*, *54*(3), 842–845. https://doi.org/10.1007/s13197-016-2474-7.

Seidlova-Wuttke, D., Ehrhardt, C., & Wuttke, W. (2010). Metabolic effects of 20-OH-ecdysone in ovariectomized rats. *Journal of Steroid Biochemistry and Molecular Biology*, *119*, 121–126. https://doi.org/10.1016/j.jsbmb.2010.01.006.

Seidlova-Wuttke, D., Jarry, H., & Wuttke, W. (2013). Plant derived alternatives for hormone replacement therapy (HRT). *Hormone Molecular Biology and Clinical Investigation*, *16*(1), 35–45. https://doi.org/10.1515/hmbci-2013-0024.

Seleem, D., Pardi, V., & Murata, R. M. (2017). Review of flavonoids: A diverse group of natural compounds with anti-*Candida albicans* activity in vitro. *Archives of Oral Biology*, *76*, 76–83. https://doi.org/10.1016/j.archoralbio.2016.08.030.

Serna, M., Hernández, F., Coll, F., Coll, Y., & Amorós, A. (2012). Brassinosteroid analogues effects on the yield and quality parameters of greenhouse-grown pepper (*Capsicum annuum* L.). *Plant Growth Regulation*, *68*(3), 333–342.

Shahzad, B., Tanveer, M., Che, Z., Rehman, A., Cheema, S. A., Sharma, A., et al. (2018). Role of 24-epibrassinolide (EBL) in mediating heavy metal and pesticide induced oxidative stress in plants: A review. *Ecotoxicology and Environmental Safety*, *147*, 935–944. https://doi.org/10.1016/j.ecoenv.2017.09.066.

Sharma, I., Kaur, N., & Pati, P. K. (2017). Brassinosteroids: A promising option in deciphering remedial strategies for abiotic stress tolerance in rice. *Frontiers in Plant Science*, *8*, 1–17. https://doi.org/10.3389/fpls.2017.02151.

Shawky, E., Abou, R. M., & Kheir, E. (2018). Rapid discrimination of different apiaceae species based on HPTLC fingerprints and targeted flavonoids determination using multivariate image analysis. *Phytochemical Analysis*, *29*, 452–462. https://doi.org/10.1002/pca.2749.

Shi, J., Arunasalam, K., Yeung, D., Kakuda, Y., Mittal, G., & Jiang, Y. (2004). Saponins from edible legumes: Chemistry, processing, and health benefits. *Journal of Medicinal Food*, 7(1), 67–78. https://doi.org/10.1089/109662004322984734.

Singh, D., & Chaudhuri, P. K. (2018). Structural characteristics, bioavailability and cardioprotective potential of saponins. *Integrative Medicine Research*, 7(1), 33–43. https://doi.org/10.1016/j.imr.2018.01.003.

Singh, S., Swain, S., Singh, D. R., Salim, K. M., Nayak, D., & Roy, S. D. (2015). Changes in phytochemicals, anti-nutrients and antioxidant activity in leafy vegetables by microwave boiling with normal and 5% NaCl solution. *Food Chemistry*, *176*, 244–253. https://doi.org/10.1016/j.foodchem.2014.12.068.

Smoliga, J. M., Baur, J. A., & Hausenblas, H. A. (2011). Resveratrol and health—A comprehensive review of human clinical trials. *Molecular Nutrition & Food Research*, *55*(8), 1129–1141. https://doi.org/10.1002/mnfr.201100143.

Sobolewska, D., Michalska, K., Podolak, I., & Grabowska, K. (2016). Steroidal saponins from the genus *Allium*. *Phytochemistry Reviews*, *15*(1), 1–35. https://doi.org/10.1007/s11101-014-9381-1.

Sok, D.-E., Cui, H. S., & Kim, M. R. (2009). Isolation and bioactivities of furfuran type lignan compounds from edible plants. *Recent Patents on Food, Nitrition and Agriculture*, *1*, 87–95. https://doi.org/10.1002/cpe.4977.

Sparg, S. G., Light, M. E., & Van Staden, J. (2004). Biological activities and distribution of plant saponins. *Journal of Ethnopharmacology*, *94*(2–3), 219–243. https://doi.org/10.1016/j.jep.2004.05.016.

Steigerová, J., Oklešt'ková, J., Levková, M., Rárová, L., Kolář, Z., & Strnad, M. (2010). Brassinosteroids cause cell cycle arrest and apoptosis of human breast cancer cells. *Chemico-Biological Interactions*, *188*, 487–496. https://doi.org/10.1016/j.cbi.2010.09.006.

Steigerová, J., Rárová, L., Oklešt'ková, J., Křížová, K., Levková, M., Šváchová, M., et al. (2012). Mechanisms of natural brassinosteroid-induced apoptosis of prostate cancer cells. *Food and Chemical Toxicology*, *50*(11), 4068–4076. https://doi.org/10.1016/j.fct.2012.08.031.

Subramanian, C., Zhang, H., Gallagher, R., Hammer, G., Timmermann, B., & Cohen, M. (2014). Withanolides are potent novel targeted therapeutic agents against adrenocortical carcinomas. *World Journal of Surgery*, *38*(6), 1343–1352. https://doi.org/10.1007/s00268-014-2532-0.

Surh, J., Kim, M.-J., Koh, E., Kim, Y.-K. L., & Kwon, H. (2006). Estimated intakes of isoflavones and coumestrol in Korean population. *International Journal of Food Sciences and Nutrition*, *57*(5–6), 325–344. https://doi.org/10.1080/09637480600802348.

Talaat, N. B., & Abdallah, A. M. (2010). Effect of 28-homobrassinolide and 24-epibrassinolide on the growth, productivity and nutritional value of two faba bean (*Vicia faba* L.) cultivars. *Archives of Agronomy and Soil Science*, *56*(6), 649–669. https://doi.org/10.1080/03650340903164249.

Tang, J., Han, Z., & Chai, J. (2016). Q&A: What are brassinosteroids and how do they act in plants? *BMC Biology*, *14*(113), 1–5. https://doi.org/10.1186/s12915-016-0340-8.

Teshima, Y., Ikeda, T., Imada, K., Sasaki, K., El-Sayed, M. A., Shigyo, M., et al. (2013). Identification and biological activity of antifungal saponins from shallot (*Allium cepa* L. Aggregatum group). *Journal of Agricultural and Food Chemistry*, *61*(31), 7440–7445. https://doi.org/10.1021/jf401720q.

Thompson, L. U., Boucher, B. A., Liu, Z., Cotterchio, M., & Kreiger, N. (2006). Phytoestrogen content of foods consumed in Canada, including isoflavones, lignans, and coumestan. *Nutrition and Cancer*, *54*(2), 184–201. https://doi.org/10.1207/s15327914nc5402_5.

Tomás-Barberán, F. A., & Espín, J. C. (2001). Phenolic compounds and related enzymes as determinants of quality in fruits and vegetables. *Journal of the Science of Food and Agriculture*, *81*(9), 853–876. https://doi.org/10.1002/jsfa.885.

Tong, Q. Y., He, Y., Zhao, Q. B., Qing, Y., Huang, W., & Wu, X. H. (2012). Cytotoxicity and apoptosis-inducing effect of steroidal saponins from *Dioscorea zingiberensis* wright against cancer cells. *Steroids*, 77, 1219–1227. https://doi.org/10.1016/j.steroids.2012.04.019.

Touillaud, M. S., Pillow, P. C., Jakovljevic, J., Bondy, M. L., Singletary, S. E., Li, D., et al. (2009). Effect of dietary intake of phytoestrogens on estrogen receptor status in premenopausal women with breast cancer. *Nutrition and Cancer*, *51*(2), 162–169. https://doi.org/10.1207/s15327914nc5102.

Tousson, E., El-Moghazy, M., Massoud, A., & Akel, A. (2012). Histopathological and immunohistochemical changes in the testes of rabbits after injection with the growth promoter boldenone. *Reproductive Sciences*, *19*(3), 253–259. https://doi.org/10.1177/1933719111418126.

Tripathi, Y. C., Tiwari, S., Anjum, N., Tewari, D., & Division, C. (2015). Phytochemical, antioxidant and antimicrobial screening of roots of *Asparagus racemosus* willd. *World Journal of Pharmaceutical Research*, *4*(4), 709–722.

Tseng, T. H., Kao, E. S., Chu, C. Y., Chou, F. P., Lin Wu, H. W., & Wang, C. J. (1997). Protective effects of dried flower extracts of *Hibiscus sabdariffa* L. against oxidative stress in rat primary hepatocytes. *Food and Chemical Toxicology*, *35*(12), 1159–1164. https://doi.org/10.1016/S0278-6915(97)85468-3.

Tzortzakis, N., Chrysargyris, A., & Petropoulos, S. (2018). Phytochemicals content and health effects of cultivated and underutilized species of the cucurbitaceae family. In S. A. Petropoulos, I. C. F. R. Ferreira, & L. Barros (Eds.), *Phytochemicals in vegetables: A valuable source of bioactive compounds* (pp. 99–165). Sharjah: Bentham Science Publishers.

Ujváry, I. (2010). Pest control agents from natural products. In R. Krieger (Ed.), *Hayes' handbook of pesticide toxicology* (3rd ed., pp. 119–229). New York: Academic Press. https://doi.org/10.1016/B978-0-12-374367-1.00003-3.

Vrchotová, N., Será, B., & Tříska, J. (2007). The stilbene and catechin content of the spring sprouts of *Reynoutria* species. *Acta Chromatographica*, *19*(July), 21–28.

Wachsman, M. B., & Castilla, V. (2012). Antiviral properties of brassinosteroids. In M. B. Wachsman & V. Castilla (Eds.), *Brassinosteroids: Practical applications in agriculture and human health* (pp. 57–71). Sharjah: Bentham Science Publishers. https://doi.org/10.2174/97816080529811120101 0057.

Wallace, P. A., Marfo, E. K., & Plahar, W. A. (1998). Nutritional quality and antinutritional composition of four non-conventional leafy vegetables. *Food Chemistry*, *61*(3), 287–291.

Wang, T., Liu, Y., Li, X., Xu, Q., Feng, Y., & Yang, S. (2018). Isoflavones from green vegetable soya beans and their antimicrobial and antioxidant activities. *Journal of the Science of Food and Agriculture*, *98*(5), 2043–2047. https://doi.org/10.1002/jsfa.8663.

Wang, Y. Q., Luo, W. H., Xu, R. J., & Zhao, Y. J. (1994). Effect of epibrassinolide on growth and fruit quality of watermelon. *Plant Physiology Communications*, *30*(6), 423–425.

Wang, D., Pan, H., Deng, X., Xiang, H., & Gao, H. (2007). Cucurbitane and hexanorcucurbitane glycosides from the fruits of *Cucurbita pepo* cv *dayangua*. *Journal of Asian Natural Products Research*, *9*(6), 525–529. https://doi.org/10.1080/10286020600782538.

Wang, S., Panter, K. E., Gaffield, W., Evans, R. C., & Bunch, T. D. (2005). Effects of steroidal glycoalkaloids from potatoes (*Solanum tuberosum*) on in vitro bovine embryo development. *Animal Reproduction Science*, *85*(3–4), 243–250. https://doi.org/10.1016/j.anireprosci.2004.06.002.

Wang, F.-Q., Yao, K., & Wei, D.-Z. (2011). From soybean phytosterols to steroid hormones. In H. El-Shemy (Ed.), *Soybean and health* (pp. 231–252). London: IntechOpen. https://doi.org/10.5772/18808.

Wani, A. S., Tahir, I., Ahmad, S. S., Dar, R. A., & Nisar, S. (2017). Efficacy of 24-epibrassinolide in improving the nitrogen metabolism and antioxidant system in chickpea cultivars under cadmium and/or NaCl stress. *Scientia Horticulturae*, *225*, 48–55. https://doi.org/10.1016/j.scienta.2017.06.063.

Whitten, P. L., Lewis, C., Russell, E., & Naftolin, F. (1995). Potential adverse effects of phytoestrogens. *The Journal of Nutrition*, *125*(Suppl. 3), 771S–776S. https://doi.org/10.1093/jn/125.3_Suppl.771S.

Wilkinson, A. P., Wähälä, K., & Williamson, G. (2002). Identification and quantification of polyphenol phytoestrogens in foods and human biological fluids. *Journal of Chromatography B*, *777*(1–2), 93–109. https://doi.org/10.1016/S1570-0232(02)00095-8.

Wu, J. J., Cheng, K. W., Zuo, X. F., Wang, M. F., Li, P., Zhang, L. Y., et al. (2010). Steroidal saponins and ecdysterone from *Asparagus filicinus* and their cytotoxic activities. *Steroids*, *75*(10), 734–739. https://doi.org/10.1016/j.steroids.2010.05.002.

Wu, Y.-D., & Lou, Y.-J. (2007). Brassinolide, a plant sterol from pollen of Brassica napus L., induces apoptosis in human prostate cancer PC-3 cells. *Pharmazie*, *62*, 392–395. https://doi.org/10.1691/ph.2007.5.6660.

Xia, X.-J., Wang, Y.-J., Zhou, Y.-H., Tao, Y., Mao, W.-H., Shi, K., et al. (2009). Reactive oxygen species are involved in brassinosteroid-induced stress tolerance in cucumber. *Plant Physiology*, *150*(2), 801–814. https://doi.org/10.1104/pp.109.138230.

Xiang, L., Wang, Y., Yi, X., & He, X. (2016). Antiproliferative and anti-inflammatory furostanol saponins from the rhizomes of *Tupistra chinensis*. *Steroids*, *116*, 28–37. https://doi.org/10.1016/J.STEROIDS.2016.10.005.

Xu, Y. M., Wijeratne, E. M. K., Brooks, A. D., Tewary, P., Xuan, L. J., Wang, W. Q., et al. (2018). Cytotoxic and other withanolides from aeroponically grown *Physalis philadelphica*. *Phytochemistry*, *152*, 174–181. https://doi.org/10.1016/j.phytochem.2018.04.018.

Yadav, P., Kaur, R., Kanwar, M. K., Sharma, A., Verma, V., Sirhindi, G., et al. (2018). Castasterone confers copper stress tolerance by regulating antioxidant enzyme responses, antioxidants, and amino acid balance in *B. juncea* seedlings. *Ecotoxicology and Environmental Safety*, *147*, 725–734. https://doi.org/10.1016/j.ecoenv.2017.09.035.

Yang, C.-X., Huang, S.-S., Yang, X.-P., & Jia, Z.-J. (2004). Nor-lignans and steroidal saponins from *Asparagus gobicus*. *Planta Medica*, *70*(5), 446–451. https://doi.org/10.1055/s-2004-818974.

Yang, C. R., Zhang, Y., Jacob, M. R., Khan, S. I., Zhang, Y. J., & Li, X. C. (2006). Antifungal activity of C-27 steroidal saponins. *Antimicrobial Agents and Chemotherapy*, *50*(5), 1710–1714. https://doi.org/10.1128/AAC.50.5.1710-1714.2006.

Yashin, A. Y., Yashunskii, D. B., Vedenin, A. N., Nifant'ev, N. E., Nemzer, B. V., & Yashin, Y. I. (2018). Chromatographic determination of lignans (antioxidants) in food products. *Journal of Analytical Chemistry*, *73*(5), 399–406. https://doi.org/10.1134/S106193481805012X.

Yoder, S. C., Lancaster, S. M., Hullar, M. A. J., & Lampe, J. W. (2015). *Gut microbial metabolism of plant lignans: Influence on human health. In K. Tuohy & D. Del Rio (Eds.), Diet-microbe interactions in the gut* (pp. 103–117). San Diego: Academic Press. https://doi.org/10.1016/B978-0-12-407825-3.00007-1.

Zhang, H., Birch, J., Xie, C., Yang, H., Dias, G., & Kong, L. (2018). Optimization of extraction parameters of antioxidant activity of extracts from New Zealand and Chinese *Asparagus officinalis* L root cultivars. *Industrial Crops and Products*, *119*, 191–200. https://doi.org/10.1016/j.indcrop.2018.03.066.

Zhang, H.-J., Sydara, K., Tan, G. T., Ma, C., Southavong, B., Soejarto, D. D., et al. (2004). Bioactive constituents from *Asparagus cochinchinensis*. *Journal of Natural Products*, *67*(2), 194–200. https://doi.org/10.1021/np030370b.

Zhang, J. D., Xu, Z., Cao, Y. B., Chen, H. S., Yan, L., An, M. M., et al. (2006). Antifungal activities and action mechanisms of compounds from *Tribulus terrestris* L. *Journal of Ethnopharmacology*, *103*(1), 76–84. https://doi.org/10.1016/j.jep.2005.07.006.

Zhou, L.-B., Chen, T.-H., Bastow, K. F., Shibano, M., Lee, K.-H., & Chen, D.-F. (2007). Filiasparosides A-D, cytotoxic steroidal saponins from the roots of *Asparagus filicinus*. *Journal of Natural Products*, *70*(8), 1263–1267. https://doi.org/10.1021/np070138w.

Zhu, N., Kikuzaki, H., Vastano, B. C., Nakatani, N., Karwe, M. V., Rosen, R. T., et al. (2001). Ecdysteroids of quinoa seeds (*Chenopodium quinoa* willd.). *Journal of Agricultural and Food Chemistry*, *49*(5), 2576–2578. https://doi.org/10.1021/jf0014462.

CHAPTER EIGHT

Terpene core in selected aromatic and edible plants: Natural health improving agents

Jovana Petrović, Dejan Stojković, Marina Soković*
Department of Plant Physiology, Institute for Biological Research "Siniša Stanković", University of Belgrade, Belgrade, Serbia
*Corresponding author: e-mail address: mris@ibiss.bg.ac.rs

Contents

Abstract

Aromatic plants synthesize and produce aromatic molecules, among these compounds some of them belong to terpenes and terpenoids. Plant species have specific genes involved in secondary metabolism which allows them to synthesize various compounds with terpene core. These kinds of plant species are also known as herbal drugs and they are primarily used as components in medicinal products or simply as health foods. This chapter will focus on terpene and terpenoid compounds found in selected edible and aromatic plants belonging to several plant families. Selected plant species are briefly discussed. Biologically active compounds with terpene core are most frequently found in essential oils of the edible and aromatic species, as well as they are separately isolated and identified from the extracts. Health beneficial effects coming from terpene compounds found in edible and aromatic plants are further presented and include anti-microbial, antiviral, cytotoxic, anticancer, anti-inflammatory and many other pharmacological activities.

Advances in Food and Nutrition Research, Volume 90
ISSN 1043-4526
https://doi.org/10.1016/bs.afnr.2019.02.009

1. An introduction to selected edible and aromatic plants

Since time immemorial humans have used different parts of wild growing plants as food and medicine, and this particular knowledge gave them the advantage to survive harsh natural environment. All cultures in the world have a comprehensive, traditional knowledge on plants with nutritive and medicinal properties, which is used for the sake of preserving health in the absence of modern medicines as well as prolonging life expectancy. Aromatic plants are plants which contain essential oils responsible for their odor and aroma. Due to this fact, they are widely used in perfumery, to improve food aroma, but also for their health beneficial effects (often applied topically and rarely internally) (Ali et al., 2015; Soković, Ćirić, Glamočlija, & Skaltsa, 2017; Talapatra & Talapatra, 2015). Literature sources indicate that over 7000 wild growing plants have been used as food at some point of human civilization, with this number continuously declining throughout time indicating slow, but constant loss of data on usefulness of certain plant species. An interesting fact is that people once consumed wild growing plants for food and remedy as a necessity, due to the scarcity of food sources, but nowadays they have developed consciousness on healthy habits in eating, so as result, some of the almost forgotten plants have been reevaluated (Pinela, Carvalho, & Ferreira, 2017; Zouari, 2015). With additional scientific confirmation of what's already known in ethnopharmacology of underdeveloped countries and what's most likely lost from the developed ones, many plants are expected to acquire status of candidates for functional food. The importance of this traditional knowledge has been even recognized by the World Health Organization (WHO) which summarized it into a worldwide strategy; the fact seems to be of great importance especially when one has in mind that almost 80% of human civilization relies on natural sourced substances as medicine at some point of their life (Vender et al., 2005).

2. Terpene core in edible and aromatic plants: Terpenes and terpenoids

Plants produce various types of biologically active compounds, generally classified into three large groups: terpene-core compounds, alkaloids and phenolic compounds, among which terpenes/isoprenoids take an important place (Zwenger & Basu, 2008). The terms terpene and terpenoid are often used interchangeably in literature, even though they are not quite the same. According to Pereira, Severino, Santos, Silva, and Souto (2018), terpenoids are actually modified terpenes that contain oxygen in form of other groups, such as: ether, hydroxyl, aldehyde, ketone, and carboxylic. Literature sources indicate that terpenes are among the largest and diversified group of low molecular weight compounds, with nearly 30,000–40,000 identified products in natural sources, such as plants, insects, microorganisms, mushrooms, and animals (Aharoni, Jongsma, & Bouwmeester, 2005; Cho et al., 2017). The most recent data even indicate that as much as 64,000 terpene-core compounds have been identified so far (Abdallah & Quax, 2017).

The theories on why there are as many terpenes have changed the concept over time. What is now a known fact is that diversification of terpenes used to occur and continues to do so, as a result of coexistence with other organisms (with an emphasis on herbivores) (Pichersky & Raguso, 2016). Plants are capable of producing a high number of these compounds thanks to the chemical nature of enzymes included in their biosynthesis which enable high number of products resulting from one single substrate. Additionally, potential changes that occur via mutation in numerous genes coding for terpene compounds (approximately 30–100) as well as post-chemical modifications which include glycosylation, acylation or addition of fatty acids, benzyl and hydroxy rings to the basic terpene core may end up increasing number of compounds (Pichersky & Raguso, 2016). In agreement with a high number of these compounds is also the immense number of their functions in organisms, to mention a few: building blocks of cell membranes, participants of cellular essential biochemical processes, such as electron transfer and photosynthesis (Pichersky & Raguso, 2016). From the aspect of their health beneficial effects on human health, multiple positive effects have been noticed, which will be elaborated throughout this chapter. Nevertheless, this fact alone doesn't mean they are safe to use no matter the consequences. Some terpenes are considered potentially toxic: for example, triterpenoid saponin identified in *Silene vulgaris* (Moench) Garcke (silenosides A, B and C) or some monoterpenes identified in *Mentha pulegium* L. (pulegone, menthofuran, thujones, 1,8-cineole) (Pinela et al., 2017).

Biochemical pathway of the terpene biosynthesis is completely elucidated, demonstrating that chemical core of vast terpene compounds includes isoprene (2-methyl-1,3-butadiene) units containing 5 C atoms

(Zuzarte & Salgueiro, 2015). According to Pereira et al. (2018), terpene precursors isopentenyl diphosphate (IPP) and its allylic isomer dimethylallyl diphosphate (DMAPP) condense, resulting in the formation of prenyl diphosphates, geranyl diphosphate (GPP), farnesyl diphosphate (FPP), and geranyl geranyl diphosphate (GGPP). These compounds are practically linear precursors of all terpenes. They come from (1) mevalonate and (2) methylerythritol phosphate (MEP) or so-called non-mevalonate pathways which have different cellular compartmentalization: mevalonate pathway occurs in cytoplasm (sesquiterpenes most likely), whereas non-mevalonate pathway takes place in chloroplasts (mono- and diterpenes) (Singh & Sharma, 2015). After terpene formation, the enzyme terpene synthase and/or cyclase steps into action, converting them into fully formed terpenes which can later undergo different modifications. Additionally, DMAPP connects with one or more IPP molecules in a "head to tail" manner (Zuzarte & Salgueiro, 2015), after which second modification of the product occurs. The modifications are responsible for specific properties of terpenes, giving them chemical and functional attributes. Most commonly, biosynthesis ends up forming mono (10 C) and sesquiterpenes (15 C), while other types are rarer to identify (Zuzarte & Salgueiro, 2015).

According to Yadav, Yadav, and Goyal (2014), the most comprehensive definition of plant terpenes is that they are hydrocarbons with formula $(C_5H_8)_n$, but also include different derivatives of basic terpene-core compounds. Depending on their structure and number of rings, terpene-core compounds may be divided into several classes that can further be subdivided. While acyclic terpenes have open structure, monocyclic contain one ring structure. Class of bicyclic terpenes includes compounds with two rings, whereas tricyclic group are comprised of three rings, and tetracyclic of four. Sesquiterpenes can also be divided into acyclic and cyclic.

The classification of terpenes based on the number of carbon atoms is as follows (Table 1): hemiterpenes (5 C), monoterpenes (10 C), sesquiterpenes (15 C), diterpenes (20 C), sesterterpenes (25 C), triterpenes (30 C), tetraterpenes (40 C), and polyterpenes (*n* C), that are comprised from a larger number of isoprene units >45 C (Fig. 1) (Pereira et al., 2018; Singh & Sharma, 2015). Which type of terpene compound will be synthesized is correlated to the biosynthesis pathway in a following manner: triterpenes and sesquiterpenes are synthesized via mevalonate pathway, whereas hemi-, mono-, di- and tetraterpenes are synthesized via MEP (Oldfield & Lin, 2012; Talapatra & Talapatra, 2015).

Table 1 Classification of terpenes according to Bhargava, Patel, and Desai (2013).

No. of isoprene units	No. of C atoms	Name of the class
1	C_5	Isoprene
2	C_{10}	Monoterpene
3	C_{15}	Sesquiterpene
4	C_{20}	Diterpene
6	C_{30}	Triterpene
8	C_{40}	Tetraterpene
N	C_n	Polyisoprene

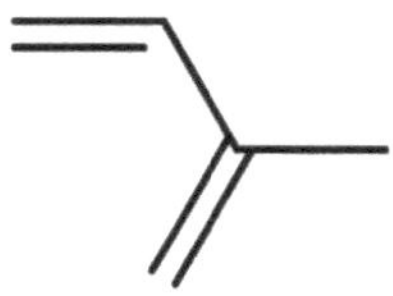

Fig. 1 Structure of isoprene unit.

When discussing whether terpenes are primary or secondary metabolites, it is safe to say that they include both types of compounds. With respect to their universal abundance among plants, metabolites are usually divided into two groups: (1) primary metabolites—compounds which are present universally through the Plant Kingdom (in this particular case), but not as numerous in comparison to the total number of identified compounds, and (2) secondary metabolites—compounds which are most often specific to a species, or its lineage. Primary metabolites are essential for growth, development and reproduction, while secondary are produced as a result of reciprocal interaction between an organism (in this case—plant) and its environment so as to increase reproductive success. Although there are perhaps a few hundred terpenes that are found in all or almost all plants—therefore they are defined as primary metabolites—the vast majority of terpenes are restricted to a given lineage or even a single species, and are therefore referred to as secondary metabolites (Aharoni et al., 2005; Chen, Tholl, Bohlmann, & Pichersky, 2011; Pichersky & Raguso, 2016; Zuzarte & Salgueiro, 2015). Plants have multiple benefits from terpenes, which are also reflected through their estimated number. Many of them improve plant's response to abiotic and biotic stress, whereas others are signal molecules to pollinators (Pichersky & Raguso, 2016; Singh & Sharma, 2015).

Some of them have been proved beneficial to human health as well, showing potent biological activities, whereas others are interesting from commercial point of view in cosmetic and food industry and biotechnology which provide opportunities for their profitable use (Aharoni et al., 2005; Talapatra & Talapatra, 2015).

2.1 Monoterpenes and derived compounds in edible and aromatic plants

Plants produce numerous types of monoterpene compounds, usually as part of essential oils where they can reach as high as 90% of the oil, while only few naturally occur as pure ones (Aharoni et al., 2005; Bakkali, Averbeck, Averbeck, & Idaomar, 2008; Cho et al., 2017; Parvin, Shahrokh, Mozafar, Hassan, & Mehrdad, 2014). Essential oils are volatile liquids that can be extracted from almost every part of the plant in different amounts: flowering parts, leaves, fruits, seeds, stems, bark, and buds (Ali et al., 2015; Yadav et al., 2014). Some of these compounds exist in enantiomeric forms, forming racemic mixtures (Talapatra & Talapatra, 2015). Commonly, few compounds, so-called main compounds, dominate over dozens of others, dictating odor and pharmacological activity of the oil. Compounds present in very low amounts (<1.00%) are referred to as trace components, but their presence in the oil is of great importance as well, especially from the aspect of its biological activity (Abad, Bedoya, Apaza, & Bermejo, 2012). A phenomenon of synergism is a known, when it comes to natural sourced substances and it includes the situation when a given compound shows greater biological potential when coincides together with other compounds rather than in an isolated form (Stojković et al., 2013).

2.1.1 Biosynthesis, classification, and isolation

Monoterpenes are widely distributed in the Plant Kingdom, especially in the following families: Asteraceae, Apiaceae, Lamiaceae (in the scope of this chapter), but may also be identified in plants belonging to other families such as: Myrtaceae, Lauraceae, Rutaceae, Poaceae, and Cupressaceae. So far, monoterpenes have been identified in plants: leaves, fruits, roots, seeds and barks; identified compounds may be somewhat different depending on the part of the plant from which they are isolated (Talapatra & Talapatra, 2015). Data indicate that their secretion takes place in schizogenous and lysogenous glands, as well as in glandular trichomes and ducts (Bhargava et al., 2013). Plants producing monoterpenes have multiple benefits from it, for example, environmental advantage reflects itself through their ability to cool down the surface of plants making their photosynthetic machinery more resistant to increasing temperatures and plants as a whole more resilient to heat stress. Additionally to this protective activity against overheating, volatile monoterpenes may protect plants against oxidative stress as well, due to their potential to interact with atmosphere particles (Parvin et al., 2014). Their function is most often ecological, with the accent on the pollinator's attraction and repellent activity toward herbivores (Paduch et al., 2016).

From the chemical point of view, monoterpenes can be divided into three groups: (1) acyclic, (2) monocyclic, and (3) bicyclic monoterpenes (Parvin et al., 2014; Pereira et al., 2018), but they may also possess double bonds and different moieties, such as hydroxyl, carbonyl, or other type of group.

Acyclic monoterpenes are derivatives of 2,6 dimethyloctane, and most commonly identified in plant sources are: geraniol, linalool, ocimene, myrcene, aloocimene, nerol, citronellol, etc. They have multiple roles, including protective against herbivores which was demonstrated after analyses of the plant parts damaged by their predators—often insects. Geraniol is, for example, known for its insect repellent activities and specific fragrance for which it has broad use in perfumery and food industry (Michaelakis et al., 2014; Müller et al., 2009; Yadav et al., 2014). Linalool, a monoterpene alcohol often identified in numerous plant species, is famous for several biological activities and has numerous industrial applications (Talapatra & Talapatra, 2015). According to Pereira et al. (2018), this acyclic monoterpene is identified in as much as 200 plant species, but in different amounts depending on the analyzed part of the plant, its harvest time and geographic/climatic conditions. It is the dominant component in the essential oils of innumerous species belonging to the families: Apiaceae (*Coriandrum sativum* L., mericarp and seeds, *Thapsia garganica* L.), Lamiaceae (*Hyssopus officinalis* L., *Lavandula angustifolia* Mill., *Lavandula latifolia* Medik., *Ocimum basilicum* L., *Origanum dictamnus* L., *Satureja montana* L.), but has also been identified in plants belonging to Betulaceae, Lauraceae, Myrtaceae, Rutaceae, Verbenaceae, etc. (Pereira et al., 2018). Ocimene is a commonly identified monoterpene identified in plant families within the scope of this chapter (Asteraceae, Apiaceae, and Lamiaceae). *cis*- and *trans*-isomers of β-ocimene are readily identified in *Lavandula multifida* L. (Lamiaceae) and sabinene in seeds of *Daucus carota* L. (Apiaceae). β-Myrcene is often identified in plants, unlike the α-isomer of the compound, which has not been identified so far in natural sources. Due to its fresh citrus-like fragrance, it is desirable in perfumery industry (Zuzarte & Salgueiro, 2015).

Monocyclic monoterpenes are biosynthesized from cyclohexane with isopropyl substituent, and some of the commonly identified in various plant families are listed below: limonene, phellandrene, terpinene, cymene, ascaridole, pulegone, etc. Limonene is a terpene hydrocarbon widely spread among citrus essential oils, very appealing for perfumes industry (Kim et al., 2013). (−)-α-Phellandrene, a compound with citrus odor and slight peppery scent, is identified among many plants including *Anethum graveolens* L. (Apiaceae), whereas (−)-β-phellandrene is identified in *Crithmum maritimum* L. Compounds belonging to monocyclic terpenes are mainly important for industrial purpose (Zuzarte & Salgueiro, 2015).

Bicyclic group of monoterpenes include: camphor, thujone, caren, sabinene, umbellulol, pinene, thujone, among others (Talapatra & Talapatra, 2015). Camphor has long been traditionally used in rituals and nowadays is important in Asian food industry. It is known for high permeability through the skin, but in high doses can be poisonous: >500 mg causes severe neurological distractions, whereas 4 g dose can be lethal (Talapatra & Talapatra, 2015). Thujone, isolated from *Artemisia absinthium* L., usually affects GABA receptors, which is the reason why absinthe, extracted from this plant is considered psychedelic agent. Nevertheless it is widely consumed in drinks, but with strictly defined maximum permissible dose of 25 mg/L (Olsen, 2000). Whereas both camphor and pinene are important allelopathic agents (identified in *Salvia leucophylla* Greene—Asteraceae). Pinene, naturally occurring in two isoforms α- and β-(making racemic mixtures) is also very important for synthesis of other compounds (Talapatra & Talapatra, 2015). High concentrations are also present in *Sideritis erythrantha* Boiss. et Heldr., *Salvia rosifolia* Sm., and *Rosmarinus officinalis* L. (Lamiaceae) (Zuzarte & Salgueiro, 2015). In addition to the mentioned ones, iridoids are among most common group of bicyclic monoterpenes. Biosynthetically, iridoids are derived from 8-oxogeranial, but are identified as glycosides in plants, since they typically bind to sugars (most often to glucose). Iridoids are often linked to numerous biological activities (sugar regulating, anticarcinogenic, immunomodulatory, etc.) but have value as taxonomic markers as well. A fine example of this is as follows: as much as five families that were prior included in Umbelliferae (Apiaceae) now have to be reclassified due to the fact that they produce iridoid compounds which makes them closer to Dipsacales (Bhargava et al., 2013). Their importance as chemotaxonomic markers is present within numerous members of the Lamiaceae family as well: aromatic group members do not produce iridoid glycosides compounds, whereas nonaromatic group members produce this compound (Bhargava et al., 2013).

Essential oils of the plants within the scope of this chapter have been chemically characterized on several occasions. Comparative analysis of the available literature data indicates that type and abundance of terepene compounds depend on multiple factors including: season of harvesting, climate, and geographical condition in which plant has grown, etc. The aforementioned will be briefly discussed with few examples (El-Zaeddi, Martínez-Tomé, Calín-Sánchez, Burló, & Carbonell-Barrachina, 2016).

Selected species harvested in Spain were rich in the following compounds: *A. graveolens*: α-phellandrene and β-phellandrene; *Petroselinum crispum* (Mill.) Fuss: 3,8-*p*-menthatriene and β-phellandrene; *C. sativum*: E-2-dodecenal, dodecanal, and octane (all members of Apiaceae family); and *Mentha piperita* L.: carvone and limonene (member of Lamiaceae family) (El-Zaeddi et al., 2016). β-Phellandrene along with sabinene, α-bisabolol, and α-phellandrene was major constituent of the essential oil of yet another member of Apiaceae family, *Selinum* sp. harvested in India (Joshi, Melkani, Nailwal, Prasad, & Bisht, 2018). For example, in samples of *Salvia officinalis* L., *L. angustifolia*, and *Mentha asiatica* Boriss. dozen compounds were identified: 23, 33 and again 33, respectively, among which terpenes: 1,8-cineole, sabinene, and linalool were the most abundant (Sonmezdag, Kelebek, & Selli, 2017). Along with essential oils, extracts have proved to be rich in terpene core compounds as well: dichloromethane extracts of some Lamiaceae members collected in Turkey were rich in different terpene compounds. Some traditionally used herbs proved to be rich source of terpene-core compounds as well, to which their biological activity may be ascribed to. An example of that is *Eryngium creticum* Lam. (Apiaceae) which has been traditionally used for treatment of various conditions. Different herbal extracts were rich in mono- and sesquiterpenes among other bioactive compounds (Kikowska, Dworacka, Kędziora, & Thiem, 2016). According to Jung et al. (2001), members of the Asteraceae family produce various derivatives of monoterpenes as well, such is the case of *Aster scaber* (Thunb.) Nees whose aerial parts are a source of two new monoterpene peroxide glycosides. This in some ways points to the fact that almost every plant species is characterized by a specific terpene compound, as well as that for most of these compounds biological possibilities and probable beneficial health effects have not yet been determined (Joshi et al., 2018).

2.1.2 Health-beneficial effects of monoterpenes and derived compounds

Numerous research studies during the years showed that monoterpenes have multiple biological activities. Several examples of medicinal potential of monoterpenes and monoterpenoids will be briefly discussed in the following section and the health beneficial effects of some compounds are summarized in Table 2.

One of the most frequently mentioned is cytotoxic/antitumor/antiproliferative activity toward different malignant cell lines. Due to the fact that synthetic cytostatics used in modern medicine have an adverse effect on non-tumor cells aside from affecting tumor cells, efforts have been made toward finding an effective and more selective cytostatics in natural products. Different types of terpenes emerged as presumptuous source of anticancer activity. Their mechanisms of action include rather different pathways than the ones achieved through the activity of commercially applied drugs. Among the described ones, the most promising is the effect of natural sourced substances on mevalonate metabolism, which raises hope for finding new effective antitumor treatment (Fernandes, 2015). Parvin et al. (2014) demonstrated terpenes have promising activity toward several malignant diseases. Namely, D-limonene and perillyl alcohol inhibited growth of several cancerous cell lines, including prostate, pancreas, skin, etc. Additionally, authors showed that metabolites of these compounds also act on malignant cell lines, for example, the activity of perillic acid, limonene-1,2-diol and few others was demonstrated. Mechanism of the activity which included suppression of posttranslational isoprenylation of proteins that regulates cell growth was described as well.

Aside from antitumor activity, some monoterpenes showed antioxidant potential which is achieved through multiple mechanisms, including lipid peroxidation prevention, scavenging of free radicals and others; thymol and carvacrol, the dominant components of *Origanum vulgare* L. (oregano) essential oil are among the most promising ones (Anastasaki, Zoumpopoulou, Astraka, & Kampoli, 2017; Parvin et al., 2014). A powerful effect of oregano essential oil has been demonstrated on several occasion; according to Chouhan, Sharma, and Guleria (2017) different types of terpene compounds are responsible for its antimicrobial activity, carvacrol and thymol to mention a few. Namely, 48.9% of thymol and 19.0% of *p*-cymene constituted >50% of the essential oil of *Thymus vulgaris* L., whereas 12.8% of carvacrol, 12.8% of α-terpinyl acetate, 11.2% of *cis*-myrtanol, and 10.4% of thymol constituted the oil of *Thymus tosevii* L. The oils in question showed antioxidant potential which may very well be attributed to these components.

Unlike for the *Thymus* species, same authors demonstrated that menthol dominated *Mentha piperita* essential oil with 37.4%, followed by menthyl-acetate with 17.4%, and menthone with 12.7%, whereas the essential oil

Table 2 Health beneficial effects of selected monoterpenes and monoterpenoids.

Monoterpenes and monoterpenoids	Source	Health beneficial effects	Reference
Geraniol	Lemongrass, roses	Antitumor, antimicrobial compound, with antioxidant and anti-inflammatory properties	Ahmad et al. (2011), Khan et al. (2013), and Thapa, Losa, Zweifel, and Wallace (2012)
Ocimene	*Ocimum basilicum*, *Artemisia absinthium* and many others	Wound healing, anti-inflammatory action, abolishes or reduced edema, hyperemia, laceration, and hemorrhage	Pravdich-Neminskaya and Kachkov (1978)
Myrcene	Myrcia, bay, rosemary, cannabis, ylang-ylang, wild thyme, parsley, cardamom, and hops	Analgesic, anti-inflammatory, antioxidant properties, antibacterial activity	Bonamin et al. (2014) and Ciftci, Ozdemir, Tanyildizi, Yildiz, and Oguzturk (2011)
Citral	Lemon myrtle, *Litsea citrate*, *Litsea cubeba*, lemongrass, lemon tea-tree, *Ocimum gratissimum*, *Lindera citriodora*	Flavor additive in foods, antimicrobial, anti-inflammatory and anticancer properties	Bhalla, Gupta, and Jaitak (2013) and Marcus, Klossek, Touraud, and Kunz (2013)
Limonene	Citrus fruits peels	Antidepressant, antinociceptive, antidiabetic, antiulcerogenic	Piccinelli et al. (2015), Piccinelli et al. (2017), Joglekar, Bavkar, Sistla, and Arvindekar (2017), and Rozza et al. (2011)
Menthol	Corn mint, peppermint, or other mint species	Antimicrobial, Radioprotective, antioxidant, analgesic	Freires, Denny, Benso, de Alencar, and Rosalen (2015), and Sueishi and Nii (2018), and Galeotti et al. (2002)

Continued

Table 2 Health beneficial effects of selected monoterpenes and monoterpenoids.—cont'd

Monoterpenes and monoterpenoids	Source	Health beneficial effects	Reference
Thymol	Oregano species, thyme species	Antifungal, anti-parasitic, antiseptic	Numpaque, Oviedo, Gil, García, and Durango (2011) and Filoche, Soma, and Sissons (2005)
Carvacrol	Oregano, thyme, pepperwort, wild bergamot	Antimicrobial, Pro- and anti-apoptotic	Wen-Xian et al. (2008) and Bhakkiyalakshmi et al. (2016)
Pinene	Pine, coniferous species, sagebrush, ironwort, sage	Antimicrobial	Stojković et al. (2008)

obtained from *Mentha spicata* L. was characterized with carvone (69.5%) and menthone (21.9%) as the main compounds. The presence and domination of these components was ascribed to the antioxidant potential of the essential oils, as well as for the exquisite antimicrobial activity of the aforementioned oils, even much better in comparison to the commercial antibiotics and antimycotics.

Prominent in vitro antifungal activity toward *Candida albicans* was also observed for *cis*- and *trans*-isomers of β-ocimene, identified in *L. multifida* (Lamiaceae) (Zuzarte & Salgueiro, 2015). Their antioxidant capacity is achieved via multiple mechanisms. A fine example on how a natural source of compounds with demonstrated in vitro activities may be practically applied onto in situ food systems is the use of *Micromeria dalmatica* Benth. essential oil as a food preservative. Namely, this essential oil is rich in monoterpene compounds (piperitone—41.46%, piperitone-oxide—19.02%, D-limonene—6.23% and *p*-menthone—5.06%) and effectively delayed the growth of pathogenic bacteria and yeasts using in vitro microdilution method. The obtained results served as a starting point for the development of in situ food system using actual pork meat rather than microbiological media to evaluate the potential of the oil to be used as natural preservative. The oil demonstrated strong activity toward *Salmonella typhimurium* with GI_{50} values of only 0.048 mg/mL which strongly indicated that this oil is a future perspective ingredient for the food industry (Bukvički et al., 2015).

The concept of using essential oils as effective food preservatives was additionally confirmed on sliced cheese with the essential oil obtained from aerial parts of *Thymus algeriensis* Boiss. & Reut. collected in Libya. The oil rich in monoterpene carvacrol (80.90%) also effectively eradicated pathogenic bacteria and micromycetes with MBC ranging from 0.05 to 0.15 mg/mL, and MFC in the range of 0.01 and 0.04 mg/mL. The oil proved to be a perspective antioxidant agent, due to the presence of high amounts of carvacrol (Bukvički et al., 2018).

A study by Stojković et al. (2008) showed that α-pinene is effective at 3.3 and 5.0 μL/mL in the growth inhibition of *Actinomadura madurae*, a causative agent of mycetoma—serious infection disease registered on the southern hemisphere.

Eucalyptol (1,8-cineole), a monoterpenoid, showed great potential in inhibiting *Aspergillus* sp. rot on wounded apple fruits (Stojković et al., 2011), showing that this monoterpenoid could be used for apple fruits preservation during storage.

Anti-inflammatory activity of terpenes is achieved through their effect on the cytokine level which usually serves as a marker of the tumor progress (Paduch et al., 2016). Different types of terpene-core compounds affect pathways and nuclear translocation of NF-κB by which they alleviate symptoms of inflammation (Paduch et al., 2016). Anti-inflammatory activity may be attributed to 1,8-cineole, linalyl acetate, (−)-linalool and few other monoterpene compounds (Parvin et al., 2014).

Some monoterpene compounds may demonstrate multiple biological activities. One of such compounds is linalool, a monoterpene alcohol with broad application in several industries, often identified in plants. It is well known for its anti-inflammatory, anti-carcinogenic, antimicrobial, anti-noceptive, analgesic, anxiolytic, and anti-depressive activities. Linalool exhibits anti-carcinogenic activity via promotion of autophagy and apoptosis of malignant cells, whereas it can also effectively retard growth of several clinically relevant pathogens, such as *Staphylococcus aureus*, *Candida albicans*, *Escherichia coli*, and others. From the ecological point of view it has an important role as attractant of the pollinators, but also as a repellent of potential herbivores (Pereira et al., 2018).

Various terpenes (mixed in form of essential oils as well as pure compounds) are known for their allelopathic potential. For example, thymol acts on *Leishmania* sp., whereas menthol derivatives have trypanocidal potential (Parvin et al., 2014). According to Pavela, Maggi, Cianfaglione, Bruno, and Benelli (2018) plants belonging to Apiaceae family may very well be a promising source of compounds with insecticidal properties, namely,

Sison amomum L., *Echinophora spinosa* L., *Heracleum sphondylium* subsp. *sphondylium* L., *Heracleum sphondylium* subsp. *ternatum* (Velen.) Brummit, and *Trachyspermum ammi* (L.) Sprague ex Turrill. Essential oils of the aforementioned species, dominated by the monoterpene phenol thymol, are held responsible for this activity. The two most active essential oils were those from *T. ammi* fruits and *E. spinosa* roots, showing LC_{50} below 20 μL/L and LD_{90} below 50 μL/L, respectively.

Available data indicate that biological activity of essential oils contains mixture of various types of compounds which have been much more thoroughly studied than the biological activity of pure compounds that constitute it. According to Fernandes (2015), it may be difficult to attribute certain biological activity to a particular terpene-core compound, especially since most of the compounds are mixed together, which can include different types of interaction among them. According to Paduch et al. (2016), what makes things even more interesting is that biological activity, aside from being dependent on the type of terpene, is dependent on the type of enantiomer and other type of modification. This basically means that while one enantiomer has potent bioactivity other may have none at all, a different affinity for the receptors is considered to be the cornerstone of this phenomenon. A true example of the aforementioned is the case of carvone, which (*S*)-(+) enantiomer has stronger effect on cell viability than it's (*R*)-(−) enantiomer. This means that by applying/consuming only one of the stereoismoers, higher specificity and efficacy will be achieved. Also, a number of cyclic skeletons as well as the type of groups is directly responsible for biological activity of terpene core compounds—for example, the number of hydroxyl groups in monoterpenes is positively correlated to anticancer activity of the compound (Paduch et al., 2016). Iridoids are also famous for their numerous biological activities and have been referred to as efficient agents for treating digestion problems, tension, headache, allergy, insomnia, etc. (Bhargava et al., 2013).

Nevertheless, even though they are isolated from natural sources, all terpene compounds are not completely safe for use. Many data indicate that these compounds are responsible for intoxicating effects after consuming plants that produce them. According to Pinela et al. (2017), monocyclic monoterpene ketone pulegone, bicyclic monoterpenes menthofuran, and thujone, monoterpene etheroxide 1,8-cineole (eucalyptol) identified in essential oil of aerial parts obtained from *M. pulegium*, as well as bicyclic monoterpene beta-thujone, monoterpene etheroxide 1,8-cineole identified in essential oils obtained from aerial parts of *O. vulgare* are considered as potential intoxicating agents, which indicates careful use of preparations on the basis of these compounds.

2.2 Sesquiterpenes and derived compounds in edible and aromatic plants

The name "sesqui" comes from Latin language; it means "one half more" and refers to the number of C atoms in these compounds, which is 15 (Talapatra & Talapatra, 2015). Depending on their structure, sesquiterpenes can be acyclic or contain rings. This group comprises a few thousand compounds, often identified in plants, micro and marine organisms. In Plant Kingdom, they are most likely to be identified in parts of the plant above the ground, i.e., flowering parts and leaves (Talapatra & Talapatra, 2015). Out of the three plant families of interest within this chapter, members of the Asteraceae family are known for producing various types of sesquiterpene lactones, whereas Apiaceae and Lamiaceae produce lower amounts.

2.2.1 Biosynthesis, classification, and isolation

All sesquiterpenes are biosynthetically derived from *trans* forms, *trans*-farnesyl pyrophosphate (FPP) in the endoplasmatic reticulum. Biosynthesis of sesquiterpenes is presented in Fig. 2. Among their acyclic group, β-farnesene, *cis*-α-farnesene and *trans*-α-farnesene are usually identified in natural sources. Group of cyclic sesquiterpenes covers much wider groups of naturally occurring compounds classified into: (1) monocyclic, (2) bicyclic

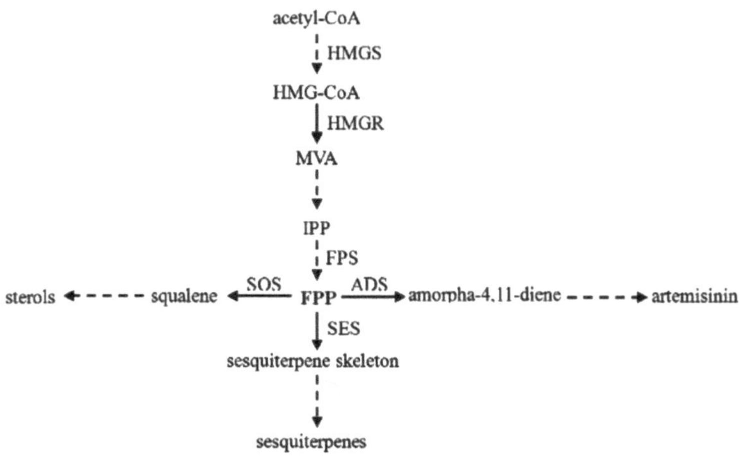

Fig. 2 Biosynthesis of sesquiterpenes.

(α-cadinene, δ-cadinene, β-eudesmol caryophyllene and isocaryophyllene), (3) tricyclic (longifolene), and (4) tetracyclic (longicyclene) with this number reaching as much as 200 (Chadwick, Trewin, Gawthrop, & Wagstaff, 2013; Talapatra & Talapatra, 2015).

Sesquiterpenes containing lactone ring are known as sesquiterpene lactones or SLs. According to Chaturvedi and Dwivedi (2016) and Chadwick et al. (2013), sesquiterpene lactones (SL) are a complex group of >5000 identified compounds in several higher plant families, among which their presence is notable in: Asteraceae (most frequent), Acanthaceae, Anacardiaceae, Apiaceae, Euphorbiaceae, Lauraceae, Cactaceae, Solanaceae, Araceae, and Euphorbiaceae. Data published by Matejić, Šarac, and Ranđelović (2010) are consistent with the above-mentioned fact and also indicate that as much as 90% of the identified SLs are identified among members of the Asteraceae family. Biosynthesis of lactones takes place according to the usual Ruzicka's "head to tail" rule of three isoprene units, followed by their cyclization, oxidative modification, and isomerization (Chaturvedi, Dwivedi, & Mishra, 2013). SLs have very diversified carbon skeleton, which is correlated with their high number of compounds, including among many: eudesmanolides (with 2 fused 6-membered rings), germacranolides (10-membered ring), guaianolides (7-membered and a 5-membered ring and a methyl group at C-4), and pseudoguaianolides (7-membered and a 5-membered ring and a methyl group at C-5). All sesquiterpene lactones contain α-methylene-γ-lactone moiety, epoxides and conjugated carbonyl groups (Chadwick et al., 2013). SLs are usually identified in leaves and glandular trichomes on leaves or flowering heads, but may very well be isolated from other plant parts (Matejić et al., 2010). The type of the compound is specific for a plant, while the content of dry matter ranges from 0.01% to 8.00% (Chaturvedi et al., 2013). People worldwide most often ingest SLs through the consummation of plain lettuce and chicory, whereas the intake of these compounds through plants traditionally used in ethno medicine is limited and specific to certain areas (Chaturvedi et al., 2013). Even though plants synthesize lactones for their own benefit, people tend to take advantage of natural strategies so as to alleviate symptoms of their own diseases/conditions, which will be elaborated throughout the following section.

2.2.2 Health-beneficial effects of sesquiterpenes and derived compounds

α-Methylene-γ-lactone group in the structure of sesquiterpenes is basically responsible for numerous biological activities. Additionally, effects of sesquiterpene lactones are dictated by their molecular geometry, chemical environment of the sulfhydryl groups, lipophilicity among other factors. According to literature data, SLs are famous mainly for their cytotoxic potential which is achieved through alkylation of thiol groups commonly found in proteins (Chadwick et al., 2013). The structural elements in question interact with nucleophiles, especially the cysteine sulfhydryl group, by means of Michael-type addition (Soković et al., 2017). Having in mind that thiol groups are present in cysteine residues in proteins and intracellular GSH, these are the targets for lactones. Their interaction results in reduction of the enzyme activity or causes the disruption of GSH metabolism and important intracellular cell redox balance. Additionally, other mechanisms include: induced rapid apoptosis, externalization of cell membrane phosphatidyl-serine and depolarization of mitochondrial membranes (Chaturvedi et al., 2013).

Other activities have been attributed to them as well, such as regulation of cardiovascular diseases, antimalarial, analgesic, antineurodegenerative, gastroprotective, and antioxidant activities (Chadwick et al., 2013). Antioxidant potential of sesquiterpene lactones has been used so as to describe cytotoxic potential of this group of compounds, but later on it was determined that this activity alone is not sufficient for pro-apoptotic effects (Chadwick et al., 2013). Lactones, produced by plants to defend themselves from several pathogenic microorganisms, may also serve as antimicrobial agents to treat human diseases. Sesquiterpenes demonstrate their activity on pathogenic microorganisms via several mechanisms including destabilization of cell membrane as the main one. They have proved active on both pathogenic bacteria and fungi (*Colletotrichum* sp., *Fusarium* sp., *Botrytis* sp., *Phomopsis* sp.), with guaianolide sesquiterpenes as the most prominent antimicrobial compounds. Volatile representatives of lactones may very well serve as primary defense strategy against pathogens, acting at the very surface of the plant part which produces it. SLs isolated from aerial parts of *Centaurea pullata* L. showed higher antimicrobial potential than those of commercial antimicrobials (Djeddi et al., 2007). Isomontanolide, montanolide, and tarolide, sesquiterpene lactones isolated from *Laserpitium* species, showed higher potential on the inhibition of *Candida albicans* and *C. kruzei* biofilm formation than fluconazole (Popović et al., 2015). Soković et al. (2017) confirm their promising antifungal potential, basing their conclusion on numerous studies dealing with the evaluation of lactones fungicidal potential toward several micromycetes, including *Aspergillus niger*, *A. versicolor*, *A. flavus*, *Cladosporium cladosporoides*, *Fusarium tricinctum*, *Trichoderma viride*, etc.

They suggest that nature of compounds has an important role in achieving antifungal activity.

SLs have also prominent allelopathic potential, one of many examples illustrating this activity is the case of the essential oil obtained from the traditionally used plant *Dendranthema indicum* (L.) Des Moul. (Asteraceae) from China. The oil demonstrated insecticidal activity toward *Tribolium castaneum* and *Stegobium paniceum*, which was attributed to sesquiterpenes content: β-caryophyllene (13.78%), germacrene D (9.11%) and β-*cis*-farnesene (6.59%) (Zhang et al., 2015). Nevertheless, even though they are generally considered as safe to consume, adverse health effects on humans have been described as well. Some of the well-known toxic lactones include repin, janerin, cynaropicrin which have been identified in *Centaurea* spp. Additionally, SLs also have tremendous chemotaxonomic value (Soković et al., 2017).

Family of Asteraceae (formerly Compositae) comprises of nearly 24,000 species which have nutritive, medicinal, and economic potential (Sülsen, Lizarraga, Mamadalieva, & Lago, 2017). Asteraceae are known for the production of bioactive SLs; parthenolide, identified in *Tanacetum parthenium* (L.) Sch. Bip., a plant with already demonstrated ethnopharmacological value, is known for its anti-inflammatory activity which is achieved through regulation of NF-κB-signaling pathway (the inhibition of interleukins and prostaglandins, known pro-inflammatory cytokines). Helenalin, a SL isolated from *Arnica* sp., showed promising anti-carcinogenic potential via multiple mechanisms: pro-apoptotic effect, suppression of NF-κB and depletion of thiol groups. Arglabin, artemisin, and its derivatives produced by *Artemisia* sp. also exhibited prominent selective in vitro as well as in vivo cytotoxic potential (Chaturvedi et al., 2013; Sülsen et al., 2017). Aside from being a perspective anti-carcinogenic agent, artemisin shows excellent anti-malarial activity with few side effects. The example of artemisin is one of the most famous examples of how a natural compound becomes the most effective and commercially available drug in the market (Abad et al., 2012; Jiang, Kempinski, & Chappell, 2017). Arylated derivatives of tourneforine, a sesquterpene lactone isolated from *A. tournefortiana* Benth., also showed promising cytotoxic activity, even better than the naturally occurring compound (Chaturvedi et al., 2013). From *Eupatorium chinense* L. eupalinin was isolated, with potential on inducing autophagy in cancerous cells, or inuviscolide identified in *Inula viscosa* (L.) Greuter, with inhibitory potential toward melanoma cells; angeloylenolin isolated from *Centipeda minima* (L.) A.Braun & Asch., profoundly inhibited proliferation of human nasopharyngeal cancer cells via mechanisms of cell-cycle arrest and pro-apoptosis (Chaturvedi et al., 2013). According to Matejić et al. (2010), numerous SLs have been proved to possess antimicrobial potential as well: tanachin and tavulin isolated from *Tanacetopsis mucronata* (Regel & Schmalh.) Kovalevsk. (Asteraceae) toward intestinal bacteria *E. coli* and *Enterobacter cloacae*; centaurepensin A, chloroyanerin, and 13-acetyl solstitialin A isolated from the plant extract of *Centaurea solstitialis* ssp. *solstitialis* L. showed weak antimicrobial potential on several clinically relevant pathogens.

Apiaceae family members are characterized with the production of bioactive guaianolides biosynthetically derived from germacrene A–D, which may also be identified in some members of the Asteraceae family. According to Chaturvedi et al. (2013) flowering parts of the *Hemisteptia lyrata* (Bunge) Fisch. & C.A.Mey. are known for the production of several bioactive SL compounds, such as aguerin B, 8α-acetoxyzaluzanin C, cynaropicrin, and deacylcynaropicrin. Using Sulforhodamine B (SRB) method, activity of aguerin B and cynaropicrin toward MCF-7 and HCT-15 cell lines has been demonstrated, while the same compounds, isolated from the member of Asteraceae family *Saussurea calcicola* Nakai, showed activity against five human tumor cell lines, A549 (nonsmall cell lung adenocarcinoma), SK-OV-3 (ovarian cancer), SK-MEL-2 (skin melanoma cancer), XF498 (central nervous system cancer), and HCT15 (colon cancer) with an effective dose ED_{50} of 0.29–1.37 and 0.23–1.72 μg/mL, respectively (Chaturvedi et al., 2013). Guaianolide compounds with potent cytotoxic activities were also isolated from *Ixeris chinensis* (Thunb.) Kitag, *Carpesium faberi* C. Winkler, and *Elephantopus mollis* Kunth (Asteraceae). Minimolide G and minimolide H (also guaianolides), isolated from *C. minima* showed anti-proliferative effect toward the cell line of human nasopharyngeal cancer, with IC_{50} of 61.4 and 28.7 μg/mL, respectively. Aside from guaianolides group, pseudoguaianolides derived biosynthetically from guaianolide, with migrated position of methyl group from C-4 to C-5 on the ring, are often identified in plants. One of such compounds is coronopilin, identified in *Ambrosia arborescens* Mill. (Asteraceae), which showed inhibitory activity on cell growth of Jurkat and U937 leukemia cells (IC_{50} values were 5.0 and 11.0 μM, respectively). Bioactive eudesmanolides are also lactones with potent anti-carcinogenic effect, from *Inula racemosa* C.B. Clarke (Asteraceae), eudesmanolide presented cytotoxic activity toward malignant diseases of gastrointestinal tract. For example, from roots of *Inula helenium* L. (Asteraceae), a germacranolide compound was isolated with cytotoxic properties toward

MK-1, HeLa, and B16F10 cell lines, whereas from *E. mollis*, a new cytotoxic compound was isolated as well (Chaturvedi et al., 2013). Various *Centaurea* species with long tradition of ethno medicinal use in Mediterranean region are known for the production of biologically active sesquiterpene lactones. A number of studies scientifically confirmed what was already known in practice, revealing that sesquiterpene lactones may very well be responsible for these activities (regulation of different types of conditions, such as arthritis, bronchitis, diabetes, fever, sprains, headaches, malaria, and inflammation). Some of the examples of bioactive lactones identified in *Centaurea* spp. include: cnicin (identified in as much as 83 species) which effectively retarded growth of pathogenic micromycetes, as well as 13-acetylsolstitialin A isolated from aerial parts of *C. solstitialis*, which showed strong anti-viral effect against Herpes simplex type-1 (Soković et al., 2017). Health beneficial effects of sesquiterpenes and derived compounds presented in this section are briefly summarized in Table 3.

2.3 Other compounds containing the terpene core and their pharmacological attributes

Aside from these "regular," numerous plants contain so-called irregular type of terpenes, whose abundance is not uniform throughout the Plant Kingdom, but rather limited to individual species. Such an example is the case of monoterpenes identified in *Chrysanthemum cinerariaefolium* (Trev) Boccone which are C 10 compounds. These compounds deviate from usual Ruzicka's "head to tail" coupling rule, and include chrysanthemic and pyrethric acid among others (Talapatra & Talapatra, 2015).

According to Lin, Ung, Feng, Huang, and Hu (2016) and Abdallah and Quax (2017) hydrocarbons containing C 20 backbone are referred to as diterpenes (so far approximately 176 compounds identified). They have complex structures and great commercial interest, since some of the potent anticancer compounds including taxol (nitrogenous diterpene ester) belong to this class of compounds (Talapatra & Talapatra, 2015). Depending on their structure, these compounds may be classified into: acyclic (phytol), monocyclic (cembrene and casbene), bicyclic (sclarene), tricyclic (rosenonolactone and rosololactone), tetracyclic (sandaracopimaradiene, ent-7-hydroxykaurenoic acid), pentacyclic, and macrocyclic compounds (Talapatra & Talapatra, 2015). Depending on the core type, they can be classified into: kaurane-type, abietane-type, nemorallisin-type, labdane-type, clerodane-type, cassane-type, dolabrane-type, pimarane-type, icetexane-type, guanacastane- type, cembrane-type, rosane-type, or a combination of

Table 3 Health beneficial effects of selected sesquiterpenes and derived compounds.

Sesquiterpenes and derivatives	Source	Health beneficial effects	Reference
Sesquiterpene lactones	*Centaurea pullata* L.	Antimicrobial	Djeddi et al. (2007)
Isomontanolide Montanolide Tarolide	*Laserpitium* sp.	Antimicrobial, antibiofilm	Popović et al. (2015)
β-Caryophyllene Germacrene D β-*cis*-farnesene	*Dendranthema indicum*	Allelopatic potential	Zhang et al. (2015)
Parthenolide	*Tanacetum parthenium* (L.) Sch.Bip.	Antiinflammatory activity	Sülsen et al. (2017)
Helenalin	*Arnica* sp.	Anticancer	Chaturvedi et al. (2013)
Arglabin Artemisin	*Artemisia* sp.	Anticancer, antimalarial	
Tourneforine	*Artemisia tournefortiana*	Cytotoxic	
Eupalinin	*Eupatorium chinense* L.	Anticancer	
Angeloylenolin	*Centipeda minima*	Antiproliferative	
Aguerin B 8α-Acetoxyzaluzanin C Cynaropicrin	*Hemisteptia lirata*	Anticancer	
Coronopilin	*Ambrosia arborescens* Mill.	Anticancer	
Tanachin Tavulin	*Tanacetopsis mucronata*	Antibacterial	Matejić et al. (2010)
Centaurepensin A Chlorocyanerin 13-acetyl solstitialin	*Centaurea solstitialis* L.	Antimicrobial, antiviral	Matejić et al. (2010) and Soković et al. (2017)
Cnicin	*Centaurea*spp.	Antimicrobial	Soković et al. (2017)

two from the above structural types (Lin et al., 2016). Literature data indicate they exhibit a myriad of biological activities, as in the case of salvinorin A, identified in *Salvia divinorum* Epling & Játiva which has been known for its psychoactive activities and is used for pain relief (Lin et al., 2016).

The aforementioned compounds are not as abundant in plants as other terpenes, and they have been identified in only 18 plant families, including ones within the scope of this chapter: Asteraceae (*Baccharis* sp.) and Lamiaceae (*Ballota* sp., *Clerodendrum* sp., *Isodon* sp., *Plectranthus* sp., *Premna* sp., *Salvia* sp., and *Teucrium* sp.). To our knowledge, these compounds are distributed in almost every part of the plant: roots, stems, leaves, fruits, and seeds, as well as bulbs.

In *Baccharis lejia* Phil. four clerodanoid dimers were identified: bacchalejins numbered 1–4, which are known as the only diterpenes in this family. As for the members of Lamiaceae family, numerous compounds have been identified—60 compounds fully described, and their biological activities were demonstrated. From the genus *Isodon* sp. 29 compounds have been identified, whereas in *Plectranthus* sp., *Teucrium* sp., *Salvia* sp., and *Clerodendrum* sp. the number goes as much as 16 abietane-type compounds. In the members of the latter two mentioned genera, four more diterpenoid dimer compounds were identified, but of clerodane type. Members of the *Premna* sp. are characterized with three icetexane-type diterpenoid compounds, whereas members of the *Ballota* sp. are characterized with one labdane type diterpenoid dimer compound. Plants belonging to *Isodon* sp. are very rich in odd type diterpene compounds, to mention a few: four totarene and labdane type, three abietane type, and one kaurane type. Persianone—dimer containing two labdane type terpenoids was also identified and structurally characterized within the members of *Ballota* sp. members, while inermes A and B, containing two clerodane units, were identified in plants belonging to *Clerodendrum* sp. A rarely occurring terpenoid type compound, named trichotomone was also isolated from *Clerodendrum* sp. members—namely, *Clerodendrum trichotomum* Thunb., and the compound proved to have potent in vitro cytotoxic activity toward several malignant cell lines, including A549, Jurkat, BGC-823, and others. Plants belonging to *Isodon* sp. genus aside from being rich in monoterpene compounds are also a rich source of structurally diverse diterpenes, and to the statement contributes the fact that as much as 36 compounds with kaurane-type core were isolated from the plants. Hispidanins A–D are appreciated due to their unique heterodimeric structure including totarane-type and labdane-type diterpenes and represent the unique naturally occurring heterodimeric compounds. Some of the members of this family have potent cytotoxic potential with low IC_{50} values. Members of the *Salvia* sp. are characterized with 14 diterpenoid dimeric compounds: 12 abietane-type and two clerodane-type. From the abietane type group, most often are mentioned hongencaotone, bisprioterones A–C (*S. prionitis* Hance), rosmanoyl carnosate (*S. canariensis* L.), salviwardins A and B (*S. wardii* E. Peter), salvialeriicone (*S. leriifolia* Benth), broussonetones A and B (*S. broussonetii* Benth.) to mention a few, and from the second group both compounds were identified in *S. wagneriana* Pol (Lin et al., 2016).

Naturally occurring diterpene, dimeric compounds, are known for their numerous biological activities, but the obstacle in turning them into powerful pharmacological tools is most likely the fact they are commonly produced in low amounts. Due to the fact that these compounds have very diverse structures, they can serve very well as taxonomic markers in genera in which morphological markers cannot lead to a final decision (Lin et al., 2016). Diterpenes may also be effective in certain aspects of antimicrobial activity. Thus, phytol, a diterpene alcohol, showed promising anti-quorum sensing activity toward opportunistic pathogen *Pseudomonas aeruginosa* PAO1. At sub-inhibitory level, phytol reduced formation of biofilm in the range of 74.00–84.33%, whereas it also successfully affected twitching and flagella motility, and inhibited production of pyocyanin, pigment that has been considered as virulence factor of *Pseudomonas aeruginosa* PAO1. This compound also showed very strong antifungal effect against *Aspergillus niger* and *A. fumigates* (Fig. 3; Pejin, Ćirić, Glamočlija, Nikolić, & Soković, 2015).

According to Abdallah and Quax (2017), triterpenes are C 30 hydrocarbons containing six isoprene units classified into two groups: steroidal type with C 27 and pentacyclic type with C 30 hydrocarbons. They are not as abundant in nature, but among those which are identified, cucurbitacins are probably the most famous ones. Concerning their chemical structure, they are tetracyclic triterpenes with a cucurbitane skeleton, but unlike other triterpenes, they encompass numerous groups such as hydroxy, keto and others. Even though they are fairly toxic, symptoms of intoxication are not to be expected after the consummation of food rich in them, since they

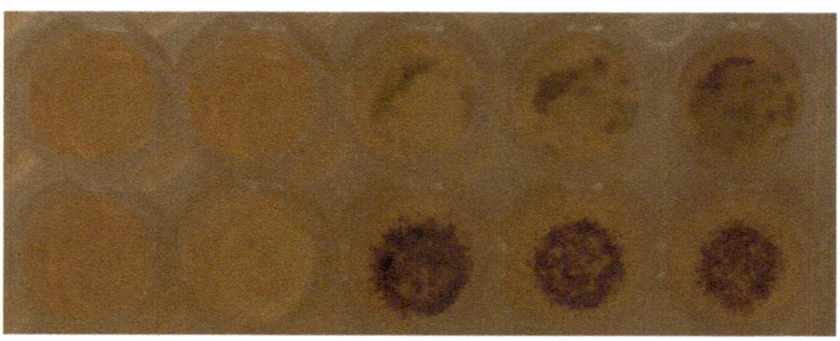

Fig. 3 Antifungal activity of phytol toward *Aspergillus niger* and *A. fumigatus.*

are also responsible for bitter and unpleasant taste of the plants producing them, which discourages people from consummation. Saponins are also triterpene core compounds which have one or few sugar chains connected to triterpene or steroid aglycone. Representatives of >100 plant families produce these type of compounds which, depending on the type, may act beneficially or harmful after ingestion (Pinela et al., 2017). The chemical structure is considered responsible for their cytotoxic potential as well as for their potential to disrupt cell membrane after the interaction with cholesterol embedded within. Additionally, saponins exhibit anti-inflammatory potential as well (Pinela et al., 2017). Antimicrobial activity of selected terpene compounds identified in ethyl acetate and dichloromethane extracts of traditionally used plant *Platostoma rotundifolium* (Briq.) A. J. Paton (Lamiaceae) was also demonstrated. Namely, extracts of this plant were rich in three terpene-core compounds, identified as cassipourol (diterpene), β-sitosterol (triterpene), and α-amyrin (triterpene), which were responsible for the aforementioned activity. Inhibition processes toward opportunistic pathogen *Pseudomonas aeruginosa* PAO1 were achieved through alteration of quorum-sensing regulated functions, such as biofilm formation, for example (Rasamiravaka et al., 2017). Tetraterpenes are C 40 hydrocarbons formed of two C 20 units coupled by "head to head" rule. Carotenoids may be the most famous compounds of this group (Abdallah & Quax, 2017). Brief presentation of health beneficial effects of selected terpenes presented in here are presented in Table 4.

Table 4 Health beneficial effects of selected terpene compounds.

Terpenic compounds	Source	Health beneficial effects	Reference
Taxol	*Taxus baccata* L.	Anticancer	Talapatra and Talapatra (2015)
Salvinorin A	*Salvia divinorum* Epling & Játiva	Pain relief	Lin et al. (2016)
Trichotomone	*Clerodendrum trichotomum* Thunb.	Cytotoxic on malignant cells	
Phytol	*Cannabis* spp., *Jasminum officinale* L., *Olea europaea* L.	Antifungal, Anti-quorum sensing	Pejin et al. (2015)
Cassipourol β-Sitosterol α-Amyrin	*Platostoma rotundifolium* (Briq.) A. J. Paton	Antimicrobial	Rasamiravaka et al. (2017)

3. Conclusions

People used and continue to use natural strategies so as to prolong and increase quality of life. This is where consummation of plants which produce bioactive terpenes becomes important. The fact that aside from fulfilling basic nutrient needs, they also benefit one's health has reached public worldwide (Ceccanti, Landi, Benvenutti, Pardosi, & Guidi, 2018).

This chapter showed that terpene and terpenoid compounds exert a wide variety of biological activities. Health beneficial effects were shown on specific chosen compounds isolated or identified in some plant species. The major effects of compounds with terpene core are oriented toward antimicrobial and cytotoxic activities. These compounds are potent antibiotics which need to be further analyzed in some clinical trials to confirm their efficacy in vivo and to exclude potential hazardous effects of these compounds. Based on the activity level and on the lowest level of toxicity, the compounds containing terpene or terpenoid structure should be chosen as potential candidates for the development of effective drugs in pharmaceutical industry, or as effective food preservatives which should be used by the food industry. Nevertheless, further clinical studies are necessary in this field as well. Compounds showing no cytotoxic potential against primary cell lines (non-tumor) and cytotoxic potential against certain cancer cell lines should be further tested as potential anticancer drugs.

Despite the fact, of how little is actually known about the true potential of all the identified compounds (several tens of thousands), and the possibility that numerous compounds have not even been discovered, terpenes are tremendously important. Their true potential is yet to be revealed, both for the benefit of comprehension basic biochemical processes and for the practical application of the obtained results.

References

Abad, M. J., Bedoya, L. M., Apaza, L., & Bermejo, P. (2012). The *Artemisia* L. Genus: A review of bioactive essential oils. *Molecules*, *17*, 2542–2566.

Abdallah, I. I., & Quax, W. J. (2017). A glimpse into the biosynthesis of terpenoids. *KnE Life Sciences*, *3*(5), 81.

Aharoni, A., Jongsma, M. A., & Bouwmeester, H. J. (2005). Volatile science? Metabolic engineering of terpenoids in plants. *Trends in Plant Science*, *10*(12), 594–602.

Ahmad, S. T., Arjumand, W., Seth, A., Nafees, A., Rashid, S., Ali, N., et al. (2011). Preclinical renal cancer chemopreventive efficacy of geraniol by modulation of multiple molecular pathways. *Toxicology*, *290*, 69–81.

Ali, B., Al-Wabel, N. A., Shams, S., Ahamad, A., Khan, S. A., & Anwar, F. (2015). Essential oils used in aromatherapy: A systemic review. *Asian Pacific Journal of Tropical Biomedicine, 5*(8), 601–611.

Anastasaki, E., Zoumpopoulou, G., Astraka, K., & Kampoli, E. (2017). Phytochemical analysis and evaluation of the antioxidant and antimicrobial properties of selected herbs cultivated in Greece. *Industrial Crops and Products, 108*, 616–628.

Bakkali, F., Averbeck, S., Averbeck, D., & Idaomar, M. (2008). Biological effects of essential oils—A review. *Food and Chemical Toxicology, 46*, 446–475.

Bhakkiyalakshmi, E., Suganya, N., Sireesh, D., Krishnamurthi, K., Saravana Devi, S., Rajaguru, P., et al. (2016). Carvacrol induces mitochondria-mediated apoptosis in HL-60 promyelocytic and Jurkat T lymphoma cells. *European Journal of Pharmacology, 772*, 92–98.

Bhalla, Y., Gupta, V. K., & Jaitak, V. (2013). Anticancer activity of essential oils: A review. *Journal of the Science of Food and Agriculture, 93*, 3643–3653.

Bhargava, V. V., Patel, S. C., & Desai, K. S. (2013). Importance of terpenoids and essential oils. *International Journal of Herbal Medicine, 1*(2), 14–21.

Bonamin, F., Moraes, T. M., Dos Santos, R. C., Kushima, H., Faria, F. M., Silva, M. A., et al. (2014). The effect of a minor constituent of essential oil from *Citrus aurantium*: The role of β myrcene in preventing peptic ulcer disease. *Chemico-Biological Interactions, 212*, 11–19.

Bukvički, D., Giweli, A., Stojković, D., Vujisić, L., Tešević, V., Nikolić, M., et al. (2018). Short communication: Cheese supplemented with *Thymus algeriensis* oil, a potential natural food preservative. *Journal of Dairy Science, 101*, 3859–3865.

Bukvički, D., Stojković, D., Soković, M., Nikolić, M., Vannini, L., Montanari, C., et al. (2015). Potential application of *Micromeria dalmatica* essential oil as a protective agent in a food system. *LWT—Food Science and Technology, 63*, 1–6.

Ceccanti, C., Landi, M., Benvenutti, S., Pardosi, A., & Guidi, L. (2018). Mediterranean wild edible plants: Weeds or "new functional crops"? *Molecules, 23*, 2299.

Chadwick, M., Trewin, H., Gawthrop, F., & Wagstaff, C. (2013). Sesquiterpenoids lactones: Benefits to plants and people. *International Journal of Molecular Sciences, 14*, 12780–12805.

Chaturvedi, D., & Dwivedi, P. K. (2016). Sesquiterpene lactones: A versatile class of structurally diverse biologically potent scaffolds in drug discovery research. *Research & Reviews: Journal of Chemistry, 5*(1), 5–9.

Chaturvedi, D., Dwivedi, P. K., & Mishra, M. (2013). Sesquiterpene lactones: A versatile class of structurally diverse natural products and their semisynthetic analogs as potential anticancer agents. In G. Brachmachari (Ed.), *Bioactive natural compounds, chemistry and biology* (pp. 321–342): Wiley VcH.

Chen, F., Tholl, D., Bohlmann, J., & Pichersky, E. (2011). The family of terpene synthases in plants: A mid-size family of genes for specialized metabolism that is highly diversified throughout the kingdom. *Plant Journal, 66*(1), 212–229.

Cho, K. S., Lim, Y. R., Lee, K., Lee, J., Lee, J. H., & Lee, I. S. (2017). Terpenes from forests and human health. *Toxicological Research, 33*(2), 97–106.

Chouhan, S., Sharma, K., & Guleria, S. (2017). Antimicrobial activity of some essential oils—Present status and future perspectives. *Medicine, 8*(4), 1–21.

Ciftci, O., Ozdemir, I., Tanyildizi, S., Yildiz, S., & Oguzturk, H. (2011). Antioxidative effects of curcumin, β-myrcene and 1,8-cineole against 2,3,7,8-tetrachlorodibenzop-dioxin-induced oxidative stress in rats liver. *Toxicology and Industrial Health, 27*(5), 447–453.

Djeddi, S., Karioti, A., Soković, M., Stojković, D., Seridi, R., & Skaltsa, H. (2007). Minor sesquiterpene lactones from *Centaurea pullata* and their antimicrobial activity. *Journal of Natural Products, 70*(11), 1796–1799.

El-Zaeddi, H., Martínez-Tomé, J., Calín-Sánchez, Á., Burló, F., & Carbonell-Barrachina, Á. (2016). Volatile composition of essential oils from different aromatic herbs grown in mediterranean regions of Spain. *Food, 5*(2), 41.

Fernandes, J. (2015). Antitumor monoterpenes. Chapter 8. In D. P. de Sousa (Ed.), *Bioactive essential oils and cancer* (pp. 175–200). Switzerland: Springer International Publishing.
Filoche, S. K., Soma, K., & Sissons, C. H. (2005). Antimicrobial effects of essential oils in combination with chlorhexidine digluconate. *Oral Microbiology and Immunology, 20*(4), 221–225.
Freires, I. A., Denny, C., Benso, B., de Alencar, S. M., & Rosalen, P. L. (2015). Antibacterial activity of essential oils and their isolated constituents against cariogenic bacteria: A systematic review. *Molecules, 20*(4), 7329–7358.
Galeotti, N., Mannelli, L. D. C., Mazzanti, G., Bartolini, A., Ghelardini, C., & Di Cesare, M. (2002). Menthol: A natural analgesic compound. *Neuroscience Letters, 322*(3), 145–148.
Jiang, Z., Kempinski, C., & Chappell, J. (2017). Engineering triterpene and methylated triterpene production in plants provides biochemical and physiological insights into terpene metabolism. *Plant Physiology, 170*, 345–358.
Joglekar, M. M., Bavkar, L. N., Sistla, S., & Arvindekar, A. U. (2017). Effective inhibition of protein glycation by combinatorial usage of limonene and aminoguanidine through differential and synergistic mechanisms. *International Journal of Biological Macromolecules, 99*, 563–569.
Joshi, D., Melkani, A. B., Nailwal, M. K., Prasad, R., & Bisht, L. S. (2018). Terpenoid composition of essential oil from a new chemotype of *Selinum wallichianum* raizada & saxena. *Natural Product Research, 32*(3), 362–365.
Jung, C. M. J., Kwon, C. K., Seo, J. S., Ohizumi, Y. O., Matsunaga, K. M., & Saito, S. S. (2001). Two new monoterpene peroxide glycosides from Aster scaber. *Chemical and Pharmaceutical Bulletin, 49*(7), 912–914.
Khan, A. Q., Khan, R., Qamar, W., Lateef, A., Rehman, M. U., Tahir, M., et al. (2013). Geraniol attenuates 12-O-tetradecanoylphorbol-13-acetate (TPA)-induced oxidative stress and inflammation in mouse skin: Possible role of p38 MAP kinase and NF-κB. *Experimental and Molecular Pathology, 94*, 419–429.
Kikowska, M., Dworacka, M., Kędziora, I., & Thiem, B. (2016). *Eryngium creticum*—Ethnopharmacology, phytochemistry and pharmacological activity. *Brazilian Journal of Pharmacognosy, 26*(3), 392–399.
Kim, Y. W., Kim, M. J., Chung, B. Y., Bang, D. Y., Lim, S. K., Choi, S. M., et al. (2013). Safety evaluation and risk assessment of *d*-limonene. *Journal of Toxicology and Environmental Health, Part B, Critical Reviews, 16*(1), 17–38.
Lin, L. G., Ung, C. O. L., Feng, Z. L., Huang, L., & Hu, H. (2016). Naturally occurring diterpenoid dimers: Source, biosynthesis, chemistry and bioactivities. *Planta Medica, 82*(15), 1309–1328.
Marcus, J., Klossek, M. L., Touraud, D., & Kunz, W. (2013). Nano-droplet formation in fragrance tinctures. *Flavour and Fragrance Journal, 28*, 294–299.
Matejić, J., Šarac, Z., & Ranđelović, V. (2010). Pharmacological activity of sesquiterpene lactones. *Biotechnology & Biotechnological Equipment, 24*, 95–100.
Michaelakis, A., Vidali, V. P., Papachristos, D. P., Pitsinos, E. N., Koliopoulos, G., Couladouros, E. A., et al. (2014). Bioefficacy of acyclic monoterpenes and their saturated derivatives against the West Nile vector *Culex pipiens*. *Chemosphere, 96*, 74–80.
Müller, G. C., Junnila, A., Butler, J., Kravchenko, V. D., Revay, E. E., Weiss, R. W., et al. (2009). Efficacy of the botanical repellents geraniol, linalool, and citronella against mosquitoes. *Journal of Vector Ecology, 34*(1), 2–8.
Numpaque, M. A., Oviedo, L. A., Gil, J. H., García, C. M., & Durango, D. L. (2011). Thymol and carvacrol: Biotransformation and antifungal activity against the plant pathogenic fungi *Colletotrichum acutatum* and *Botryodiplodia theobromae*. *Tropical Plant Pathology, 36*, 3–13.
Oldfield, E., & Lin, F. (2012). Terpene biosynthesis: Modularity rules. *Angewandte Chemie International Edition, 51*(5), 1124–1137.

Olsen, R. W. (2000). Absinthe and γ-aminobutyric acid receptors. *Proceedings of the National Academy of Sciences of the United States of America*, *97*(9), 4417–4418.

Paduch, R., Trytek, M., Król, S. K., Kud, J., Frant, M., Kandefer-Szerszeń, M., et al. (2016). Biological activity of terpene compounds produced by biotechnological methods. *Pharmaceutical Biology*, *54*(6), 1096–1107.

Parvin, R., Shahrokh, K. O., Mozafar, S., Hassan, E., & Mehrdad, B. (2014). Biosynthesis, regulation and properties of plant monoterpenoids. *Journal of Medicinal Plant Research*, *8*(29), 983–991.

Pavela, R., Maggi, F., Cianfaglione, K., Bruno, M., & Benelli, G. (2018). Larvicidal activity of essential oils of five Apiaceae taxa and some of their main constituents against *Culex quinquefasciatus*. *Chemistry & Biodiversity*, *15*.

Pejin, B., Ćirić, A., Glamočlija, J., Nikolić, M., & Soković, M. (2015). Short communication: In vitro anti-quorum sensing activity of phytol. *Natural Product Research*, *29*(4), 374–377.

Pereira, I., Severino, P., Santos, A. C., Silva, A. M., & Souto, E. B. (2018). Linalool bioactive properties and potential applicability in drug delivery systems. *Colloids and Surfaces B: Biointerfaces*, *171*, 566–578.

Piccinelli, A. C., Morato, P. N., Dos Santos Barbosa, M., Croda, J., Sampson, J., Kong, X., et al. (2017). Limonene reduces hyperalgesia induced by gp120 and cytokines by modulation of IL-1 β and protein expression in spinal cord of mice. *Life Sciences*, *174*, 28–34.

Piccinelli, A. C., Santos, J. A., Konkiewitz, E. C., Oesterreich, S. A., Formagio, A. S., Croda, J., et al. (2015). Antihyperalgesic and antidepressive actions of (R)-(+)-limonene, α-phellandrene, and essential oil from *Schinus terebinthifolius* fruits in a neuropathic pain model. *Nutritional Neuroscience*, *18*, 217–224.

Pichersky, E., & Raguso, R. A. (2016). Why do plants produce so many terpenoid compounds? *New Phytologist*, *220*, 692–702.

Pinela, J., Carvalho, A. M., & Ferreira, I. C. F. R. (2017). Wild edible plants: Nutritional and toxicological characteristics, retrieval strategies and importance for today's society. *Food and Chemical Toxicology*, *110*, 165–188.

Popović, V., Stojković, D., Nikolić, M., Heyerick, A., Petrović, S., Soković, M., et al. (2015). Extracts of three *Laserpitium* L. species and their principal components laserpitine and sesquiterpene lactones inhibit microbial growth and biofilm formation by oral *Candida* isolates. *Food & Function*, *6*(4), 1205–1211.

Pravdich-Neminskaya, T. V., & Kachkov, A. P. (1978). Effect of allo-ocimene on the healing of experimental soft tissue wounds. *Bulletin of Experimental Biology and Medicine*, *85*, 57.

Rasamiravaka, T., Ngezahayo, J., Pottier, L., Ribeiro, S. O., Souard, F., Hari, L., et al. (2017). Terpenoids from *Platostoma rotundifolium* (Briq.) A. J. Paton alter the expression of quorum sensing-related virulence factors and the formation of biofilm in *Pseudomonas aeruginosa* PAO1. *International Journal of Molecular Sciences*, *18*(6), 1270.

Rozza, A. L., Moraes, T. M., Kushima, H., Tanimoto, A., Marques, M. O., Bauab, T. M., et al. (2011). Gastroprotective mechanisms of *Citrus lemon* (Rutaceae) essential oil and its majority compounds limonene and β-pinene: Involvement of heat-shock protein-70, vasoactive intestina peptide, glutathione, sulfhydryl compounds, nitric oxide and prostaglandin E2. *Chemico-Biological Interactions*, *189*, 82–89.

Singh, B., & Sharma, R. A. (2015). Plant terpenes: Defense responses, phylogenetic analysis, regulation and clinical applications. *3 Biotech*, *5*(2), 129–151.

Soković, M., Ćirić, A., Glamočlija, J., & Skaltsa, H. (2017). Biological activities of sesquiterpene lactones isolated from the genus *Centaurea* L. (Asteraceae). *Current Pharmaceutical Design*, *23*(19), 2767–2786.

Sonmezdag, A., Kelebek, H., & Selli, S. (2017). Identification of aroma compounds of lamiaceae species in Turkey using the purge and trap technique. *Food*, *6*(2), 10.

Stojković, D., Glamočlija, J., Ćirić, A., Nikolić, M., Ristić, M., Šiljegović, J., et al. (2013). Investigation on antibacterial synergism of *Origanum vulgare* and *Thymus vulgaris* essential oils. *Archives of Biological Sciences*, *65*(2), 639–643.

Stojković, D., Soković, M., Glamočlija, J., Džamić, A., Ćirić, A., Ristić, M., et al. (2011). Chemical composition and antimicrobial activity of *Vitex agnus-castus* L. fruits and leaves essential oils. *Food Chemistry*, *128*(4), 1017–1022.

Stojković, D., Soković, M., Glamočlija, J., Džamić, A., Ristić, M., Fahal, A., et al. (2008). Susceptibility of three clinical isolates of *Actinomodura madurae* to α-pinene, the bioactive agent of *Pinus pinaster* turpentine oil. *Archives of Biological Sciences*, *60*(4), 697–701.

Sueishi, Y., & Nii, R. (2018). Monoterpene's multiple free radical scavenging capacity as compared with the radioprotective agent cysteamine and amifostine. *Bioorganic and Medicinal Chemistry Letters*, *28*(18), 3031–3033.

Sülsen, V. P., Lizarraga, E., Mamadalieva, N. Z., & Lago, J. H. G. (2017). Potential of terpenoids and flavonoids from Asteraceae as anti-inflammatory, antitumor, and antiparasitic agents. *Evidence-Based Complementary and Alternative Medicine*, *2017*, 6–8.

Talapatra, S. K., & Talapatra, B. (2015). Chemistry of Plant Natural Products: Stereochemistry, conformation, synthesis, biology, and medicine, Springer-Verlag Berlin Heidelberg. https://doi.org/10.1007/978-3-642-45410-3.

Thapa, D., Losa, R., Zweifel, B., & Wallace, R. J. (2012). Sensitivity of pathogenic and commensal bacteria from the human colon to essential oils. *Microbiology (Reading, England)*, *158*, 2870–2877.

Vender, C., Aiello, N., Fusani, P., Scartezzini, F., Zini, E., Eggher, P., et al. (2005). Edible, aromatic and medicinal plants of the alps: A resource to value regenerated plants. In *Proceedings of the XLIX Italian society of agricultural genetics annual congress potenza, Italy—12/15 september, 2005*, ISBN: 88-900622-6-6.

Wen-Xian, D., Olsen, C. W., Avena-Bustillos, R. J., McHugh, T. H., Levin, C. E., & Friedman, M. (2008). Storage stability and antibacterial activity against *Escherichia coli* O157:H7 of carvacrol in edible apple films made by two different casting methods. *Journal of Agricultural and Food Chemistry*, *56*(9), 3082–3088.

Yadav, N., Yadav, R., & Goyal, A. (2014). Chemistry of terpenoids. *International Journal of Pharmaceutical Sciences Review and Research*, *27*(2), 272–278.

Zhang, W., You, C., Yang, K., Wang, Y., Su, Y., Geng, Z., et al. (2015). Bioactivity and chemical constituents of the essential oil from *Dendranthema indicum* (L.) Des Moul. against two stored insects. *Journal of Oleo Science*, *560*(19), 553–560.

Zouari, N. (2015). Eating wild edible plants can be a good alternative? *Medicinal & Aromatic Plants*, (1). S1-e001.

Zuzarte, M., & Salgueiro, L. (2015). Essential oils chemistry. In D. P. de Sousa (Ed.), *Bioactive essential oils and cancer* (pp. 19–28). Switzerland: Springer International Publishing.

Zwenger, S., & Basu, C. (2008). Plant terpenoids: Applications and future potentials. *Biotechnology and Molecular Biology Reviews*, *3*(1), 1–7.

Further reading

Soković, M., Vukojević, J., Marin, P. D., Brkić, D., Vajs, V., & Griensven, L. J. L. D. (2009). Chemical composition of essential oils of *Thymus* and *Mentha* species and their antifungal activities. *Molecules*, *14*, 238–249.

FSC
www.fsc.org
MIX
Papier | Fördert
gute Waldnutzung
FSC® C083411

Kolibri 360 GmbH
Ferdinand-Jühlke-Straße 7
99095 Erfurt, Deutschland
produktsicherheit@kolibri360.de